DYNAMICS OF SHOCK WAVES, EXPLOSIONS, AND DETONATIONS

Edited by
J. R. Bowen
University of Washington
Seattle, Washington

N. Manson
Universite de Poitiers
Poitiers, France

A. K. Oppenheim
University of California
Berkeley, California

R. I. Soloukhin
Institute of Heat and Mass Transfer
BSSR Academy of Sciences
Minsk, USSR

Volume 94
PROGRESS IN
ASTRONAUTICS AND AERONAUTICS

Martin Summerfield, Series Editor-in-Chief
Princeton Combustion Research Laboratories, Inc.
Princeton, New Jersey

Technical papers selected from the Ninth International Colloquium on Gasdynamics of Explosions and Reactive Systems, Poitiers, France, July 1983, and subsequently revised for this volume.

Published by the American Institute of Aeronautics and Astronautics, Inc.
1633 Broadway, New York, NY 10019

American Institute of Aeronautics and Astronautics, Inc.
New York, New York

Library of Congress Cataloging in Publication Data
Main entry under title:

International Colloquium on Gasdynamics of Explosions and Reactive Systems (9th:1983:Poitiers, France) Dynamics of shock waves, explosions, and detonations.

(Progress in astronautics and aeronautics; v. 94)
Revised technical papers selected from the Ninth International Colloquium on Gasdynamics of Explosions and Reactive Systems, Poitiers, France, July 1983.
Includes bibliographies and index.
1. Explosions—Congresses. 2. Shock waves—Congresses I. Bowen, J.R. (J. Ray) II. Title. III. Series. TL507.P75 vol. 94   629.1s
[621,402'3]   84-21564
[QD516]
ISBN 0-915928-91-4

Copyright © 1984 by the American Institute of Aeronautics and Astronautics, Inc. All rights reserved. Printed in the United States of America. No part of this publication may be reproduced, distributed, or transmitted, in any form or by any means, or stored in any data base or retrieval system, without the prior written permission of the publisher.

Progress in Astronautics and Aeronautics

Series Editor-in-Chief
Martin Summerfield
Princeton Combustion Research Laboratories, Inc.

Series Associate Editors

Burton I. Edelson
*National Aeronautics
   and Space Administration*

Allen E. Fuhs
Naval Postgraduate School

J. Leith Potter
Vanderbilt University

Assistant Series Editor

Ruth F. Bryans
Ocala, Florida

Norma J. Brennan
Director, Editorial Department
AIAA

Jeanne Godette
Series Managing Editor
AIAA

Table of Contents

Preface xv

Chapter I. Detonations in Gaseous Mixtures 1

Direct Initiation of Planar Detonation Waves in Methane/ Oxygen/Nitrogen Mixtures 3
    S. Ohyagi, T. Yoshihashi, and Y. Harigaya, *Saitama University, Saitama-Ken, Japan*

Measurements of Cell Size in Hydrocarbon-Air Mixtures and Predictions of Critical Tube Diameter, Critical Initiation Energy, and Detonability Limits 23
    R. Knystautas, C. Guirao, J.H. Lee, and A. Sulmistras, *McGill University, Montreal, Canada*

Power-Energy Relations for the Direct Initiation of Gaseous Detonations 38
    K. Kailasanath and E.S. Oran, *Naval Research Laboratory, Washington, D.C.*

Detonation Length Scales for Fuel-Air Explosives 55
    I.O. Moen, J.W. Funk, S.A. Ward, and G.M. Rude, *Defence Research Establishment Suffield, Ralston, Canada* and P.A. Thibault, *University of Toronto Institute for Aerospace Studies, Ontario, Canada*

The Influence of Yielding Confinement on Large-Scale Ethylene-Air Detonations 80
    S.B. Murray, *Defence Research Establishment Suffield, Ralston, Canada,* and J.H. Lee, *McGill University, Montreal, Canada*

Cellular Structure in Detonation of Acetylene-Oxygen Mixture 104
    M. Vandermeiren and P.J. Van Tiggelen, *Université Catholique de Louvain, Louvain-la-Neuve, Belgique*

The Influence of Initial Pressure on Critical Diameters of Gaseous Explosive Mixtures 118
    P. Bauer, C. Brochet, and H.N. Presles, *Université de Poitiers, Poitiers, France*

"Galloping" Gas Detonations in the Spherical Mode............ 130
   J.E. Elsworth, P.J. Shuff, and A. Ungut, *Shell Research Ltd., Chester, Great Britian*

Chemical Kinetics of Propane Oxidation in Gaseous
   Detonations... 151
   C.K. Westbrook, W.J. Pitz, and P.A. Urtiew, *Lawrence Livermore National Laboratory, University of California, Livermore, California*

High-Speed Deflagration with Compressibility Effects........... 175
   J.F. Clarke, *Cranfield Institute of Technology, Bedford, England,* and D.R. Kassoy, *University of East Anglia, Norwich, England*

Numerical Simulations on the Establishment of Gaseous
   Detonation.. 186
   S. Taki, *Fukui University, Fukui, Japan,* and T. Fujiwara, *Nagoya University, Nagoya, Japan*

A Shock Tube Study of the Chlorine Azide Decomposition........ 201
   C. Paillard and G. Dupré, *C.N.R.S. et Université d'Orléans, France,* and N.A. Fomin, *Academy of Sciences, Minsk, USSR*

Chapter II.   Detonations in Two-Phase Systems 219

Dust, Hybrid, and Dusty Detonations 221
   C.W. Kauffman, P. Wolański, A. Arisoy, P.R. Adams, B.N. Maker, and J.A. Nicholls, *University of Michigan, Ann Arbor, Michigan*

The Structure of Dust Detonations......................... 241
   P. Wolański, D. Lee, M. Sichel, C.W. Kauffman, and J.A. Nicholls, *University of Michigan, Ann Arbor, Michigan*

"Double-Front" Detonations in Gas-Solid Particles Mixtures 264
   B. Veyssière, *Centre National de la Recherche Scientifique, Poitiers, France*

Unconfined Aluminum Particle Two-Phase Detonation in Air..... 277
   A.J. Tulis and J.R. Selman, *Illinois Institute of Technology, Chicago, Illinois*

Dynamics of Dispersion and Ignition of Dust Layers by a
   Shock Wave... 293
   V.M. Boiko and A.N. Papyrin, *Institute of Theoretical and Applied Mechanics, Novosibirsk, USSR,* and M. Woliński and P. Wolański, *Technical University of Warsaw, Poland*

Detonations in Explosive Foams 302
    J.P. Saint-Cloud and O. Peraldi, *Université de Poitiers,*
    Poitiers, France

Propagation Velocity and Mechanism of Bubble Detonation 309
    T. Hasegawa, *Nagoya Institute of Technology, Japan,* and
    T. Fujiwara, *Nagoya University, Japan*

Nonsteady Shock Wave Propagating in a Bubble-Liquid System ... 320
    T. Sugimura, *Meijo University, Nagoya, Japan,* K. Tokita, *Nippon*
    Oil & Fats Co. Ltd., Taketoyo, Chita, Aichi, Japan, and
    T. Fujiwara, *Nagoya University, Nagoya, Japan*

Ignition of Dust Suspensions Behind Shock Waves 332
    A.A. Borisov, B.E. Gel'fand, E.I. Timofeev, S.A. Tsyganov, and
    S.V. Khomic, *Academy of Sciences, Moscow, USSR*

Chapter III. Condensed Explosives 341

**Characterization of an Overdriven Detonation State in
Nitromethane** 343
    L. Hamada, H.N. Presles, C. Brochet, and R. Bouriannes,
    Université de Poitiers, Poitiers, France, and R. Cheret, *C.E.A.,*
    Vaujours, Sevran, France

**The Effects of Grain Size on Shock Initiation Mechanisms in
Hexanitrostilbene (HNS) Explosive** 350
    R.E. Setchell and P.A. Taylor, *Sandia National Laboratories,*
    Albuquerque, New Mexico

**Theoretical Modeling of Converging and Diverging Detonation
Waves in Solid and Gaseous Explosives** 369
    C.M. Tarver and P.A. Urtiew, *Lawrence Livermore National*
    Laboratory, University of California, Livermore, California

**Model Similarity Solutions for Shock Initiation Containing a
Realistic Constitutive Relationship for Condensed Explosive** 387
    M. Cowperthwaite, *SRI International, Menlo Park, California*

**The Simulation of Shock-Induced Energy Flux in Molecular
Solids** 405
    A.M. Karo, F.E. Walker, and T.M. DeBoni, *Lawrence Livermore*
    National Laboratory, University of California, Livermore,
    California, and J.R. Hardy, *Behlen Laboratory of Physics,*
    University of Nebraska, Lincoln, Nebraska

Detonation Temperatures of Nitromethane Aluminum Gels....... 416
  Y. Kato, *Fukui Institute of Technology, Fukui, Japan,*
  and C. Brochet, *Université de Poitiers, Poitiers, France*

Chapter IV. Explosions 427

**Theory of Vorticity Generation by Shock Wave and Flame
Interactions 429**
  J.M. Picone, E.S. Oran, J.P. Boris, and T.R. Young Jr., *Naval
  Research Laboratory, Washington, D.C.*

**The Interaction of Explosively Produced Shock Waves with
Internal Discontinuities and External Objects 449**
  M.A. Fry, *Science Applications, Inc., McLean, Virginia,* and
  D.L. Book, *Naval Research Laboratory, Washington, D.C.*

**Flame Propagation and Pressure Buildup in a Free Gas-Air
Mixture Due to Jet Ignition............................. 474**
  M. Schildknecht, W. Geiger, and M. Stock, *Battelle-Institut e.V.,
  Frankfurt, Federal Republic of Germany*

Flame Acceleration by a Postflame Local Explosion 491
  M. Stock, M. Schildknecht, and W. Geiger, *Battelle-Institut e.V.,
  Frankfurt, Federal Republic of Germany*

**Flame Acceleration of Propane-Air in a Large-Scale Obstructed
Tube... 504**
  B.H. Hjertager, K. Fuhre, S.J. Parker, and J.R. Bakke,
  Chr. Michelsen Institute, Bergen, Norway

**Initiation of Unconfined Gaseous Detonation by Diffraction
of a Detonation Front Emerging from a Pipe................ 523**
  A. Ungut, P.J. Shuff, and J.A. Eyre, *Shell Research Ltd., Chester,
  Great Britain*

**Large-Scale Experiments on the Transmission of Fuel-Air
Detonations from Two-Dimensional Channels 546**
  W.B. Benedick, *Sandia National Laboratories, Albuquerque,
  New Mexico,* and R. Knystautas and J.H. Lee, *McGill University,
  Montreal, Canada*

Air Blast from Unconfined Gaseous Detonations............... 556
  J. Brossard, *University of Orleans, Bourges, France,* J.C. Leyer,
  D. Desbordes, and J.P. Saint-Cloud, *University of Poitiers, Poitiers,
  France,* S. Hendrickx, *E.D.F., Paris, France,* J.L. Garnier,
  C.E.A., Fontenay aux Roses, France, A. Lannoy, *E.D.F.,
  Saint-Denis, France,* and J. Perrot, *C.E.A., Le Barp, France*

Chapter V. Interactions 567

Collapse of Gas-Filled Cavities in Water 569
  M. Holt and M.J. Djomehri, *University of California,*
  Berkeley, California

Crack Propagation in Burning Solid Propellants 575
  J.G. Siefert and K.K. Kuo, *The Pennsylvania State University,*
  University Park, Pennsylvania

Author Index for Volume 94 596
List of Series Volumes 597

Table of Contents for Companion Volume 95

Introduction to Modern Laminar Flame Theory 1
   P. Clavin, *Université de Provence, Marseille, France*

Chapter I. Premixed Flames 35

Effects of Chemical Equilibrium on the Structure and Extinction of Laminar Diffusion Flames 37
   N. Peters, *Rheinisch-Westfälische Technische Hochschule, Aachen, Federal Republic of Germany*, and F.A. Williams, *Princeton University, Princeton, New Jersey*

Stretch Effects in Planar Premixed Hydrogen-Air Flames 61
   J. Warnatz, *Universität Heidelberg, Federal Republic of Germany*, and N. Peters, *Rheinisch-Westfälische Technische Hochschule, Aachen, Federal Republic of Germany*

Slowly Varying Flames with Chain-Branching/Chain-Breaking Kinetics 75
   G.S.S. Ludford, *Cornell University, Ithaca, New York*, and N. Peters, *Rheinisch-Westfälische Technische Hochschule, Aachen, Federal Republic of Germany*

Effect of Dissociation on the Near-Stoichiometric Burning of Non-Dilute Mixtures 92
   G.S.S. Ludford, *Cornell University, Ithaca, New York*, and A.K. Sen, *Purdue School of Sciences, Indianapolis, Indiana*

The Feedback of a Flame Front on Turbulent Flows 103
   G. Searby, F. Sabathier, J. Monreal, P. Clavin, and L. Boyer, *Université de Provence, Marseille, France*

Flame Front Stability with General Intermolecular Interaction Potential 115
   P.G. Ybarra, *Université de Rouen, Mont-Saint-Aignan, France*

Stability Limits and Critical Size of Structures in Premixed Flames 129
   J. Quinard, G. Searby, and L. Boyer, *Université de Provence, Marseille, France*

Nonsteady Gasdynamic Effects in the Induction Behind a Strong Shock Wave 142
   J.F. Clarke and R.S. Cant, *Cranfield Institute of Technology, Bedford, England*

Structure of Premixed Laminar Methanol-Air Flames: Experimental and Computational Results 164
   L.L. Andersson, *Chalmers University of Technology, Göteborg, Sweden*, and B. Christenson, A. Höglund, J.O. Olsson, and L.G. Rosengren, *Volvo, Göteborg, Sweden*

**Burning Velocities of Ethanol-Air and Ethanol-Water-Air
Mixtures** 181
    O.L. Gülder, *National Research Council of Canada, Ottawa,
Ontario, Canada*

**Computer Modeling Study of Acetylene-Oxygen Ignition
and Flames Using a Truncated Reaction Mechanism** 198
    S.-M. Hwang and W.C. Gardiner Jr., *University of Texas,
Austin, Texas*, and J. Warnatz, *University of Heidelberg,
Federal Republic of Germany*

**Prediction of Laminar Flame Properties of Propane-Air
Mixtures**.. 211
    C.K. Westbrook and W.J. Pitz, *Lawrence Livermore National
Laboratory, University of California, Livermore, California*

**Homogeneity and Propagation of Autoignited Cool and Blue
Flames**.. 236
    Y. Ohta, *Nagoya Institute of Technology, Nagoya, Japan*, and
H. Takahashi, *Meijo University, Nagoya, Japan*

**Stability of Solid Propellant Combustion Subject to
Nonplanar Perturbations**.. 248
    F.J. Higuera and A. Liñán, *Escuela Técnica Superior de
Ingenieros Aeronauticos, Madrid, Spain*

Chapter II. Diffusion Flames 259

**Laminar Diffusion Flames with Cylindrical Symmetry, Arbitrary
Values of Diffusion Coefficients and Inlet Velocities, and
Chemical Reactions in the Approach Streams** 261
    S.S. Penner, M.Y. Bahadori, and E.M. Kennedy, *University
of California, San Diego, La Jolla, California*

**Transition and Transport in the Initial Region of a Turbulent
Diffusion Flame** 293
    A.R. Masri, S.H. Stårner, and R.W. Bilger, *The University
of Sydney, Sydney, Australia*

**Predicted Structure of Stretched and Unstretched Methane-Air
Diffusion Flames**... 305
    S.K. Liew and J.B. Moss, *Cranfield Institute of Technology,
Bedford, England*, and K.N.C. Bray, *University of
Southampton, Southampton, England*

An Experimental Study of Turbulent Jet Diffusion Flames............................ 320
    O.K. Sønju and J. Hustad, *Norwegian Institute of Technology,
Trondheim, Norway*

Chapter III. Turbulent Combustion 341

On Sound Sources in Turbulent Combustion.. 343
    N. Kidin and V. Librovich, *Institute for Problems in Mechanics,
Moscow, USSR*, and J. Roberts and M. Vuillermoz, *Polytechnic
of the South Bank, London, United Kingdom*

**Comparisons of Experimental and Computed Length Scales
and Velocities in Turbulent Combustion**.. 356
    A.Y. Abdalla, D. Bradley, S.B. Chin, and C. Lam, *University
of Leeds, Leeds, United Kingdom*

Flow Rate and Equivalence Ratio Influences on the Thermal
Field of a Turbulent Cool Flame 367
    I. Gökalp, *C.N.R.S., Orléans, France,* and N. Zarrad,
    G.M.L. Dumas, and R.I. Ben Aïm, *Université
    Pierre et Marie Curie, Paris, France*

Turbulent Reacting Concentric Jets: Comparison Between
pdf and Moment Calculations.. 384
    P. Givi, J.I. Ramos, and W.A. Sirignano, *Carnegie-Mellon
    University, Pittsburgh, Pennsylvania*

Chapter IV. Constant Volume Combustion 419

Influence of Turbulent Motion on Spark Ignition 421
    P.S. Tromans and S.J. O'Connor, *Shell Research Ltd.,
    Chester, Great Britain*

Vibratory Combustion Triggered by a Small Cavity in the Wall
of a Constant Volume Combustion Chamber... 433
    A. Girard, F. Fisson, and J.C. Leyer, *Université de Poitiers,
    Poitiers, France*

Direct Measurement of the Head-on Flame Quenching
Distance in Closed Chambers 443
    A. Girard and J.C. Leyer, *Université de Poitiers,
    Poitiers, France*

Chapter V. Spray Combustion 453

Timed Ignition of Explosives and Flammables from
Desensitized Solutions... 455
    M. Gerstein and P.R. Choudhury, *University of Southern California,
    Los Angeles, California*

Comparative Study of Droplet Heating and Vaporization at
High Reynolds and Peclet Numbers 464
    H.A. Dwyer and B.R. Sanders, *Sandia National Laboratories,
    Livermore, California*

Comparisons of Computed and Measured Dense Spray Jets............................ 484
    L. Martinelli and F.V. Bracco, *Princeton University, Princeton,
    New Jersey,* and R.D. Reitz, *General Motors Technical Center,
    Warren, Michigan*

A Study of the Motion of Vaporizing Droplets in a
Turbulent Flow 513
    A.A. Mostafa and S.E. Elghobashi, *University of California,
    Irvine, California*

Simulations of Two-Dimensional Fuel Droplet Flows 540
    M.J. Fritts, D.E. Fyfe, and E.S. Oran, *Naval Research
    Laboratory, Washington, D.C.*

Induction Time Measurements for Ignition of Liquid Fuel Jets
in Air at High Temperatures and Pressures 554
    V.K. Baev, A.N. Bazhaikin, A.A. Buzukov, *University of
    Novosibirsk, Academy of Sciences, Novosibirsk, USSR*

Spray Characteristics of Simplex Swirl Atomizers 563
    N.K. Rizk and A.H. Lefebvre, *Purdue University,
    West Lafayette, Indiana*

Chapter VI. Nonequilibrium Flows 581

Flows in Laval Nozzles with a High-Temperature Diatomic Gas 583
N.K. Mitra and M. Fiebig, *Ruhr-Universität, Bochum,
Federal Republic of Germany*

**Unsteady Aerodynamics of Chemically Reacting Flows Past
Oscillating Thin Bodies** 593
L. Librescu, *Tel-Aviv University, Tel-Aviv, Israel*

**Uniform Solutions for Characteristics and Weak Shock Waves
in a Reactive Medium** 610
B.D. Pandey and D.C. Chou, *University of New Mexico,
Albuquerque, New Mexico*

Chapter VII. Combustion Diagnostics 629

CARS Instrument for Practical Combustion Measurements 631
G.M. Dobbs, J.H. Stufflebeam, and A.C. Eckbreth,
United Technologies Research Center, East Hartford, Connecticut,
and P.A. Tellex, *Pratt & Whitney Aircraft Group,
West Palm Beach, Florida*

**Study of OH-Saturated Laser-Induced Fluorescence in
Low-Pressure Flame** 642
D. Stepowski and M.J. Cottereau, *Faculté des Sciences de Rouen,
Mont Saint Aignan, France*

**Flame Concentrations and Temperatures by Spontaneous
Raman Spectroscopy** 658
R. Michael-Saade, J.P. Sawerysyn, L.-R. Sochet, G. Buntinx,
M. Crunelle-Cras, F. Grase, and M. Bridoux, *Université des Sciences
et Techniques de Lille, France*

**The Application of Rotational Raman Spectroscopy to
Dynamic Measurements in Gas Flowfields** 672
P.P. Yaney, R.J. Becker, P.T. Danset, M.R. Gallis, and
J.I. Perez, *University of Dayton, Dayton, Ohio*

**Flash X-Ray Tomographic System for Diagnostics of
Microsecond Phenomena** 700
C.K. Zoltani and K.J. White, *Ballistic Research Laboratory,
Aberdeen Proving Ground, Maryland*

**Two-Dimensional Imaging of Flame Temperature Using
Laser-Induced Fluorescence** 714
R.J. Cattolica and D.A. Stephenson, *Sandia National Laboratories,
Livermore, California*

LDV Measurements of Gas Flow Behind Reflected Shocks 722
M. Frenklach and C.K. Li Kwok Cheong, *Louisiana State University,
Baton Rouge, Louisiana,* and E.S. Oran, *Naval Research
Laboratory, Washington, D.C.*

Droplet Size Distributions from Diffracted Light Intensities 736
A. Tardieu, S.M. Candel, and E. Esposito, *l' Ecole Centrale
des Arts et Manufactures, Châtenay-Malabry, France*

**Measurement of NO in Methane-Air Flames by Tunable Atomic
Line Molecular Spectroscopy** 750
E. Cuellar and N.J. Brown, *University of California
Berkeley, California*

Preface

This and a companion volume include revised and edited versions of papers presented at the Ninth International Colloquium on the Dynamics of Explosions and Reactive Systems held in Poitiers, France, in July 1983.

These Colloquia originated in 1966 as a result of the widely-held belief among leading researchers that revolutionary advances in the understanding of detonation wave structure warranted a forum for the discussion of important findings in the gasdynamics of flows associated with the exothermic process—the essential feature of detonation waves—as well as for the discussion of other, associated phenomena.

The contributions to this, the Ninth Colloquium, have been assembled into two volumes: *Dynamics of Shock Waves, Explosions, and Detonations* and *Dynamics of Flames and Reactive Systems.* The dynamics of explosions, which the former addresses, is concerned principally with the interrelationship between the rate processes of energy deposition in a compressible medium and the concurrent nonsteady flow as it occurs typically in explosion phenomena. The dynamics of reactive systems, which is the focus of the latter volume, is a broader area encompassing the processes of coupling between the dynamics of fluid flow and molecular transformations in reactive media occurring in any combustion system. The Colloquium, then, in addition to embracing the usual topics of explosions, detonations, shock phenomena, and reactive flow, included the presentation of papers that dealt primarily with the gasdynamic aspects of nonsteady flow in combustion systems, the fluid mechanics aspects of combustion, with particular emphasis on the effects of turbulence, and with diagnostic techniques employed in the study of combustion phenomena.

In this volume, *Dynamics of Shock Waves, Explosions, and Detonations,* the papers have been arranged into chapters on gaseous detonations, heterogeneous detonations, condensed explosives, explosions, and interactions. Many of the 82 papers comprising these two volumes provoked interesting discussions during the Colloquium. While the brevity of this Preface does not permit the editors to do justice to all of the stimulating papers, highlights of some of the more noteworthy contributions among them follow.

Chapter I, Detonations in Gaseous Mixtures, presents significant new results on gaseous detonations. A topic of central concern has been the detonable mixtures and the relationship of cellular structure to chemical and physical parameters such as induction delay and the geometry of confinement. This work is an extension and an enrichment of discoveries in the mid-60s on the cellular structure of detonations. *Brochet* and coworkers report that the critical diameters (below which a detonation is quenched when the front propagates into a sudden enlargement) decrease as the initial pressure of the detonable gas increases. *Vandermeiren* and *Van Tiggelen* report on the relationship of cellular structure in acetylene-oxygen and discuss the relationship of that structure to underlying chemical kinetics. *Murray* and *Lee* report the results of their observations of cell size for detonation mixtures of hydrogen and cell size for detonation mixtures of hydrogen and light paraffinic hydrocarbons with air, and they verify the empirical law that the critical diameter is about 13 times the cell size. *Moen* and coworkers report similar results for their investigations, which have employed tubes of larger diameters, but the researchers question the validity of the empirical law for less sensitive explosive mixtures. One important parameter used in these calculations is the induction delay or induction length. *Westbrook* and coworkers report a theoretical calculation of this parameter by a detailed chemical kinetic mechanism with a one-dimensional model for several explosive gas mixtures. These computed lengths are used to discuss available experimental data for critical energy, critical tube diameter, and detonation limits. *Kailasanath* and *Oran* present theoretical results for the dependence of initiation of explosive mixtures on energy depositions and power.

The recent history of accidental explosions of dusts in large buildings or enclosed industrial facilities with consequent property damage and injury has stimulated research activity on fuel particulates-gaseous oxidizer detonations, a subject of Chapter II, Detonations in Two-Phase Systems. *Nicholls* and coworkers at the University of Michigan report on their characterization studies of detonations that involve particulates. Their study, "Dust, Hybrid, and Dusty Detonations," includes experiments on dust detonations supported by combustion of particulates; hybrid detonations, supported by combustion of a gaseous *and* a particulate fuel; and dusty detonations supported by combustion of gaseous fuel in the presence of inert particulates. For dust detonations, the combustion wave velocity is near the CJ detonation velocity, and the wave structure consists of a leading shock wave followed first by an ignition delay zone and then by a reaction zone. Reactive dusts in hybrid detonations can alter detonation limits, and inert particulates may either quench or enhance detonations. In a related work, "The Structure of Dust Detonations," the Michigan group reports the results of a model study that is designed to make *ab initio* calculation of detonation properties and finds good agreement between predicted and experimental wave velocities. A collaboration between Polish and Russian investigators is reported in "Dynamics of Dispersion and Ignition of Dust Layers by a Shock Wave." The study indicates that dispersion of a dust layer on a wall is influenced by shock wave Mach number, particulate density, and particle size and that ignition of such dusts composed of combustible particulates is possible if the shock wave is sufficiently strong. *Borisov* and coworkers report in "Ignition of Dust Suspension behind Shock Waves" that ignition delays observed behind reflected shock waves for powders do not differ greatly from those observed for kerosene sprays. The hazards of detonation of explosive gaseous mixture dispersed as bubbles in nonreactive liquids or in foams was also a subject of discussion in the meeting. Detonation propagation in foams is similar to gaseous detonation when the characteristic bubble diameter exceeds a critical value for a given foam density. For characteristic bubble diameters less than the critical value, the shock and reaction zones are clearly separated and the detonation velocity is lower than that observed for the gaseous mixture. *Saint-Cloud* and *Peraldi* report on an investigation of the effects of nitrogen dilution on the phenomena. *Fujiwara* and coworkers report on their experimental and theoretical investigations on the propagation of detonations through foams.

In Chapter III, Condensed Explosives, *Setchell* and *Taylor* discuss their work that showed that the shock sensitivity of hexanitrostilbene depends on shock amplitude and duration as well as grain size. Their analysis indicates that pore collapse is the likely source of hot spots; it also identifies the microstructural parameters that influence shock sensitivity. *Tarver* and *Urtiew* report on the application of the "ignition and growth reactive flow model of shock initiation and detonation" to calculate the properties of converging and diverging detonation waves. *Cowperthwaite* reports an interesting application of Lie Groups to determine general similarity equations that may be used to predict detonation phenomena in condensed explosives. The response of diatomic molecules embedded in a monatomic host lattice upon shock loading was analyzed by *Walker* and coworkers with the aid of computer molecular dynamics. The results suggest that for more complex systems, shock loading may generate molecular fragments that could initiate explosions. *Brochet* and coworkers report on the use of optical pyrometers with rapid response times to observe brightness temperatures for explosions in liquid nitromethane.

Chapter IV, Explosions, is concerned with the effects of combustion of large volumes of explosive gaseous mixtures with and without confinement. Several contributions report on investigations of the ignition by either a free jet or a detonation issuing from a small tube into the volume. *Ungut* and coworkers report their observations of the transition (or failure phenomena) in ethane or propane in oxygen-nitrogen mixtures and correlated critical pipe diameters to the observed detonation cell sizes for a detonation passing from a circular tube. *Benedick* and coworkers show that the passage of a detonation through a rectangular orifice is influenced by orifice dimensions and that, below critical dimensions, detonation failure occurs. *Schildknecht* and coworkers report that pressure buildup and flame propagation in an external cloud depends strongly on the properties of the free jet ignition sources. *Hjertager* and coworkers' study was concerned with the effects of turbulence generated by obstacles on a flame propagating in explosive gas mixtures. Their results suggest that the largest rates of combustion are found in the turbulence shear layer. *Brossard* and coworkers summarize the results of several large-scale French tests on air blasts generated by unconfined gaseous detonations. *Stock* and coworkers report on an experimental investigation of shock-flame interactions, while *Oran* and coworkers report on a model study of vorticity generation by the interaction of a weak shock with a flame. An investigation by *Fry* and *Book* deals with modeling the blast effects of large quantities of explosives of finite dimensions with the flux corrected transport convective-equation solver. The results provide insight into the structure of the phenomena.

The companion volume includes papers on premixed flames, diffusion flames, turbulent combustions, constant combustion, spray combustion, nonequilibrium flows, and combustion diagnostics (Volume 95 in the *AIAA Progress in Astronautics and Aeronautics* series). Both volumes, we trust, will help to satisfy that need first articulated in 1966 and will continue the tradition of contribution to our understanding of the dynamics of explosions and reactive systems begun the following year in Brussels with the first Colloquium. Subsequent Colloquia have been held on a biennial basis since then (1969 in Novosibirsk, 1971 in Marseilles, 1973 in La Jolla, 1975 in Bourges, 1977 in Stockholm, 1979 in Göttingen, 1981 in Minsk, and 1983 in Poitiers). They have now achieved the status of a prime international meeting on these topics and attract contributions from scientists and engineers throughout the world. The *Proceedings* of the First through the Sixth Colloquia have appeared as part of the journal, *Acta Astronautica,* or its predecessor, *Astronautica Acta.* With the publication of the Seventh Colloquium, the *Proceedings* now appear as part of the *AIAA Progress in Astronautics and Aeronautics* series.

Acknowledgments

The Ninth Colloquium was held under the auspices of the Ecole Nationale Supérieure de Mécanique et d'Aérotechnique, Université de Poitiers, France, July 3-8, 1983. Arrangements in Poitiers were made by Dr. J. C. Bellet. The publication of the *Proceedings* has been made possible by grants from the National Science Foundation (USA) and the Army Research Office (USA).

Preparations for the Tenth Colloquium are under way. The meeting is scheduled to take place in August 1985 at the University of California, Berkeley, California.

J. Ray Bowen
Numa Manson
Antoni K. Oppenheim
R.I. Soloukhin
May 1984

Chapter I.   Detonations in Gaseous Mixtures

Direct Initiation of Planar Detonation Waves in Methane/Oxygen/Nitrogen Mixtures

Shigeharu Ohyagi,* Teruo Yoshihashi†
and
Yasuo Harigaya‡
Saitama University, Urawa-Shi, Saitama-Ken, Japan

Abstract

A study of the direct initiation of planar detonation waves in stoichiometric methane-oxygen mixtures diluted with nitrogen is described. The initiation is accomplished by a shock wave produced from detonation of stoichiometric hydrogen-oxygen mixtures at various initial pressures in a driver section. The initiation energy is estimated by measurements of the decay of blast waves in air produced by the initiators. The range of the initiation energy is 0.2 to 0.8 MJ/m^2. In experiments close to the limit of detonability, records from pressure transducers and ion probes show that the wave systems consist of the leading shock wave and the combustion wave behind it. The test gas is ignited by initiator gaseous detonation products. The following combustion wave produces pressure waves which enhance the leading shock wave until transition to a detonation wave occurs. For a particular dilution ratio with nitrogen, reduction of the initiation energy results in ignition failure of the test gas and decay of the initiator produced shock wave as a nonreactive blast wave. The critical initiation energy for a planar detonation wave in a stoichiometric mixture of methane and oxygen diluted with nitrogen is measured as a function of the dilution ratio. The values obtained are extrapolated to the methane-air

---

Presented at the 9th ICODERS, Poitiers, France, July 3-8, 1983. Copyright © American Institute of Aeronautics and Astronautics, Inc., 1984. All rights reserved.
*Lecturer, Department of Mechanical Engineering.
†Research Technician, Department of Mechanical Engineering.
‡Research Assistant, Department of Mechanical Engineering.

case and are found to be about 1 MJ/m^2 corresponding to 80 MJ for the spherical case.

Introduction

Most of the recent studies of the initiation mechanism of gaseous detonation in an unconfined medium have been concerned with spherical detonation waves initiated by solid explosives or electric sparks. Matsui and Lee (1979) adopted the linear detonation tube method of initiation used by Zel'dovich et al.(1956) to obtain the critical initiation energy for various fuels. The initiation energy is estimated as the effective work done by the combustion products behind the planar CJ detonation wave as they expand from the linear tube into the test chamber. The critical energies obtained by Matsui and Lee are within an order of magnitude of data obtained by other method of initiation of spherical detonations.

Edwards et al.(1976) and Bull et al.(1978) have used a microwave interferometry technique to observe the transition from strong blast waves produced by solid explosive charges to detonation under spherical confinement. Their results show that, under the critical conditions, propagation velocity of the blast wave decays to subCJ velocity at first and then reaccelerates to the CJ velocity of the medium. Bull and coworkers suggested that the kernel theory of Lee et al. (1976) could not be applied to model the critical behavior of the transition from blast to detonation, since the separation between the shock and reaction zone was too large. Wolanski et al.(1981) detonated methane-air mixtures with a shock wave produced by detonation of hydrogen-oxygen in a linear tube. The initiation energy was defined as the chemical energy of the initiator gas per unit crosssectional area of the tube. The explosion radius calculated at this energy was equated to that calculated for the spherical wave point source energy, to compare their results with the spherical experimental data. The critical initiation energy for a spherical detonation wave in a methane-air mixture was estimated to be 47.6 GJ, which is considerably larger than the 88 MJ reported by Bull et al.(1978). Wolanski et al. concluded that not all of the chemical energy contained in the initiator tube was transferred to the acceptor gas.

The previous works indicate the need to define initiation energy precisely to know the efficiency of energy transfer to the test gas, and to know also the details of transition process from a blast to detonation wave. This paper describes experimental results on the blast initiation of planar detonation waves in stoichiometric mixtures of methane-oxygen-nitrogen with variable dilution ratio. The classical type simple detonation tube is used, in which a driver section is filled with stoichiometric mixture of hydrogen and oxygen at various pressures and a driven section is filled with the test gases at an atmospheric pressure. The transitional process is observed by pressure transducers and ionization probes. The initiation energy is estimated from information on the decay process of a planar blast wave generated in air by the initiator system. The critical energy for the developed detonation is roughly estimated as a function of the nitrogen dilution ratio.

Experimental

A schematic diagram of the present experimental apparatus is shown in Fig. 1. The linear detonation tube is 3475-mm long and 30-mm in diameter. The tube is made of stainless steel and is divided by Mylar (0.03-mm thick) into a driver section of 595 mm and a driven section of normal 2880-mm length but extendable to a length of 3470 mm. The gases are prepared in small mixing tanks and composition controlled by adjustment of partial pressure of each component. Ignition is by an electric spark at the closed end of the driver tube. In the driven section, wall-mounted piezo-electric pressure transduces (PCB 113A24) are placed so as to record pressure variations. In this way the arrival time of the shock wave can be

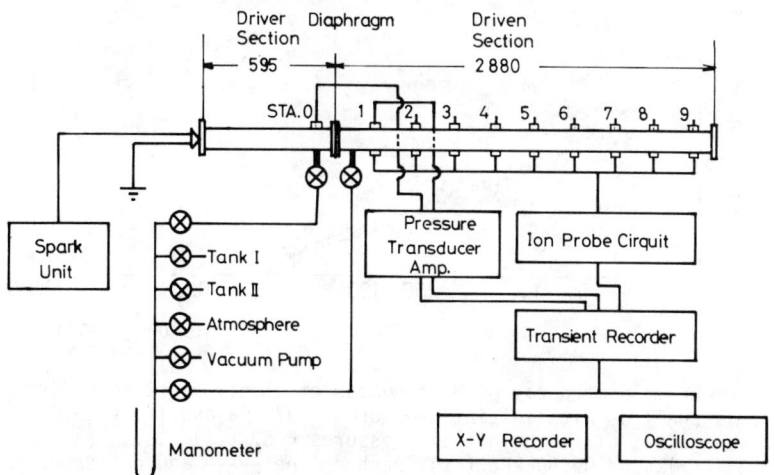

Fig. 1 Schematic diagram of experimental apparatus.

measured. Arrival times of the reaction zone, i.e., the flame, are detected by ionization probes mounted along the tube wall. The arrival time of the driver gas detonation wave at the point 495-mm downstream from the igniter is considered as a standard time. There are 9 stations located 400-, 700-, 1000-, 1300-, 1500-, 1800-, 2100-, 2400-, and 2700-mm downstream from the standard station, STA.0. At each station, the pressure transducer and the ionization probe can be mounted simultaneously at opposite sides of the tube wall to detect the shock-reaction front structure of the wave.

In the present experiments, the driver gas is a stoichiometric hydrogen-oxygen mixture with initial pressures equal to 101.3, 79.97, 53.33, and 39.99 kPa. The mixtures are referred to as driver No. 1, 2, 3, and 4, respectively. A Shchelkin spiral is inserted into the driver tube to ensure an early transition to detonation. That the transition has occurred is checked from the pressure profiles detected at station 0. Details of the experiments to determine the initiation energy of the driver gas has been described elsewhere(Ohyagi et al. 1982). The decay of Mach number and peak overpressure of a planar blast wave formed in air by the initiators are observed at several positions and several initial air pressures (99.99, 53.33, 39.99, 13.33, and 6.67 kPa) to obtain decay curves. These results are compared with theoretical curves for point source blast waves and for various initiation

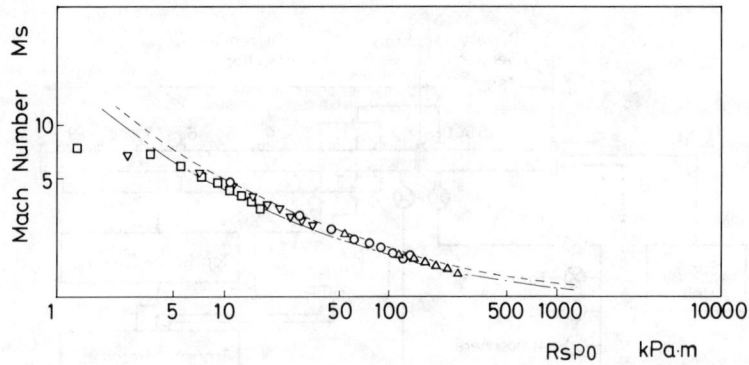

Fig. 2 Example of decay of Mach number of planar blast wave. Driver gas is $2H_2 + O_2$ with initial pressure 79.97 kPa and l/d = 19.8; driven gas is air with initial pressures 6.67 ($\square$), 13.33 ($\triangledown$), 53.33 ($\bigcirc$), and 99.99 kPa ($\triangle$). Dashed line ----- and dot-dash line —·—·— denote the results from the quasisimilar theory with assumed initiation energies equal to 616 and 490 kJ/m^2, respectively.

Table 1 List of test runs of the present experiments with driver gas $2H_2 + O_2$ at $\bar{p}_0$ kPa and test gas $CH_4 + 2O_2 + 2\beta N_2$ at 99.99 kPa

Run	Driver	$\bar{p}_0$(kPa)	β
1	No. 1	101.3	3.76
2	No. 1	101.3	3.0
3	No. 1	101.3	2.0
4	No. 2	79.97	3.0
5	No. 2	79.97	2.5
6	No. 3	53.33	3.0
7	No. 3	53.33	2.5
8	No. 3	53.33	2.0
9	No. 3	53.33	1.5
10	No. 3	53.33	1.0
11	No. 3	53.33	0.5
12	No. 3	53.33	0.0
13	No. 4	39.99	2.0
14	No. 4	39.99	1.5
15	No. 4	39.99	1.0
16	No. 4	39.99	0.5

energies. The method of solution is the quasisimilar technique developed by Oshima (1965). The results of the determination of the initiation energy are briefly described in the following section.

The experiments on methane were performed with the nitrogen dilution as a parameter. The equivalence ratio and the initial pressure were fixed at unity and 99.99 kPa, respectively. Table 1 lists combinations of driver and test gases in the present experiments, where β is the dilution ratio defined as the volume ratio of nitrogen to oxygen gases.

Results and Discussion

Initiation Energy

A plot of the Mach number M_S of blast waves in air as a function of the product of initial pressure p_0 and shock wave position R_S is shown in Fig. 2. For point source energy deposition, the Mach number M_S of a planar blast wave is a function of only two parameters: the specific heat ratio γ and the nondimensional wave position R_S/R_0 where R_0 is the explosion radius defined as

$$R_0 = E_0/2\gamma p_0 \qquad (1)$$

E_0 is the initiation energy per unit area of the planar source and p_0 the initial pressure. For fixed initiation energy and specific heat ratio, M_S depends only on the product $R_S p_0$ so that the experimental values should lie on a single curve in this type of figure. The broken line and the dashed line in Fig. 2 represent the theoretically obtained decay curve using Oshima's method for two extreme

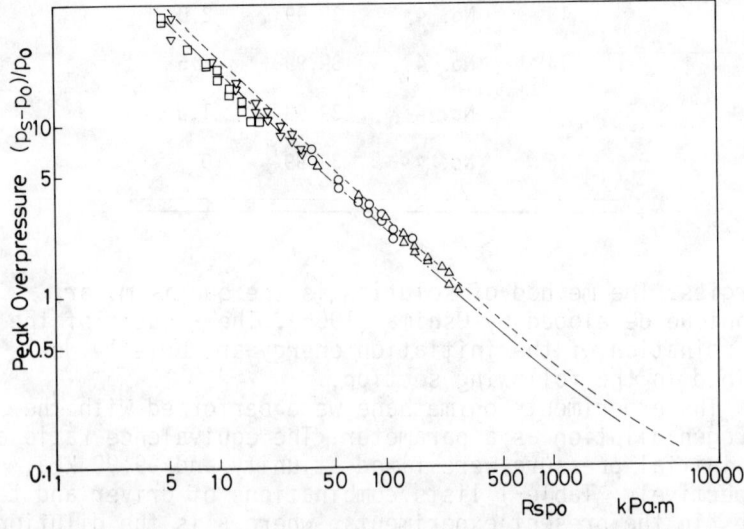

Fig. 3 Example of decay of peak overpressure of planar blast wave. Condition and keys are same as those in Fig. 2.

values of the initiation energy; nearly all of the experimental points lie between these lines.

Agreement between experiments and theory is quite good except for the range where M_S exceeds 5. In this region, the blast wave may be much affected by the detonation wave of the driver gas. Even for the case of air experiments, ionization probes detect the arrival of an ionized zone in the region where $R_S p_0$ is small. This ionized zone represents a contact surface between the driven gas and detonation products of the driver gas. For M_S larger than 5, the contact surface exists close behind the shock front so that the motion of the shock wave is dominated by this surface. Peak overpressure of the blast wave as a function of $R_S p_0$ is shown in Fig. 3. The symbol p_S denotes the pressure of the shock front. The average of these two extreme values of initiation energy is adopted as the initiation energy of the driver at 79.97 kPa. Similarly, the initiation energy can be estimated for other driver conditions. The list of the initiation energies obtained for the present experiments is shown in Table 2. The values of the explosion radius R_0 calculated for $\gamma = 1.4$ and $p_0 =$

Table 2 The properties of the present drivers at 298 K

Driver	$\bar{p}_0$(kPa)	E_0(MJ/m^2)	R_0(m)	E_0''(MJ)	Tetryl(kg)
No. 1	101.3	0.812	2.9	42.9	10.7
No. 2	79.97	0.553	1.98	13.7	3.43
No. 3	53.33	0.308	1.10	2.34	0.585
No. 4	39.99	0.185	0.66	0.506	0.140

Driver	M_{CJ}	p_{CJ}(MPa)	Q_{CJ}(MJ/kg)	η
No. 1	5.331	1.938	6.619	0.420
No. 2	5.307	1.519	6.507	0.369
No. 3	5.265	1.000	6.318	0.317
No. 4	5.235	0.739	6.194	0.259

99.99 kPa are also listed. For comparison with spherical energy, E_0'' is calculated on the assumption that for the same explosion radius the planar wave and the spherical wave have equivalent initiation energy. The Chapman-Jouguet values of the driver gas are listed, i.e., M_{CJ}(CJ Mach number), p_{CJ}(CJ wave front pressure), Q_{CJ}(CJ chemical heat release), all of which are calculated under the assumption of chemical equilibrium. The efficiency with which energy is transferred to the acceptor gas per unit amount of chemical energy contained initially in the driver η is

$$\eta = E_0/\bar{\rho}_0 Q_{CJ} l \qquad (2)$$

where $\bar{\rho}_0$ and l denote the initial density of the driver gas and the length of the driver section, respectively. The efficiency increases as M_{CJ} of the driver gas increases, so that the energy release rate affects the blast wave propagation. It should be noted that the initiation energy can be

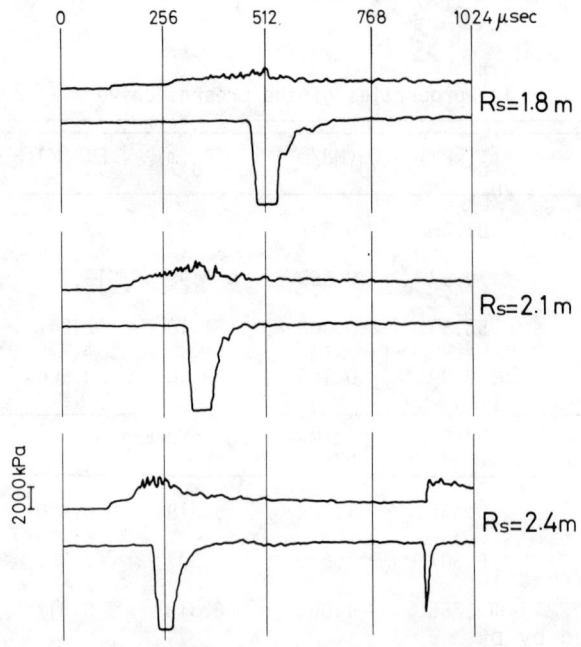

Fig. 4 Examples of pressure traces (upper) and ionization current records (lower) for each station R_s; driver gas is $2H_2 + O_2$, 79.97 kPa and driven gas is $CH_4 + 2O_2 + 5N_2$ ($\beta = 2.5$), 99.99 kPa.

INITIATION OF PLANAR DETONATION WAVES

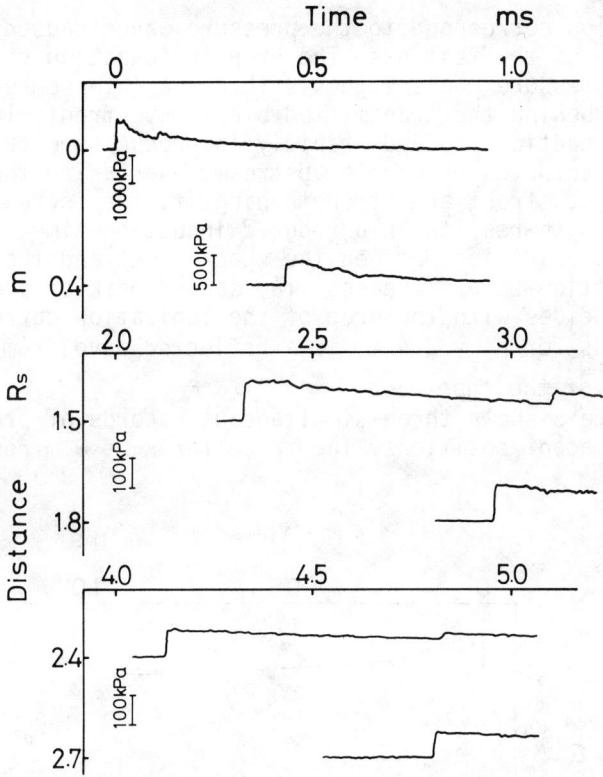

Fig. 5 Examples of pressure traces for six stations when detonation is not initiated; driver gas is $2H_2 + O_2$, 53.33 kPa and driven gas is $CH_4 + 2O_2 + 6N_2$ ($\beta = 3$), 99.99 kPa. The trace for $\dot{R}_s = 0$ m is caused by driver detonation.

obtained from the point of view of the acceptor gas rather than the donor gas, i.e., initiator. The energy thus obtained can be used as a universal value independent of the type of the initiator.

Wave Transition Process

The blast transmitted from the driver to the driven section filled with a combustible mixture initiates chemical reactions, if the initiation energy is sufficient. The wave system — the blast wave with chemical reaction — is elucidated by pressure traces and ionization current records, as shown in Fig. 4. The very small pressure increases at the outset correspond to the transmitted blast wave, and the larger peaks with considerable oscillations which follow correspond to the pressure waves caused by combustion of the test gas. The drop in ionzation currents near the pressure peaks suggests that the intense reaction zone lags behind the transmitted blast wave front. In this zone the reaction proceeds rapidly to produce pressure waves. These waves propagate upstream, overtaking the leading shock front and strengthening it. The increased shock strength results in a reduced induction time, and hence the separation between the shock front and the intense reaction zone decreases. The second peak of pressure, which coincides with the drop of the ionization current in the records for $R_S = 2.4$ m, is a reflected wave from the end plate of the tube.

Figure 5 shows three simultaneous records of pressure at two adjacent positions. The trace for $R_S = 0$ m repre-

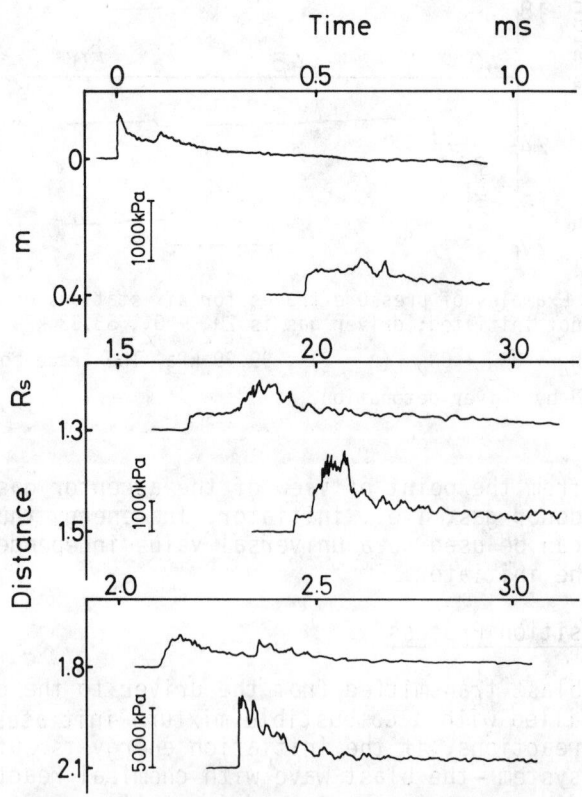

Fig. 6   Examples of pressure traces for six stations when detonation is initiated; driver gas is $2H_2 + O_2$, 39.99 kPa and driven gas is $CH_4 + 2O_2 + 2N_2$ ($\beta = 1$), 99.99 kPa. Detonation in the test gas is established at $R_S = 2.1$ m.

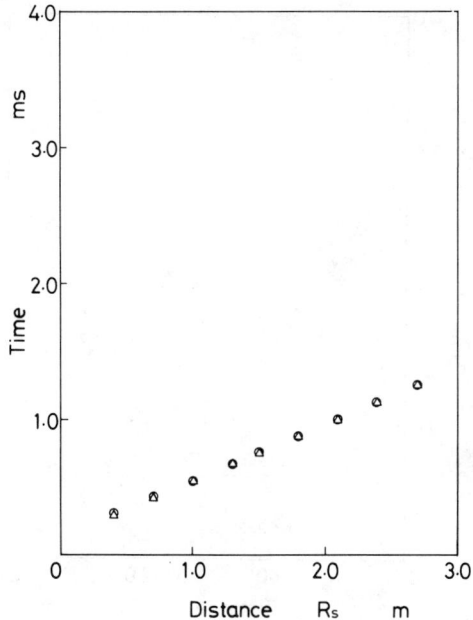

Fig. 7 Movements of shock and reaction fronts in mixture $CH_4 + 2O_2$ ($\beta = 0$), 99.99 kPa using $2H_2 + O_2$, 53.33 kPa as a driver. ○ denotes shock front and △ denotes reaction front.

sents a pressure profile of the detonation in the driver gas. In this case, the transmitted shock wave cannot be influenced by the chemical reaction of the test gas. At first, pressure profiles are affected slightly by the driver detonation. As the wave propagates downstream, the peak pressure and the pressure fluctuations decay monotonically as an inert blast wave. Figure 6 shows the case when detonation of the test gas has been initiated. The combustion induced pressure wave appears to be significant at R_S = 1.3 m and is strengthened as the wave propagates and merges with the preceding shock wave at R_S = 1.8 m. The overdriven detonation is seen at R_S = 2.1 m. Figures 7-10 are typical distance-time diagrams. For $\beta = 0$, at every measuring station the arrival of the shock front and flame front coincides and the transition to steady-state propagation is achieved rapidly. As β increased to unity, the two fronts do not merge until they reach a point 2.1 m from the standard point. Figure 9 shows the case for $\beta = 2$ where the two fronts never merge in the test tube. In this case,

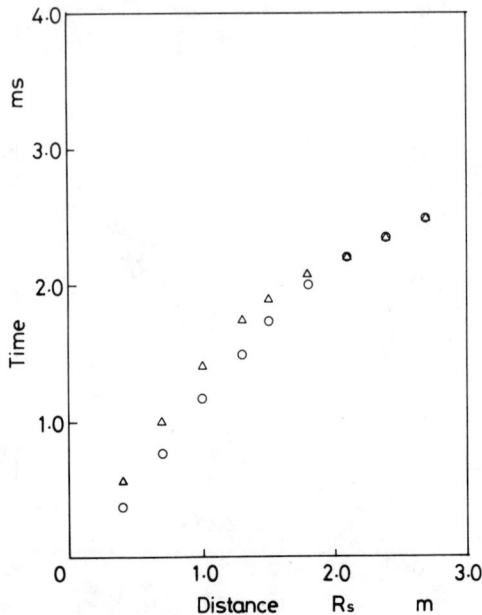

Fig. 8 Movements of shock and reaction fronts in mixture CH_4 + $2O_2$ + $2N_2$ (β = 1), 99.99 kPa using $2H_2$ + O_2, 53.33 kPa as a driver. ○ denotes shock front and △ denotes reaction front.

the phenomenon is very unstable so that among the several test runs there are some cases where the flame front does not appear. The case for β = 3 is shown in Fig. 10 which corresponds to Fig. 5. In this case, no reaction occurs in the test gas, and the reaction front appearing in this figure is the contact surface between the driven gas and the combustion products of the driver gas.

Figures 11-16 show the Mach number variation of the wave propagation. These Mach numbers are defined as the propagation velocities of the first peak of the pressure traces divided by the sound velocity of the undistubed medium. For β = 0, Fig. 11 shows that M_S decays at first and soon develops to M_{CJ}. Figure 12 shows the case for β = 0.5 where the transmitted shock wave decays at first to about M_S = 2, re-accelerates to overdiven states, and decays to the CJ state. M_{CJ}'s in these figures are the calculated values. Experiments at higher initiation energy were not performed because detonation will occur for these cases. The distance where the overdriven detonation appears

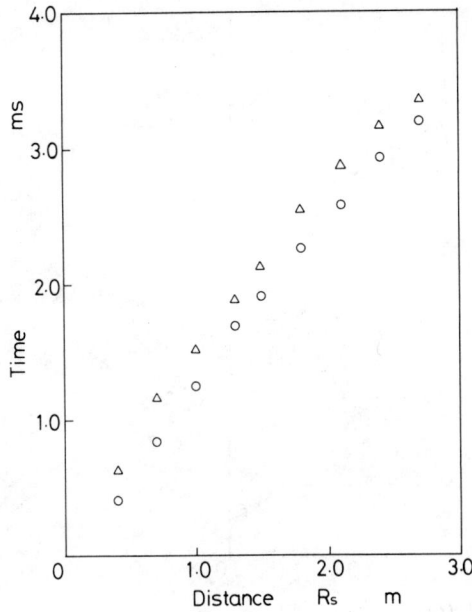

Fig. 9 Movements of shock and reaction fronts in mixture $CH_4 + 2O_2 + 4N_2$ ($\beta = 2$), 99.99 kPa using $2H_2 + O_2$, 53.33 kPa as a driver. $\circ$ denotes shock front and $\triangle$ denotes reaction front.

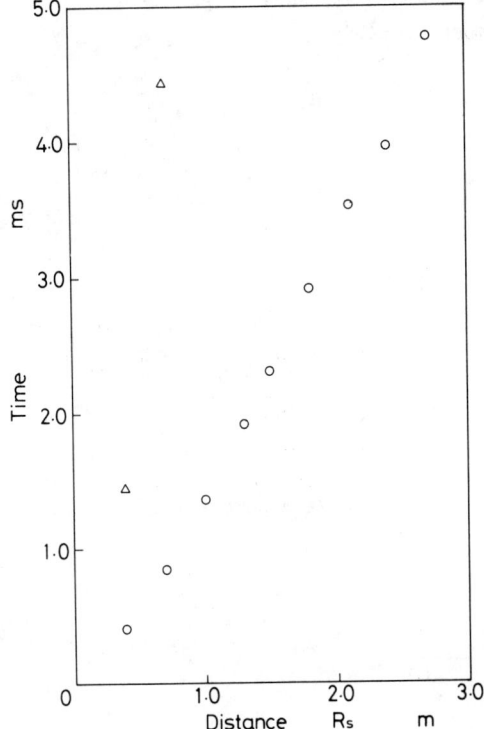

Fig. 10 Movements of shock and reaction fronts in mixture $CH_4 + 2O_2 + 6N_2$ ($\beta = 3$), 99.99 kPa using $2H_2 + O_2$, 53.33 kPa as a driver. $\circ$ denotes shock front and $\triangle$ denotes reaction front.

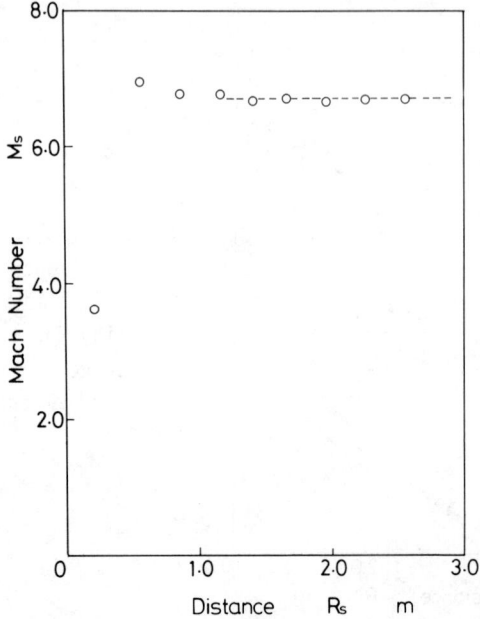

Fig. 11 Variation of wave Mach number with distance in mixture $CH_4 + 2O_2$ ($\beta = 0$), 99.99 kPa using $2H_2 + O_2$, 53.33 kPa as a driver. Dashed line denotes CJ Mach number 6.697.

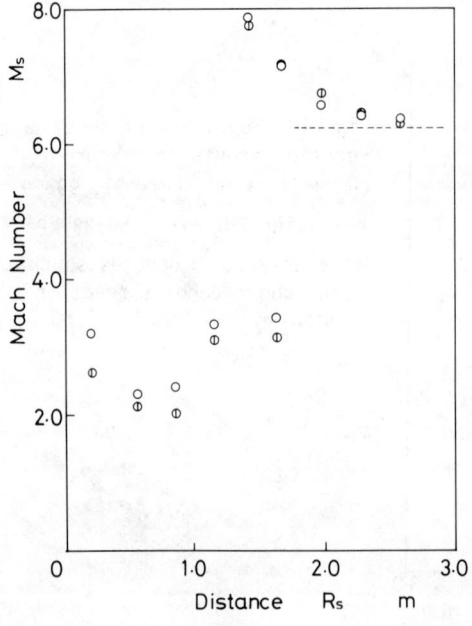

Fig. 12 Variation of wave Mach number with distance in mixture $CH_4 + 2O_2 + N_2$ ($\beta = 0.5$), 99.99 kPa using $2H_2 + O_2$, 53.33 kPa (O), and 39.99 kPa (Φ) as drivers. Dashed line denotes CJ Mach number 6.257.

INITIATION OF PLANAR DETONATION WAVES

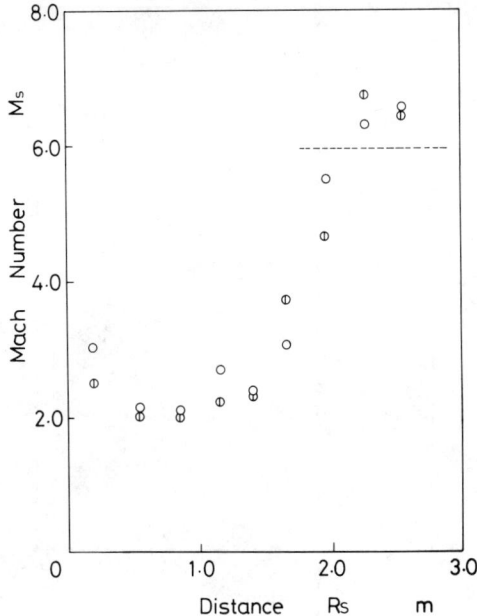

Fig. 13 Variation of wave Mach number with distance in mixture $CH_4 + 2O_2 + 2N_2$ ($\beta = 1$), 99.99 kPa using $2H_2 + O_2$, 53.33 kPa (o), and 39.99 kPa (⊕) as drivers. Dashed line denotes CJ Mach number 5.972.

for $\beta = 1$ (Fig. 13) is greater than that for $\beta = 0.5$ (Fig. 12), while the minimum Mach number in the cource of propagation remains unchanged at 2. Increasing β to 1.5 (Fig. 14), the situation becomes very marginal. M_S for driver No. 4 fluctuates markedly because the phenomenon is very unstable. For driver No. 3, overdriven detonation appears at the last measuring point. Further increase of β results in suppression of initiation of detonation even for a high-energy initator such as driver No. 1 (Fig. 15). In these cases where the wave Mach number remains nearly constant at a value around 2 or 3, the wave consists of the shock front and the following flame front, and the phenomena are very unstable so that, if the tube were considerably longer than this tube, M_S would evolve to M_{CJ} or it would decay to unity. However, it is also possible that this quasisteady propagation continues. In these quasisteady regimes, the wave system is sustained by the net of chemical heat release and the heat losses to the tube in the induction zone. The phenomena in this case depend strongly on the tube diameter. In future work the effect of the tube

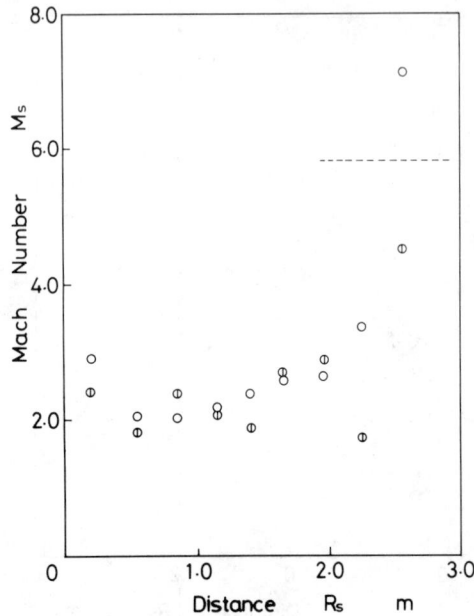

Fig. 14 Variation of wave Mach number with distance in mixture $CH_4 + 2O_2 + 3N_2$ ($\beta = 1.5$), 99.99 kPa using $2H_2 + O_2$, 53.33 kPa (O), and 39.99 kPa (Φ) as drivers. Dashed line denotes CJ Mach number 5.758.

diameter of the driven section will be studied. Figure 16 shows the case $\beta = 2.5$ in which the wave decays to a sound wave for driver No. 3, while remaining quasisteady for driver No. 2.

Figure 17 is a summary of the present experiments. In this figure a circle denotes the case in which a detonation wave is formed and the symbol x denotes an inert blast wave which decays monotonically. The dashed line denotes the probable limit of initiation. The phenomena near this line are very unstable and may be affected by the tube diameter. The limit for methane-air detonation initiation cannot be obtained for the present initiators which have very small energy. The critical initiation energy for methane-air is estimated to be 1 MJ/m^2 by an extrapolation of the results computed from the experimental data for various values of β to the value of β (= 3.76) for air. Conversion of this value into an equivalent spherical initiation energy gives 80 MJ, which is in good agreement with the results of Bull et al. (1978).

INITIATION OF PLANAR DETONATION WAVES

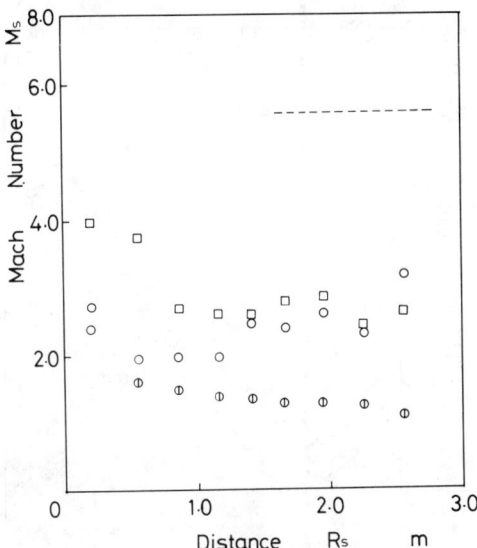

Fig. 15 Variation of wave Mach number with distance in mixture $CH_4 + 2O_2 + 4N_2$ ($\beta = 2$), 99.99 kPa using $2H_2 + O_2$, 101.3 kPa (□), 53.33 kPa (○), and 39.99 kPa (Φ) as drivers. Dashed line denotes CJ Mach number 5.585.

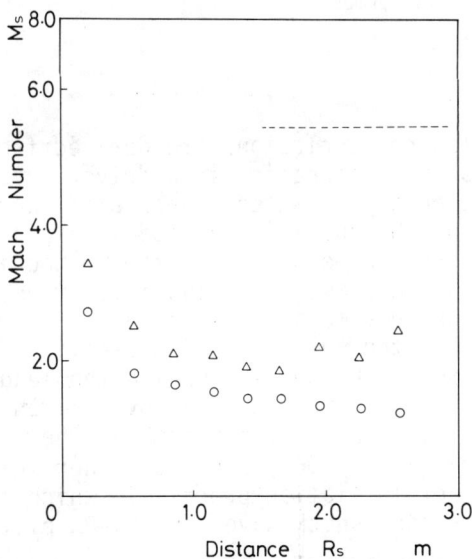

Fig. 16 Variation of wave Mach number with distance in mixture $CH_4 + 2O_2 + 5N_2$ ($\beta = 2.5$), 99.99 kPa using $2H_2 + O_2$, 79.97 kpa (△), and 53.33 kPa (○) as drivers. Dashed line denotes CJ Mach number 5.439.

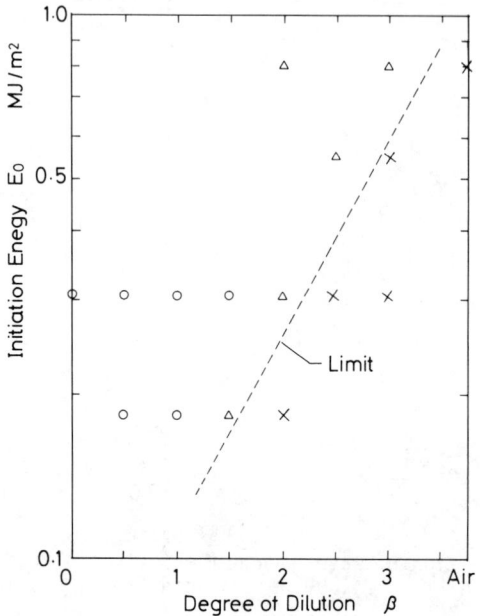

Fig. 17 The summary of the present experiments in E_0-β diagram.
O denotes the case when detonation is established, △ the case when shock-reaction wave complex is formed, and x the case when no reaction wave is formed.

Conclusions

The following conclusions are derived from this work. As hydrogen-oxygen detonation in a driver tube produces in a driven tube an approximated planar blast wave whose decay is described by the quasisimilar model of a planar explosion, the theory can be used to estimate the initiation energy. For such a blast wave, the initiation process of a detonation in a methane-oxygen-nitrogen mixture is observed. Before the detonation is established, the flame propagates along with the flow behind the leading shock wave and cause the formation of an overdriven detonation wave. This initiation process resembles the defragration-detonation-transition (DDT) process rather than the so-called direct initiation process in which the mixture is ignited by a strong shock wave. The M_s-R_s plots obtained in this experiment are similar to those obtained by Bull et al. (1978) who initiated detonation in hydrocarbon-air mixtures by solid explosives in unconfined situations. When the initiation energy is near a critical value, the direct initiation processes (even in the spherical cases) may resemble those obtained in these tube experiments. The mixture might be ignited by core (or kernel) of the initiator and generate pressure waves which enhance the leading shock wave. In these processes, the behavior of the core is more important than that of the preceding blast wave. However, the behavior of the core determines the blast wave propagation and the initiation energy. In this sense, the initiation energy of the driver determines the transition process.

The critical initiation energies were roughly evaluated for dilution ratios of nitrogen β ranging from 1.5 to 3, giving an extrapolated value for methane-air of about 1 MJ/m^2. This value is very close to that obtained by Bull et al.(1978) if it is converted to the spherical initiation energy.

Acknowledgments

This work represents a portion of the results obtained in a research program under the financial support of Grant-in-Aid for Scientific Research (Nos. 56350005, 57750043, and 57306016) from the Ministry of Education of Japan.

The authors acknowledge the helpful advice of Professor T. Asaba (University of Tokyo) and Dr. H. Matsui (Ministry of Labor of Japan) in performing the experiments and acknowledge also the discussion of their group.

References

Bull, D. C., Elsworth, J. E., and Hooper, G. (1978) Initiation of spherical detonation in hydrocabon/air mixures. Acta Astronautica 5, 997-1008.

Edwards, D. H., Hooper, G., and Morgan, J. H. (1976) An experimental investigation of the direct initiation of spherical detonations. Acta Astronautica 3, 117-130.

Lee, J. H. and Ramamurthi, K. (1976) On the concept of the critical size of a detonation kernel. Combustion and Flame 27, 331-340.

Matsui, H. and Lee, J. H. (1978) On the measure of the relative detonation hazards of gaseous fuel-oxygen and air mixtures. Seventeenth Symposium (International) on Combustion, pp. 1269-1280. The Combustion Institute, Pittsburgh, Pa.

Ohyagi, S., Yoshihashi, T., and Harigaya, Y. (1982) On the initiation energy of planar blast waves produced by gaseous detonations, (in Japanese) Proceedings of Symposium on Shock Technology, pp. 20-27. The Institute of Space and Aeronautical Science, Tokyo.

Oshima, K. (1960) Blast waves produced by exploding wire. Rept. of Aeronautical Research Institute, Univ. Tokyo, No. 358, 137-194.

Wolanski, P., Kauffman, C. W., Sichel, M., and Nicholls, J. A. (1981) Detonation of methane-air mixtures. Eighteenth Symposium (International) on Combustion, pp. 1651-1660. The Combustion Institute, Pittsburgh, Pa.

Zel'dovich, I. B., Kogarko, S. M., and Simonov, N. N. (1956) An experimental investigation of spherical detonation of gases. Soviet Physics-Technical Physics 1, Pt. 8, 1689-1713.

Measurements of Cell Size in Hydrocarbon-Air Mixtures and Predictions of Critical Tube Diameter, Critical Initiation Energy, and Detonability Limits

R. Knystautas,* C. Guirao,† J. H. Lee,‡ and A. Sulmistras§
McGill University, Montreal, Canada

Abstract

Experimental measurements of the detonation cell size in mixtures of H_2, C_2H_2, C_2H_4, C_2H_6, C_3H_8, and C_4H_{10} with air over a range of fuel concentrations have been carried out in three cylindrical tubes of diameters 5, 15, and 30 cm. The cell size has been determined from the signatures on smoked aluminum foils placed inside the tube as well as from the frequency of the pressure fluctuations recorded by piezoelectric transducers. Based on the cell size data obtained, estimates of the critical tube diameter using the empirical law of Soloukhin and Mitrofanov ($d_c \simeq 13\lambda$) have been found to be in agreement with experimental data from direct measurement of the critical tube itself. Hence, the important empirical law $d_c \simeq 13\lambda$ is thus verified. Estimates of the critical charge weight from the cell size data using the surface energy theory proposed by Lee have been found to agree reasonably well with the experimental results of Elsworth. Based on the criteria for stable propagation in tubes ($d^* \simeq \lambda/\pi$) and in two-dimensional channels ($W^* \simeq 3\lambda$), detonability limits can also be predicted from a knowledge of the cell size λ. Based on Westbrook's kinetic calculations, it is found that the cell size data are directly proportional to the induction time of the oxidation process which confirms qualitatively Shchelkin's model. However, on a quantitative basis, Shchelkin's model predicts cell size an order-of-magnitude larger than the present experimental data.

Presented at the 9th ICODERS, Poitiers, France, July 3-8, 1983. Copyright 1984 by the American Institute of Aeronautics and Astronautics, Inc. All rights reserved.
 *Associate Professor, Department of Mechanical Engineering.
 †Senior Research Scientist, Department of Mechanical Engineering.
 ‡Professor, Department of Mechanical Engineering.
 §Undergraduate Student, Department of Mechanical Engineering.

Introduction

An assessment of the relative detonation sensitivity of various fuel-oxygen-nitrogen mixtures was carried out by Matsui and Lee (1978) using the critical initiation energy for spherical detonations as a basis for comparison. The critical initiation energy itself depends on the energy time characteristics as well as the geometry of the source (Bach et al. 1971; Lee et al. 1974) and is thus a rather complex parameter. Furthermore, for the relatively insensitive fuel-air mixtures, the critical initiation energy is usually measured in terms of an equivalent weight of a high explosive charge typically of the order of 100 g or more. This requires rather large volumes of gases for a long enough detonation travel to conclusively determine the stability of the detonation in the case of successful initiation. The use of long rectangular bags rather than spherical balloons simplifies the experiments somewhat but suffers from a certain degree of lateral confinement of the plastic walls as shown by Murray et al. (1981). Thus, a rectangular bag does not simulate a truly unconfined spherical detonation. For insensitive fuels or for near limit mixtures, the critical cross-sectional dimension of the rectangular bag required for stable propagation is also very large and is not known a priori.

In recent years, the critical tube diameter required for the successful transformation of a confined planar detonation into an unconfined spherical wave has been shown to be easily measurable. Since the reinitiation process for successful transformation of a planar into a spherical wave occurs within a distance of the order of one tube diameter from the exit of the tube, even the large-scale experiments required are of rather modest dimensions. Although the critical tube diameter has been related to the critical energy by Lee and Matsui (1977) using a "work-done" concept, a far more important fundamental relationship between the critical tube diameter d_c and the cell size or transverse wave spacing λ has been established in recent years. First observed by Mitrofanov and Soloukhin (1965) and confirmed later by Edwards et al. (1979), it was found that $d_c \simeq 13\lambda$ for C_2H_2-O_2 mixtures at subatmospheric pressures of about 80 Torr. Following the suggestion of Edwards that this empirical law should have a wider applicability, Knystautas et al. (1982) carried out extensive experiments in fuel-oxygen-nitrogen mixtures to measure both the critical tube diameter d_c as well as the cell size λ over a wide range of initial pressures and confirmed the validity of this empirical law of $d_c \simeq 13\lambda$ for the fuels tested (H_2, C_2H_2, C_2H_4,

C_3H_6, C_3H_8, CH_4, C_2H_6, C_4H_{10}, and MAPP). With the critical tube diameter linked to the cell size through this simple empirical law, it is then possible to perform laboratory scale experiments in confined tubes to determine the cell size λ and deduce the critical tube diameter from it. A program for the systematic measurement of the detonation cell size in atmospheric fuel-air mixtures has been carried out. The present paper reports the results obtained and the correlation of the cell size to existing data from actual measurements of critical tube diameter as well as critical initiation energies.

Experimental Details

Three steel tubes (5, 15, and 30 cm in diameter) of length $\approx$ 20 m were used in the present study. The tubes were equipped with ionization probes for the measurement of the detonation velocity and piezoelectric transducers for pressure measurements. Commercial grades of C_2H_2, H_2, C_2H_4, C_2H_6, C_3H_8, and C_4H_{10} and bottled compressed air were used. In the 5- and 15-cm-diam tubes, the mixture of the desired composition was prepared in a continuous flow system with flow rates monitored through standard calibrated rotameters. The tube was purged with the premixed gases for at least 5 tube volumes prior to the experiment. In the larger diameter (30-cm) tube, the mixing procedure was as follows: the tube was first evacuated and the desired volumes of air and fuel were then introduced into the tube via the method of partial pressures. A bellows-type pump was then used to recirculate the gases from one end to the other to permit thorough mixing of the components. Initiation of the detonation was via an exploding wire or a blasting cap depending on the sensitivity of the mixture. In the 30-cm-diam tube, it was often required to use an additional booster charge with the blasting cap for direct initiation. A short length of wire spiral was usually placed at the initiation end to guarantee the rapid formation of the detonation wave. To record the detonation cell signatures, smoked aluminum foils were used. The aluminum foils were usually of width πd and as long a length (in the direction of propagation) as experimentally possible for a more easier estimation of the averaged cell size. For cell sizes large compared to the diameter of the pressure transducer, periodic pressure fluctuations could be seen superimposed on the main pressure trace. The periods of these pressure fluctuations t_c can be used to estimate the cell length L_c ($L_c \simeq t_c V_{CJ}$) which can then be converted into cell size via the approximate geometrical raltionship $\lambda \simeq 0.6\ L_c$.

Thus the pressure trace as well as the smoked foil record provide two independent means for estimating cell sizes. In general, a few experiments had to be carried out for each mixture composition in order to conclusively establish the cell size for the particular mixture.

Results

The averaged detonation cell diameter λ for six gaseous fuels (H_2, C_2H_2, C_2H_4, C_2H_6, C_3H_8, and C_4H_{10}) is plotted against the equivalence ratio ϕ in Fig. 1. Except for hydrogen, the minimum cell size (or the most sensitive composition) corresponds to a mixture composition slightly on the fuel-rich side ($\phi > 1$) rather than stoichiometric ($\phi = 1$). As can be observed, estimates of the cell size from pressure fluctuations are in good agreement with those obtained from smoked foil records. Except for methane (CH_4), all the alkanes (C_3H_8, C_2H_6, C_4H_{10}) appear to have the same sensitivity in that their cell sizes are practically identical. For methane, the use of 50 g of explosive charge failed to cause direct initiation for the stoichiometric composition although Kogarko (1958) reported detonations in fuel-lean (from 6.9 to 8.2% CH_4) and fuel-rich (from 11.1 to 13.5% CH_4) CH_4-air mixtures in a similar tube using 50 and 70 g of explosive charges, respectively. However, we feel that a 50-g charge is the upper limit for safe operation in a university laboratory. Furthermore, Moen (1982) recently performed experiments in a 6-ft-diam tube and estimated from his smoked foil record a cell size of about 33 cm for stoichiometric methane-air mixtures. This would correspond to about 6 times the minimum cell size of about 5.35 cm for the other alkanes (i.e., C_3H_8, C_2H_6, and C_4H_{10}). In increasing order of sensitivity (or decreasing cell size), ethylene (C_2H_4) follows the alkanes and hydrogen (H_2) is slightly more sensitive than C_2H_4. As expected, acetylene is found to be the most sensitive fuel with a minimum cell size $\lambda \simeq 0.565$ cm as compared to $\lambda \simeq 1.5$ cm for H_2, $\lambda \simeq 2.6$ cm for C_2H_4, and $\lambda \simeq 5.35$ cm for the alkane group (C_3H_8, C_2H_6, and C_4H_{10}). For the stoichiometric compositions of the mixtures studied in the present work, cell sizes were determined by Bull et al. (1982) in a detonation tube of rectangular cross section (7.6 X 3.8 cm) using smoked stainless steel foils. Bull's results are $\lambda \simeq 0.92$, 1.0, 2.4, 5.4, and 5.6 cm for C_2H_2, H_2, C_2H_4, C_2H_6, and C_3H_8, respectively, and are in good agreement with the present observations. It is also of importance to note that the width of the typical U-shaped curves for cell size λ vs fuel composition ϕ decreases with decreasing sensitivity of the fuel.

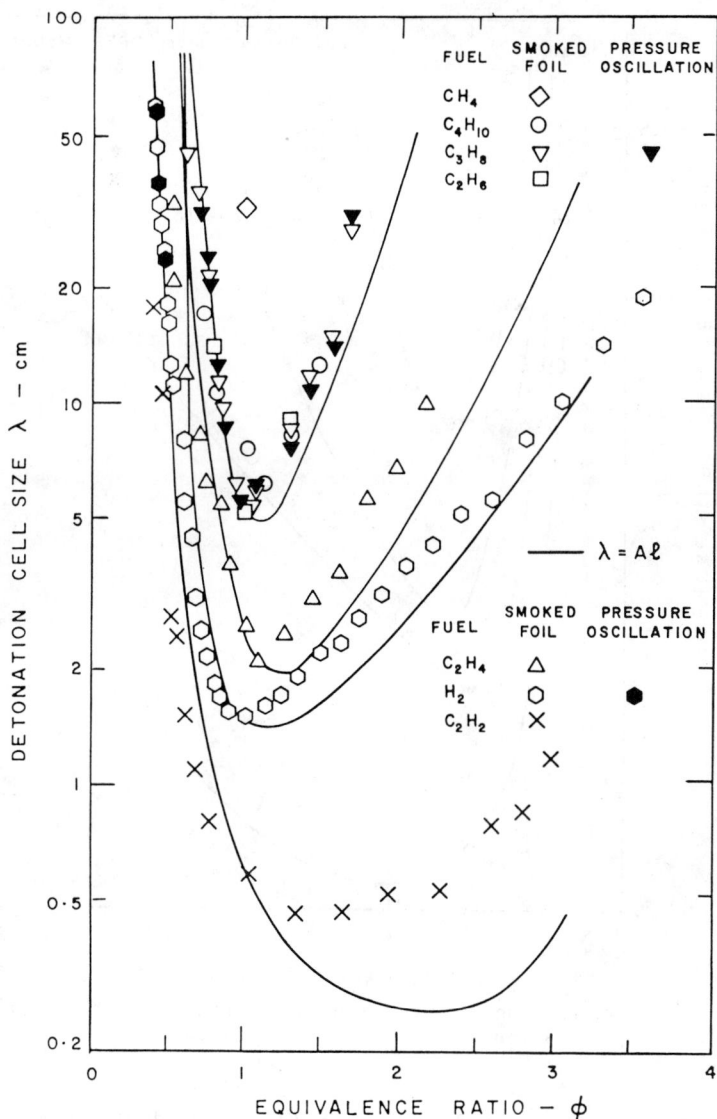

Fig. 1 The variation of measured detonation cell size λ with fuel-air composition and correlation with induction length.

Also, the increase in cell size as the equivalence ratio approaches the rich limit ($\phi > 1$) is in general much slower than towards the lean limit ($\phi < 1$). This universal behavior is reflected in practically all dynamic detonation parameters that are dependent on the induction time of the mixture itself.

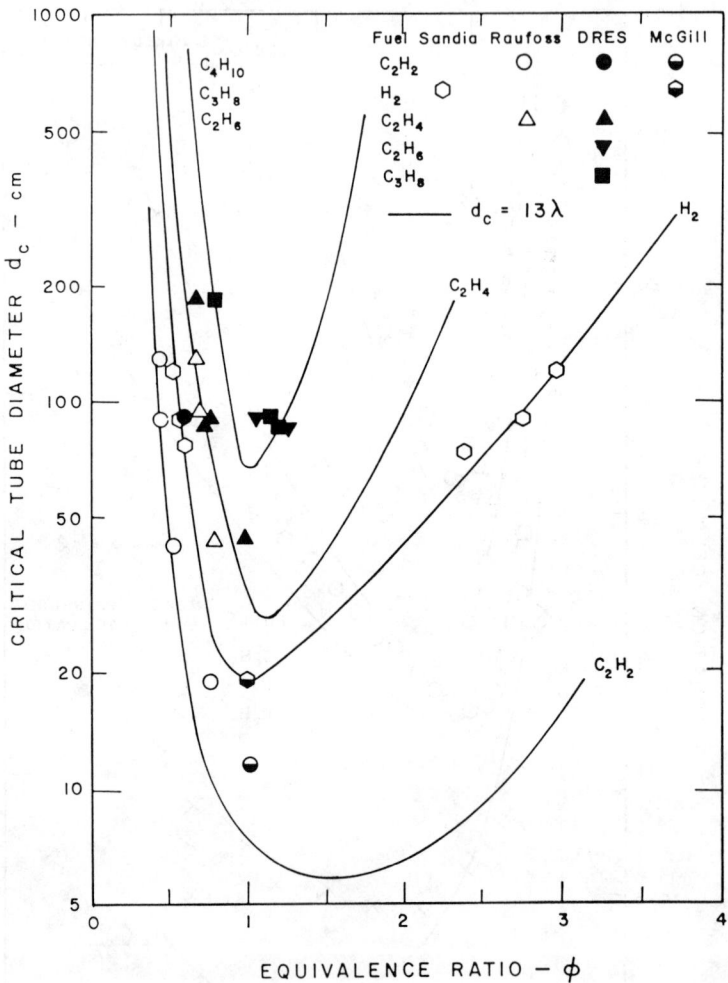

Fig. 2 Correlation of critical tube diameter d_c with the empirical law $d_c \simeq 13\lambda$ for a range of fuel-air mixtures.

Based on cell size data, the critical tube diameter d_c can be estimated from the empirical relationship $d_c \simeq 13\lambda$. The solid curves shown in Fig. 2 represent critical tube diameters estimated from the present data obtained for λ. Although an extensive experimental measurement of the critical tube diameter for fuel-oxygen-nitrogen mixtures was carried out by Knystautas et al. (1982), relatively few data exist for d_c in fuel-air mixtures, in particular for off-stoichiometric compositions. Some of the data that are currently available from large-scale field tests are also shown in Fig. 2. Because of the interest in H_2-air mixtures in connection with nuclear reactor safety, large-scale experiments on the measurements of d_c for both lean and rich H_2-air mixtures were carried out recently by Benedick at Sandia National Laboratory (New Mexico) (Guirao et al. 1982). As can be observed from Fig. 2, the agreement with the estimates of d_c from cell size data via the empirical law $d_c \simeq 13\lambda$ is extremely good for the case of H_2. For ethylene (C_2H_4), a limited number of experiments were carried out at Raufoss (Norway) (Rinnan 1982) and at DRES (Moen et al. 1982) for lean C_2H_4-air mixtures. The results are also in accord with the present estimates from cell size data. For lean acetylene-air mixtures, some direct measurements of d_c were also carried out at Raufoss (Rinnan 1982). Although no special attempt has been made to narrow down the range of mixture compositions between success and failure for transmission in a given tube, the results are also found to be in good agreement with the predictions from cell size data. For stoichiometric C_2H_2-air and H_2-air mixtures, direct measurements of d_c were carried out by Knystautas et al. (1982). For stoichiometric H_2-air mixtures, the agreement is almost perfect. For C_2H_2-air mixtures, the measurements of d_c are slightly higher than those estimated from λ. However, "bottle-to-bottle" variations of the purity of the commercial C_2H_2 used can account for the observed deviation in view of the extreme sensitivity of the C_2H_2-air mixture itself. It would be of interest to extend the critical tube diameter measurements to fuel-rich hydrocarbon-air mixtures as in the case of hydrogen-air mixtures in order to verify the validity of the $d_c \simeq 13\lambda$ correlation over the entire range of fuel compositions between the two limits.

Although the critical initiation energy has been deduced from a knowledge of the critical tube diameter from energy considerations using a "work done" concept, (Lee and Matsui 1977; Urtiew and Tarver 1981) a better model was recently developed by Lee (Lee et al. 1982; Guirao et al. 1982). The model postulates that there exists a minimum surface energy per unit area of the wave front for successful transformation of a planar detonation into an unconfined spherical wave. This minimum "surface energy" is then equated directly to the area of the critical tube (i.e., $\pi d_c^2/4$). Assuming the requirement of a minimum size of a detonation kernel for direct initiation, (Lee and Ramamurthi 1976) the surface area of the critical detonation kernel is then equated to the area of the critical tube (i.e., $4\pi R_s^{*2} = \pi d_c^2/4$). Using strong blast theory to relate the blast energy E_c to the kernel radius R_s^* where the detonation is assumed to be of Chapman-Jouguet strength, the following simple expression was obtained by Lee et al. (1976)

$$E_c = 4\pi\gamma_0 p_0 M_{CJ}^2 I(d_c^3/4) \tag{1}$$

or equivalently

$$E_c = (2197/16)\pi\gamma_0 p_0 M_{CJ}^2 I \lambda^3 \tag{2}$$

where use has been made of the empirical relationship discussed previously $d_c \simeq 13\lambda$. In Eqs. (1) and (2), γ_0 and p_0 denote the specific heat ratio and the initial pressure of the mixture, M_{CJ} is the Chapman-Jouguet detonation Mach No. and I is a dimensionless constant representing the energy integral in strong blast theory (Guirao et al. 1982). Based on the cell size data obtained, the critical initiation energies for the various mixtures studied can be evaluated from Eq. (2). Figure 3 shows a comparison of the predicted results using Eq. (2) and those obtained by Elsworth from direct experimental measurements of the critical weight. For C_2H_4, Murray et al. (1981) have also obtained a few critical initiation charge weights for lean mixtures. The critical energy in Fig. 3 has been expressed as an equivalent charge weight of tetryl (1-g tetryl is equivalent to 4270 J) for easy comparison with Elsworth's experimental data. The agreement in general is reasonably good in view of the simplicity of the model. No experimental data of critical energies have been reported for C_2H_2-air mixtures to permit a comparison with the present prediction from Eq. (2). For most practical situations, Eq. (2) can be used to predict critical initiation energies (or charge weights) with quite acceptable accuracy.

The detonability limits in tubes have also been related directly to the cell size by Lee et al. (1982). Based on the experimental work of Donato (1982) and Moen et al. (1981), it was found that the onset of single-headed spinning detonations corresponds to the limit for stable propagation in the given tube. Donato showed that for mixtures outside the single-head limit, a finite perturbation would cause the detonation to fail and the combustion wave will continue to propagate as a deflagration thereafter. Only for mixtures within the single-head limit would regeneration into the detonation mode occur shortly downstream of the finite perturbation that causes the wave to fail. It should be emphasized that using a strong enough initiator, spinning detonation can always be initiated in a given tube over a wide range of composition beyond the onset limit when spinning first occurs. However, it has been shown that these spinning waves will decay when encountering finite perturbations and after failure will not transit again to

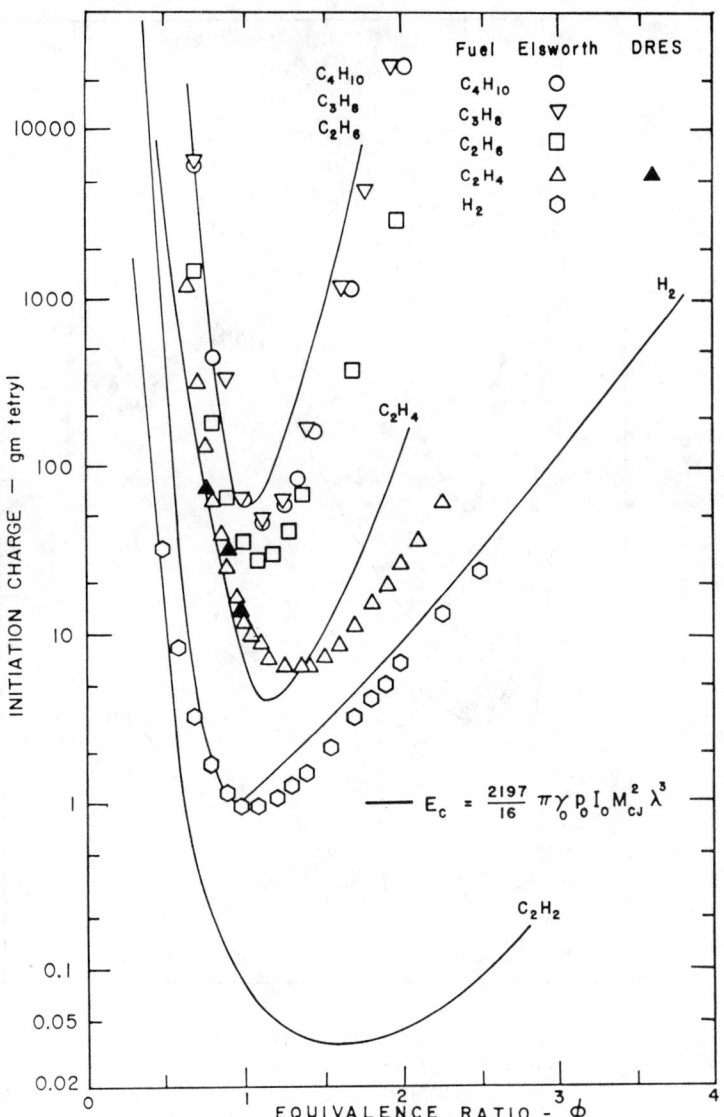

Fig. 3 Correlation of initiation energy E_c with theoretical predictions based on the surface energy model for a range of fuel-air mixtures.

the detonation mode without the support of a strong initiator. However, without a finite perturbation (such as a couple of turns of a Shchelkin spiral), these spinning waves may persist for as much as 100 tube diameters without showing any sign of decay.

Adopting the criterion that the onset of single-head spinning waves in a given tube should correspond to the de-

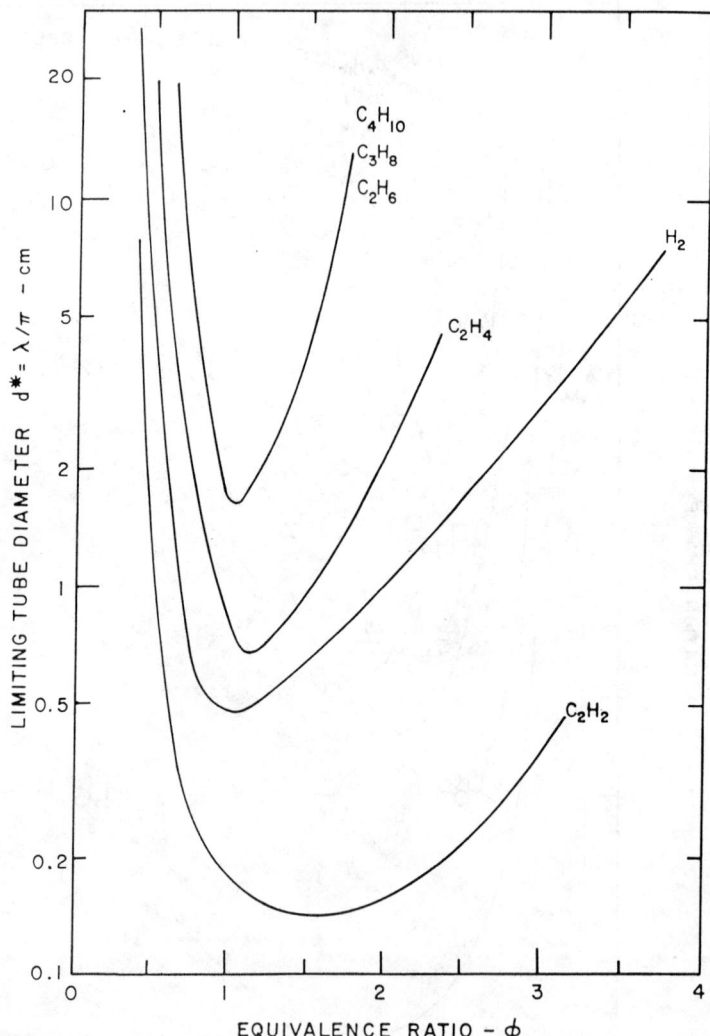

Fig. 4 The variation of the minimum limiting tube diameter $d^* \simeq \lambda/\pi$ for stable propagation of detonation waves in tubes as a function of fuel-air composition.

tonability limits for that tube, the limits in circular tubes can be estimated when cell size data are available. Since the onset of single-head spin corresponds to a cell size of the order of the tube circumference (i.e., $\lambda \simeq \pi d$), the composition limits for a given circular tube can be estimated if λ is known. Figure 4 gives the limiting tube diameter d^* as function of equivalence ratio for the various fuels studied. For a given fuel composition (i.e., a given ϕ), stable detonations cannot be propagated in tubes with a diameter $d < d^*$. Thus, for the case of stoichiometric H_2-air mixtures, $d^* \simeq 0.5$ cm when compared to $d^* \simeq 0.2$ cm for C_2H_2 and $d^* \simeq 1.7$ cm for the alkanes. It is of interest to note that according to the cell size reported by Moen for methane (i.e., $\lambda \simeq 33$ cm), a minimum tube diameter for stable detonation propagation in stoichiometric methane-air mixtures would be of the order of at least 10 cm. Thus, the results reported by Wolanski et al. (1981) in a 5-cm tube should correspond to overdriven transient waves only. This is supported by the fact that they observed single-headed spinning waves over the whole range of fuel concentrations studied. For two-dimensional channels, the limit should correspond to a composition where the channel height W is of the order of the cell size. This is supported by the recent experiments of Vasiliev (1982) who found that stable detonation propagation corresponds to $W \simeq \lambda$.

From the practical point of view, the detonation limits for a thin horizontal layer of explosive mixture bounded by one solid surface (i.e., ground) is of interest. Recent experiments by Liu et al. (1984) for transmission of planar detonations through rectangular slits into unconfined media as well as large scale transmission experiments from two-dimensional channels by Benedick et al. (1983) showed that the critical channel width $W^* \simeq 3\lambda$. Thus it is reasonable to assume that the minimum thickness h of a layer of explosive gas that can support a detonation should be of the order of one and a half cell size (i.e., $h^* \simeq 1.5\lambda$). Thus, for stoichiometric H_2-air mixtures, $h^* \simeq 3$ cm as compared to the alkanes C_3H_8, C_2H_6, and C_4H_{10} for which $h^* \simeq 8$ cm. For CH_4, $h^* \simeq 0.5$ m based on Moen's estimate of $\lambda \simeq 0.33$ m, which is much smaller than commonly expected for methane. Direct experiments must be performed to verify the limit postulate of an unconfined thin layer of explosive.

As suggested by Shchelkin and Troshin (1965), the cell size should be directly proportional to the induction zone thickness. However, no theory exists at present whereby the cell size λ can be predicted a priori from a knowledge of the induction kinetics. However, Westbrook (1980, 1982a, b) has made extensive correlations between experimentally measured λ (or equivalent d_c) and induction times computed from detailed kinetics of the oxidation processes. In general, it can be shown that $\lambda \sim \ell$ where ℓ is the induction length but the constant of proportionality has to be determined by matching with an experimental datum. The proportionality constant depends on how the induction time is computed. For example, Westbrook based his induction time on a constant volume explosion process, while Lee et al. (1982b) integrated the kinetic rate equations along a Rayleigh line as in the classical ZDN model for the detonation structure.

There is no special justification for either method since both have to use an experimental point to find the proportionality constant itself. The solid curves in Fig. 1 represent the estimated cell sizes from kinetic data matching at the stoichiometric composition point. The results are relatively in good agreement with the data despite some discrepancies as the mixture composition departs from the stoichiometric composition which may result from the choice of the matching point. Nevertheless, the reasonable agreement confirms Shchelkin's postulate. However, useful models for linking the cell size λ to the kinetics of the reaction must somehow take into consideration the complex transient gasdynamics processes of the cell interactions.

Conclusions

From the cell size data obtained in the present investigations, critical tube diameter, critical initiation energy (charge weight) as well as detonability limits have been estimated. Comparison with available data for the critical tube from direct experimental measurements confirms the general validity of the empirical law $d_c \simeq 13\lambda$ first observed by Mitrofanov and Soloukhin. However, more extensive experiments should be carried out particularly in fuel-rich mixtures to consolidate the validity of this empirical expression. Due to its fundamental significance, it is also of importance to carry out experiments in fuels other than the hydrocarbons to verify the universality of the $d_c \simeq 13\lambda$ law. The critical initiation energies predicted by the minimum surface energy model using experimental cell size data agree reasonably well with direct experimental measurements. Thus, apart from the practical usefulness of Eq. (2), the linkage of the cell size to the critical initiation energy E_c should elucidate the fundamental mechanisms of the failure and reinitiating processes in the critical tube experiment, the blast wave initiation processes as well as the detailed cell dynamics of the transient shock interactions. This may lead to a theory for the empirical law $d_c \simeq 13\lambda$ itself. Prediction of detonability limits (or minimum tube diameters for stable propagation) from cell size data is an important contribution. However, extensive experiments must be carried out further to verify the postulates for stable propagation (i.e., $d^* \simeq \lambda/\pi$ for circular tubes, $W^* \simeq \lambda$ for two-dimensional channels and $h^* \simeq 1.5\lambda$ for a thin layer bounded by one solid surface). Thus far, the experiments in circular tubes by Donato and two-dimensional channels by Vasiliev are of limited scope. Although it has been shown that all the dynamic detonation parameters (i.e., critical tube diameter, critical initiation energy and detonability limits) can be linked to the cell size, a theory for the prediction of the cell size λ from basic information on the physical chemical properties of the explosive mixture is still lacking.

Acknowledgments

The authors wish to thank I. O. Moen for informing them of the results of his work at DRES and C. Westbrook for providing them with his kinetic calculations. The work is sponsored by US-AFOSR Contract 72-2387, DRES Contract 8SU80, FCAC Grant and NSERC Grant A3347, A7091, and A6819.

References

Bach, G. G., Knystautas, R., and Lee, J. H. (1971) Initiation criteria for diverging gaseous detonations. 13th Symposium (International) on Combustion, pp. 1097-1110. The Combustion Institute, Pittsburgh, Pa.

Benedick, W., Knystautas, R., and Lee, J. H. (1983) Large-scale experiments for the transmission of detonations from a two-dimensional channel, presented at the 9th International Colloquium on Dynamics of Explosions and Reactive Systems, Poitiers, France, July 3-8, 1983.

Bull, D. C., Elsworth, J. E., Shuff, P. J., and Metcalfe, E. (1982) Detonation cell structures in fuel/air mixtures. Combustion and Flame 45(1), 7-22.

Donato, M. (1982) The influence of confinement on the propagation of near-limit detonation waves. Ph.D. Thesis, McGill University, Montreal, Canada.

Edwards, D. H., Thomas, G. O., and Nettleton, M. A. (1979) The diffraction of a planar detonation wave at an abrupt area change. J. Fluid Mechanics 95(1), 79-96.

Guirao, C. M., Knystautas, R., Lee, J. H., Benedick, W., and Berman, M. (1982) Hydrogen-air detonations. 19th Symposium (International) on Combustion, pp. 583-590. The Combustion Institute, Pittsburgh, Pa.

Knystautas, R., Lee, J. H., and Guirao, C. M. (1982) The critical tube diameter for detonation failure in hydrocarbon-air mixtures, Combustion and Flame 48(1), 63-83.

Kogarko, S. M. (1958) Detonation of methane-air mixtures and the detonation limits of hydrocarbon-air mixtures in a large-diameter pipe. Soviet Physics-Tech. Physics 28, 1904-1916.

Lee, J. H., Knystautas, R., and Guirao, C. M. (1974) Critical power density for direct initiation of unconfined gaseous detonations. 15th Symposium (International) on Combustion, pp. 53-67. The Combustion Institute, Pittsburgh, Pa.

Lee, J. H., Knystautas, R., and Guirao, C. M. (1982) The link between cell size, critical tube diameter, initiation energy and detonability limits. Fuel-Air Explosions (edited by J. H. S. Lee and C. M. Guirao) pp. 157-187. University of Waterloo Press, Waterloo, Canada.

Lee, J. H., Knystautas, R., Guirao, C. M., Benedick, W. B., and Shepherd, J. E. (1982b) Hydrogen-air detonations, 2nd International Workshop on the Impact of Hydrogen on Water Reactor Safety, Albuquerque, New Mexico.

Lee, J. H. and Matsui, H. (1977) A comparison of the critical energies for direct initiation of spherical detonations in acetylene-oxygen mixtures. Combustion and Flame 28, 61-66.

Lee, J. H. and Ramamurthi, K. (1976) On the concept of the critical size of a detonation kernel. Combustion and Flame 27, 331-340.

Liu, Y. K., Lee, J. H., and Knystautas, R. (1984) The effect of geometry on the transmission of detonation through an orifice. Combustion and Flame (in print).

Matsui, H. and Lee, J. H. (1978) On the measure of the relative detonation hazards of gaseous fuel-oxygen and air mixtures. 17th Symposium (International) on Combustion, pp. 1269-1280. The Combustion Institute, Pittsburgh, Pa.

Mitrofanov, V. V. and Soloukhin, R. I. (1965) The diffraction of multifront detonation waves. Soviet Physics-Doklady 9(12), 1055-1058.

Moen, I. O. (1982) Private communication.

Moen, I. O., Donato, M., Knystautas, R., and Lee, J. H. (1981) The influence of confinement on the propagation of detonations near the detonability limits. 18th Symposium (International) on Combustion, pp. 1615-1622. The Combustion Institute, Pittsburgh, Pa.

Moen, I. O., Murray, S. B., Bjerketvedt, D., Rinnan, A., Knystautas, R., and Lee, J. H. (1982) Diffraction of detonation from tubes into a large fuel-air explosive cloud. 19th Symposium (International) on Combustion, pp. 635-644. The Combustion Institute, Pittsburgh, Pa.

Murray, S. B., Moen, I. O., Gottlieb, J. J., Lee, J. H., Coffey, C., and Remboutsikas, D. (1981) Direct initiation of detonation in unconfined ethylene-air mixtures. 7th Symposium on Military Applications of Blast Simulation, Medicine Hat, Alberta, Canada.

Rinnan, A. (1982) Transmission of detonation through tubes and orifices. Fuel-Air Explosions (edited by J. H. S. Lee and C. M. Guirao) pp. 553-564. University of Waterloo Press, Waterloo, Canada.

Shchelkin, K. I. and Troshin, Ya. K. (1965) *Gasdynamics of Combustion*, Mono Book Corp., Baltimore, Md.

Urtiew, P. A. and Tarver, C. M. (1982) Effects of cellular structure on the behavior of gaseous detonation waves under transient conditions. *Gasdynamics of Detonations and Explosions: Progress in Astronautics and Aeronautics* (edited by J. R. Bowen, N. Manson, A. K. Oppenheim, and R. I. Soloukhin), Vol. 75, pp. 370-384, AIAA, New York.

Vasiliev, A. A. (1982) Geometric limits of gas detonation propagation. *Fizika Goreniya i Vzryva* 18(2), pp. 132-136.

Westbrook, C. K. (1980) Chemical kinetics of hydrocarbon oxidation in gaseous detonations. *Combustion and Flame* 49, pp. 191-210.

Westbrook, C. K. (1982) Chemical kinetics in gaseous detonations. *Fuel-Air Explosions* (edited by J. H. S. Lee and C. M. Guirao) pp. 189-242. University of Waterloo Press, Waterloo, Canada.

Westbrook, C. K. and Urtiew, P. A. (1982) Chemical kinetic predictions of critical parameters in gaseous detonations. *19th Symposium (International) on Combustion*, pp. 615-623. The Combustion Institute, Pittsburgh, Pa.

Wolanski, P., Kaufmann, C. W., Sichel, M., and Nicholls, J. A. (1981) Detonation of methane-air mixtures. *18th Symposium (International) on Combustion*, pp. 1651-1660. The Combustion Institute, Pittsburgh, Pa.

Power-Energy Relations for the Direct Initiation of Gaseous Detonations

K. Kailasanath* and E. S. Oran†
Naval Research Laboratory, Washington, D.C.

Abstract

Recent studies on the direct initiation of gaseous detonations have shown that initiation depends not only on the energy deposited but also on the rate at which it is deposited, namely, the power. In this paper, a theoretical model has been used to determine the relation between the power and the energy required for the initiation of planar, cylindrical, and spherical detonations in a detonable gas mixture. The results from the model show that the qualitative differences in the power-energy relations obtained from two different experimental arrangements are due to differences in the geometry. Also shown is that the minimum power requirement corresponds to a shock of minimum Mach number only in the case of planar detonations. Finally, the effect on the power-energy relation of the ratio of specific heats and the experimental uncertainties in the determination of the induction times have been studied for an acetylene-oxygen-nitrogen mixture.

Introduction

The early studies of direct initiation of gaseous detonations (Zel'dovich et al. 1956; Litchfield et al. 1963; Freiwald and Koch 1963) established the importance of the magnitude of the source energy. More recent studies (Oppenheim 1967; Bach et al. 1971; Meyer et al. 1973; Lee et al. 1975; Knystautas and Lee 1976) have shown the importance not only of the energy but also of the rate at which the energy is deposited, namely, the power. The

---

Presented at the 9th ICODERS, Poitiers, France, July 3-8, 1983. Copyright © American Institute of Aeronautics and Astronautics, Inc., 1984. All rights reserved.
*Research Scientist, Science Applications, Inc., McLean, VA.
†Senior Research Scientist, Laboratory for Computational Physics.

experimental results of Lee et al. (1975) indicate that there is a minimum detonation energy E_m below which a detonation would not occur no matter what the power is, and that there is a minimum power P_m below which a detonation would not occur no matter what the total energy is. Later, Knystautas and Lee (1976) noted that the requirement for a minimum value for the power of the source indicates that the source must be capable of generating a shock wave of certain minimum strength (Mach number). They also concluded that the minimum energy requirement implied that the shock wave must be maintained at or above this minimum strength for a certain minimum duration.

Recently, these ideas have been used by Dabora (1980, 1982) to obtain a relation between the power and energy required for the direct initiation of hydrogen-air detonations in a shock tube. However, this power-energy relation is very different qualitatively from those of Knystautas and Lee (1976). More recently, Abouseif and Toong (1982) have proposed a simple theoretical model to determine the power-energy relation and predict their respective threshold values. The predictions based on their model are in qualitative agreement with the experiments of Knystautas and Lee (1976).

In this paper, we have modified and extended the basic model proposed by Abouseif and Toong (1982) and have used it to determine the relation between the power and the energy required for the initiation of planar, cylindrical, and spherical detonations in a detonable gas mixture. Specifically, its application to an acetylene-oxygen-nitrogen mixture is discussed. We have used the results from the model to explain the qualitative differences between the experimental results of Knystautas and Lee (1976) and Dabora (1980). The relation between the minimum power requirement and the Mach number of the shock wave has also been examined. Some of the limitations of the model are discussed, and several applications are described.

The Theoretical Model

In principle, the direct initiation of detonations can be studied by performing detailed numerical simulations of the flowfield generated by a given source of energy. In general, such a calculation is a complicated, multidimensional, multispecies, time-dependent problem. Part of the complication and cost of such calculations arises from the solution of the conservation equations, and part of it arises from integrating the large number of ordinary differential equations describing the chemical reactions. This latter factor is further complicated by the fact that we usually do not have an adequate representation of the chemical reactions with which to work. Thus, a convenient, inexpensive way to evaluate the relative tendency of different explosive mixtures to detonate would be very useful. Below we develop and expand a simple theoretical model proposed earlier by Abouseif and Toong (1982). Although this approach is not as precise as solving the full set of equations numerically, it offers a number of important insights and gets around the requirement of knowing the detailed chemical kinetics.

The model considers the flow generated by the motion of a constant velocity shock wave in planar, cylindrical, and spherical geometries. As this shock wave passes through a gas mixture, the gas temperature and pressure increases. Due to this increase in temperature and pressure, ignition can occur in the shock heated gas mixture after the elapse of a certain time, and this may lead to detonation.

A constant velocity shock wave can be formed in each of the three geometries by the motion of a constant velocity piston (Taylor 1946; Kailasanath and Oran 1983). Furthermore, it has been shown (Chu 1955) that a pressure and velocity field identical to that ahead of a constant velocity piston can be generated by appropriate energy addition. For example, a flowfield bounded by a constant velocity planar piston and a constant velocity planar shock wave can be generated by a planar energy source with a constant rate of energy deposition. An example of such an energy source is the high pressure driver in a uniform shock tube. In general, the source power P_s required to generate a constant velocity piston in planar, cylindrical, and spherical geometries can be written as (Abouseif and Toong 1982; Kailasanath and Oran 1983)

$$P_s(t) = \frac{\gamma}{\gamma-1} C_\alpha p_p u_p^\alpha t^{\alpha-1} \qquad (1)$$

where $C_\alpha = 1, 2\pi, 4\pi$ for $\alpha = 1, 2, 3$ corresponding to the planar, cylindrical, and spherical geometries respectively; p_p and u_p are the pressure and velocity at the piston surface, and t is the duration of energy deposition. The energy deposited is given by the time integral of the power, that is,

$$E_s(t) = \frac{\gamma}{\gamma-1} \frac{C_\alpha}{\alpha} p_p u_p^\alpha t^\alpha \qquad (2)$$

Equations (1) and (2) give the source power and the source energy required to generate a constant velocity piston in the three geometries. As shown elsewhere (Kailasanath and Oran 1983), if the piston velocity is steady, a constant velocity shock wave could be generated ahead of it. If the piston velocity is reduced (by altering the energy deposition rate), rarefaction waves will be generated ahead of it and these, on catching up with the shock wave, will reduce the shock velocity. However, if the shock has been in motion for a sufficiently long time, chemical reactions would begin in the shock heated gas mixture. Then, even if the piston decelerates and produces rarefaction waves, these will have very little effect on the motion of the shock. In this case we could have a detonation.

Let us call the minimum time of shock travel required to initiate a detonation t_{cr}. Using this in Eqs. (1) and (2),

$$(E_s)_{cr} = \frac{\gamma}{\gamma-1} \frac{C_\alpha}{\alpha} p_p u_p^\alpha t_{cr}^\alpha \qquad (3)$$

and

$$(P_s)_{cr} = \frac{\gamma}{\gamma-1} C_\alpha p_p u_p^\alpha t_{cr}^{\alpha-1} \qquad (4)$$

In the planar case, the pressure p_p and fluid velocity u_p at the piston surface are the same as those just behind the shock. However, in the cylindrical and spherical cases, the flowfield between the shock and the piston surface is nonuniform and can be obtained by solving the governing partial differential equations. However, the solution procedure is considerably simplified if a similarity solution is sought. Then the system of partial differential equations can be reduced to a system of coupled ordinary differential equations:

$$\frac{u-L}{\rho} \frac{d\rho}{dL} + \frac{du}{dL} + (\alpha-1) \frac{u}{L} = 0 \qquad (5)$$

$$(u-L) \frac{du}{dL} = -\frac{1}{\rho} \frac{dp}{dL} \qquad (6)$$

$$\frac{dp}{dL} = \frac{\gamma p}{\rho} \frac{d\rho}{dL} \qquad (7)$$

In the above system of equations, the density ρ, the velocity u, and the pressure p are all functions of the similarity variable L, which is equal to the radial location r divided by the time t. The pressure and the velocity at the piston surface that are required in Eqs. (3) and (4) can be obtained by solving Eqs. (5-7) in the following manner. For a shock of a given Mach number, the flow condition just behind the shock can be calculated using normal shock relations. Equations (5-7) can then be integrated from just behind the shock to the piston surface to obtain P_p and u_p, which are needed in Eqs. (3) and (4). The procedure is further simplified by appropriately combining Eqs. (5-7) into two equations and normalizing them. This is discussed in detail elsewhere (Kailasanath and Oran 1983).

In order to determine the power-energy relation using Eqs. (3) and (4), t_{cr} must also be known. This time must at least be equal to the time at which ignition first occurs in the flowfield (Abouseif and Toong 1982). As noted by Urtiew and Oppenheim (1967), ignition usually occurs first at the contact surface (i.e., at the piston surface here), since the temperature and pressure is highest at this location. So, a first estimate of the time t_{cr} would be the induction delay time corresponding to the conditions at the piston surface.

Results and Discussion

The model described in the previous section has been used to determine the power-energy relations for the initiation of planar, cylindrical, and spherical detonations in an acetylene-oxygen-nitrogen mixture. The initial temperature and pressure of the mixture were taken to be 300 K and 100 Torr (0.1316 atm) to correspond to the initial conditions in the experiments of Knystautas and Lee (1976). As a first approximation, the time duration necessary for successful initiation was assumed to be equal to the chemical induction time of the mixture corresponding to the conditions at the piston surface.

The critical source power given by Eq. (4) is time dependent for the cylindrical and spherical cases. In order to relate the critical source energy to a critical source power, we need to define an average or "effective" power. Following Abouseif and Toong (1982), an average critical source power is defined as

$$(P_s)_{av} = \frac{(E_s)_{cr}}{t_{cr}} \qquad (8)$$

This power also corresponds to the critical peak averaged power of the source as defined by Knystautas and Lee

(1976). For the discussion below, the terms power and energy have been used to refer to the average critical source power [Eq. (8)] and the critical source energy [Eq. (3)], respectively.

Cylindrical Detonations in an Acetylene-Oxygen-Nitrogen Mixture

The power-energy relation for the initiation of cylindrical detonations has been determined by using Eqs. (3) and (8). The induction time data used were those obtained by Edwards et al. (1981) for an acetylene-oxygen-nitrogen (2:5:4) mixture and are given by:

$$\text{Log } (\tau[O_2]) = -9.41 \ (\pm 0.2) + \frac{71.35 \ (\pm 3.34)}{19.14 \ T} \quad (9)$$

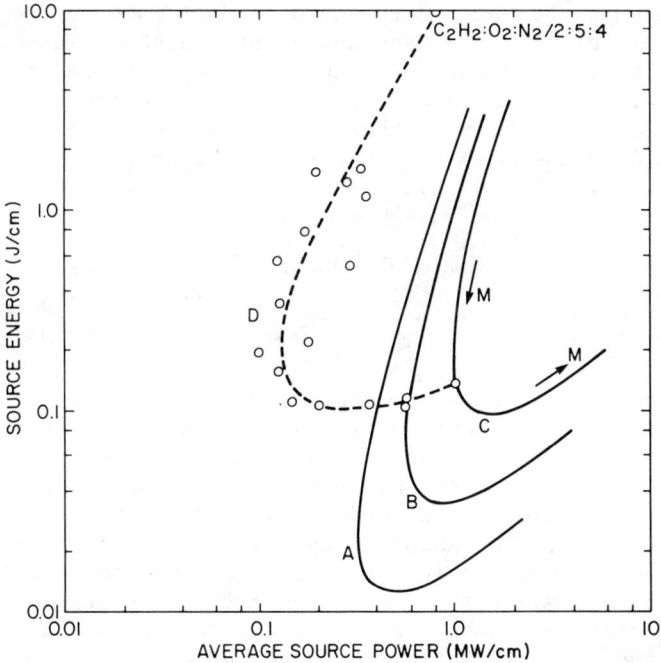

Fig. 1 Power-energy relations for the initiation of cylindrical detonations in an acetylene-oxygen-nitrogen mixture (2:5:4) at 0.1316 atm and 300 K. The data for curve D were obtained from spark ignition experiments (Knystautas and Lee 1976). Curves A, B, and C are explained in the text. The arrows on curve C indicate the direction of increasing Mach number.

where τ is the induction time in seconds; $[O_2]$ is the concentration in mole/liter, and T is the temperature in thousands of degrees K. Three different power-energy relations obtained from the theoretical model are shown in Fig. 1. Curve A was obtained by using the smallest value of the induction time given by Eq. (9), that is, by choosing the negative signs. Curve B was obtained by using the mean values, and curve C by using the largest value of the induction time (by choosing the positive signs). The arrows on curve C indicate the direction of increasing Mach number. First, note that each curve has a minimum power and a minimum energy. Also observe that as the Mach number decreases below the Mach number corresponding to the minimum power, both the average source power and the source energy increase. However, when the Mach number increases above the Mach number corresponding to the minimum power, the energy first decreases to the minimum energy and then increases again. All three curves exhibit these same qualitative trends.

The shape of these curves can be explained in the following manner: As the Mach number of the shock wave decreases, the pressure and the temperature behind it decrease. This decrease also results in a decrease of the pressure and velocity at the piston surface. This would tend to decrease both the power and the energy, since, as seen in Eqs. (1) and (2),

$$P \sim p_p u_p^2 t \qquad (10)$$

$$E \sim p_p u_p^2 t^2 \qquad (11)$$

This tendency is, however, opposed by the tendency of the induction time to increase with decreases in the pressure and the temperature. For low Mach numbers (i.e., low temperatures behind the shock), a small decrease in the Mach number of the shock wave leads to a large increase in the induction time. The shape of the curves in Fig. 1 implies that this increase in induction time is more than sufficient to compensate for the decrease in the pressure and the velocity for Mach numbers below that corresponding to the minimum power. Therefore, both the power and the energy increase with decreasing Mach number. Since the energy is proportional to the product of the power and the induction time [Eqs. (10) and (11)], the energy increases faster with induction time than the power does. As the Mach number increases above that corresponding to the minimum power, the increase in the pressure and velocity is larger than the decrease in the induction time. Therefore, the power increases. However, for a certain range of Mach numbers, the increase in the pressure and velocity is not sufficient to compensate for the decrease in the square of the induction time. Therefore, the energy decreases until it attains a minimum value, even though the power increases. Finally, for Mach numbers above that corresponding to the minimum energy, the increase in the pressure and velocity is easily able to overcome the decrease in the induction time with increasing Mach number, and both the power and the energy increase. This occurs because the rate of decrease of the induction time with temperature is small for high temperatures (i.e., high Mach numbers) according to Eq. (9).

The power-energy curve obtained using data from the spark ignition experiments of Knystautas and Lee (1976) has also been included in Fig. 1 as curve D. The data for curve D are the same as those used by Abouseif and Toong (1982) for their Fig. [1], and was originally presented in Fig. [4] of Knystautas and Lee (1976). Curve D exhibits the same qualitative trends as those of the theoretical curves discussed above. However, observe that the values of the minimum power and the minimum energy from the four curves are very different from each other. The differences in the values of these parameters from the three "theoretical" curves (A, B, and C) indicate that the experimental uncertainties in the values of the induction times used have a significant effect on the value of the minimum power and the minimum energy. The minimum power varies from about 0.3 to about 1 MW/cm, and the minimum energy varies from about 0.012 to about 0.1 J/cm. The experimentally determined minimum power (from curve D) is about 0.13 MW/cm, which is lower than the calculated values; and the minimum energy is about 0.1 J/cm, which is at the top of the range of calculated values.

The quantitative differences between the experimental and theoretical values could be due to a variety of factors, a few of which are now discussed. As observed from curves A, B, and C, uncertainties in the induction time data can have a significant effect on the values of the minimum power and the minimum energy. Expressions such as Eq. (9) for the induction time are obtained by fitting to a limited range of experimental data. However, here Eq. (9) was used for a range of temperatures and pressures far greater than that over which it was determined. Furthermore, for obtaining the theoretical results, a constant value of 1.2 had been assumed for γ, the ratio of specific heats. For high Mach numbers, the γ of the shocked gas could be very different from that ahead of the shock wave because of the large temperature difference across the shock wave. The effect of γ on the power-energy relations is discussed below.

Effect of γ on Power-Energy Relations

The power-energy calculations were repeated using different values for γ, and the results are shown as curves A and C in Figs. 2 and 3. In Fig. 2, the average source power is shown as a function of the shock Mach number, and in Fig. 3, the source energy is shown as a function of the

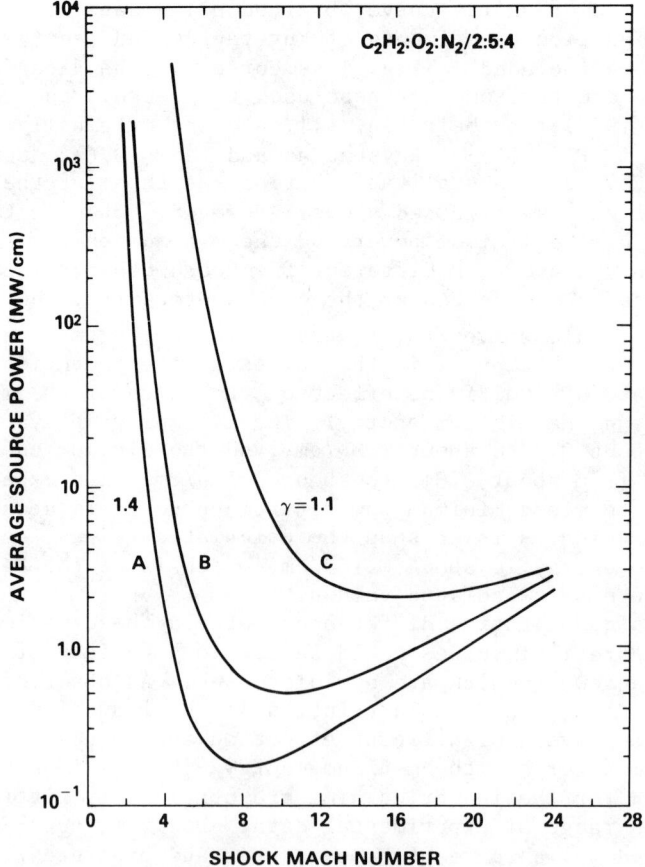

Fig. 2 The average source power as a function of the shock Mach number. Curves A and C were obtained assuming γ to be constant across the shock wave. Curve B was obtained assuming γ to be variable, as explained in the text.

shock Mach number. From these figures, it is observed that
γ does indeed have a significant effect on the minimum
power and the minimum energy. When γ is changed from 1.1
to 1.4, the minimum source power decreases from 2.0 to 0.18
MW/cm, and the minimum energy decreases from 0.065 to about
0.02 J/cm. The Mach number at which the shock must travel
to attain the minimum power is also very different, as seen
in Fig. 2. Changing γ from 1.4 to 1.1 doubles the Mach
number corresponding to the minimum power from 8 to 16.
The effect of γ on the power-energy relation arises partly
from the factor $\gamma/(\gamma-1)$ in Eqs. (3) and (4) and partly from
the fact that the temperature behind a shock of given Mach
number is very different for different γ's.

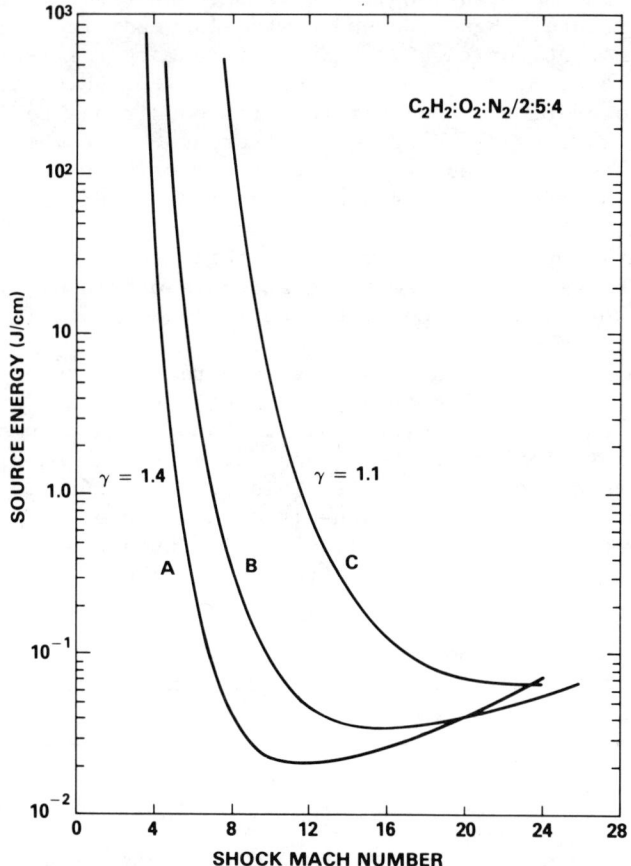

Fig. 3 The source energy as a function of the shock Mach
number. Curves A and C were obtained assuming γ to be constant
across the shock wave. Curve B was obtained assuming γ to be
variable, as explained in the text.

The effect of the factor $\gamma/(\gamma-1)$ is to change quantitatively the values of the source power and the source energy corresponding to the shock of a given Mach number and is the same for all Mach numbers. The changes in the temperature behind a shock wave due to assumed differences in γ is, however, a function of the shock Mach number. Let us consider a shock wave of the Mach number 10. Table 1 shows the pressure ratio, the temperature ratio, and the temperature behind this shock wave for different values of γ. The case where γ is different across the shock wave is included as case 3. For obtaining case 3, the normal shock relations have been rederived but with variable γ across the shock wave. The derivation of these "modified" shock relations has been presented in detail elsewhere (Kailasanath and Oran 1983). It is seen that for case 3 the temperature behind the shock wave is significantly lower than that for case 2. Case 3 is a more realistic case than case 2, since γ is generally lower behind the shock. However, the appropriate γ for conditions behind the shock wave is different for different Mach numbers, since the temperatures are different. Thus, a better approach is first to guess a γ for each Mach number and use it to calculate the temperature behind the shock. This new temperature implies a new γ. Using this new γ in the modified shock relations a new temperature is obtained. This iterative procedure is continued until convergence is achieved.

The power-energy calculations were repeated using the correct γ (as described above), and the results are presented as curve B in Figs. 2 and 3. For low Mach numbers, curve B lies close to curve A, and for very high Mach numbers, it tends toward curve C. This is not surprising, since for the acetylene-oxygen-nitrogen mixture being studied here, γ varies from 1.31 to 1.16 when the Mach number changes from 2 to 24. From curve B in Figs. 2 and

Table 1 Effect of the ratio of specific heats

Case	γ_o[a]	γ_s	P_s/P_o	T_s/T_o	T_s
1	1.2	1.2	109.000	10.900	3270.0
2	1.3	1.3	112.913	15.710	4712.89
3	1.3	1.2	118.426	11.454	3436.26

[a] The conditions ahead of the shock wave are denoted by o and those behind by s.

3, also note that the minimum power and the minimum energy conditions occur at Mach numbers of 10.0 and 15.5, respectively.

From the above discussion, it is clear that the effect of using the correct γ is mainly to alter the Mach number corresponding to the minimum power and the minimum energy condition. However, the calculated values of the minimum power and the minimum energy are still different from those obtained experimentally. Therefore, we examine another possible reason for the differences between the experimental and the theoretical values: the uncertainty in the appropriate time to be used for t_{cr} in Eqs. (3) and (8).

Critical Time for Energy Deposition

As a first approximation, it was assumed that energy must be deposited until ignition occurs at some point in the flow field between the shock and the piston surface. Since, in general, the temperature and pressure are highest at the piston surface, the chemical induction time corresponding to these conditions was used as the appropriate time for energy deposition. However, when there is fluid motion, ignition can occur before the time corresponding to the constant volume, homogeneous chemical induction time. For example, for a certain range of temperatures and pressures, oxy-hydrogen mixtures with small perturbations could have significantly reduced ignition times. The specific effect of this phenomenon on the power-energy relations will be reported in a subsequent paper. In gas mixtures that are not particularly sensitive to perturbations, the shortest induction time in the shocked region seems to be the necessary condition for the initiation of detonations. However, it must be consider whether this is a sufficient condition also.

Shock tube simulations have indicated that the time at which a detonation wave is first observed is only very slightly longer than the time at which ignition first occurs. That is, the time between ignition and the formation of a detonation wave is small when compared to the induction time. This is not surprising when considering the fact that for many reactive systems, the reaction time is very small compared to the induction time. The results of Abouseif and Toong (1982) on the initiation of planar detonations also support this observation. However the effect of geometry on the time between ignition and detonation has not been studied. It could very well be that, due to the volume change in spherical and cylindrical

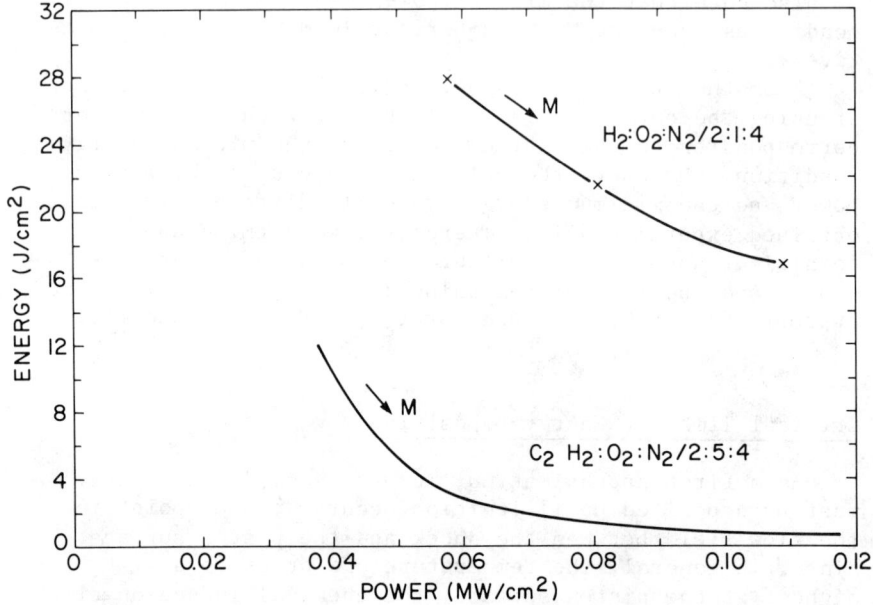

Fig. 4 Power-energy relations for the initiation of planar detonations. The x's are data obtained from shock tube experiments (Dabora 1980).

geometries, this time is significant when compared to the induction time. This needs to be studied before one can confidently use the induction time as the appropriate time for t_{cr}.

The results from the theoretical model for the case of cylindrical detonations has been compared with the experimental results of Knystautas and Lee (1976) because in both cases the amount of energy deposited was proportional to the second power of the time. However, it is important to note that in the theoretical model only constant velocity shock waves have been considered, and it was this that made it possible to assign a single induction time to each shock wave. If the velocity of the shock wave is not constant, it is not possible to assign a single induction time to it, since the flowfield behind the shock wave would be time dependent. Thus, shock waves of different time histories can deposit the same amount of energy but at different average source powers. This could be an important factor in the quantative differences between the experimental and theoretical values. Another factor which is different between the experiments and the model is the amount of diluent. The effect of diluents on the power-energy relation is currently being investigated.

Initiation of Planar Detonations

The derived power-energy relation for the initiation of planar detonations in the same oxyacetylene mixture is shown in Fig. 4. In this figure, the shock tube data of Dabora (1980) on the direct initiation of detonations in a stoichiometric hydrogen-air mixture is also shown. The point to notice is that both curves exhibit the same qualitative behavior. Unlike the cylindrical case, each value of the power corresponds to an unique value of energy. The direction of increasing shock strength (as determined by the Mach number) is also shown in Fig. 4. In the planar case, it is observed that as the Mach number decreases, the power always decreases. As noted earlier in the cylindrical case, as the Mach number decreases, the power decreases only up to the minimum power. Then the power increases with a decrease in the Mach number of the shock wave. Therefore, the qualitative difference in the experimental data of Knystautas and Lee (1976) (shown in Fig. 1) and Dabora (1980) (shown in Fig. 4) is due to the difference in the geometry of the two experiments.

It is also observed in Fig. 4 that as the Mach number decreases, more and more energy is needed to initiate a detonation. The trend of the curves indicates that there is a minimum Mach number below which a detonation will not occur (i.e., would require an infinite amount of energy). The value of the power corresponding to this minimum Mach number is the minimum power. This agrees with the observation made by Knystautas and Lee (1976) that the requirement for a minimum value of the source power indicates that the source must be capable of generating a shock wave of a certain minimum Mach number. However, it is observed from Fig. 1 (see also Fig. 2) that for the case of cylindrical detonations, the minimum power does not correspond to the shock wave of minimum Mach number. In the cylindrical case, it is possible to initiate a detonation with a shock wave of lower Mach number than that corresponding to the minimum power. Such a shock will have to be maintained for a longer time than the shock corresponding to the minimum power and, hence, will require a larger amount of energy.

Initiation of Spherical Detonations

The power-energy curve for the initiation of spherical detonations is similar to the curve for the cylindrical case. However, for the case of spherical detonations, the power is

$$P \sim p_p u_p^3 t^2 \qquad (12)$$

but the energy is still

$$E \sim P t \qquad (13)$$

Since the power and energy are proportional to higher powers of the time t, uncertainties in t will have a greater effect on the value of the minimum power and the minimum energy. Further work is being carried out currently to study the initiation of spherical detonations in hydrogen-air mixtures and to compare this to experimental data.

Summary and Conclusions

In this paper, a theoretical model has been used to determine the relation between the power and the energy required for the initiation of planar, cylindrical, and spherical detonations in a gas mixture. The results discussed above show that although the simple theoretical model has significant limitations, it can still be used to explain the qualitative differences in the power-energy relations obtained from different experimental arrangements. Another result from the model is that the minimum power requirement corresponds to a shock of minimum Mach number only in the case of planar detonations.

The results from the model on the initiation of cylindrical detonations in an acetylene-oxygen-nitrogen mixture qualitatively agree with experimental data. Some of the reasons for the quantitative differences have been examined. One of the important parameters in the model is the critical time for energy deposition. This time is related to the induction time, and the results presented above show that uncertainties in the induction time data used can have a significant effect on the power-energy relations. The results also indicate that further work needs to be done to determine the effect of the geometry on the critical time for energy deposition.

The quantitative differences between the experimental and theoretical results may also arise because of the model assumption that the velocities of the shock waves are constant. This may not be so in the experiments. Furthermore, the model considers only the minimum power and energy required to initiate a detonation wave. It was not examined whether this would result in a self-sustained propagating detonation wave. Detonation propagation is characterized by complicated interactions among incident shock waves, transverse waves, and Mach stems that form detonation cells. These must be described by multidimensional theories and simulations. The results from such studies need to be considered to extend the work presented here to the study of self-sustained detonation waves.

One application of the model presented here is to determine the relative tendency of different explosives to detonate, since the limitations of the model would then be less critical. This would be particularly useful for studying the effect of additives on the detonability of condensed phase explosives. Further work is being carried out to modify the model for such applications.

Acknowledgements

The authors gratefully acknowledge suggestions, useful conversations with and help from J. P. Boris and T. R. Young. The authors also acknowledge the editorial assistance of F. Rosenberg. This work has been supported by the Office of Naval Research through the Naval Research Laboratory.

References

Abouseif, G. E. and Toong, T. Y. (1982) On direct initiation of gaseous detonations. Combust. Flame 45, 39-46.

Bach, G. G., Knystautas, R., and Lee, J. H. (1971) Initiation criteria for diverging gaseous detonations. Thirteenth Symposium (International) on Combustion, pp. 1097-1110, The Combustion Institute, Pittsburgh, Pa.

Chu, B. T. (1955) Pressure waves generated by addition of heat in a gaseous medium. NACA TN-3411.

Dabora, E. K. (1980) Effect of additives on the lean detonation limit of kerosene sprays. UCONN0507-129-F, The University of Connecticut, Storrs, Conn.

Dabora, E. K. (1982) The relation between energy and power for direct initiation of hydrogen-air detonations. Paper presented at the Second International Workshop on the Impact of Hydrogen on Water Reactor Safety, Albuquerque, N. M.

Edwards, D. H., Thomas, G. O., and Williams, T. L. (1981) Initiation of detonation by steady planar incident shock waves. Combust. Flame 43, 187-198.

Freiwald, H. and Koch, H. W. (1963) Spherical detonations of acetylene-oxygen-nitrogen mixtures as a function of nature and strength of initiation. Ninth Symposium (International) on Combustion, pp. 275-281, Academic Press, New York.

Kailasanath, K. and Oran, E. S. (1983) The relation between power and energy in the shock initiation of detonations--I. NRL Memorandum Report 5179, Naval Research Laboratory, Washington, D. C.

Knystautas, R. and Lee, J. H. (1976) On the effective energy for direct initiation of gaseous detonations. Combust. Flame 27, 221-228.

Lee, J. H., Knystautas, R., and Guirao, C. M. (1975) Critial power density for direct initiation of unconfined gaseous detonations. Fifteenth Symposium (International) on Combustion, pp. 53-67. The Combustion Institute, Pittsburgh, Pa.

Litchfield, E. L., Hay, M. H., and Forshey, D. R. (1963) Direct electrical initiation of freely expanding gaseous detonation waves. Ninth Symposium (International) on Combustion, pp. 282-286, Academic Press, New York.

Meyer, J. W., Cohen, L. M. and Oppenheim, A. K. (1973) Study of exothermic processes in shock ignited gases by the use of laser shear interferometry. Combust. Sci. Tech. 8, 185-197.

Oppenheim, A. K. (1967) The no-man's land of gasdynamics of explosions. Appl. Mech. Rev. 20, 313-319.

Taylor, G. I. (1946) The air wave surrounding an expanding sphere. Proc. Roy. Soc. London, Ser. A 186, 273-292.

Urtiew, P. A. and Oppenheim, A. K. (1967) Detonative ignition induced by shock merging. Eleventh Symposium (International) on Combustion, pp. 665-670, The Combustion Institute, Pittsburgh, Pa.

Zel'dovich, Y. B., Kogarko, S. M. and Simonov, N. N. (1956) An experimental investigation of spherical detonation of gases. Soviet Phys.-JETP 1, 1689-1713.

Detonation Length Scales for Fuel-Air Explosives

Ingar O. Moen,* John W. Funk,† Stephen A. Ward,†
and Gerry M. Rude†
Defence Research Establishment Suffield, Ralston, Alberta, Canada
and
Paul A. Thibault‡
University of Toronto Institute for Aerospace Studies, Toronto, Ontario, Canada

Abstract

This paper reports on a series of large-scale field experiments performed to measure the critical tube diameters and detonation cell sizes for fuel-air mixtures. The results for ethylene-air reported previously are extended to include results for acetylene-, ethane-, propane- and methane-air for tube diameters up to 1.83 m. Detonation cell sizes of about 280 mm are obtained for stoichiometric methane-air mixtures. Based on the results of this investigation and results from other investigations, proposed relations between the critical initiation energy (E_c), the critical tube diameter (d_c) and the detonation cell size (S), are examined. Although, the empirical relation $d_c = 13S$ appears to provide a good correlation for much of the data, the usefulness of this relation for predicting the detonation behavior in less sensitive fuel-air mixtures is questioned. For these less sensitive mixtures, the cellular structure is highly irregular with many modes of cellular patterns, so that the identification of a dominant mode is subject to considerable interpretation. The relations of the critical tube diameter, the cell size and the critical explosion

---

Presented at the 9th ICODERS, Poitiers, France, July 3-8, 1983. Copyright 1983 by the Government of Canada. Published by the American Institute of Aeronautics and Astronautics with permission.
 *Defence Scientist, Military Engineering Section.
 †Technologist, Military Engineering Section.
 ‡Research Associate, Institute for Aerospace Studies.

length to the induction zone length, calculated from a one-dimensional Zel'dovich-von Neumann-Döring model of the detonation, are also examined. In general, linear relationships tend to underestimate the detonation sensitivity of lean fuel-air mixtures relative to that of stoichiometric mixtures. Results from numerical calculations of one-dimensional detonation instabilities which illustrate the irregularity of the structure for less sensitive mixtures are described. The link between the wavelengths of these instabilities and the induction length is also examined. The trends of the numerical results are in agreement with experimental observations for the more complex cellular fuel-air detonations.

Introduction

The detonability of an explosive mixture can be characterized by the critical conditions required to initiate a detonation and by the minimum size of an unconfined explosive cloud required to support a self-sustained detonation. The initiation requirements are usually specified by the minimum mass, or the minimum energy E_c, of high explosive required to initiate a detonation in the explosive cloud. A measure of the minimum size of a detonable cloud is provided by the critical tube diameter d_c for the transformation of a confined planar detonation wave to an unconfined spherical wave. Both the critical energy and the critical tube diameter depend on the nature of the cellular detonation structure and on the detailed chemical kinetic processes within the cellular detonation front.

The cellular structure of gaseous detonations can be attributed to the onset of chemical-gasdynamic instabilities. These instabilities result in the formation of detonation cells which are considerably larger than the induction zone thickness predicted by the steady-state Zel'dovich-von Neumann-Döring (ZND) model. For such long wavelengths, the level of instability generally increases with the activation energy of the mixture, and decreases with the heat of reaction and the degree of overdrive. One- and two-dimensional numerical simulations of detonations have revealed regular oscillatory patterns when the level of instability is relatively mild (see Fickett and Wood 1966; Abouseif and Toong 1982; Taki and Fujiwara 1978; Oran et al. 1981). For highly unstable systems, however, the linear theory of Erpenbeck (1964) and the approximate theory of Abouseif and Toong (1982) indicate that a large number of modes contribute to the overall detonation structure. These modes interact in a nonlinear manner to produce a complex irregular detonation structure. At the present time, there is no theory capable of describing the coupled chemical hydrodynamic structure of detonations responsible for this complex structure. Numerical simulations of two-dimensional cellular detonations which incorporate many of the chemical kinetic processes have been performed (Taki and Fujiwara 1978; Oran et al. 1981). However, the extension of these numerical calculations to more unstable situations, corresponding to the less sensitive fuel-air mixtures, will be difficult. It is therefore important to identify the key parameters and to establish links between these parameters so that the calculations can be simplified.

Theoretical and empirical links between detonation parameters have been proposed by various authors. Mitrofanov and Soloukhin (1964) first observed that a minimum of 10-13 detonation cells are required for successful transmission of detonations in oxy-acetylene from a tube. Edwards et al. (1979) and Edwards and Thomas (1981) have provided further evidence for the simple universal relation $d_c = 13S$ between the characteristic cell size S and the critical tube diameter d_c. This empirical relation has been confirmed for a variety of oxygen enriched fuel-air mixtures (Knystautas et al. 1982), and for selected fuel-air mixtures (see Lee et al. 1982; Guirao et al. 1983; Moen et al. 1982). Relations between other detonation parameters have also been proposed. Based on a work-done concept, Lee and Matsui (1979) have obtained a simple relation between the critical energy E_c and the critical tube diameter d_c, whereby the critical explosion length $R_{oc} = (E_c/p_0)^{1/3}$, where p_0 is the ambient pressure, is linearly related to d_c. A more recent model, based on a surface energy concept, (Lee et al. 1982) also gives a linear relationship between R_{oc} and d_c, but with a different proportionality factor.

In order to complete the link of these length scales to the chemical kinetic rates for the mixture, the criterion first proposed by Zel'dovich (1956) has been adopted. An induction length characteristic of the explosive mixture is calculated using the one-dimension ZND model of a detonation. This induction zone length is given by $\Delta = \tau u$, where τ is the chemical induction time behind a shock wave propagating at CJ velocity and u is the particle velocity relative to the shock. All other detonation length scales are then assumed to be proportional to this chemical length. Westbrook (1982a,

b) and Westbrook and Urtiew (1982) have performed extensive calculations of the chemical kinetics of gaseous detonations based on the above assumptions. Their correlations of detonation parameters for a variety of mixtures are quite impressive. However, disagreement with experimental results has been observed for some lean fuel-air mixtures (Moen et al. 1982).

The combination of the above relations between length scales which characterize; 1) the detonability properties (d_c and R_{oc}), 2) the cellular detonation structure (S), and 3) the chemical kinetics (Δ) provides a good framework for predicting the behavior of detonations in gaseous mixtures. Unfortunately, the reliability of these relations for less sensitive fuel-air mixtures with large and irregular cellular structure has not been critically examined. In fact, Edwards and Thomas (1981) have presented some evidence that detonations in systems with large and irregular cellular structure require more than 10-13 cells for the successful transmission of a diffracted wave at an abrupt expansion. It is therefore important to critically examine these links, so that their limitations can be discovered.

The present paper examines the structure and behaviour of detonations in fuel-air mixtures. The paper is divided into three parts. Results from a series of large-scale field experiments in which the critical tube diameters and cell sizes for detonations in fuel-air mixtures were measured are first reported. These results extend the results for ethylene-air reported previously (Moen et al. 1982) to include acetylene-, ethane-, propane-, and methane-air mixtures for tube diameters up to 1.83 m. Based on these results and results from other investigations, the relations between the various detonation length scales are then critically examined. Finally, numerical calculations of one-dimensional detonation instabilities, performed to further clarify the irregular instability structure observed for less sensitive mixtures and to examine the link between induction length and instability wavelengths, are described.

Experiments

Experimental Details

The experimental test facility at DRES has been described in detail elsewhere (see Moen et al. 1982; Funk and Murray 1982; Funk et al. 1982). The facility is centered around an 18.3 x 7.6 m concrete test pad onto which the experimental apparatus can be mounted. A photograph of the test pad with a typical test section used in the critical tube diameter tests is shown in Fig. 1. The test sections consist of steel tubes connected to large plastic bags constructed from 0.13-mm-thick polyethylene. The dimensions of the tube-bag configurations used both in the present tests and in previous tests (Moen et al. 1982) are shown in Fig. 1.

The test gases (CP grade) were mixed with the initial air in the test section by a multipath recirculating system using a high-capacity centrifugal blower. The composition and mixture homogeneity in the test volume were verified by continuously analyzing samples from four ports in the test section using a Wilks-Miran 80 infrared gas analyzer. This system was adequate to guarantee the concentration and homogeneity in the test volume to within ±0.05% fuel. Detonation was initiated by using either a Detasheet initiator charge or a slug of acetylene-oxygen at the end of the tube.

The progress and pressure profiles of the detonation waves were monitored at up to 14 positions along the tube-bag configuration using piezo-electric pressure transducers together with a multichannel time of arrival counter. Cinematographic records were obtained using a Hycam camera (~18,000 half fps) looking normal to the direction of propagation, a Photec camera (~18,000 half fps) looking at a 45-deg angle and a Fastax camera (~5,000 fps) looking head on at the advancing detonation wave. Smoked foils, consisting of polished steel sheets covered with a thin layer of carbon soot, were mounted in the tube near the exit to the bag to record the detonation structure.

Experimental Results

Selected frames from high-speed photographic records illustrating successful transmission and failure to transmit are shown in Fig. 2. The first frames of both sequences show the planar detonation wave just as it emerges from the tube. For successful transmission, reignition is seen to occur at a nucleus in the center of the bag in Frame 2. The subsequent formation of a detonation wave which sweeps through the preshocked region just outside the central core of burned gas can be seen in Frames 3 and 4. As seen in Frame 5, this detonation wave eventually engulfs the whole region. In the case of detonation failure, no ignition nuclei are formed. The initial planar detonation core shrinks (Frames 2 and 3) as

Fig. 1 Photograph showing the DRES testing facility with a detonation tube (1.83-m-diam) and a polyethylene bag (3.66-m-diam). Dimensions of configurations that have been used are given in Table 1 (d = tube diameter, D = bag diameter, L = tube length). DRES (1982) - present tests; DRES (1981) - Moen et al. (1982); Raufoss (1982) - Jenssen (1982).

d (m)	L/d	D/d	TEST SITE	FUELS
0.31	12.5	2.9	DRES (1981)	ETHYLENE
0.35	34.3	5.4	RAUFOSS (1982)	ETHYLENE ACETYLENE
0.45	13.4	3.9	DRES (1981)	ETHYLENE
0.89	8.6	3.0	DRES (1981)	ETHYLENE
0.89	8.6	2.0	DRES (1982)	ACETYLENE ETHANE PROPANE
1.83	6.7	1.0	DRES (1982)	ETHYLENE PROPANE
1.83	6.7	2.0	DRES (1982)	METHANE

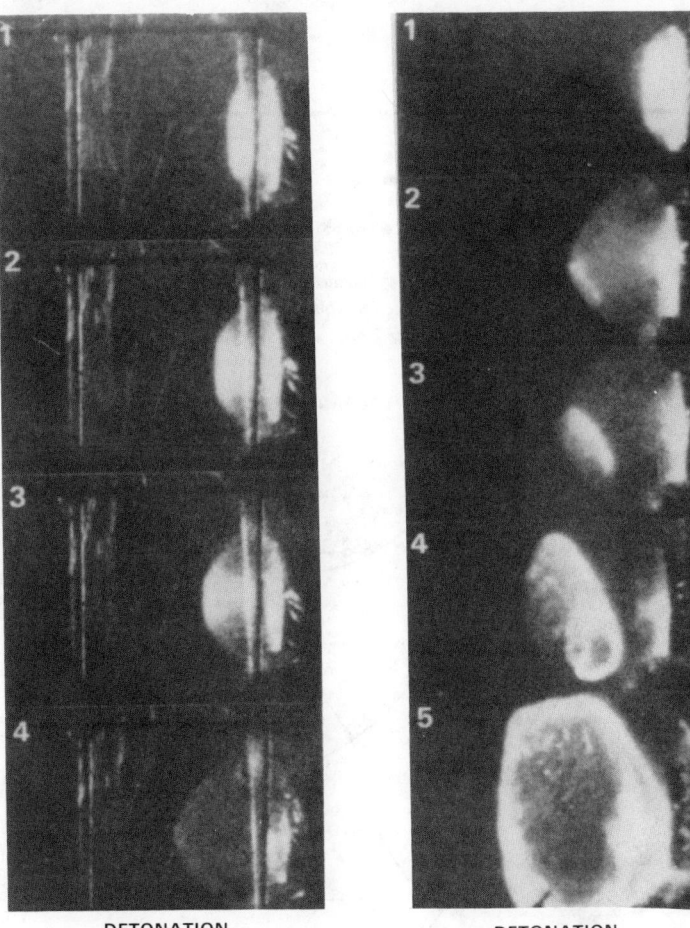

Fig. 2 Selected frames from high-speed cinematographic records showing successful transmission of detonation and failure to transmit from a 0.89-m tube.

the expansion waves sweep in towards the center and disappears completely when the head of the expansion wave reaches the center at about one tube diameter from the tube exit (Frame 4).

As described by Moen et al. (1982) the critical mixture composition for re-establishment of detonation in the bag for a given tube diameter is well defined provided that the detonation in the tube is properly established at the tube exit, and provided that the plastic bag does not influence the transmission process. The detonation

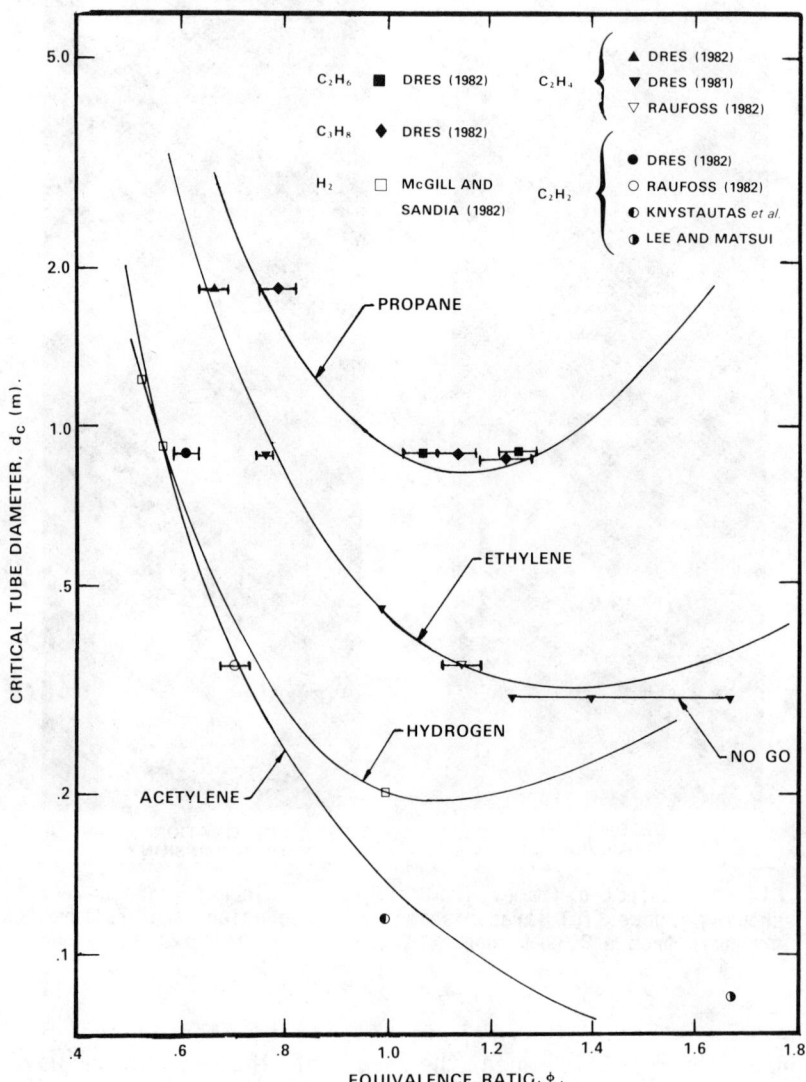

Fig. 3 Critical tube diameter d_c vs equivalence ratio ϕ for various fuel-air mixtures. The curves are fits to the data based on an induction length formula. DRES (1982) - present tests; DRES (1981) - Moen et al. (1982); Raufoss (1982) - Jenssen (1982); Knystautas et al. (1982); Lee and Matsui (1979); McGill and Sandia (1982) - Lee et al. (1982) and Guirao et al. (1982).

pressures and velocities in the tube were carefully monitored and the high-speed film records were analyzed to ensure that these conditions were satisified.

The critical tube diameter results are plotted vs equivalence ratio ϕ in Fig. 3. The uncertainty in ϕ for a given tube diameter represents go and no-go results for this tube diameter. Also shown in Fig. 3 are previous results obtained at DRES and results from other investigations. The curves shown in Fig. 3 are fits obtained using an induction length formula for the critical tube diameter. These correlations are discussed in the next section.

Detonation cell widths S were obtained from the smoked foil records by averaging over a large number of cells and by identifying dominant diagonal bands of similarly shaped cells (Moen et al. 1982). For the more sensitive mixtures with small cells, the structure recorded on the smoked foils is quite well defined so that the measurement of the cell size is relatively ambiguous.

However, for less sensitive mixtures with large cells, the structure becomes highly irregular. Furthermore, many modes of cellular patterns are observed, so that the identification of a dominant mode is subject to considerable interpretation. Typical smoked foil records illustrating the cellular structure observed for the less sensitive mixtures are shown in Fig. 4. The records for 4.1% ethylene and 3.4% propane show very clear sub-structure. Cells ranging in sizes from 40 up to about 130 mm can easily be indentified. Our interpretation of these records gives a cell width of 120 mm for 4.1% ethylene and a width of 130 mm for 3.4% propane. In other words, we have chosen the largest regular structure as that corresponding to the cell. This choice is based on detailed analyses of these records and similar records for nearby compositions.

The smoked foil record for methane shown in Fig. 4 is virtually impossible to interpret. The recorded structure is highly irregular and faint with no obvious dominant structure. In order to obtain an estimate of the cell size for methane-air mixtures, tests were performed using a plastic bag with the same diameter as the tube. As observed by Moen et al. (1982), the width of the longitudinal plastic strips produced by the detonation can then be used as a measure of the cell size. Although the detonation in a 9.6% methane-air mixture failed about half-way down the 8-m long, 1.83-m-diam plastic bag, the plastic strips produced in the first part of the bag show a very clear dominant width between 280-310 mm.

Additional tears in these strips, corresponding to the higher transverse modes, were also observed. However, these were typically relatively short (less than 0.5-m long) and intermittent. Structure with dimensions between 250-280 mm can also be identified on the smoked foils. Based on the above results, the cell size for 9.6% methane in air is interpreted to be 280 ± 30 mm, in good agreement with the value of 310 mm predicted by Bull et al. (1982).

The cell sizes for different fuel-air mixtures are plotted vs fuel-air equivalence ratio in Fig. 5. Also included in this figure are results from previous tests with

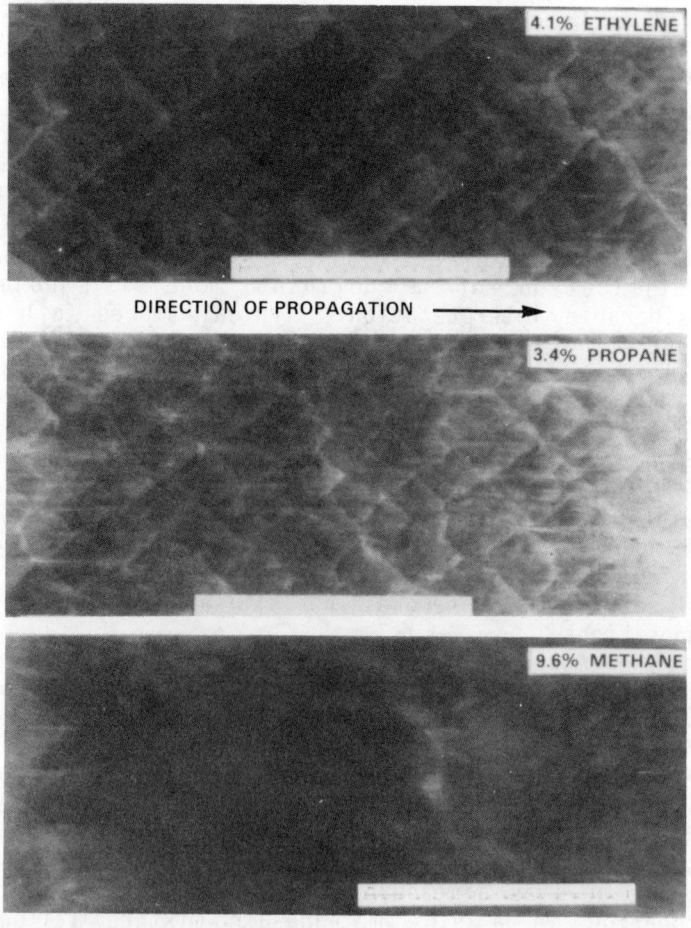

Fig. 4 Typical smoked foils illustrating the complex cellular structure observed for lean ethylene- and propane-air mixtures and for near stoichiometric methane-air.

ethylene-air and cell sizes measured in smaller tubes by other investigators. Although there is considerable scatter in the data, the agreement between results from different investigations is good, considering the difficulty in interpreting the smoked foil records. The error bars shown for the DRES(1981) results represent the spread in values obtained by repeated tests and by using different measuring techniques. The solid curves shown in Fig. 5 are fits to all the data, based on an induction length correlation formula for the cell size. These correlations are discussed in the next section.

Correlations

Critical tube diameter and cell size data are now available for a number of fuel-air mixtures, covering a wide range of compositions and mixture sensitivities. Critical initiation energy data are also available for many fuel-air mixtures. In order to correlate these data so that trends can be identified and links between the various length scales examined, we shall use a standard induction length formula of the form

$$L = uk\,[Fuel]^a\,[Oxygen]^b\,\exp(E_A/RT) \qquad (1)$$

where L is the length scale (S, d_c, or R_{oc}), and u, [Fuel], [Oxygen], and T are the postshock relative particle velocity (m/s), fuel and oxygen concentrations (moles/liter), and temperature (K) behind a shock propagating at the CJ detonation velocity of the fuel-air mixture. These properties are calculated based on frozen chemistry equilibrium states behind the shock wave. The parameters a and b are taken from proposed induction time formulae (see Kistiakowsky and Richards 1962; Hidaka et al. 1974; Burcat et al. 1972; Burcat et al. 1971; Lifshitz et al. 1971; Miyama and Takeyama 1964), but the effective activation energy E_A and the pre-exponential factor k are obtained from two parameter fits to the data. The fits to the critical tube diameter and cell size data obtained in this manner are shown by the solid curves in Figs. 3 and 5. With the exception of the rich ethylene and propane cell size results, the correlations of the data are quite good. A dashed curve is drawn through the cell size data for rich ethylene-air to indicate the trend of these data.

The parameters used for these correlations are summarized in Table 1. Also included in Table 1 are the

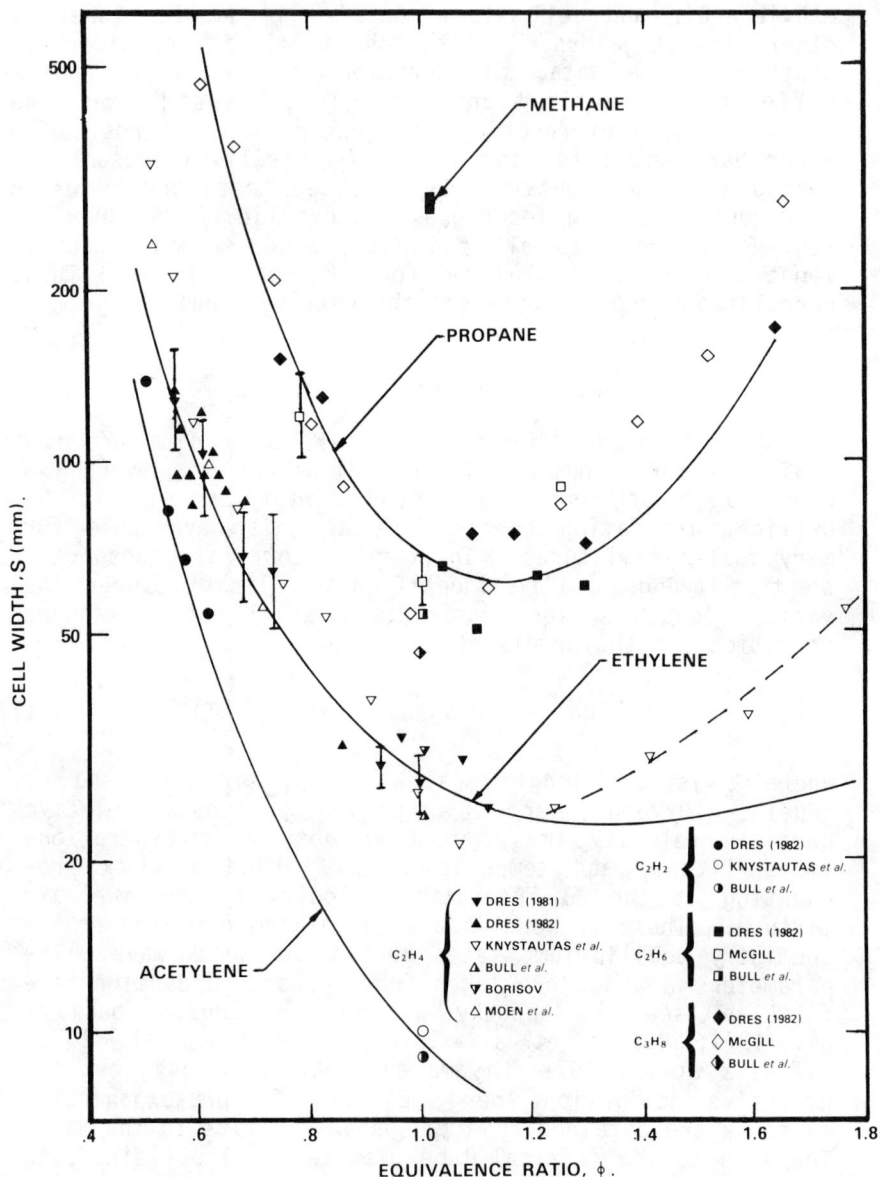

Fig. 5 Cell width S vs equivalence ratio ϕ for various fuel-air mixtures. The solid curves are fits to the data based on an induction length formula. A dashed curve is drawn through the data for rich ethylene-air mixtures. DRES (1982) - present results; DRES (1981) - Moen et al. (1982); Knystautas et al. (1982, 1983); Bull et al. (1982); McGill - Knystautas et al. (1982, 1983); Borisov (1980); Moen et al. (1982).

Table 1 Characteristic length scales for fuel-air detonations

FUEL	LENGTH SCALE		VALUE At $\phi = 1$	CORRELATION FORMULA $L = u\,k[\text{Fuel}]^a\,[\text{Oxygen}]^b\,\exp(E_A/RT)^*$		
				a, b	$k \times 10^{10}$	E_A(Kcal/mole)
ACETYLENE	Δ	(mm)	0.27	—	—	—
	S	(mm)	9.8	0, -1.0^E	138	38.9
	d_c	(m)	0.115		3.05	37.4
	R_{oc}	(m)	0.375^A	—	—	—
ETHYLENE	Δ	(mm)	2.56	0, -1.0^F	—	—
	S	(mm)	28		20,700	24.7
	d_c	(m)	0.43		5.65	37.2
	R_{oc}	(m)	$0.75 - 0.86^B$	—	—	—
ETHANE	Δ	(mm)	4.55		—	—
	S	(mm)	54 - 62		—	—
	d_c	(m)	~0.9	0.46, -1.26^G	—	—
	R_{oc}	(m)	$1.1 - 1.2^C$		114	33.5
PROPANE	S	(mm)	69		314	45.8
	d_c	(m)	0.9	0.57, -1.22^H	88.7	36.4
	R_{oc}	(m)	$1.3 - 1.5^B$	—	—	—
METHANE	Δ	(mm)	25.3	0.33, -1.03^I	—	—
	S	(mm)	280 ± 30		—	—
	R_{oc}	(m)	$9.8\,(\text{est.})^D$	—	—	—
HYDROGEN	Δ	(mm)	0.29		—	—
	S	(mm)	15	0, -1.0^J	295	32.6
	d_c	(m)	0.20		5.58	31.7
	R_{oc}	(m)	0.36		588	19.3

* Units: Concentrations, [Fuel] and [Oxygen], in moles/liter, and velocity, u, in m/sec.
A) Kogarko et al. (1965)
B) Bull et al. (1978)
C) Bull et al. (1979)
D) Bull et al. (1976)
E) Kistiakowsky and Richards (1962)
F) Hidaka et al. (1974)
G) Burcat et al. (1972)
H) Burcat et al. (1971)
I) Lifshitz et al. (1971)
J) Miyama and Takeyama (1964)

values of the characteristic length scales (Δ, S, d_c, and R_{oc}) at stoichiometric fuel-air compositions for various fuels. The induction lengths Δ are those calculated by Westbrook (1982a, b). All other results are obtained directly from experiment, or from correlations of experimental results.

At stoichiometric composition, the ratios of critical tube diameter to cell size are all between 11.7 (for acetylene) and 15.4 (for ethylene), in good agreement with the empirical relation $d_c = 13S$, obtained for more sensitive fuel-oxygen by Mitrofanov and Soloukhin (1964) and Edwards et al. (1979) and for oxygen enriched fuel-air mixtures by Knystautas et al. (1982). The variation of these length scales with composition is characterized by the effective activation energy E_A. For acetylene and hydrogen, the cell size and critical tube diameter effective activation energies are approximately the same, so that within experimental uncertainies the relation $d_c \simeq 13S$ is valid over a wide range of compositions. The cell size and critical tube diameter effective activation energies for propane and ethylene mixtures are quite different. For ethylene, for example, d_c/S varies from about 14 at $\phi = 1.2$ to almost 24 at $\phi = 0.65$. However, misinterpretation of the cellular structure for lean ethylene-air mixtures could easily account for this variation. In other words, the cell sizes measured near $\phi = 0.6$, from records such as that shown in Fig. 4, could correspond to substructure with half the dimensions of the controlling cellular structure. A similar misinterpretation of the records for propane-air, whereby twice the controlling cell size has been measured, could account for the difference in E_A in this case. Thus the existence of a simple link between the critical tube diameter and cell size cannot be ruled out. However, until a less subjective interpretation of the cellular structure is found, the usefulness of such a link for predicting the behavior of less sensitive fuel-air mixtures is questionable.

Unfortunately, there is insufficient data to draw any definitive conclusions regarding the link between the critical tube diameter and the critical initiation energy. At stoichiometric composition, the ratio R_{oc}/d_c varies from about 3.3 for acetylene-air to 1.2 for ethane-air. This can be compared with a value of about 1.3 predicted by the work-done model of Lee and Matsui (1979). The model based on a surface energy concept (Lee et al. 1982) predicts a ratio in the range 1.7-1.9. For hydrogen-air mixtures, the effective activation energy describing the variation of critical energy with composition appears to be considerably smaller than that for the cell size and critical tube diameter, indicating that the initiation and transmission of detonation may be controlled by different mechanisms.

In order to examine the links between the various detonation length scales (S, d_c, and R_{oc}) and the one-dimensional induction zone length Δ, calculated from a detailed chemical-kinetic model (Westbrook 1982a, b), we have plotted these length scales vs Δ in Fig. 6. The solid curves represent fits to the length scale data based on Eq. (1). In this figure, a straight line with slope equal to one corresponds to a linear relation between the length scale and the induction zone length. Such a linear relation appears to be valid for acetylene-air mixtures. For all other fuel-air mixtures considered, the relationship is more complex. In some cases, lean and rich composition length scales lie on two different branches of a V-shaped curve. The lower branch corresponds to lean mixtures. Furthermore, the relationship depends on the fuel-air mixture. In general, however, a linear relationship tends to underestimate the sensitivity of lean fuel-air mixtures relative to that of stoichiometric mixtures.

In view of the complex three-dimensional structure of the detonation front, involving interactions between transverse waves corresponding to many instability modes, the failure of correlations based on a one-dimensional ZND model of the detonation wave is not surprising. This structure becomes more complex for less sensitive mixtures, as more and more instability modes are observed. In fact, the identification of a dominant or controlling mode was virtually impossible for less sensitive fuel-air mixtures. This increasing complexity may be partly responsible for the failure of simple correlations which appear to hold for more sensitive mixtures.

A better understanding of the coupled chemical-gas-dynamic processes within the cellular detonation front is clearly required in order to clarify the links between detonation parameters and chemical kinetic rates. Three-dimensional analyses of cellular detonations are not feasible at the present time. However, many of the detonation instability properties can be investigated in one dimension. Results from numerical calculations of one-dimensional instabilities which illustrate many of the qualitative features of the experimental results are described in the next section.

Numerical Calculations

Although the cellular structure of the detonations described in the previous sections is intrinsically three-dimensional, the breakdown of cell regularity or symmetry can be described qualitatively by one-dimensional numerical calculations. Following Erpenbeck (1964), Fickett and Wood (1966), and Abouseif and Toong (1982), we consider, for this purpose, a mixture with constant specific heat ratio $\gamma = 1.2$, normalized activation energy $E^* = E/RT_0 = 50$, and a normalized heat of reaction $Q^* = Q/RT_0 = 50$. The level of instability is varied by changing the degree of overdrive $f=(M_S/M_{CJ})^2$, where M_S and M_{CJ} are the overdriven and CJ Mach numbers, respectively. Detonation was initiated by a piston moving at the steady-state shocked gas velocity for 3.9 half-reaction times $t_{1/2}$, followed by deceleration to the burned gas velocity during $0.1t_{1/2}$. This relatively strong initial perturbation promptly triggered the different instability modes, thus reducing the required computation time.

The numerical code initially developed by Yoshikawa (1980) was used. This code is a Lagrangian version of the McCormack (Hung and McCormack 1976) two step algorithm with the FCT antidiffusion scheme of Book et al. (1975). Mesh resolution with 50 cells in the steady-state half-reaction zone and time resolution corresponding to a Courant number of 0.5 were used.

The variation in shock pressure with distance, for degrees of overdrive corresponding to f = 1.6, 1.4, 1.3, and 1.2 are shown in Fig. 7. Distance relative to a fixed observer is measured in units of the induction length Δ, which is defined based on a steady-state ZND detonation profile. For f = 1.6 (Fig. 7a), the detonation rapidly assumes a regular oscillatory pattern with a wavelength of 160Δ (±10%). The corresponding period of $8.5t_{1/2}$ agrees well with results obtained by other investigators (see Abouseif and Toong 1982; Erpenbeck 1964; Fickett and Davis 1979). A regular pattern, with approximately the same ratio of wavelength to induction zone length, is also observed for f = 1.4 (Fig. 7b). However, at f = 1.3 (Fig. 7c), as the detonation becomes more unstable, the oscillatory pattern begins to exhibit fairly dramatic irregularities. The apparent split of a low mode into higher modes is observed at 350 and 1100Δ. Finally, at f = 1.2 (Fig. 7d), the regular symmetry of the oscillations breaks up completely. The oscillations observed are so irregular that the identification of dominant cycles is subject to considerable interpretation. If the cycles are identified as indicated in Fig. 7d, the average wavelength is 130Δ, with an uncertainty of at least ±40%. The ratio of average wavelength to induction zone length is therefore smaller than that for the more regular cycles observed with higher degrees of overdrive. It is also about 10% smaller than that reported by Abouseif and Toong (1982) based on one cycle. In view of the extreme intrinsic instability of these detonations, the details and the onset of the break-up into irregular structure are likely to depend on the numerical method and the numerical resolution. Calculations with various mesh resolutions

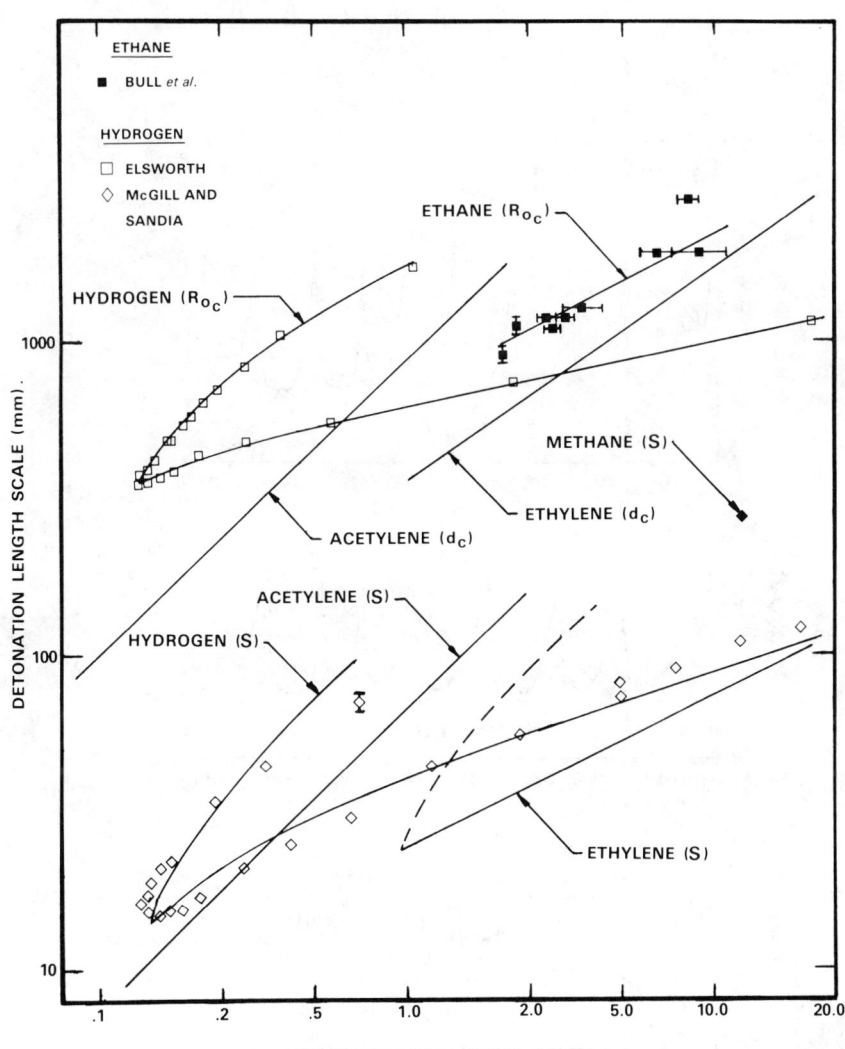

Fig. 6 Detonation length scales (d_c, R_{oc}, and S) vs induction length Δ calculated by Westbrook (1982a, b) for different fuel-air mixtures. The solid curves correspond to best fits to the length scale data. The dashed curve for ethylene corresponds to the dashed curve in Fig. 5. Bull et al. (1979); Elsworth (1982); McGill and Sandia - Lee et al. (1982) and Guirao et al. (1982).

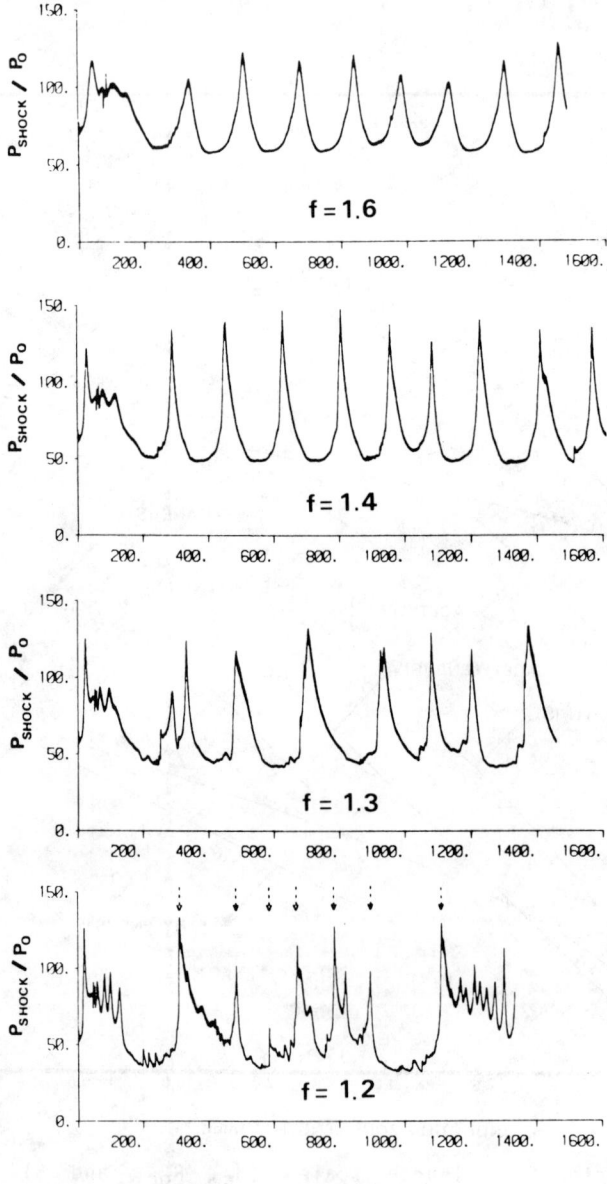

Fig. 7 Pressure traces displaying various levels of regularity for cases: a) $f = 1.6$, b) $f = 1.4$, c) $f = 1.3$, d) $f = 1.2$. P_{SHOCK} is the pressure at leading shock, P_0 the initial pressure, X_{SHOCK} the position of leading shock relative to fixed observer, and Δ the ZND induction zone length. Arrows indicate peaks selected in wavelength estimates for $f = 1.2$ (10 point averaging used in plotting).

which illustrate the delay or enhancement of the break-up have been done. Although the break-up can be delayed by using better resolution, the onset of higher order instabilities is not eliminated.

Typical pressure and reactant profiles for f = 1.2 are shown in Fig. 8. The profiles show fairly strong perturbing waves behind the leading shock. These waves interact with pockets of unburned gas causing localized explosions. Such explosions are partly responsible for the observed irregular structure.

Following Libouton et al. (1981), the above cases may be divided into three categories of good, poor, and irregular structure. Qualitatively, the one-dimensional oscillations observed for f = 1.6 and 1.4 display a regular structure analogous to the fairly regular three-dimensional structure observed for more sensitive fuel-air

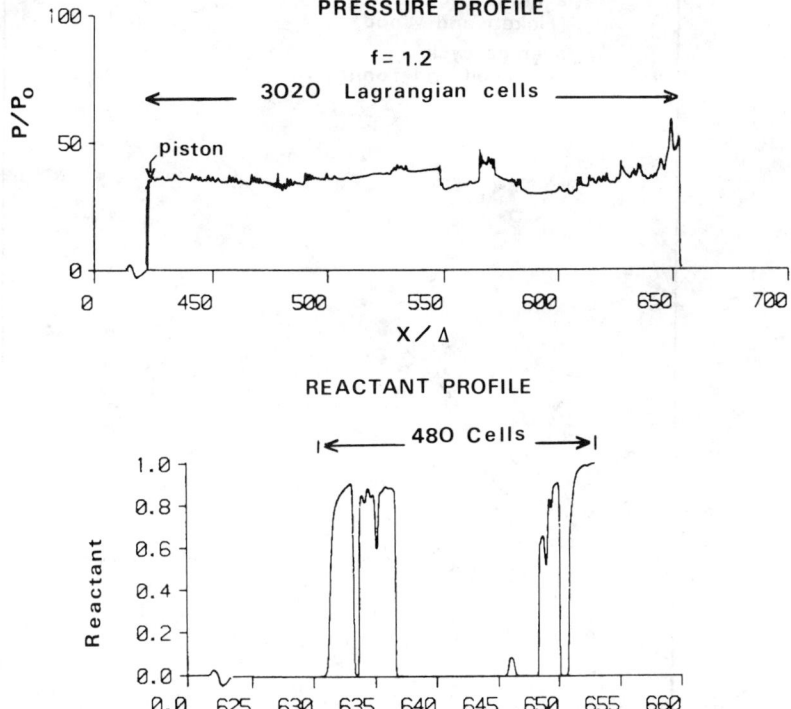

Fig. 8 Computed pressure and reactant profiles displaying secondary shocks and unburned pockets leading to local explosions for highly unstable detonations (no averaging used in plotting).

mixtures. Whereas, the intermittent wave splitting found for f = 1.3, corresponds to a relatively poor structure with more than one mode, as observed, for example, on the ethylene and propane smoked foils shown in Fig. 4. Finally, continual wave splitting, as observed for f = 1.2, results in highly irregular structures typical of methane-air mixtures.

The variation of the one-dimensional oscillation wavelength L_{osc}, with induction zone length, for various degrees of overdrive, are summarized in Fig. 9. For mildly unstable systems (i.e., f = 1.6 and 1.4), a

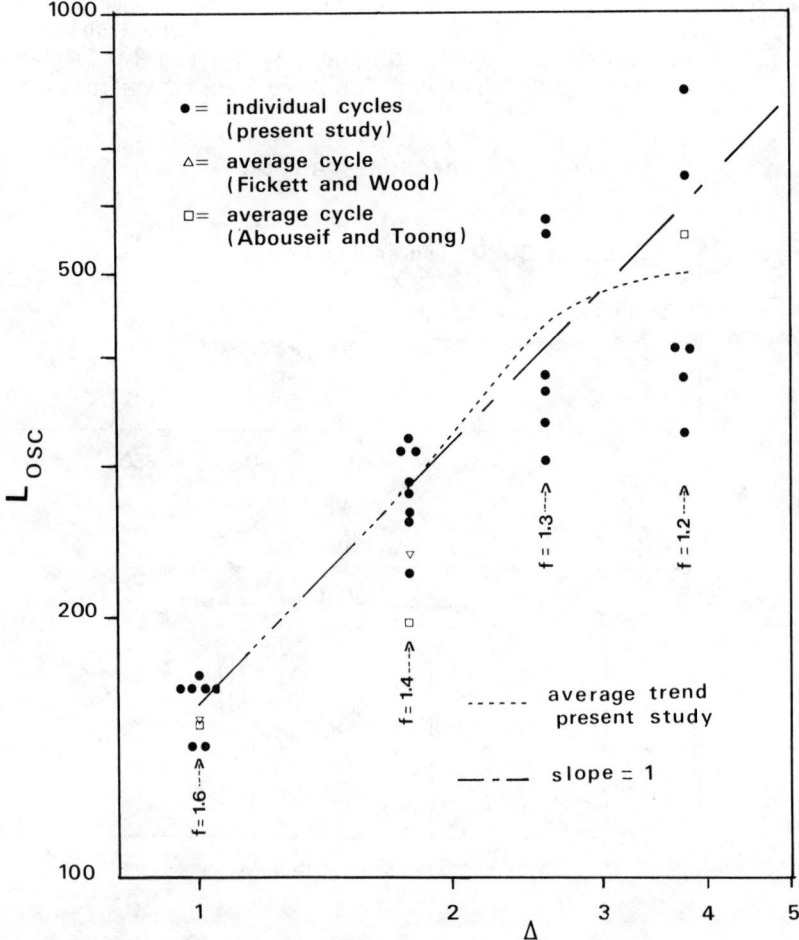

Fig. 9 Dependence of oscillation wavelength on induction zone length from numerical calculations. (ZND induction zone length set to unity for f = 1.6).

linear relationship appears to be valid. For the less stable systems, the fluctuations in wavelength are so large that a meaningful correlation is difficult. However, a correlation based on the average wavelength shows the same trend as that observed for various fuel-air detonation length scales in Fig. 6.

Conclusion

Large-scale field tests on the diffraction of detonations in fuel-air mixtures have provided key critical tube diameter and cell size data for acetylene-, ethylene-, ethane-, propane-, and methane-air mixtures for tube diameters up to 1.83 m. The detonation cell width S, for near stoichiometric methane-air mixtures, is interpreted to be 280 ± 30 mm. Based on a 13S relationship, with this cell size, a critical tube diameter of 3.6 ± 0.4 m is predicted.

These results, together with results from other investigations, now provide key detonability data for a number of fuel-air mixtures, covering a wide range of compositions and mixture sensitivities. Much of these data have been obtained by large-scale testing. Such tests are both time-consuming and expensive. It is therefore important to relate the detonation parameters of practical interest (i.e., the critical tube diameter and the critical initiation energy), to the characteristic detonation cell sizes which can be measured in the laboratory and to the parameters which characterize the chemical kinetics of the mixture. In this manner, the detonability of a gaseous mixture can be assessed without resorting to large-scale testing.

Some of the proposed relationships between detonation parameters have been critically examined using the available data for fuel-air explosives. Based on this examination, it is concluded that correlations with cell size can provide a useful assessment of detonability for more sensitive mixtures with well-defined cellular structure. However, for less sensitive mixtures, the cellular structure becomes highly irregular, with many modes of cellular patterns, so that the identification of the "controlling" cell structure is subject to considerable interpretation. Until a less subjective interpretation of this cellular structure is found, the reliability of assessments based solely on the cell size is questionable. It is also found that a linear relationship between the induction zone length, calculated from a one-dimensional ZND model, and the critical tube diameter tends to underestimate the sensitivity of lean fuel-air mixtures relative to that of stoichiometric mixtures. A similar underestimate is obtained with a cubic relationship between the critical initiation energy and the induction length.

Many of the detonation instability properties observed for fuel-air explosives have been qualitatively illustrated by numerical simulation of one-dimensional detonation instabilities. These numerical calculations also show that simple correlations between instability wavelengths and induction zone lengths lose much of their predictive value for very unstable systems which display highly irregular oscillations. Consequently, further elucidation of the coupling between chemical kinetic rates and hydrodynamics will be required in order to clarify the links between chemical lengths, cellular instability wavelengths, and detonation parameters of practical interest.

Acknowledgments

We would like to thank the personnel of the Field Operations Section, the Electronic Design and Instrumentation Group, and the Photo Group at DRES for their valuable assistance during the DRES field tests. The help of the DRES Computer Group during calculations is also appreciated. Discussions with S. B. Murray and Professor J. H. S. Lee are gratefully acknowledged.

References

Abouseif, G. E. and Toong, T. Y. (1982) Theory of unstable detonations. Combustion and Flame 45, 67-94.

Atkinson, R., Bull, D. C., and Shuff, P. J. (1980) Initiation of spherical detonation in hydrogen/air. Combustion and Flame 39, 287-300.

Book, D. L., Boris, J. P., and Hain, K. (1975) Flux corrected transport. II: Generalizations of the method. Journal of Computational Physics 18, 248.

Borisov, A. A. (1980) private communication. Institute of Chemical Physics, USSR Academy of Sciences, Moscow.

Bull, D. C., Elsworth, J. E., Hooper, G., and Quinn, C. P. (1976) A study of spherical detonation in mixtures of methane and oxygen diluted by nitrogen. J. Phys. D. Appl. Phys. 9, 1991-2000.

Bull, D. C., Elsworth, J. E., and Hooper, G. (1978) Initiation of spherical detonation in hydrocarbon/air mixtures. Acta Astronautica 5, 997-1008.

Bull, D. C., Elsworth, J. E., and Hooper, G., (1979) Concentration limits to unconfined detonation of ethane-air. Combustion and Flame 35, 27-40.

Bull, D. C., Elsworth, J. E., and Shuff, P. J. (1982) Detonation cell structures in fuel/air mixtures. Combustion and Flame 45, 7-22.

Burcat, A., Lifshitz, A., Scheller, K., and Skinner, G. B. (1971) Shock-tube investigation of ignition in propane-oxygen-argon mixtures. 13th Symposium (International) on Combustion, pp. 745-755, The Combustion Institute, Pittsburgh, Pa.

Burcat, A., Crossley, R. W., Scheller, K., and Skinner, G. B. (1972) Shock-tube investigation of ignition in ethane-oxygen-argon mixtures. Combustion and Flame 18, 115-123.

Edwards, D. H., Thomas, G. O., and Nettleton, M. A. (1979) The diffraction of a planar detonation wave at an abrupt area change. J. Fluid Mech. 95, 79-96.

Edwards, D. H. and Thomas, G. O. (1981) Diffraction of a planar detonation wave in various fuel/oxygen mixtures at an area change. Gasdynamics of Detonations and Explosions: AIAA Progress in Astronautics and Aeronautics (edited by Bowen, Manson, Oppenheim, and Soloukhin) Vol. 75, pp. 341-357, AIAA, NY.

Elsworth, J. E. (1981) Thornton Research Center, Chester, England. As reported by Guirao et al. (1981).

Erpenbeck, J. J. (1964) Stability of idealized one-reaction detonations. Phys. Fluids 7, 684-696.

Fickett, W. and Davis, W. C. (1979) Detonation, 386 pp., University of California Press, Berkeley, Ca.

Fickett, W. and Wood, W. W. (1966) Flow calculations for pulsating one-dimensional detonations. Phys. Fluids 9, 903-916.

Funk, J. W. and Murray, S. B. (1982) The DRES large-scale fuel-air explosives testing facility, Defence Research Establishment Suffield Memorandum SM-1051, pp. 19, DRES, Ralston, Alberta, Can.

Funk, J. W., Murray, S. B., Moen, I. O., and Ward, S. A. (1982) A brief description of the DRES fuel-air explosives testing facility and current research program. Proceedings of International Conference on Fuel-Air Explosions, Montreal, Nov. 1981, SM Study No. 16, pp. 565-583, University of Waterloo Press, Ontario, Canada.

Guirao, C. M., Knystautas, R., Lee, J. H., Benedick, W., and Berman, M. (1982) Hydrogen-air detonations. 19th Symposium (International) on Combustion, pp. 583-590, The Combustion Institute, Pittsburgh, Pa.

Hidaka, Y., Kataoka, T., and Suga, M. (1974) A shock-tube investigation in ethylene-oxygen-argon mixtures. Bull. Chem. Soc., Japan 47, 2166-2170.

Hung, C. M. and MacCormack, R. W. (1976) Numerical solutions of supersonic and hypersonic laminar compression corner flows. AIAA Journal 14, 475-481.

Jenssen, A. (1982) private communication. Norwegian Defence Construction Service, Oslo, Norway.

Kistiakowsky, G. B. and Richards, W. L. (1962) Emission of vacuum ultraviolet radiation from the acetylene-oxygen and the methane-oxygen reactions in shock waves. J. Chem. Phys. 36,1707.

Knystautas, R., Lee, J. H., and Guirao, C. M. (1982) The critical tube diameter for detonation failure in hydrocarbon-air mixtures. Combustion and Flame 48, 63-83.

Knystautas, R., Guirao, C., Lee, J. H., and Sulmistras, A. (1983) Measurement of cell size in hydrocarbon-air mixtures and predictions of critical tube diameter, critical initiation energy and detonability limits. Paper presented at 9th ICODERS, Poitiers, France, July 3-8.

Kogarko, S. M., Adushkin, V. V., Lyamin, A. G. (1965) Nauchno Teknicheskie Problemy Goreniya i Vzryva 2, 22.

Lee, J. H. and Matsui, H. (1978) On the measure of relative detonation hazards of gaseous fuel-oxygen and air mixtures. 17th Symposium (International) on Combustion, pp. 1269-1279, The Combustion Institute, Pittsburgh, Pa.

Lee, J. H. S., Knystautas, R., and Guirao, C. M. (1982) The link between cell size, critical tube diameter, initiation energy and detonability limits. Proceedings of the International Conference on Fuel-Air Explosions, Montreal, Nov. 1981: SM Study No. 16, pp. 157-187, University of Waterloo Press, Ontario, Canada.

Libouton, J-C., Jacques, A., and Van Tiggelen, P. J. (1981) Cinetique, structure et entretien des ondes detonation. Proceedings of First Specialist Meeting (International) of Combustion, Bordeaux, France, July, pp. 437-444, The Combustion Institute, Section Francaise.

Lifshitz, A., Scheller, K., Burcat, A., and Skinner, G. B. (1971) Shock-type investigation of ignition in methane-oxygen-argon mixtures. Combustion and Flame 16, 311.

Mitrofanov, V. V. and Soloukhin, R. I. (1964) The diffraction of multifront detonation waves. Soviet Phys. Dokl. 9, 1055.

Miyama, H. and Takeyama, T. J. (1964) Kinetics of hydrogen-oxygen reactions in shock waves. J. Chem. Phys. 41, 2287.

Moen, I. O., Donato, M., Knystautas, R., and Lee, J. H. (1981) The influence of confinement on the propagation of detonations near the detonability limits. 18th Symposium (International on Combustion, pp. 1615-1622, The Combustion Institute, Pittsburgh, Pa.

Moen, I. O., Murray, S. B., Bjerketvedt, D., Rinnan, A., Kynstautas, R., and Lee, J. H. (1982) Diffraction of detonation from tubes into a large fuel-air explosive cloud. 19th Symposium (International) on Combustion, pp. 635-644, The Combustion Institute, Pittsburgh, Pa.

Oran, E. S., Boris, J. P., Young, T. R., Flanigan, M., Picone, M., and Burks, T. (1981) Numerical simulations of detonations in hydrogen-air and methane-air mixtures. 18th Symposium (International) on Combustion, pp. 1641-1649, The Combustion Institute, Pittsburgh, Pa.

Taki, S. and Fujiwara, T. (1978) Numerical analysis of two-dimensional nonsteady detonations. AIAA 16, 73-77.

Westbrook, C. K. (1982a) Chemical kinetics of hydrocarbon oxidation in gaseous detonations. Combustion and Flame 46, 191-210.

Westbrook, C. K. (1982b) Chemical kinetics of gaseous detonations. Proceedings of International Conference on Fuel-Air Explosions, Montreal, Nov. 1981: SM Study No. 16, pp. 189-242, University of Waterloo Press, Ontario, Canada.

Westbrook, C. K. and Urtiew, P. A. (1982) Chemical kinetic predictions of critical parameters in gaseous detonations. 19th Symposium (International) on Combustion, pp. 615-622, The Combustion Institute, Pittsburgh, Pa.

Yoshikawa, N. (1980) Coherent shock wave amplification in photo-Chemical initiation of gaseous detonations, PhD. Thesis, Dept. of Mechanical Engineering, McGill University, 153 pp.

Zel'dovich, Ya. B., Kogarko, S. M., and Simonov, N. N. (1956) An experimental investigation of spherical detonation of gases. Soviet Phys. Tech. Phys. 1, 1689-1713.

The Influence of Yielding Confinement on Large-Scale Ethylene-Air Detonations

Stephen B. Murray*
Defence Research Establishment Suffield, Ralston, Canada
and
John H. Lee†
McGill University, Montreal, Canada

Abstract

This paper reports on a series of field tests performed at the Defence Research Establishment Suffield involving transmission of ethylene-air detonations from a rigid tube to a yielding polyethylene tube (1, 5, or 10 mil wall thickness), both of 0.89-m diam. High-speed cinematography shows that the phenomenon consists of an initial transient reinitiation phase followed by a steady propagation phase. The former is characterized by reignition nuclei that form near the yielding wall and give rise to detonation bubbles which sweep between the shock and decoupled reaction zone to re-establish the wave globally. In the steady propagation phase, the wave is seen to be sensitive to boundary irregularities when the mixture sensitivity for critical transmission is approached. In fact, marginal detonations are seen to quench if the spacing between boundary-induced disturbances is reduced below about 12 detonation cell widths. The data also confirm the equivalence between the critical diameters for transmission and for propagation in a "free" column of gas. The likelihood of reignition following a disturbance is predicted by first calculating the thermodynamic history of the critical particle using a cylindrical piston analogy to describe the wall motion. Assuming the relevant chemical time to be of the standard induction time form and after

---

Paper presented at the 9th ICODERS, Poitiers, France, July 3-8, 1983. Copyright 1983 by Government of Canada. Published by the American Institute of Aeronautics and Astronautics with permission.
*Defence Scientist, Military Engineering Section.
+Professor, Department of Mechanical Engineering.

applying a Shchelkin-like failure criterion, the controlling chemical length in the phenomenon is found to be on the order of the detonation cell length.

Introduction

Recently, much attention has been focused on the critical conditions for transmission of a detonation from a round tube to an unconfined region (Moen et al. 1982; Knystautas et al. 1982). The criterion for transmission can be based on the equality of a gasdynamic time scale that characterizes the quenching process to some global chemical time scale which describes the reaction processes within the cellular detonation front. Given a mixture, and hence a chemical time scale, the gasdynamic time scale can be varied by adjusting the tube diameter until the critical conditions are realized. However, it is apparent that the severity of the quenching process, which depends strongly on the boundary conditions, should also play an important role. One means of controlling the strength of the lateral rarefaction wave is to arrange transmission to a tube with yielding walls rather than to an unconfined region. The acceleration of the wall by the high-pressure gases, and thus the severity of the expansion, can be regulated by varying the wall mass. Investigations involving yielding boundaries have been carried out by many researchers (Dabora 1963; Dabora et al. 1965; Sichel 1965; Bazhenova et al. 1981; Brossard and Renard 1981; Sommers and Morrison 1962; Vasiliev et al. 1972). However, much of this work has not considered the three-dimensional structure of detonations or the details of the reinitiation process. Furthermore, much of the analysis has been based on a mean velocity-deficit approach that has met with only marginal success.

This paper reports on the results of a large-scale study on the transmission of ethylene-air detonations from a rigid tube to yielding polyethylene tubes of various wall thicknesses. Reinitiation of the disturbed wave is seen to occur via the formation of the reignition nuclei in the vicinity of the yielding tube wall. The delays to reignition predicted by a simple model correlate well with the experiment.

Experimental Details

The large-scale field facility at the Defence Research Establishment Suffield (DRES) has been described elsewhere (Funk and Murray 1982; Funk et al. 1981). A photograph of the present apparatus is shown in Fig. 1. It consists of a round steel tube 7.82 m in length and 0.89 m in diameter, connected to one of several seamless, extruded polyethylene tubes of the same diameter and typically 10-m long. Three nominal wall thicknesses were investigated. These were 1, 5, and 10 mil (1 mil = 0.0254 mm). The measured area densities are given in Table 1. These polyethylene tubes were supported by steel hoops 1.5-m apart.

The test gas (CP grade, 99.5% pure C_2H_4) was mixed with air by a multipath recirculation system. The composition and homogeneity of the mixture were continuously monitored by analyzing the gas from four sampling ports using a precalibrated "Wilks Miran 80" infrared analyzer. This system guaranteed the concentration and homogeneity to within ±0.05% C_2H_4. Direct initiation of detonation at the ignition end of the tube was achieved using either high-explosive PETN or a sensitive (i.e., easily detonable) volume of oxyacetylene mixture contained within a 1.5-m long plastic liner.

Five types of diagnostic techniques were employed. Piezoelectric pressure transducers (PCB 113A24) were positioned along the rigid tube and in the support hoops. Pressure-time signatures and velocities from these sensors were used to determine the success or failure of the transmission. Velocities were independently confirmed using an ionization-gap probe system and time-of-arrival counter. The detonation structure was recorded near the exit of the rigid tube using 350 x 500 mm smoked plates. Cinematographic records were obtained using a "Hycam" camera [~ 18,000 frames/s (fps)] looking normal to the direction of propagation and either a "Fastax" (~ 5000 fps) or a "Photec" (~ 18,000 fps) camera positioned 45 deg off axis. For some tests, an additional "Fastax" camera (~ 5000 fps) and a microwave Doppler system (10.25 GHz), both mounted in a protective shelter at the downstream end of the yielding

Table 1 Critical conditions for transmission from a rigid tube of 0.89 m diam to various yielding tubes and to an unconfined region

Nominal tube wall thickness t, mil	Measured wall surface density, ϱ_s, kg/m^2	Critical C_2H_4 concentration, %	Critical cell size, λ, mm	D/λ
0	0.0	5.05	46.9	19.0
1	0.0242	4.70	58.7	15.2
5	0.124	4.15	92.5	9.6
10	0.220	3.90	119.0	7.5

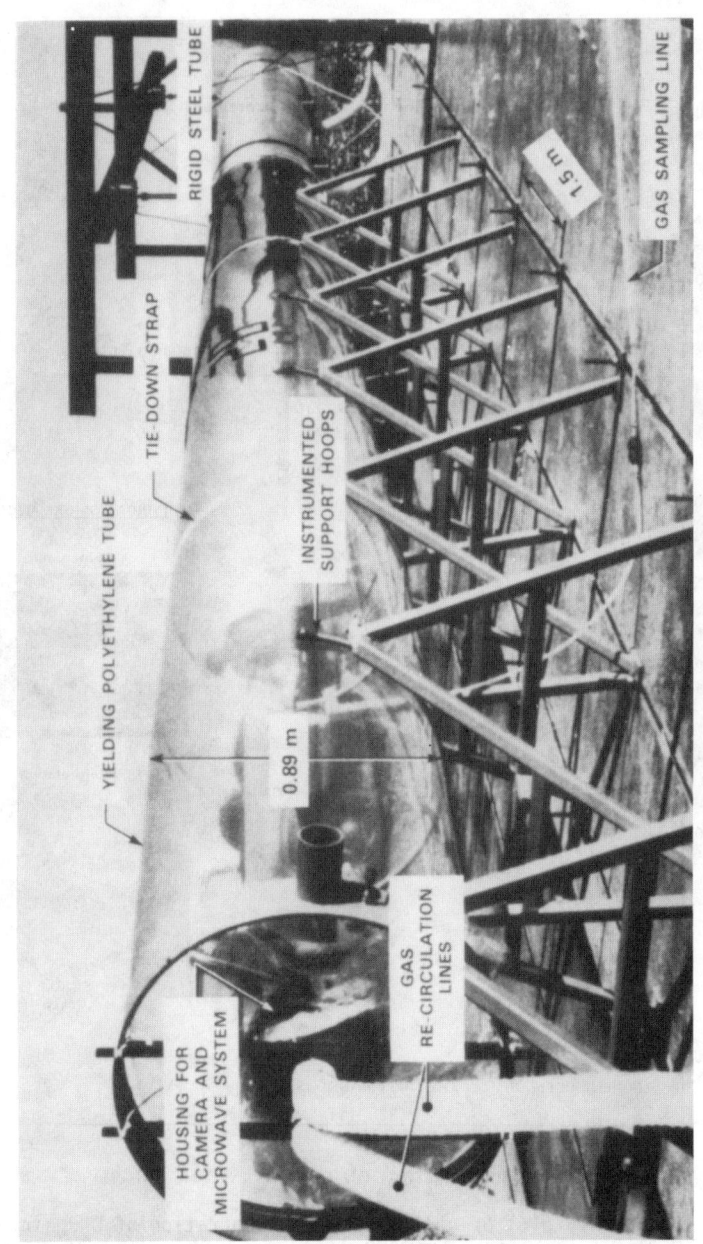

Fig. 1 Large-scale field facility showing rigid/yielding tube combination prior to testing.

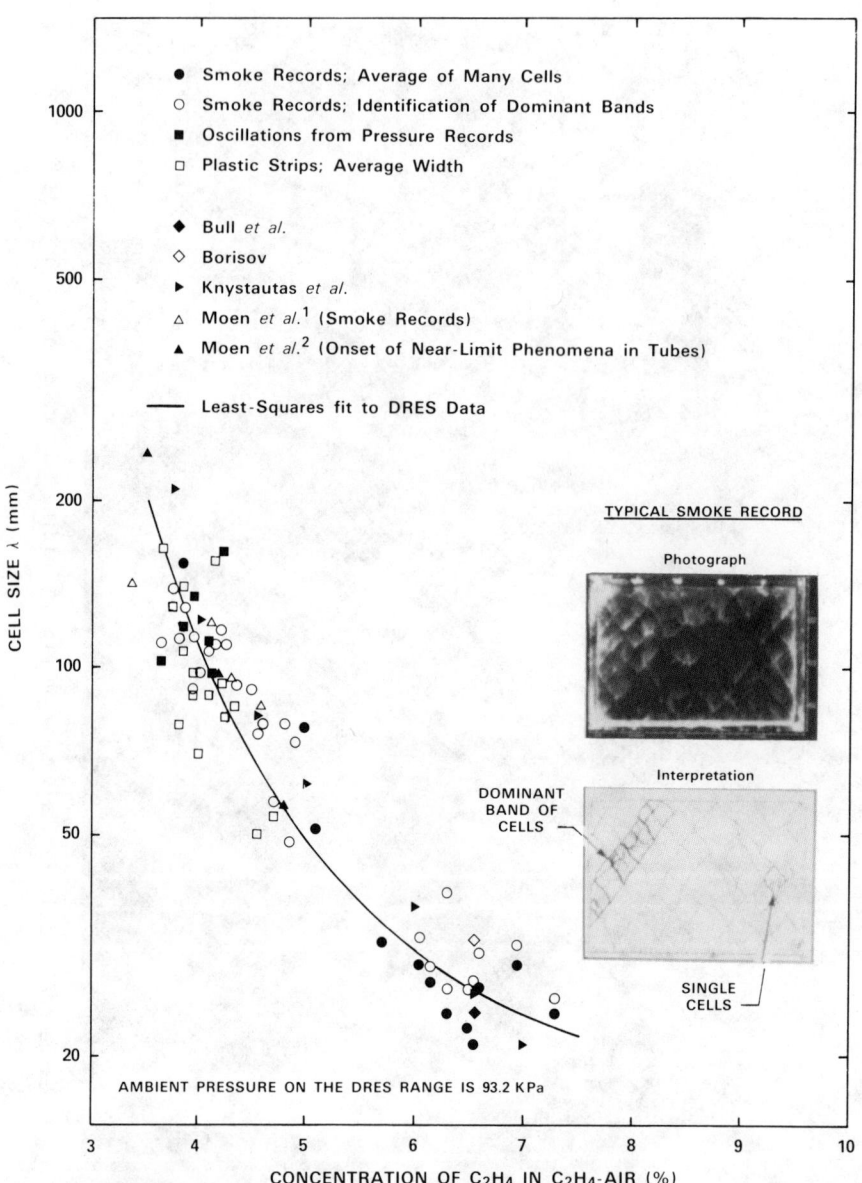

Fig. 2 Summary plot of cell size vs concentration of C_2H_4 in C_2H_4-air mixtures (Bull et al. 1982; Borisov 1982; Knystautas et al. 1982; Moen et al. 1981, 1983).

tube, were used to obtain photographic records and velocity profiles along the axis of propagation.

Experimental Results

All detonation velocities and pressures in the steel tube were within ±5% of the corresponding Chapman-Jouguet (CJ) values. Equilibrium propagation prior to transmission was further confirmed from smoke records. A summary plot of cell width λ vs mixture composition (%C_2H_4) is shown in Fig. 2. Cell widths were deduced either by averaging many individual measurements or by identifying dominant diagonal bands as illustrated in the inset of Fig. 2 (see Moen et al. 1982 for details). Also included on the graph are cell widths computed from the period of oscillations Δt on the pressure-time signatures. An experimentally observed cell aspect ratio λ/ℓ of 0.6, together with the CJ velocity V_{CJ} gives $\lambda = 0.6 \cdot V_{CJ}\Delta t$. An interesting observation is that the width of plastic strips remaining on the test bed following an experiment is often found to be quite close to the cell size. They are thought to be produced by the piercing action of the triple points followed by a longitudinal shearing process. These measurements are included on the graph. The present data are in good agreement with the direct measurements of other investigators (Borisov 1982; Bull et al. 1982; Knystautas 1982; Moen et al. 1981, 1983). The curve through the data has been calculated using a least-squares fit assuming

$$\lambda = u_s A [O_2]^{-1} \exp(E/RT)$$

where the pre-exponential factor A is the free parameter. Here E, $[O_2]$, R, and T have their usual meaning and u_s is the postshock relative particle velocity. Assuming frozen chemistry and vibrationally relaxed equilibrium conditions, the DRES data give $A = 7.15 \times 10^{-10}$ s·mole/l for $[O_2]$ in moles/l, u_s in m/s, λ in m, and T in degrees Kelvin. The postshock parameters have been computed using the chemical equilibrium computer code developed by Gordon and McBride (1976). The activation energy of 27.5 kcal/mole noted by Hidaka et al. (1974) has been used in these calculations. The curve in Fig. 2 will be used as a reference for later analysis.

The contributions of this study center around high-speed films of detonation in the yielding tubes that show the existence of both a transient reinitiation phase and a steady propagation phase. These are discussed in detail below.

Transient Reinitiation Phase

When the wave first emerges from the rigid tube, the head of the expansion propagates inward at sonic velocity C_{CJ} in the burned gases. This is felt over the entire front before it has moved a distance of about $V_{CJ}R/C_{CJ}$ from the exit, where R is the tube radius. Since a typical value of V_{CJ}/C_{CJ} from equilibrium calculations is ≈2, this distance is about one tube diameter. If the detonation wave does not act to overcome the disturbance before travelling a distance of this order, it will be quenched. A previous study has shown that, for transmission to an unconfined medium, reinitiation is seen to occur via the formation of reignition nuclei or "hot spots" near the edge of the detonation core (Murray and Lee 1983). These evolve into expanding "detonation bubbles" that eventually coalesce to form the spherically diverging wave. Transmission

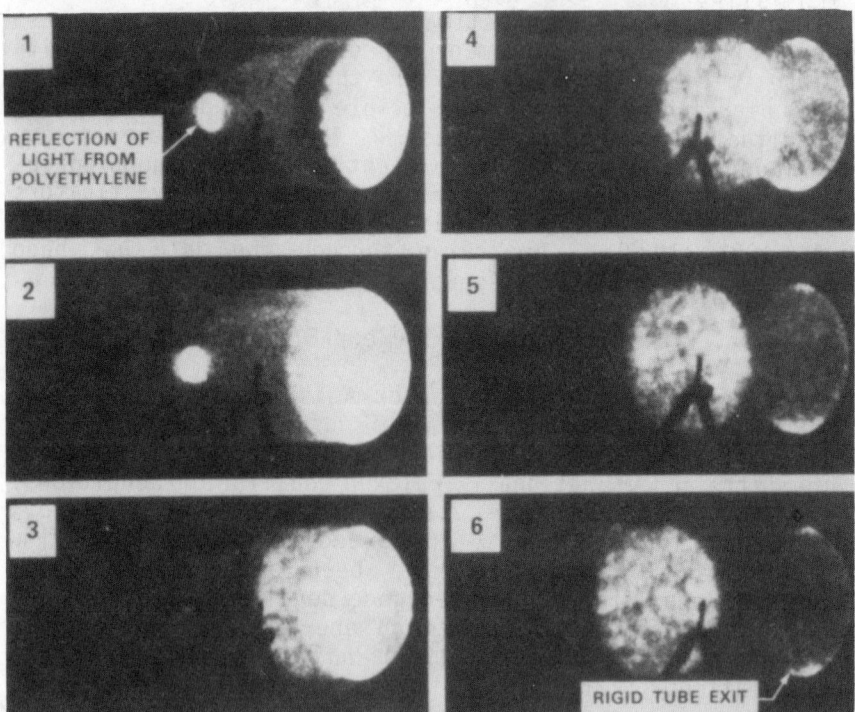

Fig. 3 Cinematographic sequence showing supercritical transmission to a yielding tube of nominal 10-mil wall thickness and 0.89-m diam (gas mixture is 4.15% C_2H_4 in C_2H_4-air; elapsed time between frames is ≈0.2 ms).

of this type was found to be possible from the present 0.89-m diam tube for an ethylene concentration of 5.05%, corresponding to a cell width λ_S of about 47 mm. This cell width gives a value for D/λ of 19.0, which is larger than that of 13 observed for most of the hydrocarbon systems (Knystautas et al. 1982). The difficulty in identifying the relevant structure on the relatively small smoked plates used in the field tests may account for this discrepancy.

For transmission to a yielding tube, the gasdynamic gradients are milder so that reinitiation should be possi-

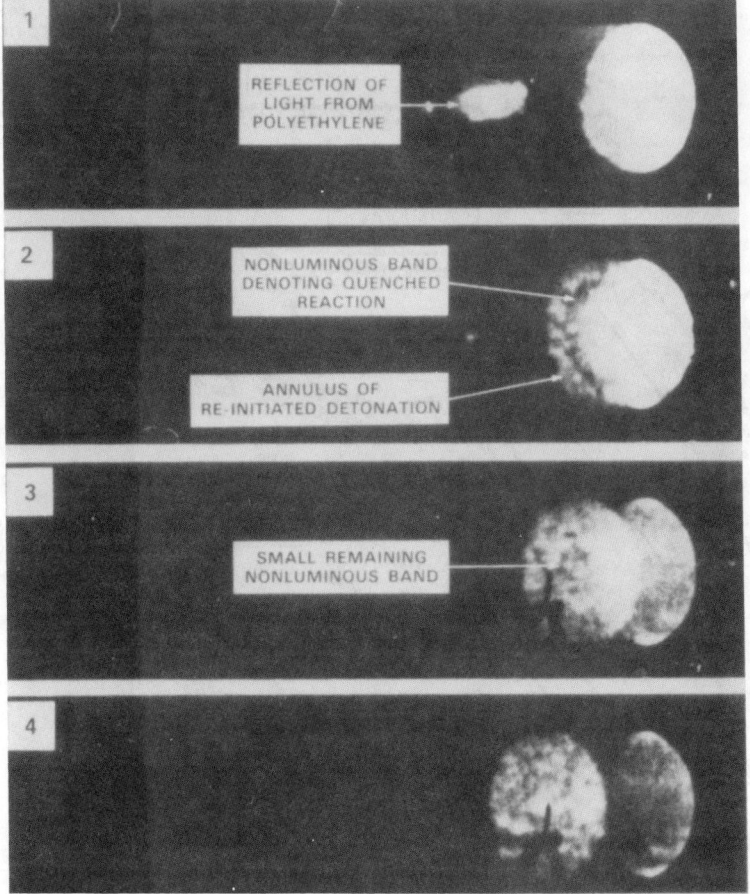

Fig. 4 Cinematographic sequence showing transmission to a yielding tube of nominal 5-mil wall thickness and 0.89-m diam [mixture of 4.25% C_2H_4 in C_2H_4-air is slightly more sensitive than the critical one (4.15% C_2H_4) for this wall thickness; elapsed time between frames is ≈ 0.2 ms].

ble in less sensitive mixtures (i.e., $\lambda_s/\lambda \leqslant 1$) than those required for transmission to an unconfined medium. This is confirmed by the high-speed cinematographic records. A sequence from one such record is shown in Fig. 3. Here, transmission to a yielding tube of 10-mil wall thickness is seen to be successful for a C_2H_4 concentration of only 4.15%, that is, for a relative sensitivity λ_s/λ of only 0.51. In fact, there is no obvious reinitiation mechanism apparent in this sequence, indicating that conditions are well removed from criticality.

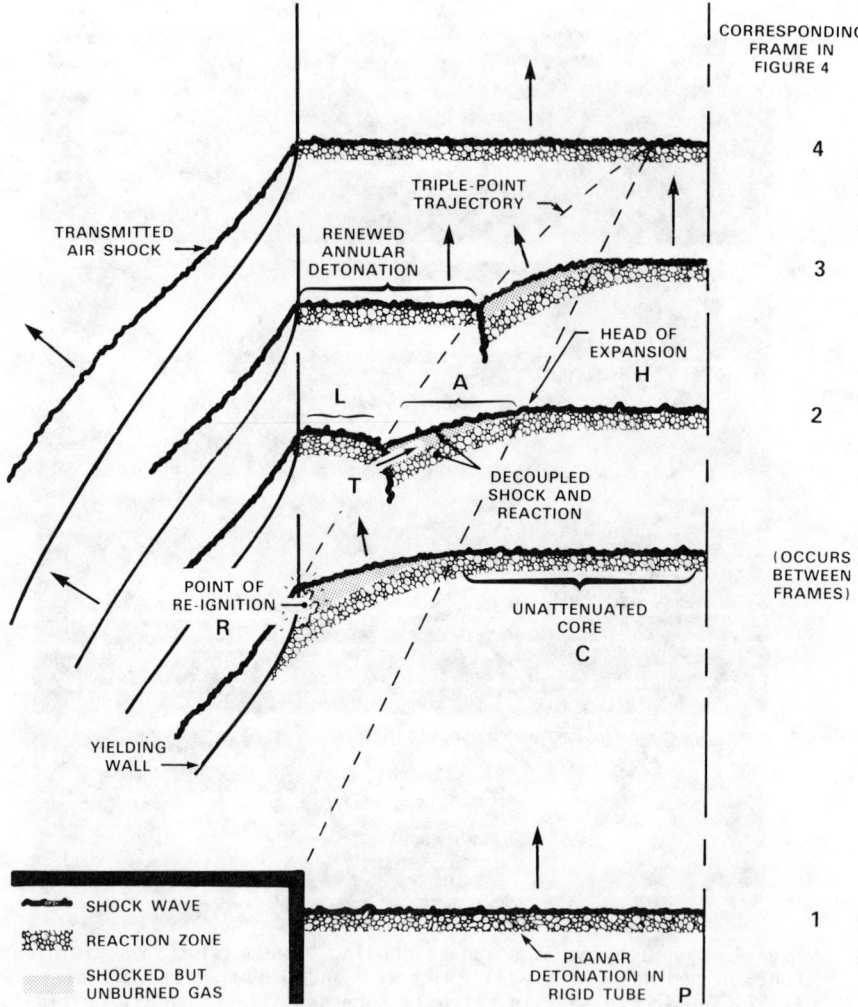

Fig. 5 Schematic diagrams showing details of the reinitiation process typical of transmission to a tube with yielding walls.

A distinct reinitiation mechanism is observed as the confinement is lessened or as the mixture is made less sensitive. Figure 4 shows a similar sequence of frames for transmission to a tube of 5-mil wall thickness for $\lambda_S/\lambda = 0.56$. The events in the sequence will be described with the aid of the schematic diagrams shown in Fig. 5. The first frame shows the planar wave (P) just prior to exiting the rigid tube. In the second frame, a dark annular band (A) surrounding a luminous circular core (C) is apparent. The inner boundary of the band marks the location of the head of the expansion (H) as it advances radially inward. A ring of luminous patches (L) bounded by the wall on the outside and dark band on the inside is seen to have formed in the wake of the expansion, indicating that a reinitiated wave has evolved from a locus of reignition centers (R) near the tube wall. Reignition near the boundary may be due to several factors: 1) it may be induced by reflection of transverse wave remnants from the wall; 2) a thicker layer of shocked but unreacted gas exists near the wall; thus, possibly satisfying a mini-

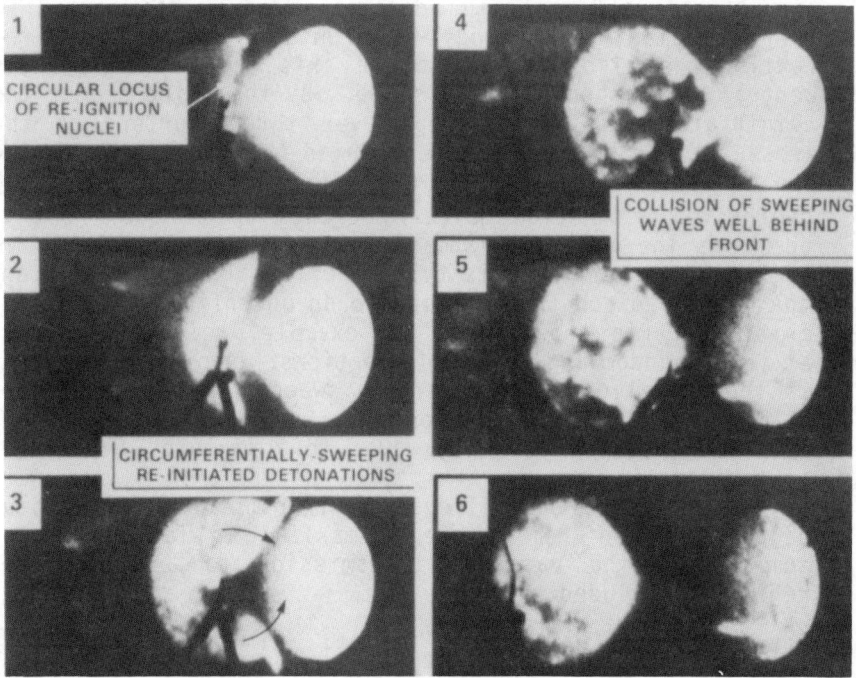

Fig. 6 Cinematographic sequence showing critical transmission to a yielding tube of nominal 1-mil wall thickness and 0.89-m diam (gas mixture is 4.70% C_2H_4 in C_2H_4-air; elapsed time between frames is ≈0.2 ms).

mum volume prerequisite for reignition; 3) a higher temperature may exist near the wall due to the presence of "x" shaped compression waves behind the slightly curved shock front (Fay and Opel 1958; Edwards et al. 1963); and 4) small ripples or other irregularities in the polyethylene surface may give rise to localized shock reflections and thus hot spots. By the third frame, the renewed annular detonation has expanded transversely (T) inward, converging on the tube axis. The final frame shows a re-established planar wave. This mode of reinitiation is similar to that for transmission to an unconfined region in that the number of reignition centers increases, while their location moves toward the tube exit as the mixture is made more sensitive.

Critical transmission to a yielding tube of 1-mil wall thickness for $\lambda_s/\lambda = 0.80$ is shown in the film sequence of Fig. 6. In the first frame, a locus of reignition centers can be seen along part of the periphery of the tube. A small circular core of unattenuated planar wave can also be seen, although much of it is obscured by light emitting from the tube. In the second frame, a pair of circumferentially sweeping detonation waves has formed in the preheated gas behind the frontal shock. The rarefaction has also arrived at the axis so that a luminous central core is no longer visible. By the third frame, the reinitiated waves have swept over three-quarters of the cross section but have fallen behind the front. Following the collision between these waves (frame 4), the resulting overdriven detonation quickly catches up to the front to complete the re-establishment process. Again, there are similarities between this type of critical reinitiation and that observed for transmission to an unconfined region. As criticality is approached, for example, reignition occurs at fewer random and isolated locations. Another similarity is the formation of transversely sweeping waves that eventually engulf the entire front. An end-on view showing the collision between two such waves in a mixture of propane-air appears in Fig. 7.

The critical conditions for transmission for various degrees of confinement are summarized in Table 1. For comparison, the case of transmission to an unconfined region is included.

Steady Propagation Phase

Velocity and pressure measurements confirm that propagation following re-establishment is steady at near CJ conditions. The front is only slightly curved and exhibits

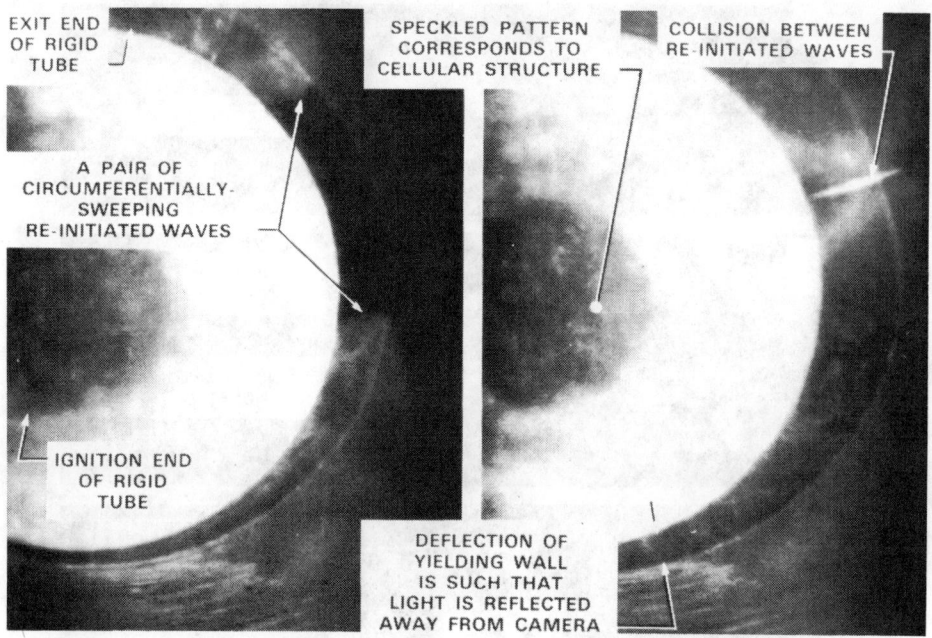

Fig. 7 Cinematographic sequence looking along the axis of propagation. Transmission of detonation is to a yielding tube of nominal 6-mil wall thickness and 1.83-m diam (gas mixture is 6.45% C_3H_8 in C_3H_8-air; elapsed time between frames is ≈0.06 ms).

no time-averaged changes in either the luminosity or scale of the cellular structure. Well above criticality, propagation is uninfluenced by the support hoops. However, as the mixture sensitivity for critical transmission is approached, the reinitiated wave becomes sensitive to perturbations such as the hoops. Figure 8 (a later continuation of the sequence in Fig. 6) shows a wave encountering a hoop 3.4 diam from the exit. Quenching is seen to occur around the periphery, followed by the formation of reignition nuclei that evolve into a renewed detonation front, in exactly the same manner as observed during transmission. This perturbation/recovery sequence repeats itself at each hoop. It would appear that recovery from a noncritical perturbation is identical to supercritical reinitiation during transmission. It may therefore be likely that recovery from a critical perturbation is identical to critical reinitiation during transmission. In this context, a test for critical transmission to a given confinement may be the equivalent to determining the limiting perturbation that can be imposed on a wave propagating in that confine-

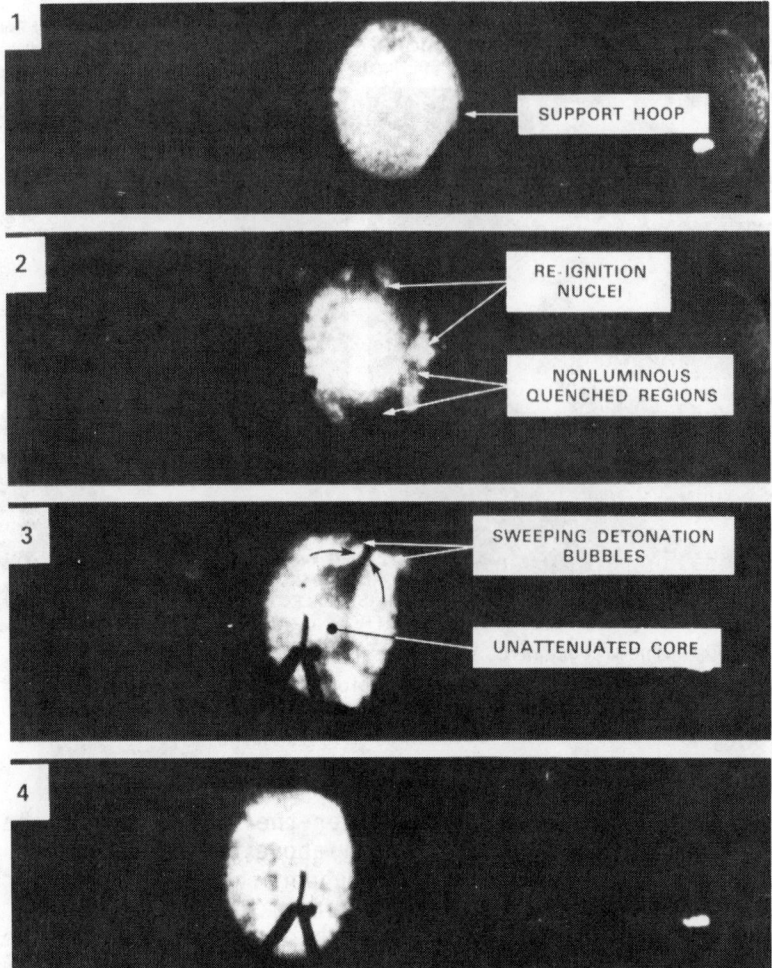

Fig. 8 Cinematographic sequence showing recovery of a detonation wave perturbed at the boundary; the perturbing obstacles are a steel hoop along the bottom and a heavy piece of tape along the top of the periphery (mixture is 4.70% C_2H_4 in C_2H_4-air; elapsed time between frames is ≈0.2 ms).

ment (i.e., to determining the propagation limits). Indeed, experimentally the critical conditions for initiation (i.e., transmission) and propagation are the same in the present tests. This is true on the scale of the cellular structure, since propagation is made possible by what appears to be critical reinitiation at the end of each cell. To demonstrate that these results were not dependent on the particular initiation source (i.e., the rigid tube in this case), an additional test was carried out using a disk of high explosive as an initiation source. This disk was fastened to a plywood backing at one end of the yielding tube. It was found that the critical mixture sensitivity for propagation did not change.

The results in Table 1 show that, as the wall mass tends to zero, the critical diameter for propagation in a free column of gas tends to the familiar critical tube diameter for transmission to an unconfined medium. This is in contrast to the results of Vasiliev (1980), which suggest that the critical diameter of a free column is some 2-8 times the critical diameter for transmission. It is quite possible that the larger column diameter observed by Vasiliev is due to a turbulent interface between the test mixture and the surrounding air caused by the rapid withdrawal of the walls containing the test mixture.

For the 1-mil wall thickness, the critical cell width (i.e., $\lambda \cong 58.7$ mm) gives a hoop spacing of 25λ that allows the perturbed wave plenty of time for recovery before a subsequent disturbance is encountered. In Fig. 8, recovery would appear to be complete somewhere between the third and fourth frames. That is, between 8.5 and 16.6λ downstream of the disturbance. As the tube wall thickness is increased to 10 mil, the critical cell width increases to 119 mm. The effective hoop spacing therefore decreases to about 12.6λ, which does not allow sufficient time for recovery between hoops. This is illustrated in the film sequence of Fig. 9 where each pair of frames (a and b) show the wave prior to and subsequent to traversing a hoop. In short, each successive interaction degenerates the wave until it becomes a low-mode spinning detonation by the end of the tube (not shown in the figure). The velocity profile from the film records shows both intense oscillations and a 400-m/s decrease in mean velocity over the 10-m length. This wave has been perturbed to extinction.

A similar observation was made by Murray et al. (1981) during a series of large-scale tests on the critical energy for direct initiation of ethylene-air mixtures. When the concentration of C_2H_4 was reduced to 3.95% (i.e., $\lambda \cong 113$ mm), it was observed that, after successful initiation using high-explosive PETN, the wave propagated the 10-m length of the test section while exhibiting the same type of oscillations and decreasing mean velocity as described above. The reason for this may have been related to the 1220-mm spacing between the cross members of the test section frame. If this was the case, the near-critical spacing between disturbances would have been about 10.8λ.

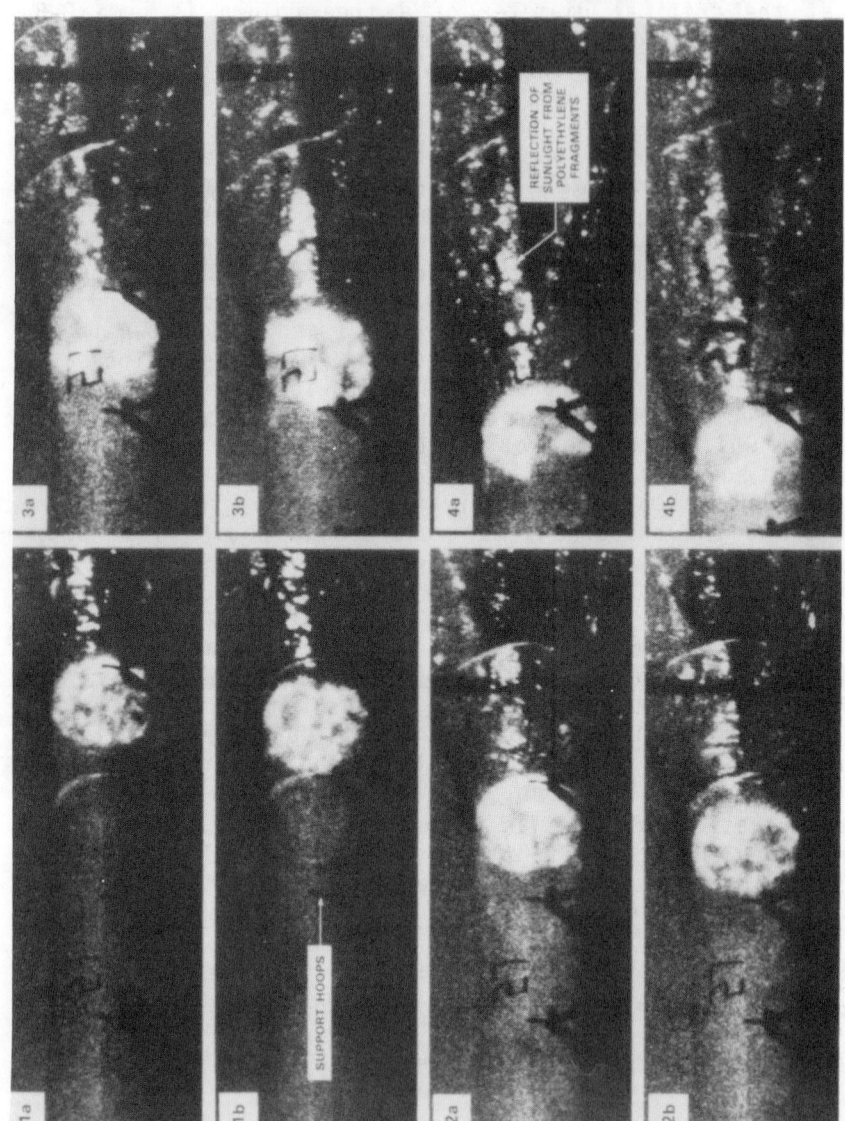

Fig. 9 Cinematographic sequence showing gradual failure of detonation induced by a series of perturbing obstacles around the periphery of a tube of nominal 10-mil wall thickness (gas mixture is 3.90% C_2H_4 in C_2H_4-air). Each pair of frames shows the wave (a) prior to and (b) subsequent to traversing a hoop (elapsed time between a and b frames is ≈0.2 ms).

All of the above observations support an earlier proposal by Murray and Lee (1983) that a disturbed wave must reinitiate itself before propagating a distance of 11.5λ, or approximately two hydrodynamic thicknesses, from the disturbance.

Analysis

Criterion for Criticality

Since reignition near the boundary appears to be favored, a criterion for criticality should be applied to particles near the tube wall that enter the shock and suffer an immediate expansion thereafter. Rather than committing ourselves to a specific chemical time (i.e., one-dimensional induction time, cell transit time, etc.), the controlling chemical time will simply be assumed to be of the form

$$\tau = K [O_2]^{-1} \exp(E/RT)$$

where all variables have been defined previously. The conditions for reignition will be two-fold: 1) a particle must complete its chemical residence before reacting, that is,

$$\int_0^t \tau^{-1} dt$$

must reach unity at some stage during the expansion; and 2) the failure criterion is similar to the one advanced by Shchelkin (1963), which simply proposes that the chemical-gasdynamic coupling will not resume if the chemical time increases by an amount on the order of itself. That is, if $\Delta \tau \cong \tau$. Hence, for a given particle thermodynamic history (i.e., a given wall thickness), the pre-exponential factor K will be varied until a chemical time is found for which the above integral reaches unity just before the failure criterion is satisfied.

Particle Thermodynamic History

The particle thermodynamic history will be calculated by first determining the wall trajectory as it accelerates radially outward in the wake of the passing shock. We will consider only a perfect, inviscid gas where the heat transfer to the wall and the skin friction are negligible. Since the history prior to reaction is of interest, the coupling between hydrodynamics and chemistry will be ignored. Resistance of the wall to motion will be due to inertia alone. Loss of the wall's integrity (tearing, for example) will be assumed to occur at times later than are of interest here. Since the wall trajectory does not vary with time to an observer attached to the shock, this is a steady two-dimensional problem with rotational symmetry. It can thus be transformed into the unsteady one-dimensional problem of an accelerating cylindrical piston. This approach is valid for small flow deflections and is identical to that used in computing hypersonic flow over slender bodies. Dabora (1963) has previously used this approach with good success. The radial position of the wall a distance x behind the shock in the two-dimensional problem will be the same as that of the piston in the one-dimensional case after time x/V_{CJ}, as shown in Fig. 10.

The one-dimensional piston motion will be calculated by the method first proposed by Meyer (1958). This author presents a means of solving the conservation equations, in characteristic form, applied to the particles immediately in front of and behind the piston. The present analysis requires three modifications to Meyer's original formulation: 1) eliminating the reflected shock in his head-on impact problem and instead beginning with a quiescent high-pressure reservoir at equilibrium conditions consistent with those behind a CJ shock; 2) allowing for different ratios of specific heats γ on either side of the wall; and 3) adding an integral term to account for cylindrical divergence. The analysis will not be repeated here, but only the modified equation for the wall motion given

$$\frac{du_r}{dt} = \frac{1}{m/A} \left\{ p_2 \left[\frac{P_2 - \int_0^t \frac{c_b u_r}{r} dt - u_r}{P_2} \right]^{2\gamma_2/(\gamma_2 - 1)} \right.$$

$$\left. - p_1 \left[\frac{Q_1 + \int_0^t \frac{c_a u_r}{r} dt + u_r}{Q_1} \right]^{2\gamma_1/(\gamma_1 - 1)} \right\}$$

where m/A is the surface density of the wall material, u_r and r the particle velocity and radius, p the pressure, γ the ratio of specific heats, and c the sonic velocity. Subscripts 1 and 2 correspond to the initial low-pressure (atmospheric air ahead of the piston) and high-pressure (postshock mixture) states, respectively. Subscripts a and

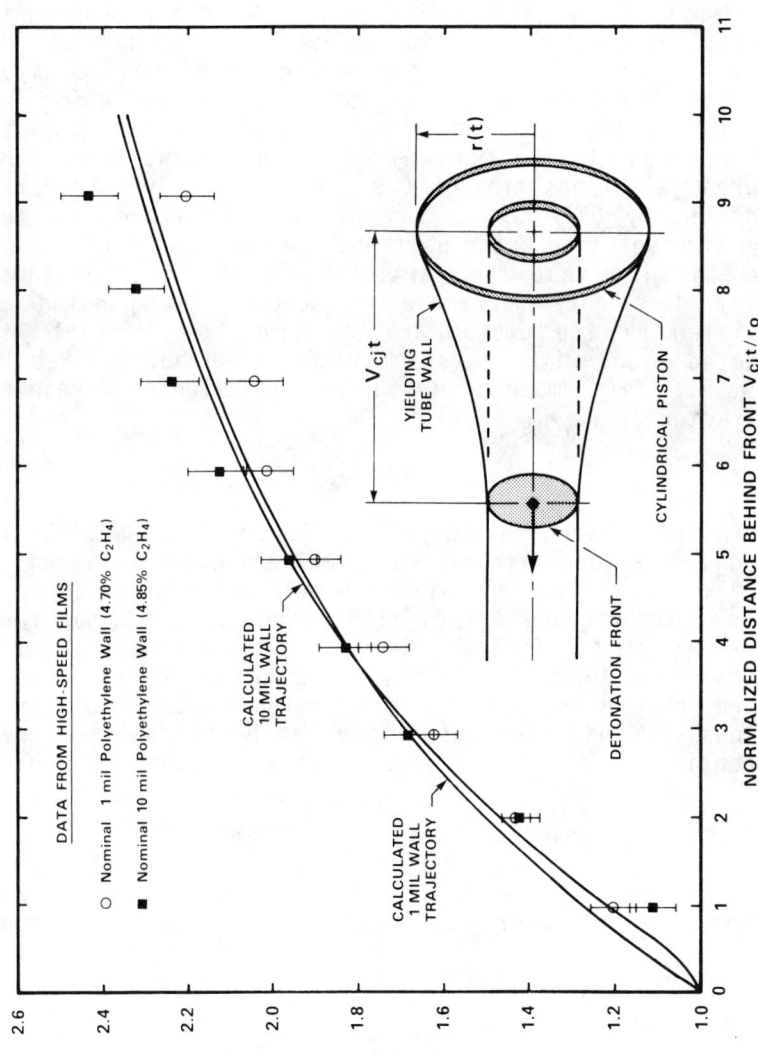

Fig. 10 Comparison between measured wall trajectories and those calculated using a cylindrical piston analogy.

b refer to conditions immediately ahead of and behind the piston, respectively. The initial Riemann invariants P_2 and Q_1 are given by $2c_2/(\gamma_2-1)$ and $2c_1/(\gamma_1-1)$. The above equation must be integrated numerically for $u_r(t)$.

It is not possible to resolve the wall motion between the shock and reaction zone from the present photographic records. However, the motion at much later times can be resolved. As a check on the method, the measured trajectories were compared to those calculated using CJ parameters rather than postshock parameters. As shown in Fig. 10, even at five tube diameters behind the front, the measured radial position of the wall differs from the predicted one by only ±8%. This gives us confidence in the early time solution using postshock parameters. It must be borne in mind that the particle in the two-dimensional problem does not experience as severe an expansion as calculated by the piston analogy since the particle is convected along with the shock rather than being fixed to the wall. This amounts to scaling the calculated results by a factor of V_{CJ}/u_s in time.

Results

The calculated thermodynamic history of a particle at the edge of a tube with a 10-mil wall thickness is shown in Fig. 11. The particle radial velocity and postshock pressure and temperature are plotted vs normalized time, t/τ_0, at the top of the figure. Here, τ_0 is the characteristic chemical time (based on initial postshock conditions) of a mixture that satisfies the reignition criteria outlined previously. This time was arrived at by varying the pre-exponential factor K until a value was found for which

$$\int_0^t \tau^{-1} dt \to 1 \text{ as } t \to 2\tau_0$$

The procedure is summarized graphically in the lower part of the figure where progress toward the chemical reaction,

$$\int_0^t \tau^{-1} dt$$

is plotted against time, t/τ_0, for a range of assumed controlling chemical times. For small chemical times the particle begins to react before it is aware of an expansion, while for large chemical times the expansion is so severe that the particle barely begins reaction. There is but one chemical time satisfying the reignition criteria.

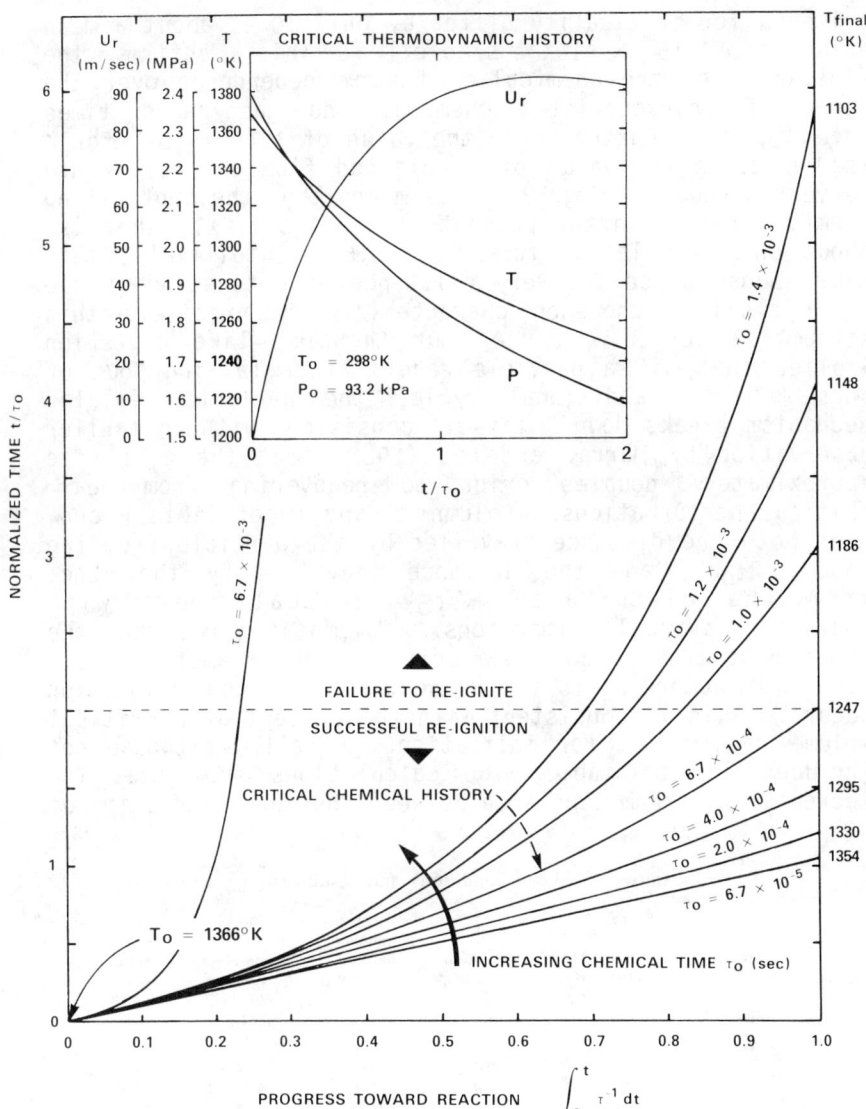

Fig. 11 Chemical histories (bottom plot) for various assumed controlling chemical times τ_0 given the expansion corresponding to the 10-mil thick polyethylene wall (critical particle thermodynamic history is shown in inset at top left).

A summary of the calculations for the three expansions investigated is shown in Table 2. Columns 3-6 list the initial postshock conditions as calculated by the Gordon-McBride (1976) chemical equilibrium computer code. Column 8 shows that the pre-exponential factors satisfying the criteria for criticality differ by only ±0.9% about a mean value of 1.15 x 10^{-9} s·mole/l. This confirms the similarity of the chemical-gasdynamic dependence over the range of characteristic chemical and gasdynamic times investigated. Furthermore, the value of 1.15 x 10^{-9} for K is 1.6 times the value of A obtained from the cell width curve fit shown in Fig. 2. This means that the controlling chemical length in the present problem is 1.6λ. That is, about one cell length for $\lambda/\ell \cong 0.6$. Intuitively, this makes sense since the very existence of detonation is due to a cyclic phenomenon characterized by precisely this streamwise length. Hence, our Shchelkin-like criterion implies that, if a particle reacts after falling "out of phase" by an additional cycle, the periodic cellular mechanism breaks down. This is consistent with an earlier observation by Murray and Lee (1983) that the cell size approximately doubles prior to recovering from near-critical perturbations. Columns 9 and 10 of Table 2 show that both the distance travelled by the particle from the shock, $2\tau_0 u_s$, and the distance travelled by the shock from the disturbance, $2\tau_0 V_{CJ}$, scale well with characteristic cell dimensions. Column 10 shows that the shock must travel about one critical tube diameter (i.e., about 19λ according to the present data) before reignition occurs. This is consistent with the concept of a critical volume requirement for initiation. Finally, although not included in the table, the calculations show that the increase in stream tube area between the shock and point of

Table 2 Initial postshock conditions and results from calculations

(1) Nominal wall thickness t, mil	(2) Critical C_2H_4 concentration, %	(3) Detonation velocity V_{CJ}, m/s	(4) Relative particle velocity u_s, m/s	(5) Temperature ratio across shock T/T_o	(6) Pressure ratio across shock P/P_o
1	4.70	1696	289.2	4.92	28.85
5	4.15	1637	287.0	4.70	26.80
10	3.90	1606	285.7	4.58	25.76

(1) Nominal wall thickness t, mil	(7) Critical cell size λ, mm	(8) Pre-exponential factor, $K \times 10^9$ s·mole/l	(9) Particle travel before reignition $2\tau_0 u_s/\lambda$	(10) Shock travel before reignition $2\tau_0 V_{CJ}/\lambda$
1	58.7	1.16	3.24	19.0
5	92.5	1.16	3.24	18.5
10	119.0	1.14	3.21	18.0
Average, %		1.15 ± 0.9	3.23 ± 0.3	18.5 ± 3

ignition (i.e., a distance $2\tau_0 u_s$ from the shock) is close to 40% in all three cases. This implies the existence of some maximum possible expansion within the reaction zone in order for propagation to be sustained.

Conclusions

The present study shows that reignition of detonation upon transmission from a rigid tube to a yielding tube is via the formation of reignition nuclei in the vicinity of the yielding tube wall. Experiments also indicate that the critical conditions for transmission and propagation in the present confinement are the same, in that a self-sustained wave is very sensitive to disturbances following near-critical transmission. In fact, it has been observed that the spacing between perturbing obstacles must exceed about 12λ if marginal propagation is to be maintained. The present results are in disagreement with those of Vasiliev (1980) in that the minimum diameter of a free column of gas appears to be identical to the critical diameter for transmission to an unconfined medium.

The delays to reignition following a disturbance have been calculated using a cylindrical piston analogy to obtain the thermodynamic history of a particle. A Shchelkin-like criterion has been used to describe when the detonative mechanism breaks down. These calculations show that the reinitiation process is similar over a range of characteristic chemical and gasdynamic times and that the controlling chemical distance is about one cell length. They also suggest the existence of some maximum possible expansion within the reaction zone if propagation is to be sustained. Further effort is required to understand the stability of the cyclic chemical-gasdynamic processes that are responsible for the cellular structure of detonations.

Acknowledgements

The valuable assistance of Mr. J. Funk, Mr. S. Ward, and Mr. G. Rude is greatly appreciated as the present field tests would not have been possible without their dedication and ingenuity. Likewise, the stimulating dialogue and suggestions provided by Dr. I. Moen and Dr. P. Thibault is acknowledged with thanks. The authors would also like to acknowledge the services of the Field Operations Section, the Chemistry Section, the Electronic Design and Instrumentation Group and the Photo Group. Finally, we wish to express our thanks to

Mrs. C. Shovar, Mrs. N. Smart, and Mrs. L. Wall for their timely assistance in typing the manuscript and preparing the graphics.

References

Bazhenova, T. V. et al. (1981) Influence of the nature of confinement on gaseous detonation. Prog. Astronaut. Aeronaut. 75, 87.

Borisov, A. A. (1982) Private communication as cited by Moen et al. (1982).

Brossard, J. and Renard, J. (1981) Effets mechaniques de la detonation sur un confinement souple. Prog. Astronaut. Aeronaut. 75, 108.

Bull, D. C., Elsworth, J. E., Shuff, P. J., and Metcalfe, E. (1982) Detonation cell structures in fuel/air mixtures. Combust. Flame 45, 7.

Dabora, E. K. (1963) The influence of a compressible boundary on the propagation of gaseous detonations. Rept. 05170-1-T, University of Michigan, Ann Arbor.

Dabora, E. K., Nicholls, J. A., and Morrison, R. B. (1965) The influence of a compressible boundary on the propagation of gaseous detonations. 10th Symposium (International) on Combustion, p. 817. The Combustion Institute, Pittsburgh, Pa.

Edwards, D. H., Jones, T. G., and Price, B. (1963) Observations on oblique shock waves in gaseous detonations. J. Fluid Mech. 17, 21.

Fay, J. A. and Opel, G. (1958) Two-dimensional effects in gaseous detonation waves. J. Chem. Phys. 29, 955.

Funk, J. W., Murray, S. B., Moen, I. O., and Ward, S. A. (1981) A brief description of the DRES fuel-air explosives testing facility and current research program. Proceedings of the International Specialist Meeting on Fuel-Air Explosions, Montreal, Nov. 1981, p. 565. SM Study 16, University of Waterloo Press, Ontario, Canada.

Funk, J. W. and Murray, S. B. (1982) The DRES large-scale fuel-air explosives testing facility. Suffield Memorandum 1051, Defence Research Establishment Suffield, Ralston, Alberta, Canada.

Gordon, S. and McBride, B. (1976) Computer program for calculation of complex chemical equilibrium compositions, rocket performance, incident and reflected shocks, and Chapman-Jouquet detonations. NASA SP-273.

Hidaka, Y., Kataoka, T., and Suga, M. (1974) A shock-tube investigation in ethylene-oxygen-argon mixtures. Bull.

Chem. Soc. Jpn. 47, p. 2166.

Knystautas, R. (1982) Private communication.

Knystautas, R., Lee, J. H., and Guirao, C. M. (1982) The critical tube diameter for detonation failure in hydrocarbon-air mixtures. Combust. Flame 48, 63.

Meyer, R. F. (1958) The impact of a shock wave on a movable wall. J. Fluid Mech. 3, 309.

Moen, I. O., Donato, M., Knystautas, R., and Lee, J. H. (1981) The influence of confinement on the propagation of detonations near the detonability limits. 18th Symposium (International) on Combustion, p. 1615. The Combustion Institute, Pittsburgh, Pa.

Moen, I. O., Murray, S. B., Bjerketvedt, D., Rinnan, A., Knystautas, R., and Lee, J. H. (1982) Diffraction of detonation from tubes into a large fuel-air explosive cloud. 19th Symposium (International) on Combustion, p. 635. The Combustion Institute, Pittsburgh, Pa.

Moen, I. O., Funk, J. W., Ward, S. A., Rude, G. A., and Thibault, P. (1983) Detonation length scales for fuel-air explosives. 9th International Colloquium on Dynamics of Explosions and Reactive Systems.

Murray, S. B., Moen, I. O., Gottlieb, J. J., Lee, J. H., Coffey, C. G., and Remboutsikas, D. (1981) Direct initiation of detonation in unconfined ethylene-air mixtures - influence of bag size. Proceedings of the 7th International Symposium on Military Applications of Blast Simulation, Medicine Hat, Alberta, Canada, Paper 6-3. Defence Research Establishment Suffield, Ralston, Alberta, Canada.

Murray, S. B. and Lee, J. H. (1983) On the transformation of planar detonation to cylindrical detonation. Combust. Flame 52, 269.

Shchelkin, K. I. and Troshin, Ya. K. (1963) Gasdynamics of combustion. NASA Tech. Translation F-231.

Sichel, M. (1965) A hydrodynamic theory for the interaction of gaseous detonation with a compressible boundary. Rept. 05170-2-T, University of Michigan, Ann Arbor.

Sommers, W. P. and Morrison, R. B. (1962) Simulation of condensed-explosive detonation phenomena with gases. J. Fluid Mech. 5, 241.

Vasiliev, A. A. (1980) Gas detonation of a free column. Rept. Lavrentyev Institute of Hydrodynamics, Novosibirsk, USSR.

Vasiliev, A. A., Gavrilenko, T. P., and Topchian, M. E. (1972) On the Chapman-Jouquet surface in multi-headed gaseous detonations. Astronaut. Acta. 17, 499.

Cellular Structure in Detonation of Acetylene-Oxygen Mixtures

M. Vandermeiren* and P. J. Van Tiggelen†
Université Catholique de Louvain
Louvain-la-Neuve, Belgique

Abstract

Preliminary results of a systematic study of detonation in acetylene-oxygen mixtures diluted with argon are presented. Detonation velocities have been measured with ionization gages. Cellular structure has been recorded by means of the smoked foil technique in a rectangular tube. The detonation velocities, D, are larger than those measured in hydrogen-oxygen mixtures at identical initial pressures. The maximum values of D are observed in fuel rich mixtures, particularly when the diluent (argon) concentration is 50% or less. The cell lengths, L, vary with the fuel content. A minimum L is observed for a fuel rich mixture equivalence ratio, $\phi = 2.0$, provided that the initial pressure is lower than 200 Torr and that the argon dilution is larger than 50%. The regularity of the transverse waves in lean acetylene-oxygen mixtures at low initial pressure and highly diluted with argon can be classified as good to excellent. In these mixtures the detonation shock Mach number < 4.7. Complete irregular patterns have never been observed for equivalence ratios $\phi \leq 2.5$. At any pressure and mixture composition the characteristic time ($t_{car} = L/D$) exhibits a similar behavior as that found for cell length. From the experimental data, an overall apparent activation energy (about 25 kcal/mole) has been deduced from the relation between characteristic and induction times. The influence of chemical kinetics on the structure and the self-sustaining mechanism in acetylene-oxygen mixtures will be briefly discussed.

Presented at the 9th ICODERS, Poitiers, France, July 3-8, 1983. Copyright © 1984 by P. J. Van Tiggelen. Published by the American Institute of Aeronautics and Astronautics, Inc. with permission.
　*Assistant-Boursier, Laboratoire de Physico-Chimie de la Combustion.
　†Professeur, Laboratoire de Physico-Chimie de la Combustion.

STRUCTURE OF ACETYLENE-OXYGEN DETONATIONS

Introduction

Gas phase detonations have been studied extensively particularly for hydrogen-oxygen (Strehlow et al. 1969, and Biller 1973) and carbon monoxide-hydrogen-oxygen mixtures (Kistiakowsky and Kydd 1956, and Dove and Wagner 1962). A systematic investigation of both systems has been reported by Libouton and Van Tiggelen (1976), Dormal et al. (1983), and Libouton et al. (1981a). Even so, there is a lack of detailed studies of mixtures such as acetylene-oxygen, ethylene-oxygen, etc., which correspond to more complex kinetic mechanisms. For those mixtures simultaneous measurements of the detonation velocity and of the characteristic shock wave structure are still missing. However, the smoked foil technique has been used by Strehlow (1968) and Strehlow and Engel (1969) to characterize cell length, and Breton (1936) and Kistiakowsky and Zinman (1955) have determined detonation velocity for undiluted acetylene-oxygen mixture at atmospheric pressure for several equivalence ratios, ϕ.

More recently, Nagaishi et al. (1979) attempted to measure cell length and detonation velocity for C_2H_2/O_2, but at low-pressure and variable ϕ. The data was collected

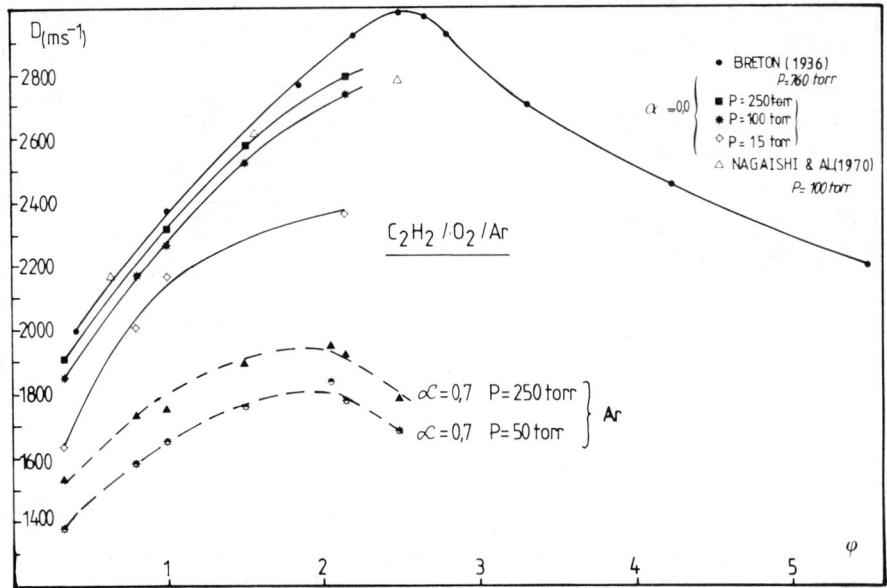

Fig. 1 Detonation velocity vs equivalence ratio for the fresh gaseous mixtures at variable initial pressure P and dilution α. Comparison with literature's data.

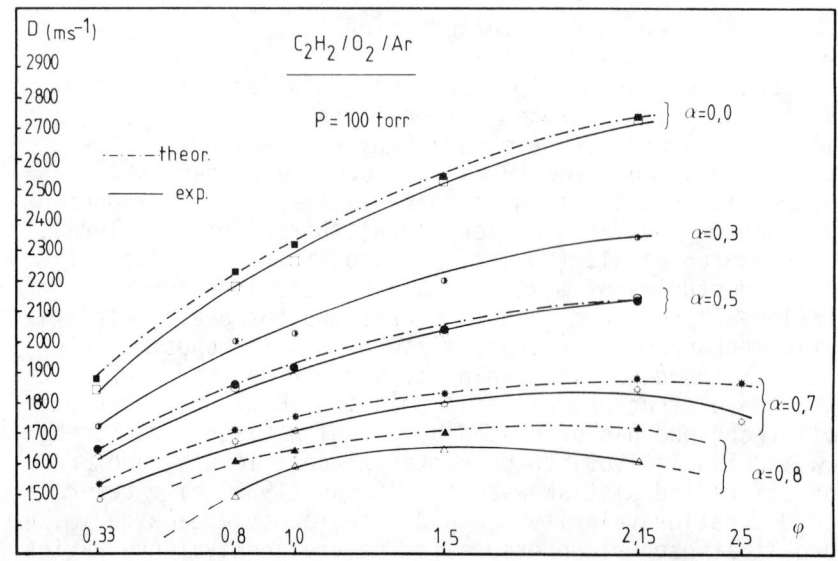

Fig. 2 Comparison of computed (dashed point line) and measured (continuous line) detonation velocities at initial of 100 Torr.

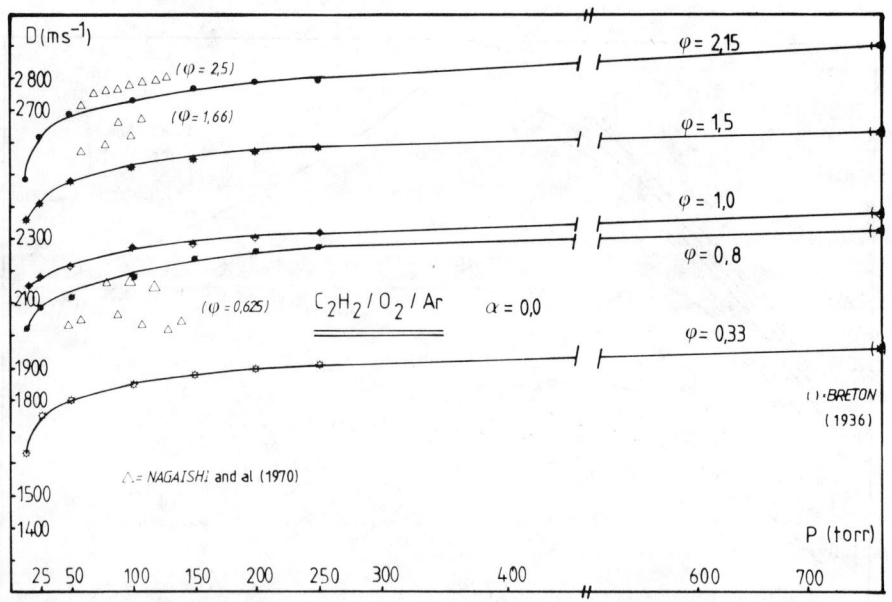

Fig. 3 Detonation velocity vs initial pressure for undiluted acetylene-oxygen-mixtures. Data of Nagaishi et al. (1970) are reported for comparison.

in a 2.5-m tube with a square cross section (15 cm) . The
fuel-oxidizer was prepared by diffusion of the constituent
into the tube, over a period of one day. Such a technique did
not guarantee a perfect homogeneous mixture. Ignition was
achieved directly in the test section by a spark plug located at one end of the tube. With this experimental setup, the
overdriven character of the detonation can persist during
the whole development of the shock wave along the tube.

This short survey of acetylene-oxygen detonation literature indicates the need for systematic study to extend the
modeling of the mechanism presented for gaseous detonation
in a H_2-CO-O_2 mixture (Libouton et al. 1981b).

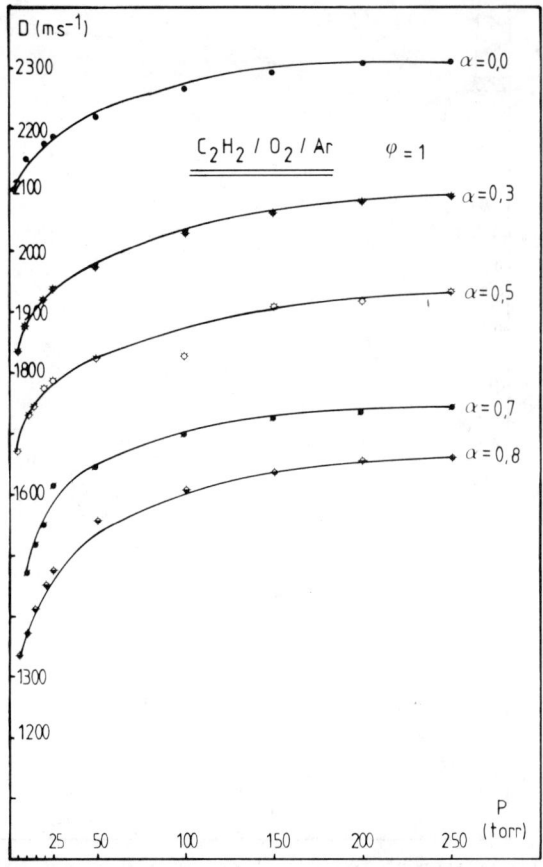

Fig. 4 Detonation velocity of stoichiometric acetylene-oxygen mixtures vs pressure at variable dilution with argon.

Experimental

A complete description of the experimental setup has been given previously (Libouton et al. 1975). An anodic aluminum tube with a rectangular cross section (32 x 92 mm) was filled by a flow method after the tube has been alternately evacuated (P = 10^{-2} Torr) with a mechanical pump and then flushed with the test mixture several times.
The composition is controlled precisely, and mixing is achieved, by choked flow of the individual gas through sonic nozzles. A better homogeneity of mixture and a more accurate composition results when this technique is used. Moreover, a large number of experiments with identical mix-

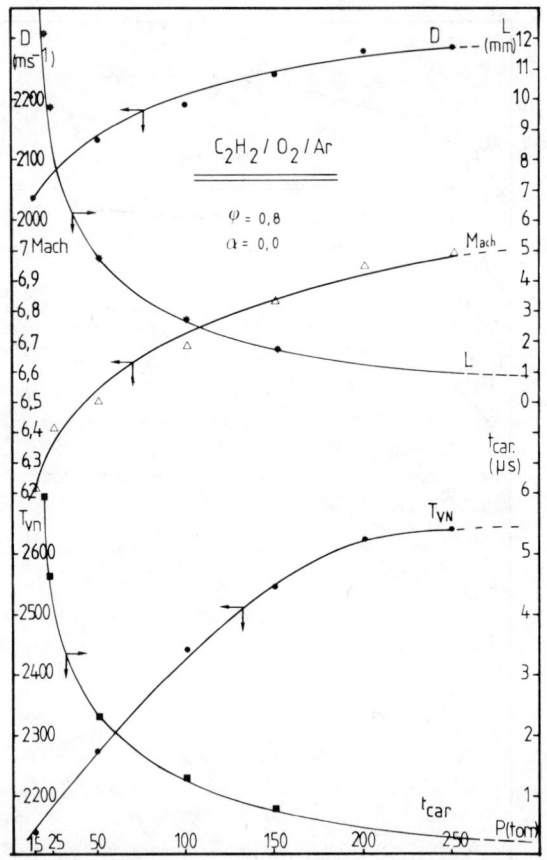

Fig. 5 Comparison of the variations with pressure of the detonation velocity D, the shock Mach number, the cell length L, the von Neumann temperature T_{vn}, and the characteristic time t_{car} in lean acetylene-oxygen undiluted mixtures.

ture composition can be achieved without danger of hazards due to storage of large quantities of explosive mixtures. Great care has been taken to reduce the overdriven character of the detonation wave. The initiation of the detonation is, in fact, achieved by a detonation of H_2/O_2 mixture in the driver section which bursts a mylar film and then ignites the test section after passage through an orifice 1 cm in diameter. The structure of the detonation has been recorded on a glass plate covered by soot. The plate is located 4-m after the diaphragm. The detonation velocities have been measured by ionization gages 2.1-m apart and located on each side of the smoked plate.

Results and Discussion

Detonation Velocity

The runs were performed in a $C_2H_2/O_2/Ar$ mixture at several equivalence ratios ($2.5 \geq \phi \geq 0.33$) with variable argon dilution ($0.8 \geq \alpha \geq 0.0$) and at initial pressures varying from 15 to 250 Torr.

For pure acetylene-oxygen mixture ($\alpha = 0.0$) the dependence of the detonation velocity on ϕ is similar to that observed by Breton (1936) and Kistiakowsky and Zinman (1955) at atmospheric pressure. The experimental data (presented in Fig. 1) indicate that the detonation velocity reaches a maximum for fuel rich mixture ($\phi > 2$). This behavior is observed at any initial pressure studied so far. For mixtures highly diluted with argon ($\alpha = 0.7$ or 0.8) the maximum is reached at about $\phi = 1.9$. The influence of large amounts of diluent (Ar) in the fuel-oxidizer mixture is exhibited by the plot of the detonation velocity vs the equivalence ratio for every dilution (Fig. 2).

Computation of the detonation velocity, D, was performed with a program based on the approach developed by Guirao et al. (1972). The computed values for D and the experimental data are compared on Fig. 2. Figure 3 exhibits the dependence of detonation velocity on initial pressure for undiluted mixtures at several equivalence ratios. At low initial pressure a decrease of detonation velocity is observed due essentially to the approach of the detonation limits, while at higher initial pressure the data converge with those reported by Breton (1936) at atmospheric pressure. The data also agree well with those of Nagaishi and al. (1970) (see also Table 1). The larger dispersion of Nagaishi's results is probably related to their ignition mode which favors the overdriven character of detonation, particularly at low velocities, i.e., for lean mixtures. For stoichiometric mixture, the dependence of the detona-

Table 1 Comparison of data of Nagaishi et al. (1970) with ours; P = initial pressure in Torr, D = measured detonation velocity in m s^{-1}, ϕ = equivalence ratio, and L = cell length in mm.

P	ϕ		0,33	0,625[a]	0,8	1,0	1,07[a]	1,5	1,66[a]	2,5[a]
40		D	...	...	...	...	...	...	...	...
		L	...	7,8	...	...	2,7	...	1,8	...
50		D	1806	2040	2123	2210	...	2481	...	...
		L	14,3	6,0	4,7	3,2	...	1,9	1,5	...
60		D	...	2060	...	...	...	...	2570	2710
		L	...	3,2	...	...	...	...	1,15	...
100		D	1856	2170	2183	2261	...	2526	2620	2275
		L	7,3	3,00	2,7	1,9	1,5	0,7	1,0	...

[a] Data of Nagaishi et al. (1970).

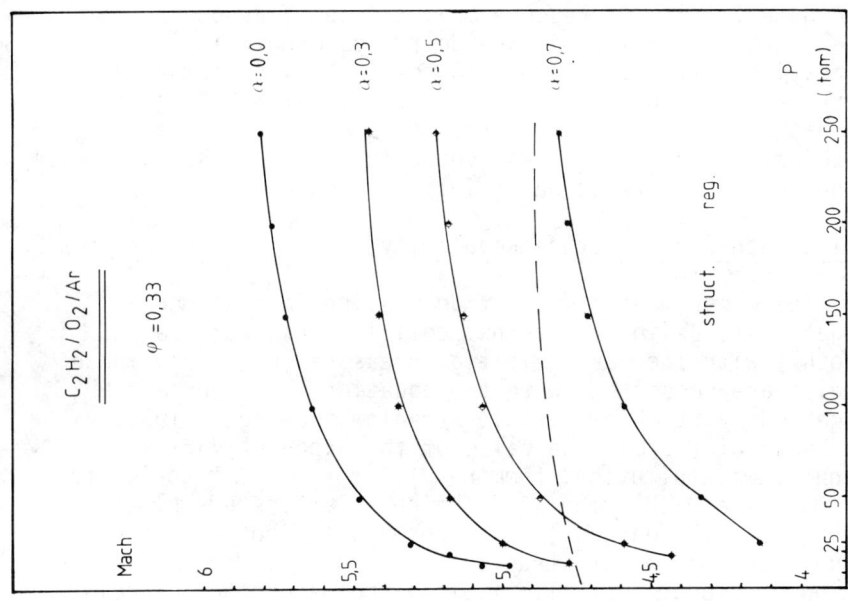
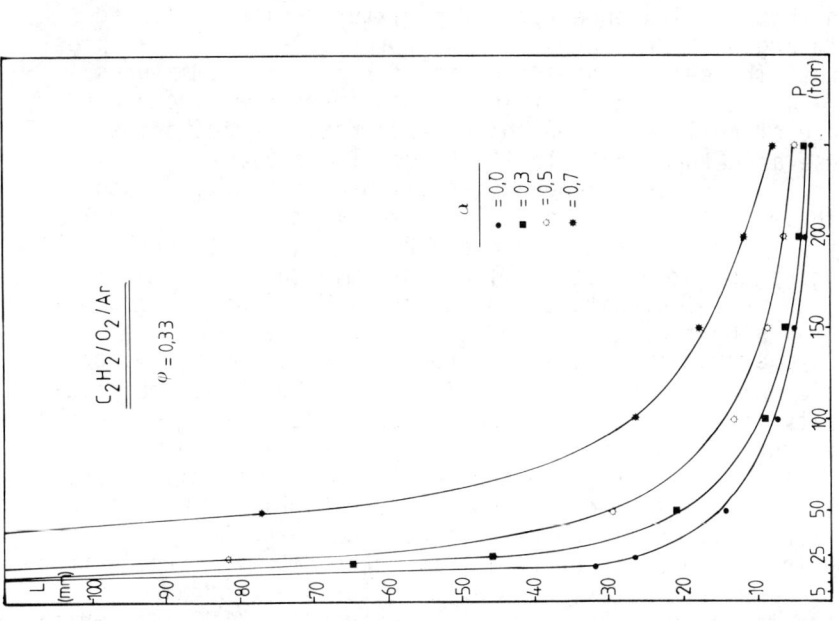

Fig. 6 Lean acetylene-oxygen detonations with variable argon dilution. a) Cell length vs pressure. b) Shock Mach number vs pressure; highly regular structure are observed below the dashed line.

tion velocity on pressure at increasing quantities of argon is presented in Fig. 4.

A similar trend is noticed at any dilution. Moreover, at constant initial pressure an increase of argon percentage of the explosive mixture decreases monotonically the detonation velocity. The Mach number, the von Neuman temperature T_{vn}, and the characteristic time, $t_{car} = L/D$, of the detonation were calculated from the experimental data D, L. These calculated parameters (shown in Fig. 5) also vary monotonically as a function of initial pressure.

Cell Length and Structure Regularity

The experimental data of detonation in acetylene-oxygen mixtures indicates that cell lengths decrease smoothly with increasing initial pressure (Fig. 6a). These results are consistent with the equation, $L = \text{constant } P^n$, suggested by Strehlow (1968), Strehlow and Engel (1969), and Nagaishi (1970). The value of the exponent varies slightly with dilution, from $n = 1.1$ for $\alpha = 0.0$ to 1.4 for $\alpha = 0.7$. The Mach numbers of the cell lengths are plotted in Fig. 6b. The dashed line represents the border between regular and irregular structures according to the classification by Libouton et al. (1981a). The dependence of cell length on equivalence ratio for mixtures highly diluted with argon is presented as Fig. 7 for several initial pressures. The cell lengths increase for fuel rich mixtures ($\phi > 1.7$), especially at low initial pressure. The dependence of cell length on equivalence ratio varies almost inverse as detonation velocity (Figs. 1 and 2).

For acetylene-oxygen mixtures slightly diluted with argon ($\alpha < 0.5$), as well as for mixtures at $\alpha = 0.7$ but with initial pressure above 200 Torr, the observed cell lengths do not exhibit well defined minima.

The structure regularity classification (suggested by Libouton and al. 1981a) has been used. For the C_2H_2/O system two distinct structures were observed. For stoichiometric and fuel-rich mixtures, the structure regularity is classified as "poor"; new transverse waves in the structure do appear on soot records, but, in lean mixture ($\phi < 1$), regularity is enhanced. A "good to excellent" regularity is observed for low-pressure highly diluted lean mixtures. It should be noted that even large amounts of argon in stoichiometric or rich mixtures do not improve structure regularity. Very regular structures are obtained for detonations propagating with velocity below 1500 m/s, which corresponds to Mach numbers ≤ 4.8 (Fig. 6b).

STRUCTURE OF ACETYLENE-OXYGEN DETONATIONS 113

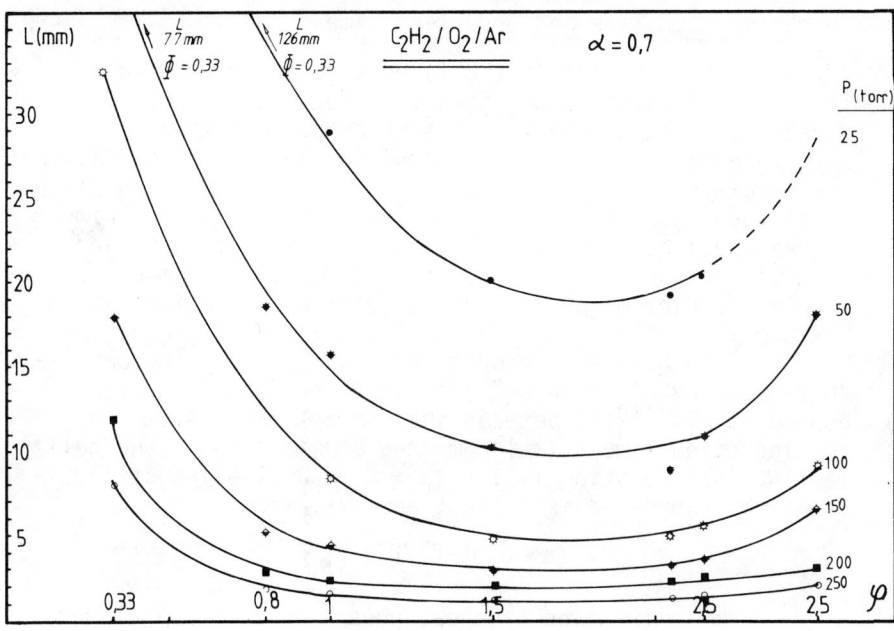

Fig. 7 Cell length vs equivalence ratio with variable initial pressure for acetylene-oxygen mixtures diluted with 70% argon.

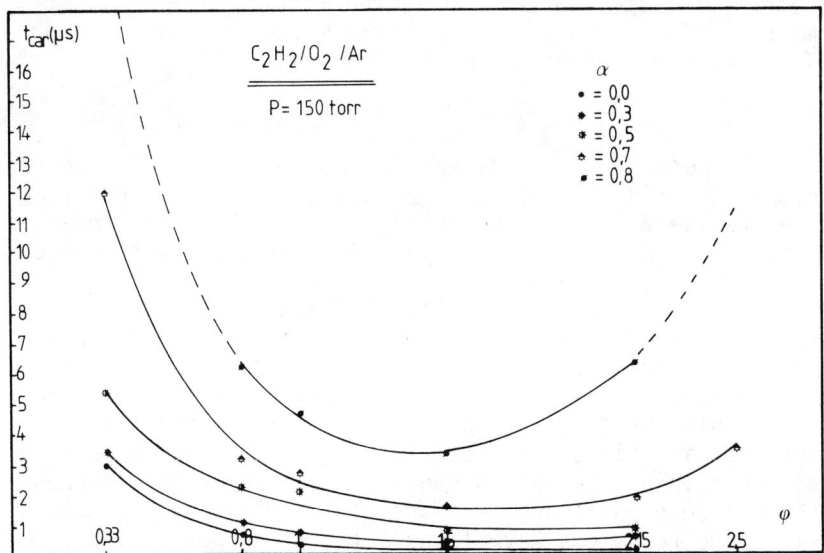

Fig. 8 Characteristic time vs equivalence ratio with variable argon dilution for acetylene-oxygen detonations at constant initial pressure.

Characteristic Time and Induction Time

The characteristic time ($t_{car} = L/D$), defined as the time the waves need to travel throughout one cell, is direct function of the kinetics which controls the sustaining mechanism of the detonation (Libouton et al. 1975). The parameter t_{car} is plotted vs equivalence ratio on Fig. 8. It turns out that, at high dilution in argon, a minimum is reached around $\phi = 1.7$. Similar behavior of D, L, and t_{car} with respect to ϕ suggests that different combustion kinetic mechanisms play a role in rich and lean initial mixture.

In H_2-O_2-CO detonations a direct relation was demonstrated by Libouton and Van Tiggelen (1976) and by Libouton et al. (1978) between the characteristic time and the induction time (t_{ind}) computed at midvalue of the cell length. That relation is : $t_{car} = K \, t_{ind}$. The induction time may be expressed as a first approximation

$$t_{ind} = (K_o \exp(-E_a/RT) \, X_{O_2})^{-1}$$

where K_o is a function of pre-exponential terms of elementary reactions, rate constants, and concentrations. X_{O_2} characterizes the oxygen molar fraction in the unburned gas. The equation is developed from the assumption that the chain branching reaction $H + O_2 \rightarrow OH + O$ remains the rate determining step. Nonetheless, the activation energy E_a remains partly a function of several elementary processes. The foregoing relations may be combined to yield

$$\ln(X_{O_2} t_{car}) = \ln \frac{K}{K_o} + \frac{E_a}{R}\left(\frac{1}{T}\right)$$

Temperature T is the von Neumann temperature calculated at the midcell point. In practice T_{vn} is obtained from the average detonation velocity (Fig. 5). From an Arrhenius plot, $\ln(t_{car} X_{O_2})$ vs $(1/T_{vn})$, an overall activation energy has been deduced.

In Fig. 9 data are plotted for mixtures at constant initial pressure with variable dilutions and equivalence ratios. ϕ varies from 0.33 to 2.15. The overall activation energy is about 24 kcal/mole. However, if one does consider the results at constant equivalence ratio, the activation energies are higher for fuel-rich mixture than for lean mixture. This observation suggests that for rich mixtures thermal decomposition reactions of acetylene play a role since the reactions have large activation energies. Decomposition reactions could become the rate-determining steps of the kinetics of the release mechanism and, therefore, for the sustaining process of the detonation wave.

Poor regularity of the structure, in fact, observed for fuel-rich mixtures, even when the mixtures are highly diluted with argon and are at low initial pressure. This observation reinforces the previous considerations (Libouton et al. 1981b) about the influence of elementary processes with high activation energy which supersede the net branching factor. This last factor is the dominating one for hydrogen-detonating mixtures where an overall activation energy of about 11 kcal/mole is observed (Libouton et al. 1975)

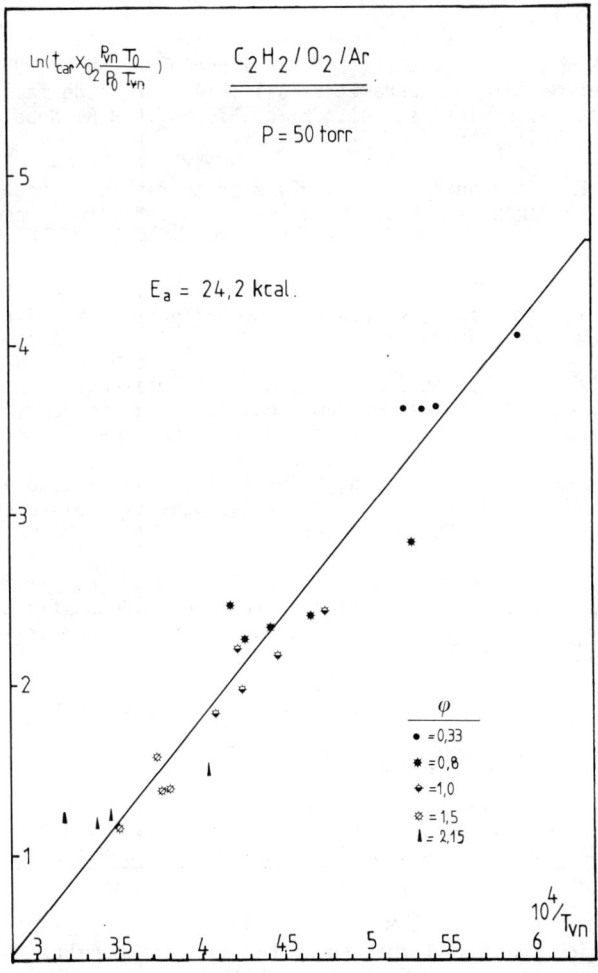

Fig. 9 Arrhenius plot as deduced from the detonation structure (see text).

Acknowledgments

This work has been supported by F.R.F.C., Belgium (Grant No. 2.9003.82). M. Vandermeiren is indebted to the I.R.S.I.A. (Belgium) for a doctoral fellowship.

References

Biller, J. R. (1973) An investigation of hydrogen-oxygen-argon detonations. Tech. Rep. AAE 73-5, UILU - ENG - 73 0505.

Breton, J. (1936) La détonation des mélanges gazeux. Ann. Comb. Liq., 11, pp. 487-546.

Dormal, M., Libouton, J. C., and Van Tiggelen, P. J. (1983) Etude expérimentale des paramètres à l'intérieur d'une maille de détonation. Explosifs, vol. 36, pp. 76-94, P.R.B. Nobel Explosifs, Bruxelles.

Dove, J. E. and Wagner H. G. (1962) A photographic investigation of the mechanism of spinning detonation. VIIIth Symposium on Combustion, pp. 589-599. Williams and Wilkins, Baltimore, Ma.

Guirao, C. M., Knystautas, R., and Lee, J. H. (1972) Detonation study in H_2/Cl_2, CS_2/O_2 and $CO/H_2/O_2/N_2$ mixtures. Merl Rept. 72-76, AFOSR Grant 69-17 52B.

Kistiakowsky, G. B. and Zinman, W. G. (1955) Gaseous detonations. VII. A study of thermodynamic equilibration in acetylene-oxygen waves. J. Chem. Phys. 23, pp. 1889-1894.

Kistiakowsky, G. B., and Kydd, P. H. (1956) Gaseous detonations IX. A study of the reaction zone by gas density measurements. J. Chem. Phys. 25, pp. 824-835.

Libouton, J. C., Dormal, M., and Van Tiggelen, P. J. (1975) The role of chemical kinetics on structure of detonations waves. XVth Symposium on Combustion, pp. 79-86. The Combustion Institute, Pittsburgh, Pa.

Libouton, J. C. and Van Tiggelen, P. J. (1976) Influence de la composition du mélange gazeux sur la structure des ondes de détonations. Acta Astron. 3, pp. 759-769.

Libouton, J. C., Jacques, A., and Van Tiggelen, P. J. (1981a) Cinétique, structure et entretien des ondes de détonation. Actes du Colloque International Berthelot-Vieille-Mallard-Le Chatelier, Tome II, pp. 437-442, Bordeaux.

Libouton, J. C., Dormal, M. and Van Tiggelen, P. J. (1981b) Reinitiation process at the end of the detonation cell. Gasdynamics of Detonations and Explosions: AIAA Progress in Astronautics and Aeronautics, (edited by J. R. Bowen, N. Manson, A. K. Oppenheim, and R. I. Soloukhin), Vol. 75, pp. 358-369. AIAA, New York.

Nagaishi, T., Yoneda, K., and Hikita, T. (1970) Studies on detonation waves. J. Chem. Soc. Japan Industrial Chem. Sect. 73, pp. 2070-2076.

Strehlow, R. A. (1968) Gas phase detonations : Recent developments. Combust. Flame 12, pp. 81-101.

Strehlow, R. A., Maurer, R.E., and Ryan, S. (1969) Transverse waves in detonations : I.Spacing in hydrogen-oxygen system. AIAA J. 7, pp. 323-328.

Strehlow, R.A. and Engel, C. D. (1969) Transverse waves in detonations : II.Structure and spacing in H_2-O_2, C_2H_2-O_2, C_2H_4-O_2 and CH_4-O_2 systems. AIAA J. 7, 492-496.

The Influence of Initial Pressure on Critical Diameters of Gaseous Explosive Mixtures

P. Bauer,* C. Brochet,† and H. N. Presles‡
Université de Poitiers, Poitiers, France

Abstract

The influence of pressure on the detonability limits of hydrocarbon-O_2-N_2 mixtures has been characterized by means of an experimental investigation of the structure of the detonation front in slightly rich C_2H_4-O_2-N_2 and CH_4-O_2-N_2 mixtures. Observations were made with a rapid response pyrometer (50 ns response time at λ = 0.657 µm). Mixtures were ignited by a low energy source in a 6.5-m-long tube with an inside diameter of 15 mm. The initial pressures ranged from 2 to 50 bars. At a nitrogen-to-oxygen ratio of $Z = n_{N2}/n_{O2}$ above Z_{lim}, no detonation could occur. These values were $Z_{lim} = 3.23$ and $Z_{lim} = 6$, respectively, for CH_4 and C_2H_4 mixtures. Thus a CH_4-air mixture could not be ignited in this tube. For Z close to Z_{lim}, a spinning detonation was obtained. These results are discussed on the basis of the critical diameter defined as the diameter below which a detonation is quenched when the front propagates into a sudden enlargement. It is shown that a rise of the initial pressure tends to decrease this parameter.

Presented at the 9th ICODERS, Poitiers, France, July 3-8, 1983. Copyright © American Institute of Aeronautics and Astronautics, Inc., 1984. All rights reserved.

*Assistant Professor, Laboratoire d'Energétique et de Détonique, E.N.S.M.A.

†Maître de Recherche, Laboratoire d'Energétique et de Détonique, E.N.S.M.A.

‡Chargé de Recherche, Laboratoire d'Energétique et de Détonique, E.N.S.M.A.

Introduction

To determine the detonability limits of gaseous mixtures, a knowledge of the parameters that influence the stability of the detonation front is required. The stability and velocity of detonations, obtained in tubes with either a square or a circular section, depend upon the diameter. As wall effects play a prominent part in the phenomenon because of mechanical and thermal energy dissipation in the boundary layer, this dependence is more pronounced near the explosibility limits. Guénoche and Manson (1952) have shown that, in narrow tubes, a decrease in the tube diameter results in a decrease in the detonation velocity. Moreover, they tried to determine conditions under which a detonation could be unstable and three-dimensional effects could occur. Manson et al. (1963) suggested that these conditions might be obtained whenever the discrepancy between the mean and the local velocity exceeded 0.2%. A less arbitrary criterion was proposed by Zel'dovich et al. (1956) based upon the ability of the detonation to propagate without quenching when it expands in a large vessel. The tube diameter beyond which the propagation becomes impossible was termed the critical diameter. Mitrofanov and Soloukhin (1964) and later Edwards et al. (1979) showed that, in cylindrical tubes, this critical diameter can be expressed as a linear function of the cell height b:

$$d_c = 13\,b \qquad (1)$$

Urtiew and Tarver (1981) studied the influence of an abrupt expansion in tube diameter on the structure of the detonation front and proposed values of the critical energy of initiation. Matsui and Lee (1970), Moen et al. (1981), and Knystautas et al. (1982) measured the critical diameters of numerous gaseous explosive mixtures at atmospheric pressure and confirmed the Zel'dovich relation [Eq. (1)] for mixtures of ethylene in oxygen and nitrogen with various ratios of $Z = n_{N_2}/n_{O_2} < 3$. Critical diameters were found to be less than 23 cm in stoichiometric mixtures. A critical diameter of 33 cm predicted for an ethylene-air mixture by extrapolation of their results agrees well with the value proposed by Bull et al. (1982). For CH_4-O_2-N_2 mixtures, Knystautas and co-workers have shown that, for $Z < 1.2$, the critical diameters for a slightly rich mixture are of the same order as those for a stoichiometric mixture but are much higher (about one order of magnitude) than those obtained in C_2H_4-O_2-N_2 mixtures. Measurement of the critical diameter for a CH_4-air mixture was impossible because the tube diameter was too small. From kinetic considerations, Knystautas and co-workers suggested that the critical diameter for CH_4-air would be of the order of meters. This result agrees with the value proposed by Bull et al. (1982) : d_c = 4 m. Matsui and Lee (1970) suggested that d_c = 10 m, based on an extrapolation of their data to the N_2 dilution value of Z = 3.76. Bauer and Brochet (1983 a) measured the critical diameters in CH_4-O_2-N_2 as well as in C_2H_4-O_2-N_2 mixtures at a high initial pressure (up to p_i = 20 bars) and found that the propagation of a detonation was possible at higher dilutions by means of a high initial pressure.

Experimental Setup

The experiments were performed in a 6.5-m cylindrical tube with an inside diameter of 15 mm. The initial pressures ranged from 0.8 to 20 bars. The experimental device has been described in detail in previous works (Bauer et al. 1981; Bauer and Brochet 1983 b). Because of the device design, blowoff occurs at initial pressures greater than 5 bars. As over this range of pressures it was not possible to use the soot trace method, an optical pyrometer with a response time less than 50 ns (Bouriannes et al. 1977) was used to measure brightness temperatures at λ = 0.657 μm associated with the triple points. From simultaneous observations of the detonation velocity, the cell size

Table 1 Characteristics of the mixtures

Mixture	Dilution $Z = n_{N2}/n_{O2}$	Equivalence ratio
A: CH_4 + 1.86 O_2 + 1.34 N_2 [a]	0.72	1.08
B: CH_4 + 1.83 O_2 + 3.45 N_2 [a]	1.89	1.09
C: CH_4 + 1.74 O_2 + 4.93 N_2 [a]	2.83	1.15
D: CH_4 + 1.84 O_2 + 5.94 N_2 [a]	3.23	1.09
E: C_2H_4 + 2.97 O_2 + 8.94 N_2 [b]	3.01	1.01
F: C_2H_4 + 2.85 O_2 + 11.68 N_2 [b]	4.09	1.05
G: C_2H_4 + 2.85 O_2 + 14.25 N_2 [b]	5	1.05

[a]The compostions were confirmed by gas chromatographic analysis. [b]For these mixtures, the cell sizes were determined in a previous work (Bauer and Brochet 1983a) on the basis of an electrical conductivity measurement.

could be determined. The optical method could not be used when the detonation front had a very fine structure (cell size less than the width of the optical beam, i.e., several millimeters). The experiments were performed in CH_4-O_2-N_2 and C_2H_4-O_2-N_2 mixtures with equivalence ratios slightly higher than 1 and for various values of $Z = n_{N2}/n_{O2}$. Table 1 shows the composition of the mixtures referenced as A, B, C, D, E, F, G that were investigated.

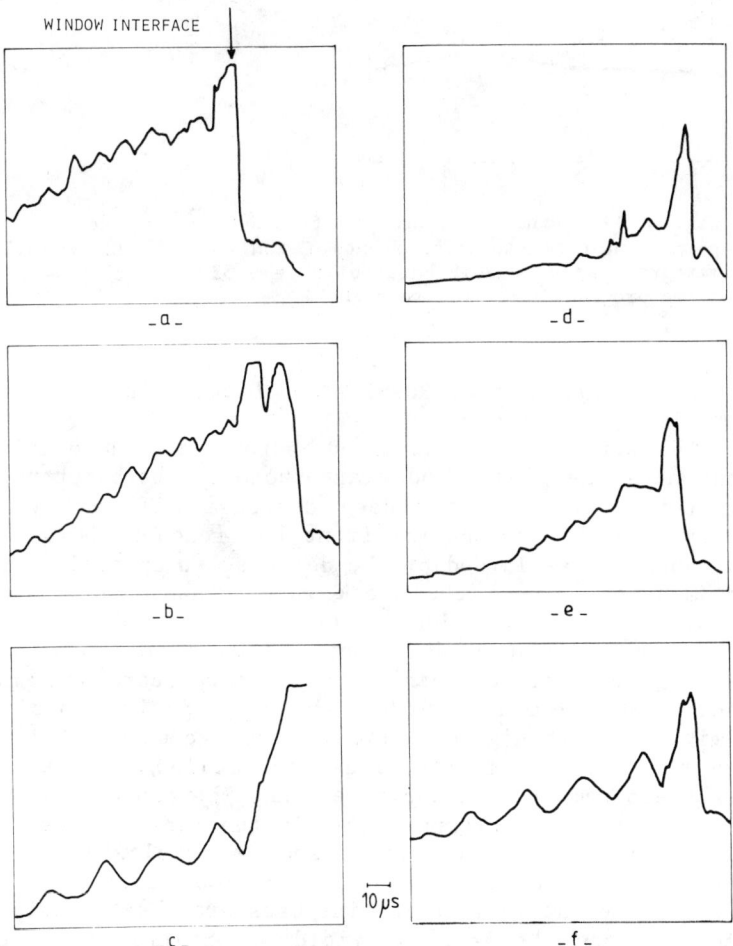

Fig. 1 Brightness temperature fluctuations due to the triple points in the detonation front of CH_4 + 1.74 O_2 + 4.93 N_2:
a) p_i = 11.13 bars; b) p_i = 8.66 bars; c) p_i = 3.6 bars and in the detonation front of CH_4 + 1.84 O_2 + 5.94 N_2; d) p_i = 19.02 bars; e) p_i = 11.49 bars; f) p_i = 6.94 bars (time proceeds from left to right).

WINDOW INTERFACE

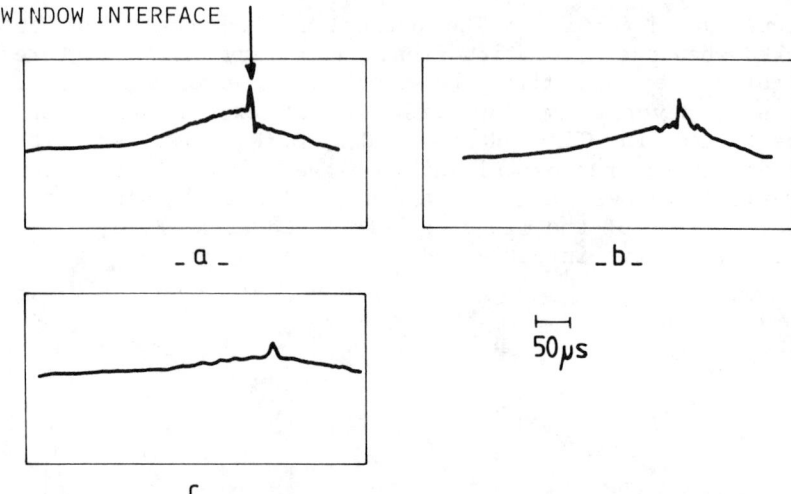

Fig. 2 Brightness temperature fluctuations due to the triple points in the detonation front of $C_2H_4 + 2.85\ O_2 + 14.25\ N_2$ mixture: a) $p_i = 9.19$ bars; b) $p_i = 7.32$ bars; c) $p_i = 3.5$ bar (time proceeds from left to right).

Experimental Results and Discussion

The oscillations observed before the detonation front hits the glass window are those of the brightness temperature peaks. The sudden decrease of the signal occurs once the detonation front has reached the glass and cannot be explained by the decrease in optical transmission due to the shock wave propagating inside glass. In that case, the signal decrease would be of a much shorter duration by about one order of magnitude (See Fig. 1). This brightness temperature decrease might also result from the window breakage as the glass is submitted to the high pressure of the products. If that were the case, the fluctuations that follow this sudden brightness temperature decrease cannot be explained.

The brightness temperature fluctuations recorded in the detonation front were used to calculate the characteristic time between two maxima. These maxima were obtained at various initial pressures as the triple points crossed the optical field. Typical records are presented on Fig. 1 ($CH_4-O_2-N_2$) and Fig. 2 ($C_2H_4-O_2-N_2$) for several pressures and dilutions. As the initial pressure decreases, 1) the cell size increases and 2) the higher Z is chosen, the earlier spinning detonation appears. The characteristic frequencies obtained for such detonations agree with the acoustic theory proposed by Manson (1946) and are only slightly dependent on the composition of the mixtures, as expected, since computed sound velocities in this range of pressures do not depend strongly on composition.

The cell length a, with an uncertainty of 10%, was determined from the records, and the cell height b was derived from the relation $b = a/2$ proposed by Urtiew and Tarver (1981). While Bull et al. (1982) have observed in H_2-air and C_2H_2-air that $a/2 \leq b \leq a$ for initial pressures lower than 1 bar, Wolanski (1983) has confirmed that $b = a/2$ for high initial pressures. As a consequence of the assumption $b = a/2$, the cell height is strongly dependent on the initial pressure of the mixture. While this assumption might lead to major inconsistencies in the critical diameters that are predicted from Eq. (1), the predicted critical diameters are consistent with those proposed by other authors.

Data on cell size for initial pressures up to 10 bars measured by soot track method are available in the literature (Manzhalei et al. 1974). Our recent results (Bauer et al. 1984) are consistent with their data. These authors also used that method to study the roughness of the detonation front. Siwiec and Wolanski (1983) used a metalic plate coated with a thin lead layer, instead of soot, to measure the cell size in hydrocarbon-O_2-N_2 mixtures at initial pressures up to 40 bars. For much lower initial pressures (3 bars), we obtained soot trace records (Bauer et al. 1984) in a C_2H_4-air mixture; the cell length, thus obtained, confirmed our optical measurements.

C_2H_4-O_2-N_2 Mixtures

The critical diameters of mixtures E, F, and G for several initial pressures are reported on Fig. 3. The values corresponding to mixture E were obtained in a previous study from observations based on electrical conductivity measurements (Bauer and Brochet 1983 a; Brochet and Veyssière 1974). Critical diameters for $Z = 3.01$, $Z = 4.09$, and $Z = 5$ fall on straight lines on a $\log p_i$ - $\log d_c$ scale. For an initial pressure of $p_i = 10$ bars, the critical diameters are, respectively, 2.5, 4.5 and 8 cm for these dilutions. Extrapolation of the correlation to $p_i = 1$ bar yields critical diameters of 25, 50, and 90 cm, respectively. A decrease of the initial pressure from 10 bars to 1 bar leads to an order of magnitude increase of the critical diameter.

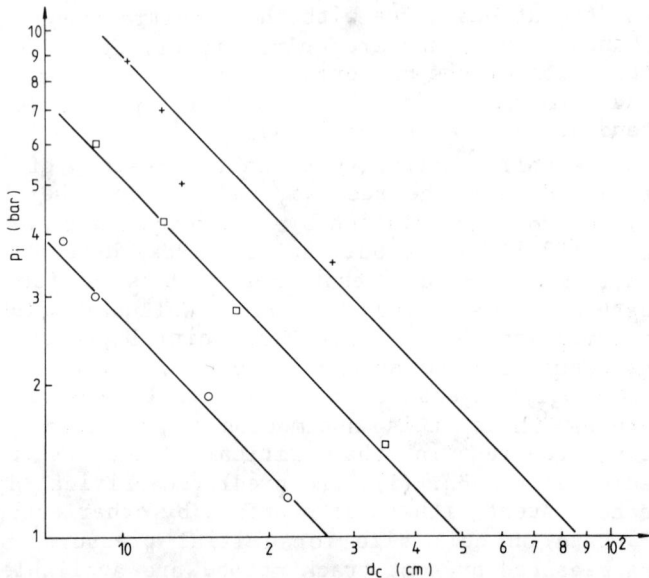

Fig. 3 Critical diameters obtained for various initial pressures in C_2H_4-O_2-N_2 mixtures: (+) $Z = 5$; (□) $Z = 4.09$; (○) $Z = 3.01$.

The dependence of the critical diameter for $p_i = 1$ bar on Z is shown in Fig. 4. The values obtained by Knystautas et al. (1982) for this C_2H_4 in stoichiometric proportions with O_2-N_2 are also shown in Fig. 4. As both sets of critical diameters are in good agreement, this work confirms the critical diameter of 33 cm obtained by Knystautas and co-workers for C_2H_4-air. Their work entails extrapolation on the basis of the N_2 dilution, whereas the present paper entails extrapolation on the basis of pressure. For the range of the experimental conditions in the present study, no detonation occurred for $Z > 6$. This limit corresponds to a critical diameter of $d_c = 1.5$ m. Also, shown on Fig. 4 are the values obtained by Matsui and Lee (1970) for a fuel-rich mixture. The value corresponding to C_2H_4-air mixture also agrees with the recent results of Bauer et al. (1984).

CH_4-O_2-N_2 Mixtures

The critical diameter for several initial pressures for mixtures A, B, C, D are plotted on Fig. 5. An important finding is that the cell length, and therefore d_c, was essentially the same for several initial

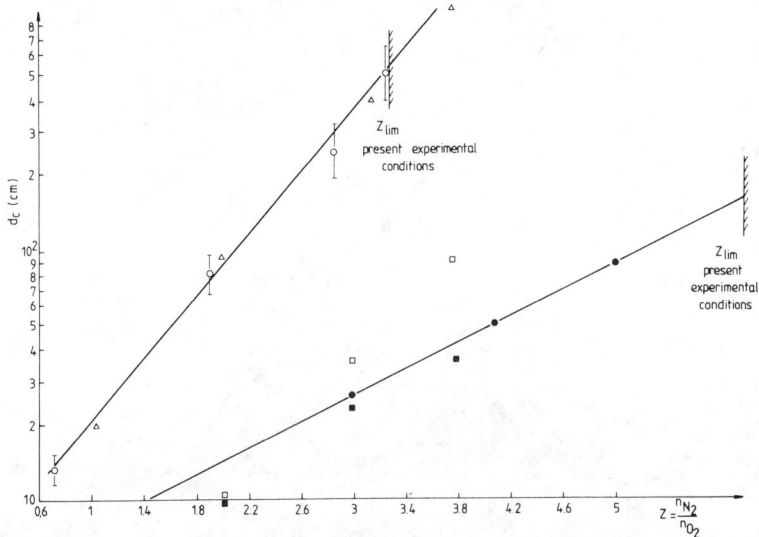

Fig. 4 Extrapolated critical diameters as a function of the dilution Z for p_i = 1 bar: (○) CH_4 + 1.85 (O_2 + ZN_2); (●) C_2H_4 + 2.85 (O_2 + ZN_2); (△) CH_4 + 1.5 (O_2 + ZN_2) (Matsui and Lee 1970); (□) C_2H_4 + 2 (O_2 + ZN_2) (Matsui and Lee 1970); (■) C_2H_4 + 3 (O_2 + ZN_2) (Knystautas et al. 1982).

pressures. These observations are explicable if a given vibratory mode exists not only for a single initial pressure but also for several initial pressures. This result implies that the evolution of these vibratory modes toward the fundamental one is not a monotonic function of the initial pressure. This appears to be particularly true in the case of CH_4-O_2-N_2 mixtures for large Z. These conditions have not been observed in the experiments with C_2H_4-O_2-N_2 mixtures or with the CH_4-O_2-N_2 mixture at the lowest dilution Z = 0.72. In the case Z = 0.72, the critical diameter is a linear function of p_i on a $\log d_c$ - $\log p_i$ plot. Since similar behavior is expected for other dilutions, d_c was plotted against p_i for each of the dilutions on a line of identical slope to that for Z = 0.72. This slope is equal to that of the C_2H_4-O_2-N_2 mixtures (Fig. 3). For p_i = 10 bars, critical diameters were found to be 1, 8, 17 and 40 cm, respectively, for mixtures A, B, C, and D. The critical diameters obtained by extrapolation of these curves to p_i = 1 bar are about an order of magnitude larger. The d_c for p_i = 1 bar for CH_4-O_2-N_2 mixtures is shown on Fig. 4. In spite of the assumptions (the uncertainty is shown in terms of uncertainty bars

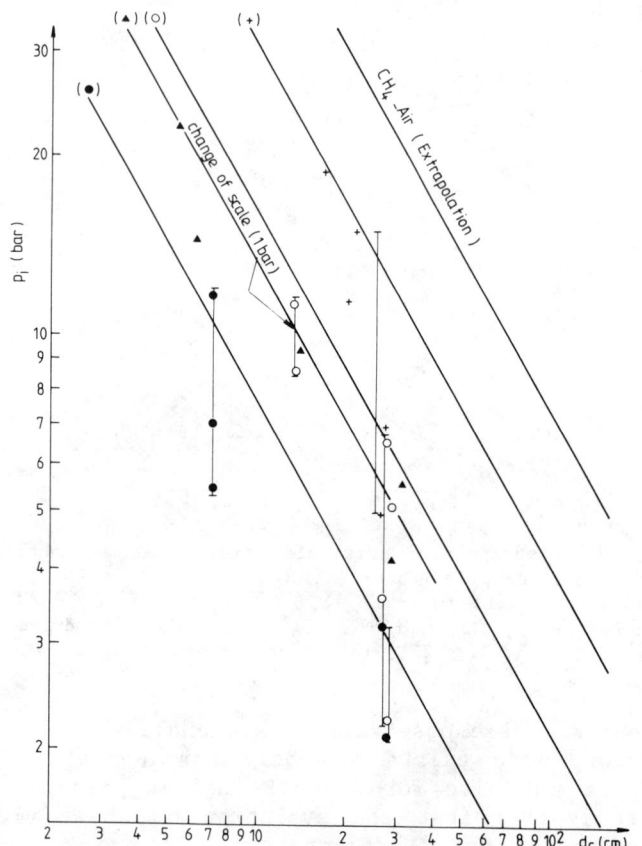

Fig. 5 Critical diameters obtained for various initial pressures in $CH_4-O_2-N_2$ mixtures: (▲) $Z = 0.72$; (●) $Z = 1.89$; (○) $Z = 2.83$; (+) $Z = 3.23$.

on Fig. 4), the critical diameter derived from Fig. 4 for $Z = 1$ is in good agreement with that obtained by Knystautas et al. (1982). This shows that studies at high initial pressures permit the determination of the critical diameters in a higher range of dilutions than had been done previously.

Extrapolation of the results to $Z = 3.76$ indicate that $d_c \simeq 10$ m, in good agreement with the value proposed by Matsui and Lee (1970). At an initial pressure of $p_i = 100$ bars, the critical diameter for a CH_4-air mixture would be about 5 cm. Bauer et al. (1983) have shown that a detonation could not occur with this mixture at p_i = 120 bars in a 55-mm-i.d. tube when a detonator served as the initiation source.

Summary and Conclusion

A rapid response time pyrometer has been used to measure the cell size of the detonation front of several $C_2H_4-O_2-N_2$ and $CH_4-O_2-N_2$ mixtures for initial pressures of up to 20 bars. Critical diameters derived from these measurements appeared to be highly dependent upon the initial pressure. A decrease of p_i from 10 bars to 1 bar yields an increase of about one order of magnitude of the critical diameter. As a consequence, high pressures can be used to determine the critical diameters of mixtures at dilutions that at low pressures would require a very large device to observe cell dimensions. As a consequence critical diameters for more diluted mixtures could be obtained than those investigated by Knystautas et al. (1982). The values derived from their study, once extrapolated toward the dilutions of this work, appeared to be in good agreement with the present results. The critical diameter in the CH_4-air mixture for an initial pressure p_i = 1 bar was found to be $\simeq$ 10 m, in good agreement with that proposed by Matsui and Lee (1970), but somewhat larger than the d_c proposed by Bull et al. (1982) ($d_c \simeq 4$ m). Since all three results are based on extrapolations on log plots at some distance from the experimental data, and since entirely different approaches were used, it is likely that the error bounds are so large that the three estimates might be in good accord. For Z = 1.5 and for very close equivalence ratios, the critical diameter of methane is approximately five times that of the ethylene and becomes 15 times this value for higher dilution (Z $\simeq$ 3). This suggests that the critical diameter of methane increases much more rapidly with the dilution than it does in the case of ethylene. The dilutions above which detonation could not occur in our experimental conditions were Z = 3.23 and Z = 6 for methane and ethylene, respectively. This confirms the results of a previous study (Bauer and Brochet, 1983 b).

References

Bauer, P., Krishnan, S., and Brochet, C. (1981) Detonation characteristics of gaseous ethylene, oxygen and nitrogen mixtures at high initial pressures. <u>Gasdynamics of Detonations and Explosions. AIAA Progress in Astronautics and Aeronautics</u>, (edited by J. R. Bowen, N. Manson, A. K. Oppenhiem, and R. I. Soloukhin), Vol. 75, pp. 408-422. AIAA, New York.

Bauer, P., Brochet, C., and Presles, H. N. (1983) Detonation study of gaseous mixtures at initial pressures reaching 100 bars. Arch. Combus. (to be published).

Bauer, P. and Brochet, C. (1983a) The structure of the detonation front in an explosive mixtures at a high initial pressure. Arch. Combus. 3 (1), 39-45.

Bauer, P. and Brochet, C. (1983b) Properties of detonation waves in hydrocarbon-oxygen-nitrogen mixtures at at high initial pressure. Shock Waves, Explosions, and Detonations. AIAA Progress in Astronautics and Aeronautics, (edited by J.R. Bowen, N. Manson, A. K. Oppenheim, and R. I. Soloukhin), Vol.87, pp. 231-243. AIAA, New York.

Bauer, P., Presles, H. N., Heuze, O., and Brochet, C. (1984) Cell and induction lengths in detonation waves of fuel-oxygen-nitrogen mixtures at high initial pressures. (unpublished).

Bouriannes, R., Moreau, M., and Martinet, J. (1977) Un pyromètre rapide à plusieurs couleurs. Rev. Phys. Appl. 12 (5),893-899.

Brochet, C. and Veyssière, M. (1974) Mesures électriques des propriétés gazodynamiques des détonations. Acta Astron. (3),267-285.

Bull, D. C., Elsworth, J. E., Shuff, P. J., and Metcalfe, E. (1982) Detonation cell structures in fuel-air mixtures. Combus. Flame 45 (1), 7-22.

Edwards, D. H., Thomas, G. O., and Nettleton, M. A. (1979) The diffraction of a planar detonation wave at an abrupt area change. J. Fluid. Mech. 95(1), 79-96.

Guénoche, H. and Manson, N. (1957) Sur la variation de la célérité des ondes explosives avec le diamètre tubes. CR Acad. Sci. (Paris) 235, 1617.

Knystautas, R., Lee, J. H., and Guirao, C. M. (1982) The critical tube diameter for hydrocarbon-air mixtures. Comb. Flame 48 (1), 63-83.

Manson, N. (1946) Sur la structure des ondes explosives dites hélicoïdales dans les mélanges gazeux. CR Acad. Sci. (Paris) 222, 46-48.

Manson, N., Brochet, C., Brossard, J., and Pujol, Y. (1963) Vibratory phenomena and instability of self sustained detonation in gases. 9th Symposium (International) on Combustion, pp. 461-469. Academic Press, New York and London.

Manzhalei, V. I., Mitrofanov, V. V., and Subbotin, V. A. (1974) Measurement of inhomogeneities of a detonation front in gas mixtures at elevated pressures. Fiz. Goreniya Vzryva, 0 (1), 102-110.

Matsui, H. and Lee, J. H. (1970) On the measure of the relative detonations hazards of gaseous fuel-oxygen and air mixtures. 17th

Symposium (International) on Combustion, pp. 1269-1280. The Combustion Institute, Pittsburgh, Pa.

Mitrofanov, V. V. and Soloukhin, R. I. (1964) The diffraction of multifront detonation waves. CR Acad. Sci. URSS 159 (5), pp. 1003-1006.

Moen, I. O., Donato, M., Knystautas, R., and Lee, J. H. (1981) The influence of confinement on the propagation of detonations near the detonability limits. 18th Symposium (International) on Combustion, pp. 615- 622. The Combustion Institute, Pittsburgh, Pa.

Siwiec, S. and Wolanski, P. (1983) Detonation cell structure in methane-air pressure. Archiv. Combus. (to be published).

Urtiew, P. A. and Tarver, C. M. (1981) Effect of cellular structures on the behavior of gaseous detonation waves under transient conditions. Gasdynamics of Detonations and Explosions. AIAA Progress in Astronautics and Aeronautics, (edited by J. R. Bowen, N. Manson, A. K. Oppenheim, and R. I. Soloukhin), Vol. 75, pp. 370-384, AIAA, New York.

Wolanski, P. (1983) Technical University of Warsaw, Poland. Private communication.

Zeldovich, Ya. B., Kogarko, S. M., and Simonov, M. M. (1956) Experimental study of spherical detonation in gases, Sov. Phys. Tech. Phys. 1(8), 1689-1713.

"Galloping" Gas Detonations in the Spherical Mode

John E. Elsworth,* Philip J. Shuff,† and Aziz Ungut‡
Shell Research Ltd., Chester, Great Britain

Abstract

Pulsating detonations, having distinct similarities to so-called "galloping" detonations previously only observed in pipes characterized by periodic velocity fluctuations and irregular cell structures, have been produced in the spherical mode in several fuel-oxidant-diluent mixtures. A detailed study of one of these mixtures, $\lambda C_2H_6 + 3.5O_2 + 3.5N_2$, where λ was varied between 0.4 to 2.5 at an initial pressure of 6.74×10^4 Pa (2/3 atm) and ambient temperature has been made in the Thornton Detonation Chamber. Galloping waves were detected only over relatively narrow compositional ranges, around both the lean and rich limbs of the deflagration-detonation boundary, in mixtures marginal with respect to initiation energy. As the stoichiometry (λ) of the mixture approached 1.3 (the most detonation-susceptible mixture) the range of compositions exhibiting galloping tended to narrow; it was not possible to determine unequivocally if galloping eventually disappeared, yielding an unambiguous deflagration-detonation boundary, or merely became narrower than the experimental resolution. At compositions where galloping detonations were observed the phenomenon was very reproducible. The unconfined nature of the pulsating-galloping spherical detonations observed makes comparison with tube studies difficult. However, the present work tends to substantiate existing theories of galloping detonation derived from planar wave studies of Moen et al. (1981) and Ul'yanitskii (1981) in tubes. Other theories of galloping detonation are also discussed.

Presented at the 9th ICODERS, Poitiers, France, July, 3-8, 1983. Copyright © 1984 by Shell Research Ltd. Published by the American Institute of Aeronautics and Astronautics with permission.
*Scientist, Thornton Research Centre.
†Senior Technician, Thornton Research Centre.
‡Senior Scientist, Thornton Research Centre.

Introduction

For any reactive gas-oxidant mixture the boundary between nondetonation (deflagration) and detonation can be delineated on a plot of gas concentration vs mass (or energy equivalent) of the initiator. Typically, these plots have a characteristic "U" shape and a comparison (e.g., Bull 1979) of such plots for several gases shows that for detonation to be achieved a certain threshold level of initiation energy, differing in amount for each gas-oxidant system is required. In general, an increase of initiator energy leads to a broadening of the effective detonation limits, and conversely reducing the initiation energy narrows the limits eventually to a point beneath which detonation becomes impossible. This point is the most susceptible composition for detonation and for most hydrocarbon-air systems is slightly fuel-rich of stoichiometric. Consequently, detonation limits cannot be defined adequately without reference to the initiator, gas composition, mode of propagation (planar, cylindrical, or spherical), and possibly also the apparatus or enclosure if detonation cell accommodation effects become limiting.

The present study is a more critical examination of the deflagration-detonation boundary and evidence is presented for the existence of a marginal type of spherically propagating waves having similarities to "galloping" detonation waves which have previously only been observed in tubes. Galloping waves are characterized by a decrease, followed by a sudden increase in the velocity of the detonation front resulting in the average velocity being close to the theoretical Chapman Jouguet (CJ) value. Spherical galloping waves have, however, one significant difference to planar galloping detonations observed in tubes. With spherical waves the galloping cycles are composed entirely of variable speed waves whereas in tubes the galloping cycles are extended, both in distance and time, by periods of constant velocity CJ propagation. This tube effect is illustrated by Edwards et al. (1974) and shows a significant fraction of a velocity-distance plot to be at CJ velocity. Although a direct comparison with the present work is not possible it may be concluded the nonequilibrium periods in both spherical and planar mode waves are roughly equal and that the tendency of planar galloping waves to stabilize at CJ velocity is due to the presence of the tube wall.

In this present study, "galloping" spherical detonations are defined as combustion waves characterized by regular velocity excursions both below and above the theoretical CJ values and which produce continuously changing cell dimensions on sooted plates.

The galloping phenomenon is not to be confused with "spinning" detonation which is also sometimes seen in tubes under marginal conditions. Spinning waves are characteristically very regular and their periodicity has been shown by several authors (e.g., Campbell and Woodhead 1927; Fay 1952) to be proportional to the tube diameter whereas galloping waves appear to be less regular and each galloping cycle may extend over several meters of tube length.

Historically, the first observation of a galloping wave (a single gallop in hydrogen-air) was reported in 1951 by Mooradian and Gordon and was regarded as a nondetonation. The name galloping was suggested by Duff in 1963 who had observed the phenomenon in acetylene. Since that time galloping detonations have been reported in propane-oxygen-nitrogen mixtures by Manson et al. (1963), Saint-Cloud et al. (1972), and Edwards et al. (1974), in hydrogen-oxygen and methane-oxygen diluted with nitrogen or argon by Ul'yanitskii (1981a) and in ethylene-air by Moen et al. (1981). A common factor which emerges from all the tube studies is that in systems where galloping detonation was observed the conditions were always marginal with respect to at least one experimental parameter. Galloping waves have not been detected in all systems, particularly in the more reactive mixtures (i.e., easily detonable), but it is not possible at the present time because of the paucity of data to determine whether or not this is a real effect or simply reflects insufficient experimental resolution.

Several theories have been advanced to explain the galloping phenomenon. Most of these require that during the "slow" period the shock and chemical reaction waves become progressively uncoupled resulting in a degradation of the detonation cell structure. Moen et al. (1981) indicate that the galloping mode of propagation is controlled by a process of transverse pressure wave amplification behind the leading shock front; the pressure oscillations amplify, and begin to catch up the leading shock, which then interact together and produce a small explosion sufficient to cause the resulting detonation wave to be overdriven. The overdriven wave has an enhanced velocity and a smaller than normal cell structure is produced. Subsequently the wave velocity gradually decreases accompanied by a broadening of the cell structure; eventually wave uncoupling occurs again and signifies the end of the galloping cycle. Manson et al. (1963) noted when recoupling occurs there is a strong association of a flame front with the reinforced shock wave, the whole phenomenon resembling the formation of detonation by an accelerating deflagration. The appearance of flame in the induction zone between the separated shock and combustion fronts initiating detonation has been termed "flashing" by Ul'yanitskii (1981a,b). The flashing phenomenon is believed to be caused by spontaneous ignition in the postshocked gas and should not be confused with the "microexplosions" produced by collision of transvere shock waves in steadily propagating detonation waves, which yield regular cellular structures.

Although studies of spherically galloping detonations have not been reported, quasisteady regimes have, however, been observed by Edwards et al. (1978) and Desbordes et al. (1981) in unconfined oxyacetylene mixtures with various diluents. These regimes were characterized by relatively small velocity pulsations below the CJ value but unlike galloping waves in tubes ultimately resulted in stable (CJ) spherical detonations.

In this work spherically propagating galloping detonations have been observed in various oxidant-diluent mixtures with ethylene oxide, hydrogen, methane, ethane, propane, n-butane, isobutane, cyclohexane, and toluene at subatmospheric pressures which showed distinct similarities with galloping waves observed in pipes. For detailed study the system:

$$\lambda C_2H_6 + 3.5O_2 + 3.5N_2$$

was examined at an initial pressure of 6.75×10^4 Pa (2/3 atm) and at ambient temperature in the Thornton Detonation Chamber. This particular mixture was chosen because ethane is the saturated hydrocarbon most susceptible to detonation (Bull et al. 1978) and has been examined previously in the study of detonation cell measurements (Bull et al. 1982). With this gas system it was found possible to vary λ over the range 0.4-2.5 using relatively small (up to 9 g) Tetryl initiation charges.

Experimental

Gas mixtures were prepared in the Thornton Detonation Chamber by the method of partial pressures to an initial pressure of 6.75×10^4 Pa (2/3 atm) at ambient temperatures over the range 286-303 K. As in previously reported vessel experiments with hydrogen mixtures (Atkinson et al. 1980) final fuel concentration accuracy was not limited by pressure measurement, but by temperature fluctuations during the mixing procedure to about ±0.1%vol. of the fuel. Mixing in the chamber, a dome-ended, vertical mild-steel

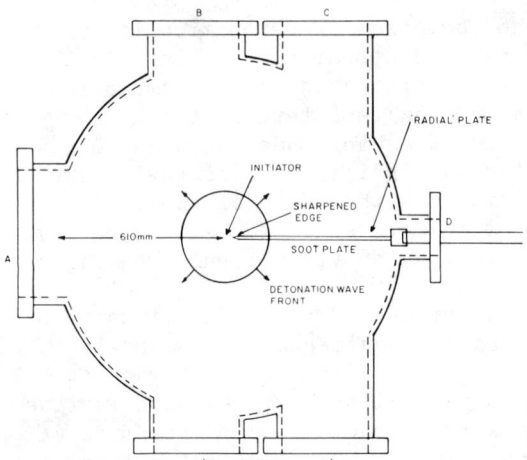

Fig. 1 Section through detonation chamber.

cylinder of volume ~1.5 m³, was accomplished by recirculation through an external loop using an oil-free diaphragm pump. The chamber has a horizontal radius of 0.61 m and, as shown in Fig. 1 is fitted with several ports which held either glass windows or blanking plates as required.

Combustion was induced in the gas mixtures by centrally located cylindrical Tetryl charges suspended by a 6-mm-diam stainless-steel tube and activated by special plastic detonators supplied by Nobels Explosives Ltd. which contained nominally the equivalent of 0.1-g Tetryl. Most experiments were at the 8-9-g Tetryl level but smaller charges of ~1 and ~0.25 g were also used.

The mode of combustion was diagnosed with reference to several measuring techniques, as follows.

Windows A and B were used for high-speed photography at framing rates of either 5×10^4 or 1×10^5 pictures s^{-1} using an IMACON 790 camera which could alternatively be used in the streak mode in the range 1-10-μs mm^{-1}. Triggering of the camera was by an optical pickup activated by light from the event, all photographs were by self-illumination.

A pair of windows C and C' were used for a multibeam laser anemometer. A He-Ne laser beam was split into 13 secondary beams 2 cm apart using a series of partially reflecting glass plates. After passing through the chamber the beams were recombined and focussed into a photomultiplier. The passage of a shock wave through the chamber caused systematic displacement of the beams which enabled the wave velocity to be deduced.

Port D could be used to mount either an X-band microwave interferometer or a sooted plate. When used for X-band interferometry a small pyramidal horn was mounted inside the port and an AEI Doppler unit (type DA 8525/6) was fixed to the outside of the metal blanking plate machined to form an integral wave guide. Alternatively, 500 × 130-mm sooted plates could be radially mounted. The plates had a sharpened leading edge and were normally located 65 mm from the initating charges.

Supplementary measurements of wave arrival times were sometimes made using stalk-mounted piezoelectric pressure transducers (Kistler type 601A). All analog signals were stored in Datalab DL922 transient event recorders. Because it was not always practicable to use all types of measuring systems in any one experiment, repeat runs were made as necessary. A total of 101 experiments were performed.

Results

Distinction between deflagration, galloping detonation, and stable detonation was made primarily by the type of cell pattern recorded on sooted plates as shown in Fig. 2. Deflagrations (a and f) were characterized by transient cell patterns followed by an area of clean plate; galloping detonations (b and d) yielded distinct areas of continuously varying cell dimensions; and stable detonations (c) gave a continuous highly regular pattern. The type of combustion was always confirmed by at least one other diagnostic. It was found that the nonincursive measurements, e.g., X-band microwave interferometry, laser-Schlieren anemometry, and photography, gave overall similar results whether or not the incursive sooted plates and/or pressure transducer stalks were present.

Figures 3-5 show typical records obtained by the various methods, the suffixes a and b refer to stable and galloping detonations, respectively. In all cases stable detonations had wave velocities close to the theoretical (CJ) values, while the galloping detonations showed wide variations. The only continuous records of wave velocity were given by the streak photographs; the example in Fig. 5b is from a two-gallop experiment and indicates a minimum wave speed $\sim$1.26 km s^{-1} immediately prior to the flash reinitiation. Immediately after the flash a very rapid increase in wave velocity occurred, a transient value of $\sim$5.5 km s^{-1} was indicated but this high value was not detected by any of the other measuring systems, possibly owing to a lack of resolution. Figure 5b also shows that the flashing occurred in discrete bursts with some degree

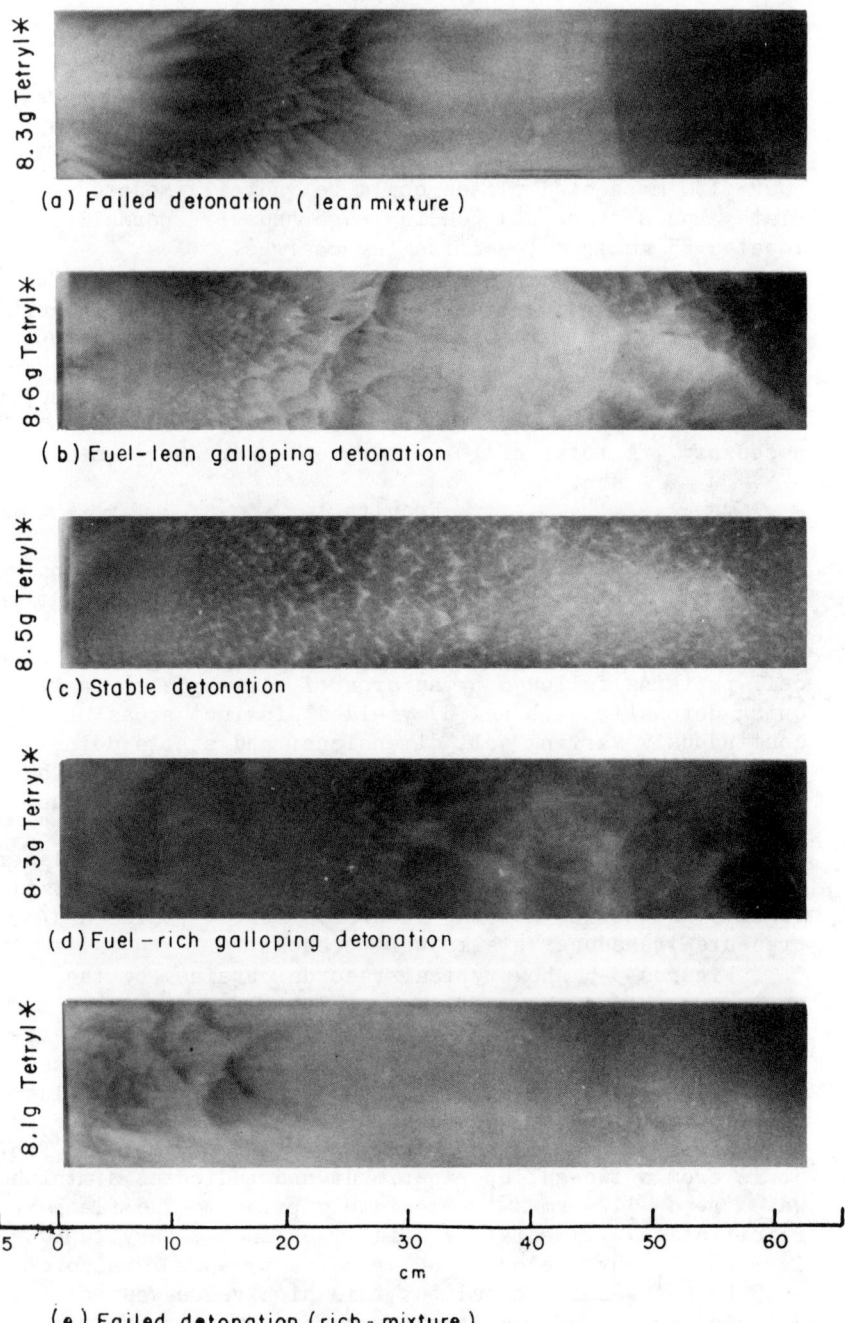

Fig. 2 Sooted plate records.

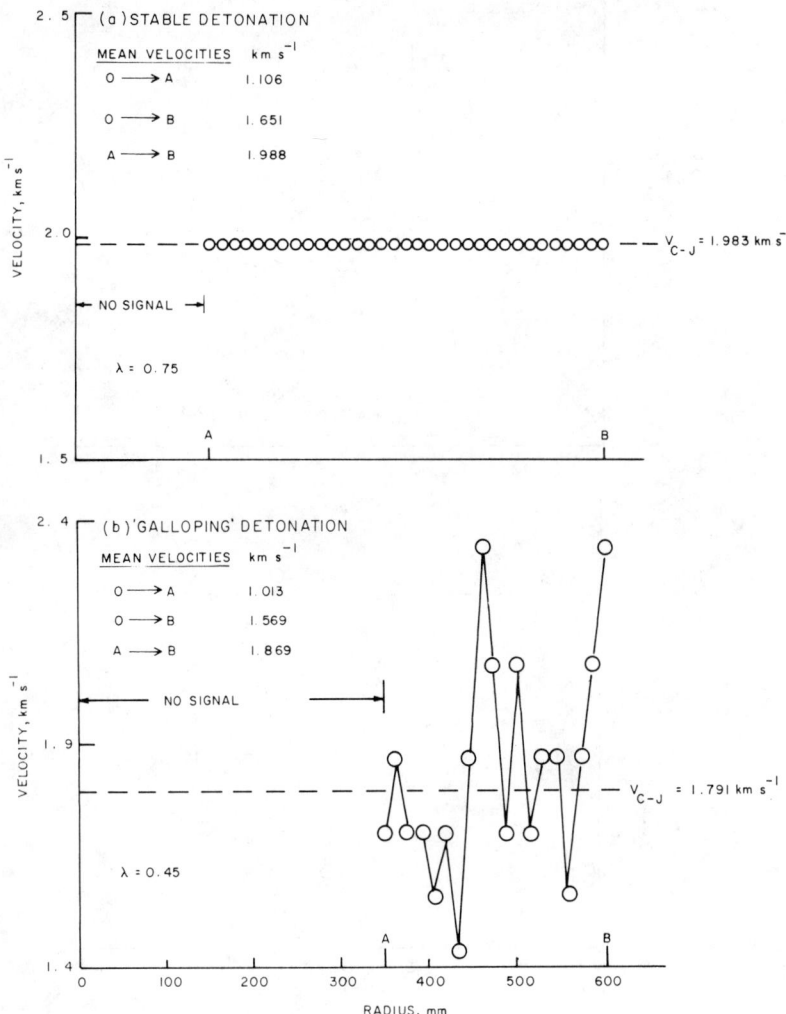

Fig. 3 X-band microwave Doppler radius-velocity plots. (Each o represents consecutive cycles.)

of symmetry, indicating that the flashing shell is not truly spherical. Some further evidence of off-spherical symmetry is shown in Fig. 6, a sequence of high-speed diametrical photographs of a galloping detonation taken at 20-μs intervals. Frames 1 and 2 show only bright light emission attributable to the Tetryl charge, but starting at frame 3 a lower luminosity gas envelope becomes evident which by frame 5 has assumed an oblate spheroid shape.

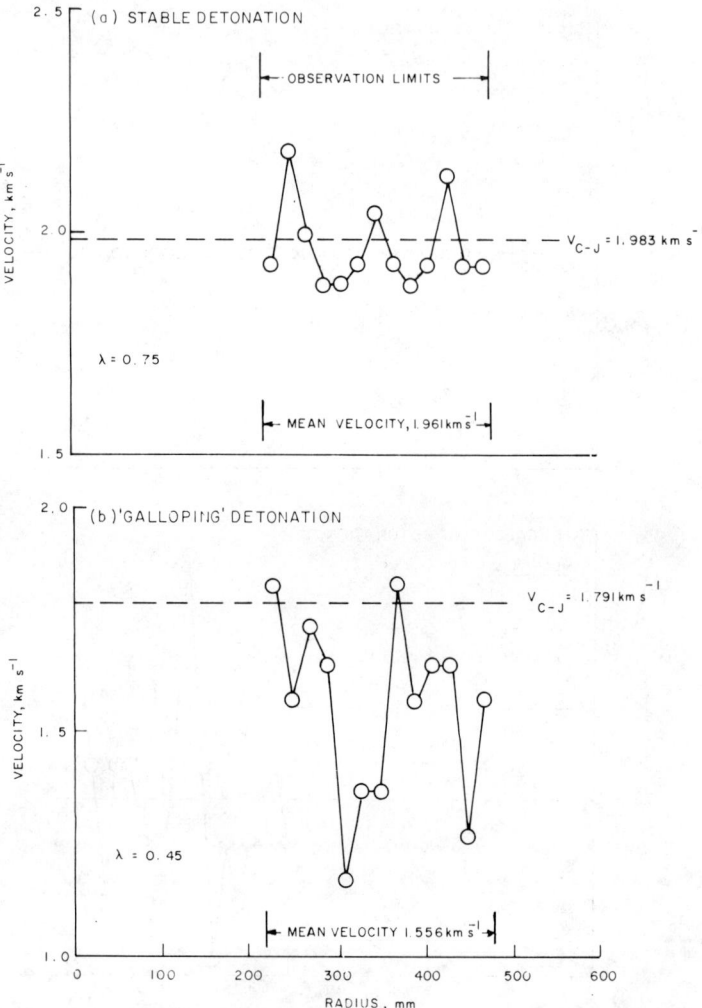

Fig. 4 Multibeam laser Schlieren radius-velocity plots. (Each o represents adjoining beams.)

Collectively, as shown in Fig. 7, galloping detonation occurred between the deflagration and stable detonation regions at both the fuel-lean and fuel-rich boundaries. At the 8-g Tetryl level, deflagrations (denoted by x) were produced at λ values <0.43 and >2.35 and galloping detonations (+) between 0.43<λ<0.5 and 2.05<λ<2.26. Stable detonation (o) was recorded between 0.5<λ<2.05. Galloping detonations were also observed with a reduced quantity of ~1-g Tetryl initiator, but further reduction of the ini-

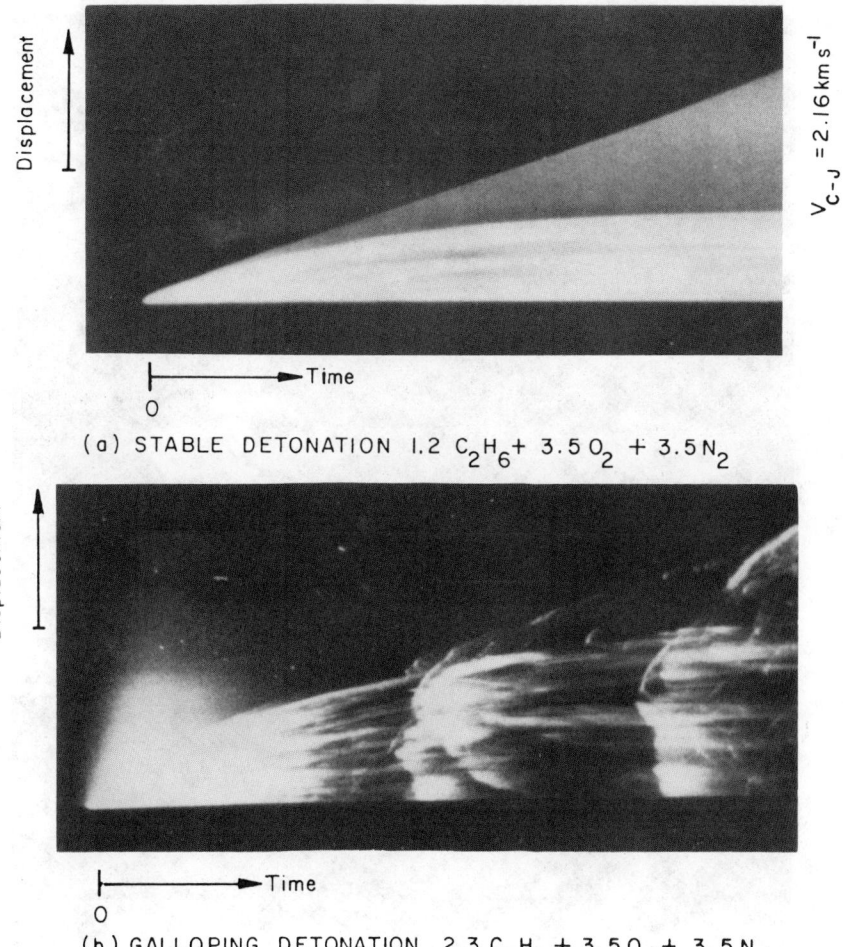

Fig. 5 Streak photographs.

tiator did not yield any further galloping detonations. At the nominal 0.1-g initiation level (single detonators) stable detonations were shown by regular cell patterns, but only after a relatively long period of nonwriting by the wave on the sooted plates. This delayed detonation phenomenon is discussed in more detail elsewhere (Elsworth and Eyre 1984).

The unbroken and broken lines in Fig. 7 are theoretical detonation susceptibility curves based on equations of Ul'yanitskii (1981b) and Bull et al. (1979), respectively, and are discussed below. Both curves give a good fit to the experimental points. The regions of galloping detonation are also indicated.

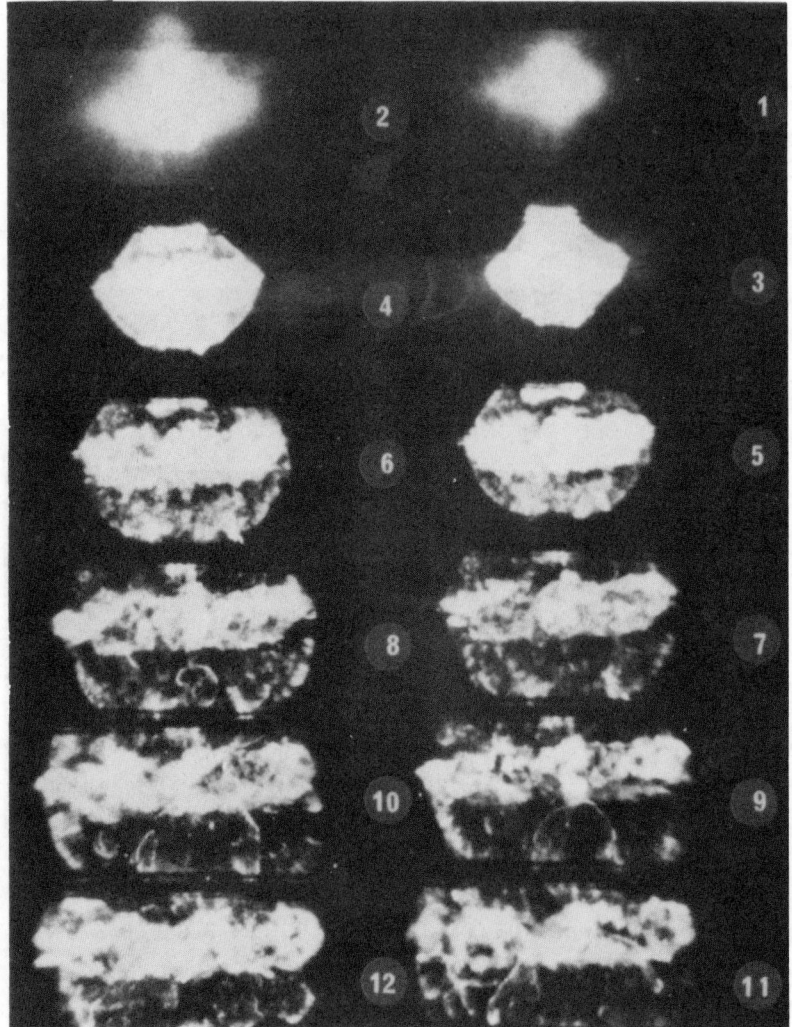

Fig. 6 Sequence of galloping detonation taken with IMACON 790 camera at framing rate 5×10^4 pictures s^{-1}.

Discussion

Galloping Zones

The experimental points in Fig. 7 define the same essentially U-shaped curve characteristic of a normal composition-detonation susceptibility plot. It can be seen, however, particularly at the 8-g Tetryl level, that the deflagration-detonation boundary (at both lean and

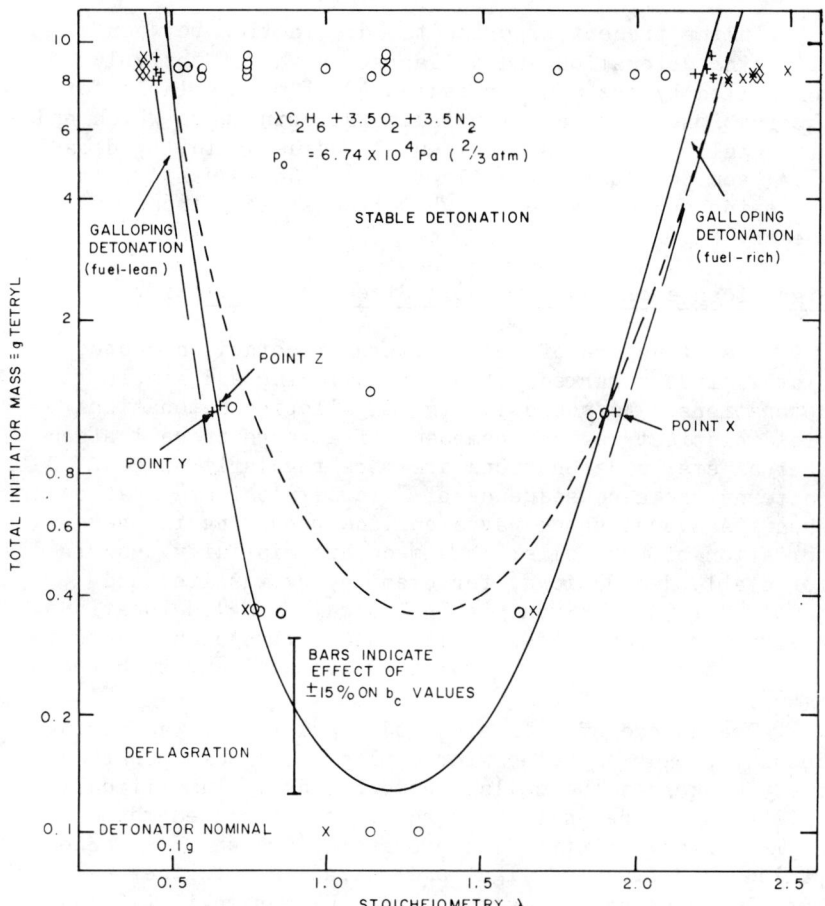

Fig. 7 Plot of stoichiometry (λ) vs total initiator mass (g) (Tetryl + detonator). o indicates stable detonation, + galloping detonation, and x deflagration. ———————— and — — — — predicted detonation susceptibility curves from the models of Ul'yanitskii and Bull et al., respectively. (Refer to text). —— —— —— approximate limit to galloping detonation.

rich mixtures) were separated by narrow zones where galloping detonations occur. At this level of initiation, 83 experimental points are represented (duplicate experiments not plotted) and show, without a single exception, that the experiments were very reproducible. At lower levels of initiation, the galloping zones became even narrower, but it was not possible to determine if the galloping zones reduced to an unambiguous deflagration-detonation boundary or simply became narrower than the experimental resolution set by uncertainties in mixture composition.

In the present experiments, distinction between galloping detonation and deflagration was effectively controlled by the chamber radius (0.61 m) available for observation. Thus a potentially galloping wave which had a first gallop exceeding the vessel radius would, by default, have been designated a deflagration. Consequently, the galloping zones may be somewhat broader than those indicated by Fig. 7.

Significance of Changing Cell Sizes and Regularity

The structure of cell patterns recorded on sooted plates differs markedly between galloping and stable detonations. As shown in Fig. 2 galloping detonations yield distinct regions composed of a range of cell sizes whereas stable detonations are more regularly spaced. Cell patterns are a consequence of "microexplosions" that occur when transverse shock waves collide resulting in the formation of new cells; this mechanism is fully described for stable detonations, for example, by Vasiliev and Nikolaev (1978). Even stable (nongalloping) detonations, however, may have somewhat irregular structures and these have been shown to vary from poor to excellent by Strehlow (1963).

The change of cell size and regularity in a failing detonation suggests the microexplosions become progressively weaker as the cells become broader. In critical initiation of detonation it has been postulated that a supplementary reinitiation process occurs which is accompanied by a flash of light and rekindles a further cycle of rapidly changing cell structures. Ul'yanitskii (1981a) attributed the flash to localized ignition in the post-shocked gas deriving from separated shock and combustion fronts. After the separation the transverse wave collisions become too weak to produce instantaneous microexplosions as in a regular structure. Reinitiation of detonation cannot occur if the chemical induction period becomes too long.

In a second paper Ul'yanitskii (1981b) developed a model based on strong "point" explosion with ensuing chemical reaction. In a galloping detonation the cell length (b_F - using Ul'yanitskii's notation) just prior to flashing was calculated to be larger than the steady cell size (b_c) and thus effectively limits the maximum cell size that can be observed.

$$b_F \simeq b_c \cdot E_a / 3RT_{10} \qquad (1)$$

where $E_a/3RT_{10}$ is the dimensionless activation energy and T_{10} is the particle temperature appropriate to a normal shock travelling at CJ velocity.

Ul'yanitskii derives relationships for both point source-initiation distance (r_{*3}) and critical energy (E_{*3}). For the spherical mode these are given as:

$$r_{*3} \simeq 1.7\, b_F \qquad (2)$$

$$E_{*3} \simeq 6\rho_o D_o^2 b_F^3 \qquad (3)$$

where ρ_o and D_o are the initial density and wave speed, respectively.

From Eqs. (1) and (2) with a b_c value of 0.04 m ± 15% (taken from Bull et al. 1982) it is estimated that a 0.45λ

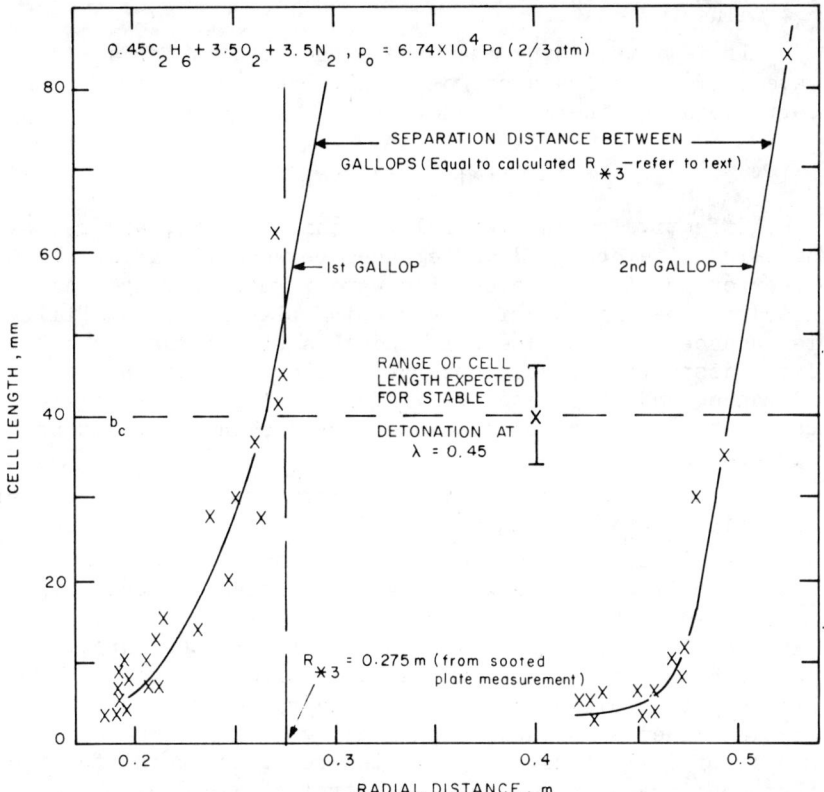

Fig. 8 Plot of radial distance (m) vs cell length (mm) showing measured r_{*3} and separation distance between consecutive gallops.

mixture would have a r_{*3} value of 0.25 m. This value compares, as shown in Fig. 8, with a value of 0.275 m for the initiator (point source)/1st reinitiation distance deduced from cell sizes in the present work. It is interesting to note that the separation distance between the first and second gallops is exactly 0.25 m indicating that the flash produced in the spherical case caused a similar galloping cycle to occur. Comparable behavior is shown in Fig. 5b, a streak photograph of a fuel-rich galloping detonation.

As $\lambda \to 1.2$ the steady cell size (b_c) becomes smaller, and it could be expected that r_{*3} should also diminish correspondingly. This was found to be so at the 1-g Tetryl initiation level (points x and y on Fig. 7) where r_{*3} was calculated to be 0.106 m compared with a mean gallop distance of 0.13 m. The present study therefore shows that the radii of subsequent galloping cycles are comparable in length with both the first (kernel) radius and calculated r_{*3} values.

It is interesting that the kernel radius for a planar wave (r_{*1}) is calculated to be only about one third of the corresponding spherical case:

$$r_{*1} \simeq 0.5 \, b_F \qquad (4)$$

Effectively, the two relationships for r_{*1} and r_{*3} highlight the observed differences between planar and spherical galloping detonation waves, respectively. As noted in the introduction, galloping planar wave stability is enhanced by the tube wall and this effect can cause a large disparity between observed gallop separation distances and calculated r_{*1}. Thus, unlike the spherical case (where separation distance $\simeq r_{*3}$) planar separation distance $\gg r_{*1}$.

With the values of b_c, taken from the curve of Bull et al. (1982) and D_o values derived by the method of Eisen et al. (1960), Eq. (3) is used to calculate the detonation susceptibility for the whole range of ethane concentrations used in this study. The results are shown as the solid line on Fig. 7 and give an excellent fit to the experimental data points. A minimum value is predicted at $\lambda = 1.2$ and compares tolerably well with the rather shallower curve (broken line) also shown representing the "phenomenological model" of Bull et al. (1979). The reference point for the phenomenological model was 8-g Tetryl at $\lambda = 0.5$.

The present experiments therefore give very good support to the spherical Ul'yanitskii model but neither confirm nor refute the observations of Moen et al. (1981)

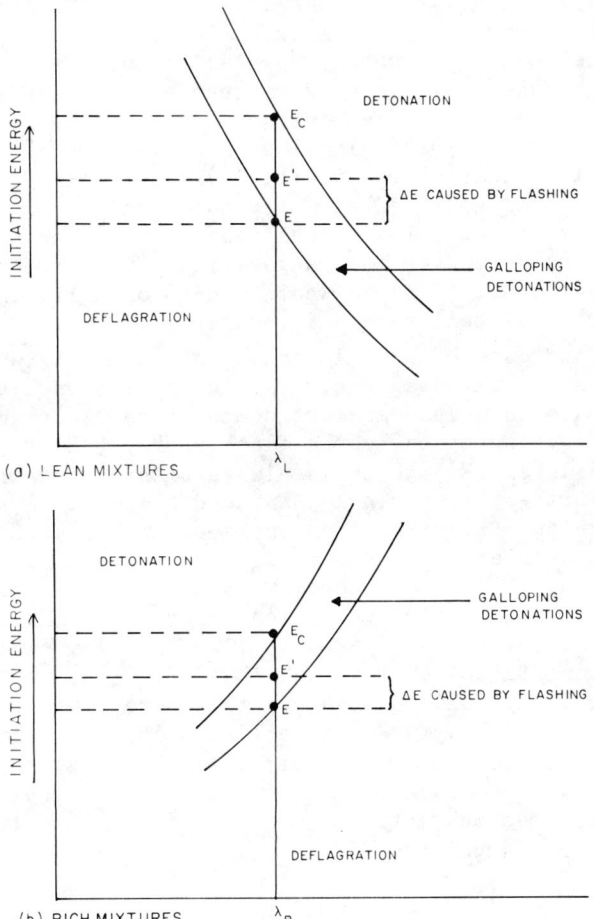

Fig. 9 Sketches showing effects of additional energy releases due to flashing, a) fuel-lean mixtures, and b) fuel-rich mixtures.

of transverse pressure wave amplification behind the leading shock front. In the present study the pressure transducers were mounted face on to the advancing wave and consequently were insensitive to transverse waves. Typically the pressure-time records showed <u>decaying</u> oscillattions in the range 120-150 kHz irrespective of combustion wave type, and are attributable to transducer resonances.

Narrowing or Disappearance of Galloping Zones

Presently available evidence suggests that galloping detonation may not occur in all gas systems. Galloping detonations are most obvious when the cellular structure is grossly irregular, and become less obvious if the cell regularity improves. It is possible, therefore, that for gas mixtures producing very regular cellular structures the galloping zones may either disappear or become narrower than the experimental resolution. In the cell measurements of Bull et al. (1982) using identical mixtures it was shown that as the chemical reactivity increased ($\lambda \rightarrow 1.2$) the cell size decreased and the cell structure became more regular. The fact that galloping detonations were not observed in the present work over the range $0.7 < \lambda < 1.8$ (corresponding to cell lengths ≤ 0.015 m) is wholly in accord with the concept of a threshold cell size and regularity.

In the case of a truly galloping wave, it is speculated, as indicated in Fig. 9a, an initiation level (E) appropriate to a lean mixture composition (λ_L) may be momentarily increased by the flash energy (ΔE) to an enhanced level (E') which is always within the galloping zone. Further, if a single ΔE is sufficiently large or if a succession of flashes progressively increases E' to above the true detonation boundary (E_c), then stable detonation would result. Such a mechanism at least qualitatively explains the quasisteady regimes observed by Edwards et al. (1978) and Desbordes et al. (1981). Should a single gallop produce sufficient energy such that $\Delta E > (E_c - E)$ then only a single gallop would be required to ensure a stable detonation and the galloping zone would effectively disappear. This supposition was given support by point z in Fig. 7 which showed a single gallop ~0.2-m long and thereafter the detonation became stable.

Similar arguments also apply to fuel-rich mixtures as shown in Fig. 9b.

Other Theories of Galloping Detonation

It may be argued that galloping detonation is simply due to unmixedness causing velocity fluctuations as the wave propagates through pockets of different compositions. However, because not even a single anomaly was observed in our 83 results at the 8-g Tetryl initiation level, this suggestion can be dismissed. Supporting evidence was also provided by the various photographic methods which indicatdicated a considerable degree of spherical symmetry.

It is not as easy, however, to dismiss the existence of oscillation between dual alternative modes of detonation, as reported in a tube study by Ubbelohde and Munday (1969). In their study, again at marginal conditions, there was steady propagation at either the CJ velocity or at about 70% lower; there was no evidence of velocities greater than CJ values. Similarly, other studies of marginal gas detonation, e.g., Adams (1978) and Edwards et al. (1978) invoke the existence of low-speed detonation waves. The latter authors suggest, as one of two possibilities, that low-speed detonation could be converted into high-speed (CJ) detonation by the interaction of secondary shock in the postshocked gas contained in the separation zone. Secondary, and also higher order, shock waves derive from collapse of the rarefaction produced by the gaseous products of solid explosive charges. Accordingly, Brode's (1958) equations have been used to estimate the behavior of the secondary blast wave that would be produced by the masses of Tetryl used in the present study.

For the purposes of computation cylindrical masses of Tetryl were assumed to behave similarly to spherical charges of TNT. At the 8-g level it can be shown that secondary shock would reach the vessel wall about 2.4 times after the arrival of the primary shock. Because the <u>mean</u> wave velocities of both stable and galloping detonations are almost identical (as shown by Fig. 3) it follows that the flashing reinitiation, characteristic of galloping detonation, could not have been triggered by secondary blast wave effects.

The second possibility enumerated by Edwards et al. (1978) for transition to fast detonation is the formation of "bubbles" of gas which explode spontaneously, the exothermocity leading to the production of a blast wave that overtakes the lead shock and transforms the wave into a normal (CJ) detonation. This second possibility, involving spontaneous ignition is clearly in accord with Ul'yanitskii's (1981a,b) flashing mechanism.

In the present study, the experimental evidence of the sooted plates and all the velocity measuring techniques, always indicated gradual rather than sudden changes in decaying wave speeds. Consequently, the present work does not support either an alternative mode theory or secondary blast wave effects controlling the galloping wave reinitiation process characterized by flashing.

Modelling of Galloping Detonation

Though the present work establishes that spherically propagating galloping detonation waves were formed very reproducibly, the precise mechanisms controlling their formation are still imperfectly understood. Undoubtedly galloping cycles are reinitiated by localized explosions characterized by flashes, in the postshocked gas. It is difficult, for spherical waves, to reconcile the formation of a local explosion with pressure amplification of transverse waves in a decaying situation as observed by Moen et al. (1981) in tubes. Consequently, a true spontaneous reinitiation mechanism, controlled by chemical kinetics is favored. In principle, induction delay periods can now be computed for several fuel gas-oxidant systems using comprehensive oxidation schemes (e.g., Westbrook 1982; Atkinson 1980, 1983), but their application depends on assigning values of temperature and pressure appropriate to the conditions prevailing in a galloping detonation.

The theoretical model of Ul'yanitskii (1981a,b) while yielding excellent predictions for r_{*3} and E_{*3}, does not quantitatively explain the dynamic "catch up" and transient over-driving of the lead-shock, which stimulates the formation of a gradually widening cell structure as the gallop degrades. While the decreasing levels of shock temperature and pressure associated with the cellular decay are qualitatively in accord with kinetic considerations it is not possible, at present, to model even the simpler dynamic-kinetic processes occurring in regular detonation cell structures. Ultimately, a comprehensive model of galloping detonation would have to include mutually compatible wave dynamics and chemical kinetics.

Conclusions

1) Pulsating-galloping detonation in the spherical mode, having distinct similarity to galloping detonations in tubes, has been shown to exist in several hydrocarbon-oxidant-diluent systems. In the mixture examined in detail, galloping was restricted to narrow compositional limits in both fuel-lean and fuel-rich mixtures provided that sufficiently large initiation energy was available.

2) The radial length of each galloping cycle is related to detonation cell dimensions and measured values agree well with theoretical values deduced from the flash reinitiation model of Ul'yanitskii.

3) The flash mechanism of reinitiation is supplementary to the normal microexplosion cell regeneration process and is only manifested after an irregular cell structure broadens and fails. Flashing is controlled by a chemical induction period induced in the postshocked gas and ultimately ceases to reinitiate a galloping cycle when the induction period becomes too long.

4) The view that galloping detonation does not occur in all gas mixtures is given some support but not confirmed.

References

Adams, T. G. (1978) Do weak detonation waves exist? *AIAA J.* 16, 1035-1040.

Atkinson, R., Bull, D. C., and Shuff, P. (1980) Initiation of spherical detonation in hydrogen/air. *Combustion and Flame* 39, 287-300.

Atkinson, R., and Bull, D. C. (1983) Kinetic modeling of ethane/air detonability. *Shock Waves, Explosions, and Detonations: AIAA Progress in Astronautics and Aeronautics* (edited by J. R. Bowen, N. Manson, A. K. Oppenheim, and R. I. Soloukhin), Vol. 87, pp. 318-332. AIAA, New York.

Brode, H. L. (1959) Blast wave from a spherical charge. *Physics of Fluids* 2, 217-229.

Bull, D. C. (1979) Concentration limits to the initiation of unconfined detonation in fuel/air mixtures. *Trans. I. Chem. E.* 57, 219-227.

Bull, D. C., Elsworth, J. E., and Hooper, G. (1978) Initiation of spherical detonation in hydrocarbon/air mixtures. *Acta Astronautica* 5, 997-1008.

Bull, D. C., Elsworth, J. E., and Hooper, G. (1979) Concentration limits to unconfined detonation of ethane-air. *Combustion and Flame* 35, 27-40.

Bull, D. C., Elsworth, J. E., Shuff, P. J., and Metcalfe, E. (1982) Detonation cell structures in fuel/air mixtures. *Combustion and Flame* 45, 7-22.

Campbell, C., and Woodhead, D. W. (1927) Striated photographic records of explosion-waves. *J. Chem. Soc.* 130, 1572-1578.

Desbordes, D., Manson, N., and Brossard, J. (1981) Pressure evolution behind spherical and hemispherical detonations in gases. *Gasdynamics of Detonations and Explosions: AIAA Progress in Astronautics and Aeronautics* (edited by J. R. Bowen, N. Manson, A. K. Oppenheim, and R. I. Soloukhin), Vol. 75, pp. 150-165. AIAA, New York.

Duff, R. E. (1963) Contribution to discussion. *Ninth Symposium (International) on Combustion*, p. 469. The Combustion Institute, Pittsburgh, Pa.

Edwards, D. H., Hooper, G., and Morgan, J. M. (1974) A study of unstable detonations using a microwave interferometer. *J. Phys. D: Appl. Phys.* 7, 242-247.

Edwards, D. H., Hooper, G., Morgan, J. M., and Thomas, G. O. (1978) The quasi-steady regime in critically initiated detonation waves. *J. Phys. D: Appl. Phys.* 11, 2103-2117.

Eisen, C. L., Gross, R. A., and Rivlin, T. J. (1960) Theoretical calculations in gaseous detonations. Combustion and Flame 4, 137-147.

Elsworth, J. E., and Eyre, J. A. (1984) The susceptibility of propene-propane/air mixtures to detonation. Combustion and Flame 55, 237-243.

Fay, J. A. (1952) A mechanical theory of spinning detonation. J. Chem. Phys. 20, 942-950.

Manson, M., Brochet, C., Brossard, J., and Pujal, Y. (1963) Vibratory phenomena and instability of self-sustained detonation in gases. Ninth Symposium (International) on Combustion, pp. 461-468. The Combustion Institute, Pittsburgh, Pa.

Moen, I. O., Donato, M., Knystautas, R., and Lee, J. H. (1981) The infleunce of confinement on the propagation of detonations near the detonability limits. Eighteenth Symposium (International) on Combustion, pp. 1615-1622. The Combustion Institute, Pittsburgh, Pa.

Mooradian, A. J., and Gordon, W. E. (1951) Gaseous detonation. I. Initiation of detonation. J. Chem. Phys. 19, 1166-1172.

Saint-Cloud, J. P., Guerraud, C. L., Brochet, C., and Manson, N. (1972) Quelques particularités des detonations très instables dans les mélanges gazeux. Acta Astronautica 17, 487-98.

Strehlow, R. A. (1963) Gas phase detonations: Recent developments. Combustion and Flame 12, 81-101.

Ubbelohde, A. R., and Munday, G. (1969) Some current problems in the marginal detonation of gases. Twelfth Symposium (International) on Combustion, pp. 809-818. The Combustion Institute, Pittsburgh, Pa.

Ul'yanitskii, V. Yu. (1981a) Galloping mode in a gas detonation. Translated from Fizika Goreniya i Vzryva 17, 118-124.

Ul'yanitskii, V. Yu. (1981b) Role of "flashing" and transverse-wave collisions in the evolution of a multi-front detonation wave structure in gases. Translated from Fizika Goreniya i Vzryva 17, 127-133.

Vasiliev, A. A., and Nikolaev, Yu. (1978) Closed theoretical model of a detonation cell. Acta Astronautica 5, 983-996.

Westbrook, C. K. (1982) Chemical kinetics of hydrocarbon oxidation in gaseous detonations. Combustion and Flame 46, 191-210.

Chemical Kinetics of Propane Oxidation in Gaseous Detonations

C. K. Westbrook,* W. J. Pitz,† and P. A. Urtiew‡
Lawrence Livermore National Laboratory,
University of California, Livermore, California

Abstract

A theoretical model including a detailed chemical kinetic reaction mechanism for hydrocarbon oxidation is used to examine detonation properties for mixtures of propane-air, propane-oxygen, and propane-oxygen diluted with varying amounts of nitrogen. Computed induction lengths are compared with available experimental data for critical energy and critical tube diameter for initiation of unconfined spherical detonation, as well as detonation limits in linear tubes.

Introduction

Gaseous detonation waves have been studied experimentally for many years. Detonation limits, propagation rates, and other properties have been examined for many fuel-oxidizer mixtures. Theoretical studies of detonations have also appeared, showing how hydrodynamic and kinetic processes interact. Recent modeling studies of chemical kinetics of hydrogen (Strehlow and Rubins 1969; Dove and Tribbeck 1970; Tsuge et al. 1970; Atkinson et al. 1980; Westbrook 1982a) and hydrocarbon (Westbrook 1982b, 1982c; Westbrook and Urtiew 1983a, 1983b) oxidation in detonation waves have significantly improved the understanding of the physical and chemical processes which take

---

Presented at the 9th ICODERS, Poitiers, France, July 3-8, 1983. Copyright © American Institute of Aeronautics and Astronautics, Inc., 1984. All rights reserved.
*Senior Physicist, Theoretical Physics Division.
†Physicist, Theoretical Physics Division.
‡Engineer, Chemistry & Materials Science Department.

place. Continued development of detailed chemical kinetic reaction mechanisms (Westbrook and Pitz 1983) has made it possible to extend this type of analysis to consider propane oxidation in detonations.

Propane is an important hydrocarbon fuel used in a wide variety of practical applications. It is also a convenient test fuel for laboratory studies of experiments ranging from flat flame burners to internal combustion engines, and has been used extensively in the field of detonations.

In this paper, a detailed kinetic mechanism is used to compute kinetic ignition delay times for propane oxidation. These calculations are combined with theoretical Chapman-Jouguet (CJ) conditions and then used to examine properties of gaseous detonations. Numerical predictions of critical detonation parameters are compared with experimental data for lean and rich limits of detonability, critical tube diameter for initiation of unconfined spherical detonation by means of a planar detonation in a linear tube, and critical energy for initiation of unconfined spherical detonation by high explosive charges and sparks. Effects of variations in fuel-oxidizer equivalence ratio, initial pressure, and dilution by nitrogen are assessed.

Chemical Kinetics

The reaction mechanism used in these calculations has been developed and validated in a series of studies (Westbrook et al. 1977, 1983; Westbrook and Dryer 1979; Westbrook and Pitz 1983). The elementary reactions and their rate expressions are summarized in Table 1. References for the individual rates can be found elsewhere (Westbrook and Pitz 1983). Reverse reaction rates are computed from the forward rates and the appropriate thermodynamic data (JANNAF 1970; Bahn 1973). This mechanism has been shown to describe the oxidation of propane, hydrogen, carbon monoxide, methane, methanol, ethane, ethylene, and acetylene over wide ranges of experimental conditions. In particular, experimental data from shock tube studies of propane oxidation (Burcat et al. 1971a, 1971b) are reproduced well by this mechanism (Westbrook and Pitz 1983). Since the physical conditions encountered during the ignition delay period in detonation waves are closely related to those in shock tube environments, this agreement provides some confidence in the adequacy of the reaction mechanism for the present study.

Table 1 Fuel oxidation mechanism. Reaction rates in cm^3-mole-s-kcal units, $k=AT^n\exp(-E_a/RT)$.

Reaction	Forward rate			Reverse rate		
	log A	n	E_a	log A	n	E_a
1. $H+O_2 \rightarrow O+OH$	14.27	0	16.79	13.17	0	0.68
2. $H_2+O \rightarrow H+OH$	10.26	1	8.90	9.92	1	6.95
3. $H_2O+O \rightarrow OH+OH$	13.53	0	18.35	12.50	0	1.10
4. $H_2O+H \rightarrow H_2+OH$	13.98	0	20.30	13.34	0	5.15
5. $H_2O_2+OH \rightarrow H_2O+HO_2$	13.00	0	1.80	13.45	0	32.79
6. $H_2O+M \rightarrow H+OH+M$	16.34	0	105.00	23.15	-2	0.00
7. $H+O_2+M \rightarrow HO_2+M$	15.22	0	-1.00	15.36	0	45.90
8. $HO_2+O \rightarrow OH+O_2$	13.70	0	1.00	13.81	0	56.61
9. $HO_2+H \rightarrow OH+OH$	14.40	0	1.90	13.08	0	40.10
10. $HO_2+H \rightarrow H_2+O_2$	13.40	0	0.70	13.74	0	57.80
11. $HO_2+OH \rightarrow H_2O+O_2$	13.70	0	1.00	14.80	0	73.86
12. $H_2O_2+O_2 \rightarrow HO_2+HO_2$	13.60	0	42.64	13.00	0	1.00
13. $H_2O_2+M \rightarrow OH+OH+M$	17.08	0	45.50	14.96	0	-5.07
14. $H_2O_2+H \rightarrow HO_2+H_2$	12.23	0	3.75	11.86	0	18.70
15. $O+H+M \rightarrow OH+M$	16.00	0	0.00	19.90	-1	103.72
16. $O_2+M \rightarrow O+O+M$	15.71	0	115.00	15.67	-0.28	0.00
17. $H_2+M \rightarrow H+H+M$	14.34	0	96.00	15.48	0	0.00
18. $CO+OH \rightarrow CO_2+H$	7.11	1.3	-0.77	9.15	1.3	21.58
19. $CO+HO_2 \rightarrow CO_2+OH$	14.18	0	23.65	15.23	0	85.50
20. $CO+O+M \rightarrow CO_2+M$	15.77	0	4.10	21.74	-1	131.78
21. $CO_2+O \rightarrow CO+O_2$	12.44	0	43.83	11.50	0	37.60
22. $HCO+OH \rightarrow CO+H_2O$	14.00	0	0.00	15.45	0	105.15
23. $HCO+M \rightarrow H+CO+M$	14.16	0	19.00	11.70	1	1.55
24. $HCO+H \rightarrow CO+H_2$	14.30	0	0.00	15.12	0	90.00
25. $HCO+O \rightarrow CO+OH$	14.00	0	0.00	14.46	0	87.90
26. $HCO+HO_2 \rightarrow CH_2O+O_2$	14.00	0	3.00	15.56	0	46.04
27. $HCO+O_2 \rightarrow CO+HO_2$	12.60	0	7.00	12.95	0	39.29
28. $CH_2O+M \rightarrow HCO+H+M$	16.52	0	81.00	11.15	1	-11.77
29. $CH_2O+OH \rightarrow HCO+H_2O$	12.88	0	0.17	12.41	0	29.99
30. $CH_2O+H \rightarrow HCO+H_2$	14.52	0	10.50	13.42	0	25.17
31. $CH_2O+O \rightarrow HCO+OH$	13.70	0	4.60	12.24	0	17.17
32. $CH_2O+HO_2 \rightarrow HCO+H_2O_2$	12.00	0	8.00	11.04	0	6.59

(Table continued on next page)

Table 1 (continued) Fuel oxidation mechanism. Reaction rates in cm^3-mole-s-kcal units, $k=AT^n \exp(-E_a/RT)$.

Reaction	Forward rate			Reverse rate		
	log A	n	E_a	log A	n	E_a
33. $CH_4+M \rightarrow CH_3+H+M$	17.15	0	88.40	11.45	1	-19.52
34. $CH_4+H \rightarrow CH_3+H_2$	14.10	0	11.90	12.68	0	11.43
35. $CH_4+OH \rightarrow CH_3+H_2O$	3.54	3.08	2.00	2.76	3.08	16.68
36. $CH_4+O \rightarrow CH_3+OH$	13.20	0	9.20	11.43	0	6.64
37. $CH_4+HO_2 \rightarrow CH_3+H_2O_2$	13.30	0	18.00	12.02	0	1.45
38. $CH_3+HO_2 \rightarrow CH_3O+OH$	13.51	0	0.00	10.00	0	0.00
39. $CH_3+OH \rightarrow CH_2O+H_2$	12.60	0	0.00	14.08	0	71.73
40. $CH_3+O \rightarrow CH_2O+H$	14.11	0	2.00	15.23	0	71.63
41. $CH_3+O_2 \rightarrow CH_3O+O$	13.68	0	29.00	14.48	0	0.73
42. $CH_2O+CH_3 \rightarrow CH_4+HCO$	10.00	0.5	6.00	10.32	0.5	21.14
43. $CH_3+HCO \rightarrow CH_4+CO$	11.48	0.5	0.00	13.71	0.5	90.47
44. $CH_3+HO_2 \rightarrow CH_4+O_2$	12.00	0	0.40	13.88	0	58.59
45. $CH_3O+M \rightarrow CH_2O+H+M$	13.70	0	21.00	9.00	1	-2.56
46. $CH_3O+O_2 \rightarrow CH_2O+HO_2$	12.00	0	6.00	11.11	0	32.17
47. $C_2H_6 \rightarrow CH_3+CH_3$	19.35	-1	88.31	12.95	0	0.00
48. $C_2H_6+CH_3 \rightarrow C_2H_5+CH_4$	-0.26	4	8.28	10.48	0	12.50
49. $C_2H_6+H \rightarrow C_2H_5+H_2$	2.73	3.5	5.20	2.99	3.5	27.32
50. $C_2H_6+OH \rightarrow C_2H_5+H_2O$	9.94	1.05	1.81	10.23	1.05	20.94
51. $C_2H_6+O \rightarrow C_2H_5+OH$	13.40	0	6.36	12.66	0	11.23
52. $C_2H_5+M \rightarrow C_2H_4+H+M$	15.30	0	30.00	10.62	0	-11.03
53. $C_2H_5+O_2 \rightarrow C_2H_4+HO_2$	12.00	0	5.00	11.12	0	13.70
54. $C_2H_4+C_2H_4 \rightarrow C_2H_5+C_2H_3$	14.70	0	64.70	14.17	0	-2.61
55. $C_2H_4+M \rightarrow C_2H_2+H_2$	16.97	0	77.20	12.66	1	36.52
56. $C_2H_4+M \rightarrow C_2H_3+H+M$	18.80	0	108.72	17.30	0	0.00
57. $C_2H_4+O \rightarrow CH_3+HCO$	12.52	0	1.13	11.20	0	31.18
58. $C_2H_4+O \rightarrow CH_2O+CH_2$	13.40	0	5.00	12.48	0	15.68
59. $C_2H_4+H \rightarrow C_2H_3+H_2$	7.18	2	6.00	6.24	2	5.11
60. $C_2H_4+OH \rightarrow C_2H_3+H_2O$	12.68	0	1.23	12.08	0	14.00
61. $C_2H_4+OH \rightarrow CH_3+CH_2O$	12.30	0	0.96	11.78	0	16.48
62. $C_2H_3+M \rightarrow C_2H_2+H+M$	14.90	0	31.50	11.09	1	-10.36
63. $C_2H_3+O_2 \rightarrow C_2H_2+HO_2$	12.00	0	10.00	12.00	0	17.87
64. $C_2H_2+M \rightarrow C_2H+H+M$	14.00	0	114.00	9.04	1	0.77
65. $C_2H_2+O_2 \rightarrow HCO+HCO$	12.60	0	28.00	11.00	0	63.65

(Table continued on next page)

Table 1 (continued) Fuel oxidation mechanism. Reaction rates in cm^3-mole-s-kcal units, $k = AT^n \exp(-E_a/RT)$.

	Reaction	Forward rate			Reverse rate		
		log A	n	E_a	log A	n	E_a
66.	$C_2H_2 + H \rightarrow C_2H + H_2$	14.30	0	19.00	13.62	0	13.21
67.	$C_2H_2 + OH \rightarrow C_2H + H_2O$	12.78	0	7.00	12.73	0	16.36
68.	$C_2H_2 + OH \rightarrow CH_2CO + H$	11.51	0	0.20	12.50	0	20.87
69.	$C_2H_2 + O \rightarrow C_2H + OH$	15.51	-0.6	17.00	14.47	-0.6	0.91
70.	$C_2H_2 + O \rightarrow CH_2 + CO$	13.83	0	4.00	13.10	0	54.67
71.	$C_2H + O_2 \rightarrow HCO + CO$	13.00	0	7.00	12.93	0	138.40
72.	$C_2H + O \rightarrow CO + CH$	13.70	0	0.00	13.50	0	59.43
73.	$CH_2 + O_2 \rightarrow HCO + OH$	14.00	0	3.70	13.61	0	76.58
74.	$CH_2 + O \rightarrow CH + OH$	11.28	0.68	25.00	10.77	0.68	25.93
75.	$CH_2 + H \rightarrow CH + H_2$	11.43	0.67	25.70	11.28	0.67	28.72
76.	$CH_2 + OH \rightarrow CH + H_2O$	11.43	0.67	25.70	11.91	0.67	43.88
77.	$CH + O_2 \rightarrow CO + OH$	11.13	0.67	25.70	11.71	0.67	185.60
78.	$CH + O_2 \rightarrow HCO + O$	13.00	0	0.00	13.13	0	71.95
79.	$CH_3OH + M \rightarrow CH_3 + OH + M$	18.48	0	80.00	13.16	1	-10.98
80.	$CH_3OH + OH \rightarrow CH_2OH + H_2O$	12.60	0	2.00	7.27	1.66	25.31
81.	$CH_3OH + O \rightarrow CH_2OH + OH$	12.23	0	2.29	5.90	1.66	8.35
82.	$CH_3OH + H \rightarrow CH_2OH + H_2$	13.48	0	7.00	7.51	1.66	15.16
83.	$CH_3OH + H \rightarrow CH_3 + H_2O$	12.72	0	5.34	12.32	0	36.95
84.	$CH_3OH + CH_3 \rightarrow CH_2OH + CH_4$	11.26	0	9.80	6.70	1.66	18.43
85.	$CH_3OH + HO_2 \rightarrow CH_2OH + H_2O_2$	12.80	0	19.36	7.00	1.66	11.44
86.	$CH_2OH + M \rightarrow CH_2O + H + M$	13.40	0	29.00	16.69	-0.66	7.58
87.	$CH_2OH + O_2 \rightarrow CH_2O + HO_2$	12.00	0	6.00	17.94	-1.66	28.32
88.	$C_2H_3 + C_2H_4 \rightarrow C_4H_6 + H$	12.00	0	7.30	13.00	0	4.70
89.	$C_2H_2 + C_2H_2 \rightarrow C_4H_3 + H$	13.00	0	45.00	13.18	0	0.00
90.	$C_4H_3 + M \rightarrow C_4H_2 + H + M$	16.00	0	60.00	11.92	1	2.54
91.	$C_2H_2 + C_2H \rightarrow C_4H_2 + H$	13.60	0	0.00	14.65	0	0.55
92.	$C_4H_2 + M \rightarrow C_4H + H + M$	17.54	0	80.00	12.30	1.0	-16.40
93.	$C_2H_3 + H \rightarrow C_2H_2 + H_2$	13.30	0	2.50	13.12	0	68.08
94.	$C_3H_8 \rightarrow CH_3 + C_2H_5$	16.23	0	84.84	10.18	1	-0.32
95.	$CH_3 + C_3H_8 \rightarrow CH_4 + iC_3H_7$	15.04	0	25.14	15.64	0	32.12
96.	$CH_3 + C_3H_8 \rightarrow CH_4 + nC_3H_7$	15.04	0	25.14	15.64	0	32.12
97.	$H + C_3H_8 \rightarrow H_2 + iC_3H_7$	6.94	2	5.00	12.89	0	15.87
98.	$H + C_3H_8 \rightarrow H_2 + nC_3H_7$	7.75	2	7.70	12.96	0	14.46

(Table continued on next page)

Table 1 (continued) Fuel oxidation mechanism. Reaction rates in cm^3-mole-s-kcal units, $k=AT^n \exp(-E_a/RT)$.

	Forward rate			Reverse rate		
Reaction	log A	n	E_a	log A	n	E_a
99. $iC_3H_7 \rightarrow H+C_3H_6$	13.80	0	36.90	13.00	0	1.50
100. $iC_3H_7 \rightarrow CH_3+C_2H_4$	10.30	0	29.50	4.66	1	4.29
101. $nC_3H_7 \rightarrow CH_3+C_2H_4$	13.98	0	31.00	8.34	1	5.79
102. $nC_3H_7 \rightarrow H+C_3H_6$	14.10	0	37.00	13.00	0	1.50
103. $iC_3H_7+C_3H_8 \rightarrow nC_3H_7+C_3H_8$	10.48	0	12.90	10.48	0	12.90
104. $C_2H_3+C_3H_8 \rightarrow C_2H_4+iC_3H_7$	11.00	0	10.40	11.12	0	17.80
105. $C_2H_3+C_3H_8 \rightarrow C_2H_4+nC_3H_7$	11.00	0	10.40	11.12	0	17.80
106. $C_2H_5+C_3H_8 \rightarrow C_2H_6+iC_3H_7$	11.00	0	10.40	10.56	0	9.93
107. $C_2H_5+C_3H_8 \rightarrow C_2H_6+nC_3H_7$	11.00	0	10.40	10.56	0	9.93
108. $C_3H_8+O \rightarrow iC_3H_7+OH$	6.70	2	3.00	5.52	2	7.41
109. $C_3H_8+O \rightarrow nC_3H_7+OH$	6.70	2	3.00	5.52	2	7.41
110. $C_3H_8+OH \rightarrow iC_3H_7+H_2O$	8.68	1.4	0.85	8.93	1.25	22.37
111. $C_3H_8+OH \rightarrow nC_3H_7+H_2O$	8.76	1.4	0.85	9.01	1.25	22.37
112. $C_3H_8+HO_2 \rightarrow iC_3H_7+H_2O_2$	12.70	0	18.00	12.01	0	8.43
113. $C_3H_8+HO_2 \rightarrow iC_3H_7+H_2O_2$	12.70	0	18.00	12.01	0	8.43
114. $C_3H_6+O \rightarrow C_2H_4+CH_2O$	13.77	0	5.00	13.76	0	86.67
115. $iC_3H_7+O_2 \rightarrow C_3H_6+HO_2$	12.00	0	5.00	11.30	0	17.48
116. $nC_3H_7+O_2 \rightarrow C_3H_6+HO_2$	12.00	0	5.00	11.30	0	17.48
117. $C_3H_8+O_2 \rightarrow iC_3H_7+HO_2$	13.60	0	47.50	12.31	0	0.00
118. $C_3H_8+O_2 \rightarrow nC_3H_7+HO_2$	13.60	0	47.50	12.31	0	0.00
119. $C_3H_6+OH \rightarrow C_2H_5+CH_2O$	12.90	0	0.00	13.66	0	17.35
120. $C_3H_6+O \rightarrow C_2H_5+HCO$	12.55	0	0.00	11.85	0	29.92
121. $C_3H_6+OH \rightarrow CH_3+CH_3CHO$	11.54	0	0.00	11.44	0	20.40
122. $C_3H_6+O \rightarrow CH_3+CH_3CO$	13.07	0	0.60	12.25	0	38.37
123. $CH_3CHO+H \rightarrow CH_3CO+H_2$	13.60	0	4.20	13.25	0	23.67
124. $CH_3CHO+OH \rightarrow CH_3CO+H_2O$	13.00	0	0.00	13.28	0	36.62
125. $CH_3CHO+O \rightarrow CH_3CO+OH$	12.70	0	1.79	12.00	0	19.16
126. $CH_3CHO+CH_3 \rightarrow CH_3CO+CH_4$	12.23	0	8.43	13.48	0	28.00
127. $CH_3CHO+HO_2 \rightarrow CH_3CO+H_2O_2$	12.23	0	10.70	12.00	0	14.10
128. $CH_3CHO \rightarrow CH_3+HCO$	15.85	0	81.78	9.58	1	0.00
129. $CH_3CHO+O_2 \rightarrow CH_3CO+HO_2$	13.30	0.5	42.20	7.00	0.5	4.00
130. $CH_3CO \rightarrow CH_3+CO$	13.48	0	17.24	11.20	0	5.97
131. $C_3H_6+H \rightarrow C_3H_5+H_2$	12.70	0	1.50	12.18	0	17.70

(Table continued on next page)

Table 1 (continued) Fuel oxidation mechanism. Reaction rates in cm^3-mole-s-kcal units, $k = AT^n \exp(-E_a/RT)$.

Reaction	Forward rate			Reverse rate		
	log A	n	E_a	log A	n	E_a
132. $C_3H_6 + CH_3 \rightarrow C_3H_5 + CH_4$	10.95	0	8.50	11.87	0	25.18
133. $C_3H_6 + C_2H_5 \rightarrow C_3H_5 + C_2H_6$	11.00	0	9.20	5.00	0	56.77
134. $C_3H_6 + OH \rightarrow C_3H_5 + H_2O$	12.60	0	0.00	7.18	0	69.69
135. $C_3H_8 + C_3H_5 \rightarrow iC_3H_7 + C_3H_6$	11.60	0	16.20	11.30	0	6.50
136. $C_3H_8 + C_3H_5 \rightarrow iC_3H_7 + C_3H_6$	11.60	0	16.20	11.30	0	6.50
137. $C_3H_5 \rightarrow C_3H_4 + H$	13.60	0	70.00	8.00	1	0.00
138. $C_3H_5 + O_2 \rightarrow C_3H_4 + HO_2$	11.78	0	10.00	11.08	0	10.00
139. $1C_4H_8 \rightarrow C_3H_5 + CH_3$	19.18	-1	73.40	13.13	0	0.00
140. $1C_4H_8 \rightarrow C_2H_3 + C_2H_5$	19.00	-1	96.77	12.95	0	0.00
141. $1C_4H_8 + O \rightarrow CH_3CHO + C_2H_4$	13.11	0	0.85	12.32	0	85.10
142. $1C_4H_8 + O \rightarrow CH_3CO + C_2H_5$	13.11	0	0.85	12.37	0	38.15
143. $1C_4H_8 + OH \rightarrow CH_3CHO + C_2H_5$	13.00	0	0.00	12.97	0	19.93
144. $1C_4H_8 + OH \rightarrow CH_3CO + C_2H_6$	13.00	0	0.00	12.99	0	32.43
145. $C_3H_4 + O \rightarrow CH_2O + C_2H_2$	12.00	0	0.00	12.03	0	81.73
146. $C_3H_4 + O \rightarrow HCO + C_2H_3$	12.00	0	0.00	10.47	0	30.82
147. $C_3H_4 + OH \rightarrow CH_2O + C_2H_3$	12.00	0	0.00	11.93	0	18.25
148. $C_3H_4 + OH \rightarrow HCO + C_2H_4$	12.00	0	0.00	11.77	0	33.81
149. $C_3H_6 \rightarrow C_3H_5 + H$	13.00	0	78.00	11.00	0	0.00
150. $C_2H_2 + O \rightarrow HCCO + H$	4.55	2.7	1.39	2.70	2.7	12.79
151. $CH_2CO + H \rightarrow CH_3 + CO$	13.04	0	3.40	12.38	0	40.20
152. $CH_2CO + O \rightarrow HCO + HCO$	13.00	0	2.40	11.54	0	33.50
153. $CH_2CO + OH \rightarrow CH_2O + HCO$	13.45	0	0.00	13.44	0	18.50
154. $CH_2CO + M \rightarrow CH_2 + CO + M$	16.30	0	60.00	10.66	0	0.00
155. $CH_2CO + O \rightarrow HCCO + OH$	13.70	0	8.00	10.86	0	8.00
156. $CH_2CO + OH \rightarrow HCCO + H_2O$	12.88	0	3.00	11.03	0	11.00
157. $CH_2CO + H \rightarrow HCCO + H_2$	13.88	0	8.00	11.39	0	8.00
158. $HCCO + OH \rightarrow HCO + HCO$	13.00	0	0.00	13.68	0	40.36
159. $HCCO + H \rightarrow CH_2 + CO$	13.70	0	0.00	13.82	0	39.26
160. $HCCO + O \rightarrow HCO + CO$	13.53	0	2.00	13.92	0	128.26
161. $C_3H_6 \rightarrow C_2H_3 + CH_3$	15.80	0	85.80	10.00	1	0.00
162. $C_3H_5 + H \rightarrow C_3H_4 + H_2$	13.00	0	0.00	13.00	0	40.00
163. $C_3H_5 + CH_3 \rightarrow C_3H_4 + CH_4$	12.00	0	0.00	13.00	0	40.00

The general features of propane consumption in this mechanism are straightforward. Initiation takes place primarily by means of Reaction 94

$$C_3H_8 = C_2H_5 + CH_3 \qquad (94)$$

The radical pool is rapidly established through the decomposition of ethyl radicals

$$C_2H_5 + M = C_2H_4 + H + M \qquad (52)$$

and by H atom abstraction reactions, including

$$C_3H_8 + H = C_3H_7 + H + M \qquad (97,98)$$

$$C_3H_8 + CH_3 = C_3H_7 + CH_4 \qquad (95,96)$$

$$C_3H_8 + C_2H_5 = C_3H_7 + C_2H_6 \qquad (106,107)$$

Reactions of H atoms and hydrocarbon radicals with O_2 molecules lead to the production of O atoms and OH radicals, which then react with the fuel

$$C_3H_8 + O = C_3H_7 + OH \qquad (108,109)$$

$$C_3H_8 + OH = C_3H_7 + H_2O \qquad (110,111)$$

A central problem in describing propane oxidation and pyrolysis is the determination of the relative rates at which the two types of propyl radicals are produced in Reactions 95-98 and 103-113. This issue is important because n-propyl radicals rapidly decompose to methyl radicals and ethylene

$$nC_3H_7 = CH_3 + C_2H_4 \qquad (101)$$

while the majority of the isopropyl radicals produce H atoms and propylene

$$iC_3H_7 = H + C_3H_6 \qquad (99)$$

Generally speaking Reaction 101 tends to retard the overall process of fuel consumption, while Reaction 99 tends to accelerate the process. These trends occur because both methyl radicals and ethylene react quite slowly and do not substantially increase the total radical population. Since a significant percentage of the methyl radicals in fact recombine (Westbrook et al. 1977; Warnatz 1981) to produce stable species such as ethane, methane, and methanol, this route actually reduces the radical levels. In contrast, H atoms produced by Reaction 99 lead to radical multiplication through the major chain branching Reaction 1

$$H + O_2 = O + OH \qquad (1)$$

This increases the radical pool and accelerates the overall oxidation rate.

Subsequent reactions of the C_2 and C_1 species have been described at length in the past and the principal reaction sequences and rates are fairly well understood. The reactions of the C_3H_6 and its products are less well established. In the present mechanism, some of the propylene reacts with O and OH through addition reactions, producing acetaldehyde and ethylene, while reactions with H and CH_3 can lead to C_3H_5. Some of these allyl radicals then react to form C_3H_4 (this mechanism does not distinguish between the isomeric forms allene and propyne), while another fraction recombines with methyl radicals to produce butene (C_4H_8). Subsequent reaction paths for C_4H_8 and C_3H_4 are somewhat speculative, and only estimated rate expressions are currently available.

Although there are a number of areas in the mechanism where substantial information is still missing, the modeling work to date has indicated that the most important steps are now quite well known. For the computations reported in this study, the initiation Reaction 94, the H atom abstraction Reactions 95-98 and 103-113, and the two propyl radical decomposition Reactions 99 and 101 are most significant and their rates are reasonably well established. As observed above, the mechanism has been shown to produce good agreement with experimental shock tube ignition data.

Detonation Model

The general approach used is the Zel'dovich-von Neumann-Doring (ZND) model in which, locally, a detonation consists of a shock wave traveling at the CJ velocity, followed by a reaction zone. The shock wave compresses and heats the fuel-oxidizer mixture which then begins to react. In most mixtures, the overall fuel oxidation consists of a relatively long induction period during which the temperature and pressure of the gas mixture remain nearly constant, followed by a rapid release of chemical energy and temperature increase.

For each C_3H_8-oxidizer mixture selected, a calculation is first made of the relevant CJ conditions, as functions of the initial pressure and temperature of the unreacted, unshocked mixture. From the resulting value of the detonation velocity D_{CJ} the conditions in the von Neumann spike, including the temperature T_1, pressure P_1, and particle velocity v_1 of the postshock, unreacted gases can be calculated from the shock conservation relations and are then used as initial conditions for the chemical kinetics model. The shock velocity within a single detonation cell varies substantially, from an initial value of about 1.6 D_{CJ} to a minimum of about 0.6 D_{CJ}, so the CJ conditions used here represent average values, and the computed ignition delay times will also be averages.

For the kinetics calculations, the reactive mixture is assumed to remain at a constant volume over its reaction time, and the ignition delay time is defined in terms of its temperature history. Most of the mixtures examined underwent a large temperature increase of 1200-2000 K, and the ignition delay time was defined as the time of maximum rate of temperature increase. This coincides roughly with the time at which the temperature has completed about half of its total increase. This is not, strictly speaking, a true induction period, often defined as the time required for a small (i.e., 1-5%) temperature or pressure increase, but here the release of substantial amounts of energy is of primary interest. In addition to the induction time τ, it is useful to define the induction length $\Delta \equiv \tau(D_{CJ} - v_1)$.

As a result of these simplifications, the computed induction times and induction lengths should be considered as characteristic time and length scales, not as a precise description of the state of a gas element through the detonation front. The evolution of the reacted gas subsequent to the induction period considered here is dominated by the fluid mechanics of the postreaction expansion of the reaction products. This expansion reduces both the pressure and density of these products. Since virtually all of the reactants have been consumed by this time, the kinetics of this phase are controlled by relatively slow radical recombination processes.

This model of the detonation is simplified and neglects several potentially significant effects arising from hydrodynamic-kinetic interactions. Variations of density, temperature, and particle velocity in the postshock unreacted mixture are not considered, changes which can affect the induction period in a number of ways. Some of these variations have been included in a recent study of H_2 oxidation in detonation waves (Lee et al. 1982) but the computed results are not qualitatively different from these obtained (Westbrook 1982a) using the present, simpler approach. Multiple shock wave reflections, rarefactions, interactions with confining walls, cellular structure, and other related effects are also not treated directly by the present model. A truly complete detonation model, incorporating these effects and including two or three spatial dimensions, is well beyond the scope of the present study.

Results

An extensive series of induction calculations was carried out for propane oxidation. The effects of variation in fuel-oxidizer equivalence ratio were examined for propane-air and propane-O_2 mixtures, initially at a temperature of 300 K and atmospheric pressure. Also for mixtures initially at 300 K and atmospheric pressure, the influence of dilution by N_2 was studied by varying the initial ratio of N_2/O_2 from zero (i.e., C_3H_8 - O_2) to 3.76 (C_3H_8 - air). In this dilution analysis, both conventional stoichiometric (initial ratio of C_3H_8 to O_2 of 1:5) and "CO stoichiometric" (C_3H_8/O_2 equal to 1:3.5) mixtures were included. Finally, the influence of initial pressure on induction properties was examined for stoichiometric C_3H_8 - O_2 and C_3H_8 - air mixtures initially at 300 K. For all these conditions, values of the induction time τ and induction length Δ were calculated as described above. In this section computed values of Δ are related to experimental detonation parameters.

Computed induction lengths for C_3H_8 - O_2 and C_3H_8 - air mixtures initially at 300 K and atmospheric pressure are given in Fig. 1, showing the effects of variations in equivalence ratio ϕ. Also shown for comparison are curves computed previously (Westbrook 1982b) for ethane and ethylene oxidation. For both fuel-O_2 and fuel-air mixtures, the analogous curves (not shown) for CH_4 lie well above the curves in Fig. 1, and

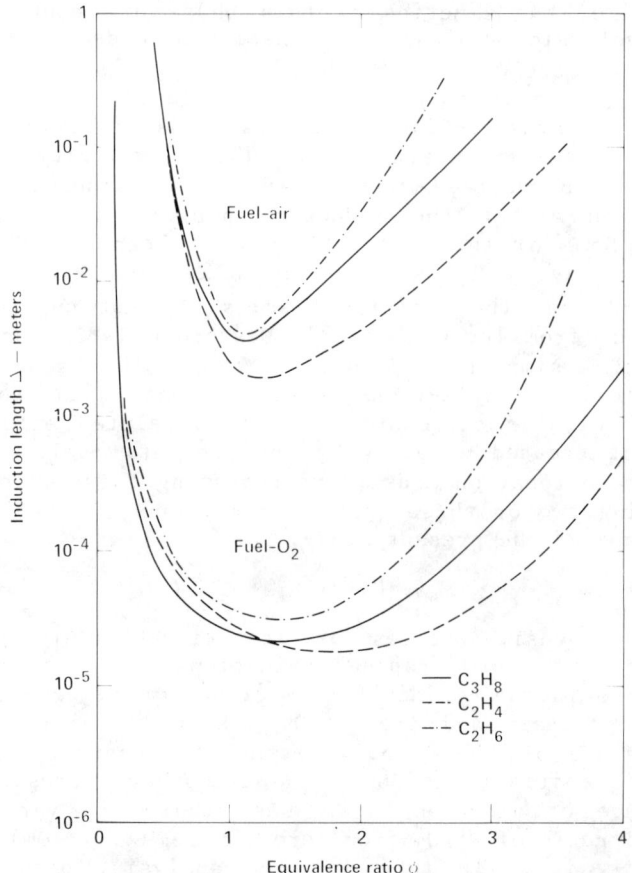

Fig. 1 Computed induction length for fuel-air and fuel-oxygen mixtures, initially at atmospheric pressure and 300 K.

those for C_2H_2 well below those in Fig. 1. For all the hydrocarbon fuels examined, including propane, these curves are roughly U-shaped, with their minimum values lying slightly on the rich side of stoichiometric. The induction lengths for fuel-O_2 are about two orders of magnitude smaller than for the fuel-air mixtures. The curves rise sharply at the lean extremes but rise much more gradually for rich mixtures.

For propagation of detonations in linear tubes, the limits of detonability depend on the diameter of the tube. A lean limit of $\phi_L = 0.128$ (2.5% C_3H_8) was observed (Michels et al. 1970; Ubbelohde and Munday 1969) for propagation of detonations in $C_3H_8 - O_2$ mixtures in a 2.54-cm diam linear tube, consistent with the computed values in Fig. 1. For propane-air, Borisov and Loban (1977) found $\phi_L \simeq 0.64$ in a 70-mm-diam tube, and a rich limit $\phi_R \simeq 1.89$. Using the data from Fig. 1, the ratio of tube diameter d to induction length Δ for propane-air is approximately 2.8 at the lean limit and 4.6 at the rich limit. If it is assumed that at the limit, single-spin detonation occurs with a transverse wave spacing $\lambda \simeq 2d$, then the ratio λ/Δ is approximately 5.6 and 9.2 at the two limits for C_3H_8-air. Analogous ratios for CH_4, C_2H_6, and C_2H_4 at the rich limit are 6, 2, and 14, respectively (Westbrook 1982b), so the present results for C_3H_8 are generally consistent with those for the other hydrocarbon-air mixtures. For C_3H_8 - O_2 in a 2.54-cm-diam tube (Michels et al. 1970; Ubbelohde and Munday 1969), $\phi_R \simeq 3.7$. Again using Fig. 1, the ratio λ/Δ for this mixture has a value of approximately 50. Earlier values (Westbrook 1982b) of 65, 70, and 30 for CH_4, C_2H_6, and C_2H_4, respectively, are again consistent with the present values for C_3H_8 - O_2. It appears to be a general feature of hydrocarbon-fueled detonations that the ratio of cell size to induction length is much larger for fuel-O_2 than for fuel-air mixtures. It is possible that the difference in this ratio is one reason why fuel-O_2 detonations display such regular and well-defined structure while fuel-air detonations leave much poorer soot records (Libouton et al. 1981; Bull et al. 1982).

A number of experimental studies have examined the effects of dilution by N_2 on detonations in C_3H_8 - oxidizer mixtures. In some of these studies (Bull et al. 1982; Knystautas et al. 1982) the cell spacings λ were measured by means of a soot record technique, while in others (Knystautas et al. 1982; Matsui and Lee 1979; Kogarko 1958) the critical tube diameter d_c for initiation of unconfined spherical detonation was determined. It has been shown (Mitrofanov and Soloukhin 1964) that $d_c \simeq 13 \lambda$ for circular tubes. Therefore, if the transverse cell spacing can be related to the induction length Δ, then the present model can be used to predict d_c. This has been done for a variety of fuels (Westbrook 1982b, 1982c; Westbrook and Urtiew 1983a, 1983b) with good success. Computed variations in induction length Δ for stoichiometric C_3H_8 + 5 O_2, with different amounts of N_2 dilution, are shown in Fig. 2. Also shown are corresponding curves for CH_4 + 2 O_2 and C_2H_2 + 2.5 O_2. Results for C_2H_4 + 3 O_2 and C_2H_6 + 3.5 O_2 are very close to the propane curve. The dashed curve in Fig. 2 represents a "CO stoichiometric" mixture of C_3H_8 + 3.5 O_2 diluted by N_2. For each

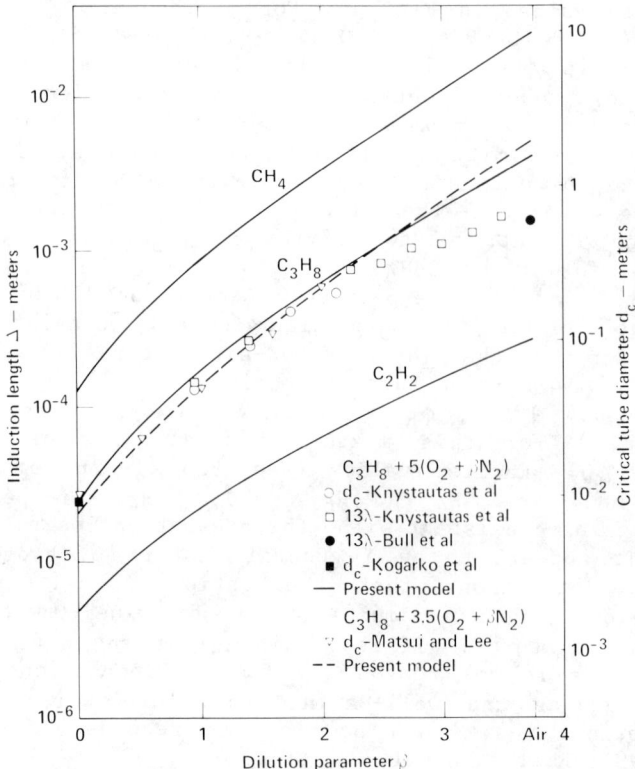

Fig. 2 Computed variation of induction length with dilution by N_2. Dilution parameter $\beta = N_2/O_2$. Solid curves are stoichiometric mixtures, dashed curve is "CO stoichiometric." Symbols represent experimental data for critical tube diameter or 13 times transverse cell spacing.

curve, the rate of variation of log Δ with dilution parameter $\beta \equiv N_2/O_2$ is greatest for β close to zero, changing smoothly to a nearly straight line for $\beta \geq 2$.

Also indicated in Fig. 2 are experimental values of d_c, either measured directly or evaluated as equal to 13 λ. When the proportionality between Δ and d_c is set to

$$d_c = 380 \, \Delta \tag{1}$$

as shown in Fig. 2, the computed curve appears to provide a fairly good representation of the experimental results. The same proportionality gives similar agreement for CH_4, C_2H_4, C_2H_6, and H_2 fuels as well (Westbrook 1982b, 1982c; Westbrook and Urtiew 1983a, 1983b). This equation is equivalent to a relationship between cell size λ and induction length

$$\lambda \equiv 29 \, \Delta \qquad (2)$$

since $29 \simeq 380/13$. This value of 29 is an average value of λ/Δ between that of about 10 for the fuel-air mixtures and 50 for fuel - O_2 mixtures already described.

The values of d_c predicted by the computed curve in Fig. 2 as $\beta \to 3.76$ are somewhat higher than the experimental data of both Knystautas et al. (1982) and Bull et al. (1982). Predicted values for C_2H_4 in the same range of β were also too high (Westbrook 1982b, 1982c; Moen et al. 1981), although the agreement for $\beta \leq 2.5$ is excellent for both fuels. It is apparent from Fig. 2 that the errors in this range are due to a fairly rapid change in the ratio λ/Δ. For $0 \leq \beta \leq 2.5$, Eqs. (1) and (2) are quite reliable, but for $\beta > 2.5$, the ratio λ/Δ falls rapidly to a value of about 10. The computed curve in Fig. 2, at $\beta = 3.76$, is higher than the experimental point of Bull et al. (1982) precisely by the ratio 29/10.

For stoichiometric C_3H_8 - O_2 and C_3H_8 - air, induction length calculations were carried out at initial pressures from 0.01 to 10.0 atmospheres. The results are summarized in Fig. 3. Also shown are computed results for the other hydrocarbon fuels reported previously (Westbrook and Urtiew 1983a). The dotted line in Fig. 3 shows the experimental values of d_c reported by Matsui and Lee (1979) for C_3H_8 - O_2, data which have since been confirmed by Knystautas et al. (1982). Over the pressure range examined, the model reproduces the experimental data well when Eq. (1) is used to relate d_c and Δ. For both C_3H_8 - O_2 and C_3H_8 - air the computed curves are essentially straight lines, and no multiple values of P_0 were found at a single value of Δ or d_c, as was predicted for H_2 - air (Westbrook and Urtiew 1983a). Again, the curves for C_3H_8 are very similar to those for C_2H_4 and C_2H_6, all widely separated from those for CH_4 and C_2H_2.

The critical energy E_c for initiation of unconfined detonation can be related to the induction length Δ by means of the Zel'dovich criterion (Zel'dovich et al. 1956)

$$E_c = k_1 \, \Delta^j \qquad (3)$$

where $j = 1$, 2, or 3 in planar, cylindrical, or spherical configurations, respectively. For initiation of spherical detonation by means of a planar detonation in a critical

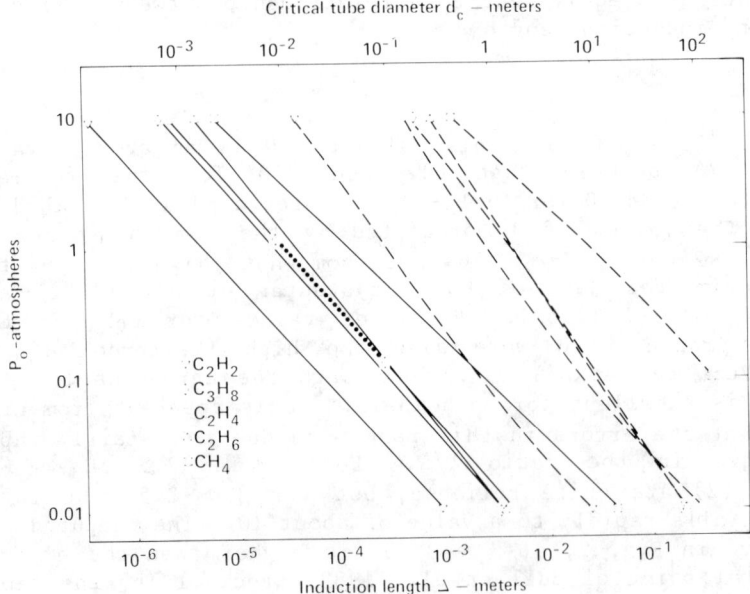

Fig. 3 Computed variation of induction length with initial pressure P_o. Solid curves show results for stoichiometric fuel-O_2, dashed curves for fuel-air. Dotted line shows experimental values of critical tube diameter d_c at different initial pressures for propane-oxygen from Matsui and Lee (1979).

tube of diameter d_c, the coefficient k_1 depends on the properties of the planar detonation (Lee and Matsui 1977; Urtiew and Tarver 1981)

$$E_c = k_o(Pu/D) d_c^3 \qquad (4)$$

where P, u, and D refer to the CJ conditions, and k_o = 0.1964. Combining Eq. (4) with Eq. (1) results in

$$E_c = k_o(Pu/D)(380)^3 \Delta^3 \qquad (5)$$

relates the critical energy E_c to the induction length.

Computed values of E_c from Eq. (5) are plotted as functions of ϕ for $C_3H_8 - O_2$ mixtures in Fig. 4, together with the similar curves for the other hydrocarbon - O_2 mixtures examined previously (Westbrook and Urtiew

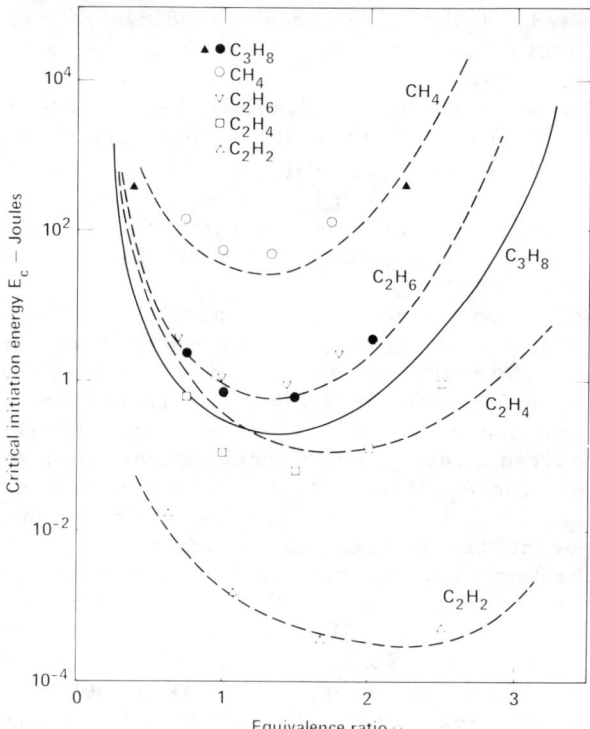

Fig. 4 Critical energy for initiation of unconfined spherical detonation. Symbols represent results of Matsui and Lee (1979), curves are computed from Eq. (4) with $k_0 = 0.1964$. Solid triangles are lean and rich limits determined by Carlson (1973).

1983a). Also shown are values of E_c computed by Matsui and Lee (1979) from their experimental values of d_c. It appears that a single relationship between E_c and Δ adequately represents the results for all of the fuels studied, although the degree of agreement varies between the different fuels. Independent calibration of the C_3H_8-O_2 curve at $\phi = 1$ would yield much better agreement between computed and experimental results at other values of ϕ than is shown in Fig. 4.

Carlson (1973) determined lean and rich limits for unconfined spherical detonation, using 400 J high explosives for initiation. The limits for C_3H_8, $\phi_L \simeq 0.35$ and $\phi_R \simeq 2.25$, are indicated by the solid triangles in Fig. 4 at 400 J. The agreement with the computed curve is good for the lean limit, but the computations give a much larger rich limit than Carlson's value. The same behavior, good agreement at ϕ_L and overprediction of ϕ_R was observed for C_2H_4,

C_2H_6, and C_2H_2 (Westbrook 1982b). For propane, Carlson determined a rich limit which is much less than that for ethylene ($\phi_R \simeq 3.25$) and even less than that for ethane ($\phi_R \simeq 2.33$). This pattern disagrees sharply with the present results indicated in Fig. 1 which shows the results for C_3H_8-O_2 falling midway between those for C_2H_4-O_2 and C_2H_6-O_2; in this light, no explanation can be provided for the qualitative differences between the computed rich limits for C_3H_8-O_2 and the data of Carlson.

In addition to detonation limits at a fixed initiator energy, Carlson determined minimum initiator energies for spherical detonations in a wide variety of fuel-O_2 mixtures, using an exploding wire technique. Since not all of the energy, which is stored capacitively, actually is deposited into the reactive gas, these energies represent conservative values. For propane-O_2, Carlson found E_{min} = 2.5 J at $\phi \simeq 1.43$. Comparison with Fig. 4 shows that the minimum in the computed curve occurs at about the same equivalence ratio but the energy scale is displaced by about an order of magnitude. The same difference in energy scale was also observed for the other fuels (Westbrook 1982b; Westbrook and Urtiew 1983a). Ratios of E_{min} measured by Carlson between different fuels agree well with the computed values, so the differences between the experimental and computed values seem to indicate that for initiating a spherical detonation, capacitively stored energy is not as efficient as a detonation from a tube with the critical diameter d_c, and that the different methods scale similarly with induction length. The only other comparable experimental data on energy of initiation for C_3H_8-O_2 are those of Brossard et al. (1972) who used an initiation energy of 105 J. However, only a small fraction of this energy actually is deposited in the reactive gases, and no variations in initiation energy with equivalence ratio were examined. As the stoichiometric C_3H_8-O_2 mixture was diluted by N_2, this energy of 105 J soon became inadequate to initiate a spherical detonation, and higher energy initiation was not pursued.

The Zel'dovich criterion [Eq. (3)] also applies to initiation of unconfined spherical detonations in fuel-air mixtures by high explosive charges. In Fig. 5 computed values of Δ^3 are plotted for C_3H_8-air mixtures at different equivalence ratios. The minimum, corresponding to maximum detonability and minimum energy for initiation, occurs slightly on the rich side of stoichiometric, in common with other hydrocarbon-air mixtures (Westbrook 1982b; Bull 1979). Also shown in Fig. 5 are available

PROPANE DETONATIONS

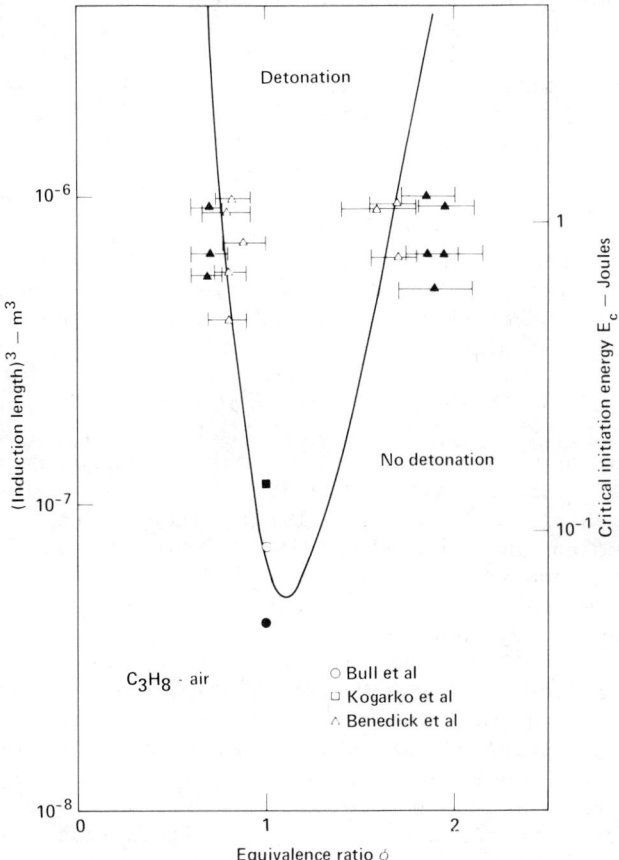

Fig. 5 Cube of induction length for propane-air mixtures initially at atmospheric pressure. Individual symbols indicate experimental data for critical initiator mass of high explosive, open symbols for successful detonation, solid symbols for no detonation.

experimental results for E_c as summarized by Bull (1979). The best agreement between the computed and experimental results was found with $k_1 = 1.22 \times 10^6$ kg Tetryl/m³ [Eq. (3)]. The same coefficient relating Δ^3 and E_c for C_2H_4 and CH_4 detonation was found to be 0.73×10^6 and 1.0×10^6, respectively (Westbrook 1982b). This indicates that the present results for propane-air mixtures are in reasonable agreement with those for most of the other hydrocarbon fuels. One exception to this rule is C_2H_6-air, for which the coefficient k_1 was found to be 0.25×10^6 kg/m³, much lower than for all the other fuels.

Furthermore, in contrast with the present results (see Fig. 1) and the results of Matsui and Lee (1979), and Bull et al. (1978, 1979) found that E_c was lower for C_2H_6-air than for C_3H_8-air. On the other hand, Bull et al. (1982) subsequently found that cell sizes in planar detonations were somewhat smaller for C_3H_8-air than those for C_2H_6-air, which suggests that E_c for C_3H_8-air should also be less than for C_2H_6-air. Considering the uncertainties in the kinetic mechanisms for both propane and ethane, together with the fact that computed induction times and lengths (Fig. 1) for the two fuels are so similar, the present modeling results cannot alone establish their relative initiation energies.

Propane in Natural Gas

Most natural gas samples contain approximately 1-4% propane, along with 80-95% CH_4 and 5-10% C_2H_6. Experimental (Bull et al. 1979; Nicholls et al., 1977; Vandermolen and Nicholls 1979; Eubank et al. 1981) and modeling analyses (Westbrook 1979) showed conclusively that methane-ethane-air mixtures have much shorter ignition delay and induction lengths than methane-air mixtures. This phenomenon, which can be termed sensitization, is a result of kinetic interactions between the reaction mechanisms for the two fuels. Because a reliable propane oxidation mechanism was not yet available, none of these earlier modeling studies considered the role of propane in the natural gas mixture.

A series of induction length calculations was carried out for fuel-air mixtures which are representative of typical natural gas. Although natural gas composition can vary, depending on its original source and its storage history, a reasonable average composition is 90% CH_4-8% C_2H_6-2% C_3H_8. Therefore, in all of the mixtures studied 90% of the fuel consisted of methane. In order to determine the importance of propane in these mixtures, the composition of the remaining 10% was varied over the range from pure ethane to pure propane.

The most important minor constitutent in natural gas is ethane. The computed induction length in a stoichiometric mixture of 90% CH_4-10% C_2H_6 is 1.07 x 10^{-2} m, which is less than half the computed value of 2.41 x 10^{-2} m when all of the fuel is methane. This reduction of 56% in the induction length is equivalent to a 91% reduction in the predicted critical initiation energy [from Eq. (3)] relative to stoichiometric methane-air. The variation of the induction length Δ

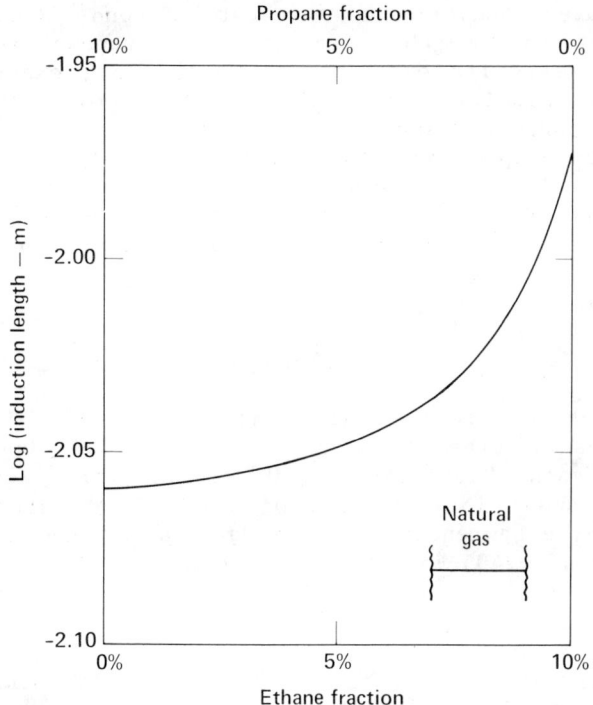

Fig. 6 Variation of computed induction length with ethane and propane content. All mixtures are stoichiometric fuel-air, with 90% of the fuel consisting of methane. Typical composition range in natural gas in indicated.

with the composition of the remaining 10% of the fuel is summarized in Fig. 6. Although complete replacement of the 10% ethane by propane reduces Δ by 18%, more than half of this reduction occurs by the time 2% C_3H_8 has been added to the fuel. Together with the Zel'dovich criterion, this predicts that the minimum initiation energy for a mixture 90% CH_4-8% C_2H_6-2% C_3H_8 is 33% less than that for a comparable mixture of 90% CH_4-10% C_2H_6.

Summary

There are two principal conclusions which can be drawn from the present modeling study. First, the reaction mechanism summarized in Table 1 provides a reliable kinetic model for the oxidation of propane under detonation conditions. The major reaction steps and the rates of the key reactions are reasonably well known, and the mechanism can be used to predict detonation parameters under related conditions of interest. Second, it has been shown that the computed results for propane oxidation correlate well with experimental data, using exactly the same constants of proportionality as those for all of the other hydrocarbon fuels studied previously. The present results for propane oxidation therefore reinforce the view that chemical kinetic processes play a dominant role in detonations and that a simplified detonation model, coupled with a detailed comprehensive kinetics model, can provide a great deal of useful information on detonation parameters.

Acknowledgments

The authors are pleased to acknowledge many valuable discussions on the subject of modeling of detonation kinetics with Professor J. H. Lee. This work was performed under the auspices of the U.S. Department of Energy by the Lawrence Livermore National Laboratory under Contract No. W-7405-Eng-48.

References

Atkinson, R., Bull, D. C., and Shuff, P. J. (1980). Initiation of spherical detonation in hydrogen/air. Combustion and Flame 39, 287.

Bahn, G. S. (1973) Approximate thermochemical tables for some C-H and C-H-O species. NASA Report NASA-CR-2178.

Benedick, W. B., Kennedy, J. D., and Morosin, B. (1970) Detonation limits of unconfined hydrocarbon-air mixtures. Combustion and Flame 15, 83.

Borisov, A. A., and Loban, S. A. (1977) Detonation limits of hydrocarbon-air mixtures in tubes. Fiz. Goreniya Vzryva 13, 729.

Brossard, J., Desbordes, D., and Manson, N. (1972) Variation de la celerite des detonations spheriques devergentes dans les melanges stricts propane-oxygene-azote et acetylene-oxygene-azote en fonction de l'abscisse radiale: influence de la dilution en inerte. C.R. Acad. Sc. Paris 274, 1198.

Bull, D. C., Elsworth, J. E., and Hooper, G. (1978) Initiation of spherical detonation in hydrocarbon/air mixtures. Acta Astr. 5, 997.

Bull, D. C. (1979) Concentration limits to the initiation of unconfined detonation in fuel/air mixtures. Trans. Inst. Chem. Eng. 57, 219.

Bull, D. C., Elsworth, J. E., and Hooper, G. (1979) Susceptibility of methane-ethane mixtures to gaseous detonation in air. Combustion and Flame 34, 327.

Bull, D. C., Elsworth, J. E., and Shuff, P. J. (1982) Detonation cell structures in fuel/air mixtures. Combustion and Flame 45, 7.

Burcat, A., Lifshitz, A., Scheller, K., and Skinner, G. B. (1971a) Shock tube investigation of ignition in propane-oxygen-argon mixtures. Thirteenth Symposium (International) on Combustion, p. 745 The Combustion Institute, Pittsburgh, Pa.

Burcat, A., Scheller, K., and Lifshitz, A. (1971b) Shock tube investigation of comparative ignition delay times for C_1-C_5 alkanes. Combustion and Flame 16, 29.

Carlson, G. A. (1973) Spherical detonations in gas-oxygen mixtures. Combustion and Flame 21, 383.

Dove, J. E., and Tribbeck, T. D. (1970) Computational study of the kinetics of the hydrogen-oxygen reaction behind steady-state shock waves. Application to the composition limits and transverse stability of gaseous detonations. Astr. Acta 15, 387.

Eubank, C. S., Rabinowitz, M. H., Gardiner, W. C., Jr., and Zellner, R. E. (1981) Shock-initiated ignition of natural gas-air mixtures. Eighteenth Symposium (International) on Combustion, p. 1767 The Combustion Institute, Pittsburgh, Pa.

JANNAF Thermochemical Tables (1970) D. R. Stull et al. eds., U.S. National Bureau of Standards NSRDS-NBS 37 and supplements.

Knystautas, R., Lee, J. H., and Guirao, C. M. (1982) The critical tube diameter for detonation failure in hydrocarbon-air mixtures. Combustion and Flame 48, 63.

Kogarko, S. M. (1958) Detonation of methane/air mixtures and the detonation limits of hydrocarbon/air mixtures in a large diameter pipe. Sov. Phys. Tech. Phys. 28, 1904.

Kogarko, S. M., Adushkin, V. V., and Lyamin, A. G. (1965) An investigation of spherical detonations of gas mixtures. Nauchno Tekh. Probl. Goren. Vzr. 2, 22.

Lee, J. H., and Matsui, H. (1977) A comparison of the critical energies for direct initiation of spherical detonations in acetylene-oxygen mixtures. Combustion and Flame 28, 61.

Lee, J. H., Knystautas, R., Guirao, C., Benedick, W. B., and Shepherd, J. E. (1982) Hydrogen-air detonations. Second International Workshop on the Impact of Hydrogen on Water Reactor Safety, Albuquerque, N.M..

Libouton, J.-C., Jacques, A., and Van Tiggelen, P. J. (1981) Cinetique, structure et entretien des ondes de detonation. Proceedings of the First International Specialists Meeting of The Combustion Institute, p. 437 Bordeaux, France.

Matsui, H. and Lee, J. H. (1979) On the measure of the relative detonation hazards of gaseous fuel-oxygen and air mixtures. Seventeenth Symposium (International) on Combustion, p. 1269. The Combustion Institute, Pittsburgh, Pa.

Michels, H. J., Munday, G. F., and Ubbelohde, A. R. (1970) Detonation limits in mixtures of oxygen and homologous hydrocarbons. Proc. Roy. Soc. Lond. A, 319, 461.

Mitrofanov, V. V. and Soloukhin, R. I. (1964) The diffraction of multifront detonation waves. Soviet Phys. Dokl. 9, 1055.

Moen, I. O., Donato, M., Knystautas, R., and Lee, J. H. (1981) The influence of confinement on the propagation of detonations near the detonability limits. Eighteenth Symposium (International) on Combustion, p. 1615. The Combustion Institute, Pittsburgh, Pa.

Nicholls, J. A., Sichel, M., Gabrijel, Z., Oza, R. D., and Vandermolen, R. (1977) Detonability of unconfined natural gas-air clouds. Seventeenth Symposium (International) on Combustion, p. 1223. The Combustion Institute, Pittsburgh, Pa.

Strehlow, R. A., and Rubins, P. M. (1969) Experimental and analytical study of H_2-air reaction kinetics using a standing-wave normal shock. AIAA J. 7, 1335.

Tsuge, S., Furukawa, H. J., Matsukawa, M., and Nakagawa, T. (1970) On the dualproperty and the limit of hydrogen-oxygen free detonation waves. Astr. Acta 15, 377.

Ubbelohde, A. R. and Munday, G. (1969) Some current problems in the marginal detonation of gases. Twelfth Symposium (International) on Combustion, p. 809. The Combustion Institute, Pittsburgh, Pa.

Urtiew, P. A. and Tarver, C. M. (1981) Effects of cellular structure on the behavior of gaseous detonation waves under transient conditions. Prog. Astr. Aero. 75, 370.

Vandermolen, R. and Nicholls, J. A. (1979) Blast wave initiation energy forthe detonation of methane-ethane mixtures. Comb. Sci. Tech. 21, 75.

Warnatz, J. (1981) The structure of laminar alkane-, alkene-, and acetylene flames. Eighteenth Symposium (International) on Combustion, p. 369. The Combustion Institute, Pittsburgh. Pa.

Westbrook, C. K., Creighton, J., Lund, C., and Dryer, F. L. (1977) A numerical model of chemical kinetics of combustion in a turbulent flow reactor. J. Phys. Chem. 81, 2542.

Westbrook, C. K. (1979) An analytical study of the shock tube ignition of mixtures of methane and ethane. Comb. Sci. Tech. 20, 5.

Westbrook, C. K. and Dryer, F. L. (1979) A comprehensive mechanism for the oxidation of methanol. Comb. Sci. Tech. 20, 125.

Westbrook, C. K. (1982a) Hydrogen oxidation kinetics in gaseous detonations. Comb. Sci. Tech. 29, 67.

Westbrook, C. K. (1982b) Chemical kinetics of hydrocarbon oxidation in gaseous detonations. Combustion and Flame 46, 191.

Westbrook, C. K. (1982c) Chemical kinetics in gaseous detonations. Proceedings of the International Specialists Meeting on Fuel-Air Explosives, McGill University, Montreal, Canada. University of Waterloo Press.

Westbrook, C. K., Dryer, F. L., and Schug, K. P. (1983) A comprehensive mechanism for the pyrolysis and oxidation of ethylene. Nineteenth Symposium (International) on Combustion, p. 153. The Combustion Institute, Pittsburgh, Pa.

Westbrook, C. K., and Urtiew, P. A. (1983a) Chemical kinetic prediction of critical parameters in gaseous detonations. Nineteenth Symposium (International) on Combustion, p. 615. The Combustion Institute, Pittsburgh.

Westbrook, C. K. and Urtiew, P. A. (1983b) Use of chemical kinetics to predict critical parameters of gaseous detonations. Fiz. Goren. Vzr. 19, 65.

Westbrook, C. K. and Pitz, W. J. (1983) A comprehensive mechanism for the pyrolysis and oxidation of propane and propene (submitted for publication).

Zel'dovich, Y. B., Kogarko, S. M., and Semenov, N. N. (1956) An experimental investigation of spherical detonation in gases. Sov. Phys. Tech. Phys. 1, 1689.

High-Speed Deflagration with Compressibility Effects

J. F. Clarke*
Cranfield Institute of Technology, Bedford, England
and
D. R. Kassoy†
University of East Anglia, Norwich, England

Abstract

A theoretical study is made of planar, steady compressible flow with significant heat addition due to distributed chemical reactions. Combustion processes in a flow, exhibiting a strong interaction between reactive and compressibility effects, are examined downstream of a specific origin from which an order-unity Mach number flow emanates. The general properties of such flows are considered when the Prandtl number is 3/4 and the Lewis number is unity. For sufficiently small initial strain rates, the reaction process is initiated weakly in a region of convection-conduction-reaction-compressibility balance. Thereafter, transport properties are of negligible importance in determining the combustion zone structure unless an imbedded shock is present in an initially supersonic flow. When the initial strain rate is sufficiently large, an inert gasdynamical adjustment zone proceeds the reaction dominated process. A detailed solution to the problem in the limit of high activation energy is given for a perfect gas with fairly general properties. After the weak reaction initiation zone, an ignition process, reminiscent of that in an adiabatic thermal explosion, occurs over an ignition-delay distance. Large variations then occur in a relatively thinner region where there is a major interaction between reaction and compressibility effects. Maximum temperatures are constrained by the latter. Chapman-

---
Presented at the 9th ICODERS, Poitiers, France, July 3-8, 1983. Copyright © American Institute of Aeronautics and Astronautics, Inc., 1984. All rights reserved.
*Professor, College of Aeronautics.
†Senior Visiting Fellow, School of Mathematics and Physics. Present address: Professor, Mechanical Engineering Department, University of Colorado, Boulder.

Jouguet processes occur when the heat addition is maximized for a given initial Mach number. All of the results are derived without resorting to an artificially defined ignition temperature.

Introduction

Most models of heat addition to a high-speed flow make use of idealizations of a rather profound character. Classical Hugoniot types of analysis certainly require, and sometimes state, that any process of heat addition is internal to the wave, which is then treated as a discontinuity linking equilibrium states in the gas that flows into and out of the front. When the energy is added by chemical combustion, the state ahead of any steady propagating wave is strictly one, not of equilibrium, but of metastable equilibrium. This is usually prescribed by simply distorting the Arrhenius rate law so that reactions are switched off under upstream conditions. Conversely, it may just be assumed that the energy addition effects have known distributions through space and time, so that they become simple known sources in the energy equation. The vital cross-couplings that exist between gasdynamical and chemical effects are therefore dismissed arbitrarily.

These deficiencies in the modeling of gas flows with combustion chemistry can be circumvented by using a "flame-holder" at a specified origin $x = 0$, as described by Hirschfelder et al. (1953). A steady state downstream of the holder is guaranteed by allowing it to extract a certain amount of thermal energy from the gas flow, so that the system is not adiabatic, and by accounting for the fractional mass fluxes of all of the chemical constituents through the holder. One may choose to forbid any back-flow of combustion products into the holder, and thence into the apparatus that supplies it. Some contamination of the apparatus can occur through this back-flow mechanism, but it is usually small in amount and extremely slow to develop to any noticeable degree; therefore, to make it zero is, as has just been stated, a rational thing to do. The use of the holder model in low-speed thermal flame studies is well known (Hirschfelder et al. 1953; Williams 1965; Buckmaster and Ludford 1982). Here we employ it to study high-speed (that is to say, order-unity Mach number) flows with combustion. In so doing, we are continuing the analysis begun by Clarke (1983a), who described the sequence of changes in flame structure that come about as the inlet flow Mach is increased from its very small ($\approx 10^{-3}$) values, which apply to thermal-flame propagation, up to values like

10^{-2}, at and beyond which all vestiges of thermal flame structure have vanished. In these latter circumstances, each fluid element is, for all practical purposes, isolated from its neighbors, and the process is predominantly one of a simple balance between convection (C) and reaction (R) effects. In fact, the entire process resembles a convecting adiabatic thermal explosion (Kassoy 1975).

General Compressibility Effects

As the initial Mach number increases toward O(1) values, one begins to encounter direct effects of gasdynamical behavior. In the first place, it is no longer possible to ignore the fact that the gas is compressible. Expansion or contraction of a fluid element, with consistent changes of pressure and, especially, temperature, will couple directly with the internal chemical activity. In addition to these factors, it will be necessary to keep track of the kinetic energy of mass motion of the system. Indeed, one may say that the system we must analyze is complicated by the single fact that it is mandatory to follow the distribution of energy between chemical, thermal, and kinetic modes, and that all three of these forms of energy storage are strongly coupled.

If the Prandtl number is 3/4 and the Lewis number is unity, the energy equation simplifies substantially and can be written as

$$\overset{(C)}{\frac{dT_s}{d\xi}} - \overset{(D)}{\frac{d^2T_s}{d\xi^2}} = \overset{(R)}{\{A_4 \theta^s M^{-2} e^{-\theta/T}\}} (T_{sb} - T_s) \qquad (1)$$

where $T_s = T + \frac{1}{2}(\gamma - 1)M^2 u^2$ is the dimensionless stagnation temperature, and T_{sb} is the value of T_s in the burnt state. In Eq. (1) T is the absolute temperature, u the flow velocity, M the inlet Mach number, and γ the specific heat ratio. In general A_4 is a number that depends weakly on T. For the sake of simplicity we assume that it is a constant here. The quantity in the curly braces {} is the local Damköhler number. Although it changes significantly from place to place within the flow on account of the Arrhenius factor $\exp(-\theta/T)$, where θ is the activation energy number, it must be noticed that it is <u>always small</u>. The length variable ξ is measured in units of a diffusion length; in particular

$$\xi = \overline{m} \int_0^x (\overline{C}_p/\overline{\lambda})dx \qquad (2)$$

where $\overline{m}$, $\overline{C}_p$, and $\overline{\lambda}$ are the dimensional mass flux (constant), specific heat, and thermal conductivity, respectively. The diffusive effects (D) that are summarized in the term $d^2T/d^2\xi$ will always be small in any place where R is significant for the fundamental reason that the local Damköhler number {} is always small. This fact is discussed in detail in Clarke (1983a) and in Clarke (1983b) where the general properties of steady planar, reactive, high Mach number flows are discussed in considerable detail.

Of course, Eq. (1) is not sufficient to enable one to solve the problem when M is O(1). It must be coupled to the momentum equation

$$\frac{du}{d\xi} - \sigma_0 = u - 1 + \frac{1}{\gamma M^2}\left(\frac{T}{u} - 1\right) \qquad (3)$$

where $\sigma_0 = du/d\xi$ at $\xi = 0 = x$. This initial strain rate σ_0 must be specified; it cannot be chosen at will but must obey certain constraints that are described by Clarke (1983b).

It is shown by general arguments in Clarke (1983b) that the structure of a high-speed flow with combustion has a distinct and unique character. When the inlet Mach number M is subsonic, the flow near the flameholder may, if σ_0 is of order unity, begin with a region of rapid, effectively inert, gasdynamical adjustment in which CD effects dominate. Following a short transitional CDR region in which all of C, D, and R play a part (R provides an essentially constant small rate of chemical energy supply in this case), the major amount of chemical energy release occurs in a CR domain; D is relegated to an insignificant role. If $|\sigma_0|$ is sufficiently small, the flow may begin with a CDR region.

If M is less than $1/(\gamma)^{1/2}$, temperature rises and an ignition event takes place. If M exceeds $1/(\gamma)^{1/2}$, the release of chemical energy causes the flow speed to increase while temperature diminishes. The consequent slowing down of reaction rates means that the processes of energy release will occur over very large distances.

When M exceeds unity, the flow usually begins with a CD domain adjacent to the holder that is, depending upon the σ_0 value, almost the whole of a conventional inert shock wave. Events behind this shock proceed subsonically through a small CDR region and into the dominant CR domain, as described above, in which ignition takes place. If M is insufficiently large [in fact, if $1 < M^2 < (3\gamma - 1)/\gamma(3 - \gamma)$], the effective subsonic entry Mach number to the CDR → CR activity is greater than $1/(\gamma)^{1/2}$ and the same kind of gasdynamically induced ignition suppression as just described takes place. For quantifiably small σ_0 values, a variety of circumstances can be encountered with, for example, CD-type shocks buried within CR-type induction domains and linked by fore and aft CDR zones. In some cases, a fully supersonic transition from inlet to final burnt states is possible.

Clarke's (1983b) analysis of the general properties of steady, planar, compressible, reactive flows is limited by the assumption of Prandtl number and Lewis number equal to 3/4 and 1, respectively. This was done primarily to simplify the energy equation as in Eq. (1) and because explicit equations for shock structure can be obtained more easily. Of course, the general study shows quite convincingly that transport effects play a role only in very narrow transition zones that include weak reaction effects (CDR-zones) and in shocks where reaction processes have a very limited effect because too few collision times are available. These facts suggest that detailed flow models can be developed for entirely general material properties, particularly if shock processes are absent.

High-Speed Deflagration Solution

In the present paper, we describe a fully quantitative model of a steady, planar deflagration downstream of a specific origin from which a compressible reactive gas emanates with an O(1) initial Mach number. This configuration allows us to emphasize that all physical processes occur on finite time scales, although some may be long compared to others. The chemistry is modeled by a global one-step decomposition reaction. An Arrhenius rate law is employed with a large activation energy parameter θ. No artificially defined ignition temperature is invoked. At the origin, where the pressure, temperature, and speed are known exactly, the reaction rate is finite, but exponentially small with respect to $\theta \to \infty$. This ensures that the thermal induction time of the gas in a conceptual delivery system upstream of the origin is very long relative to reaction times in the deflagration itself. In this model it should be recognized that the process begins at a weakly nonequilibrium initial point. In the conceptual sense, the fully interactive compressible flow with distributed, chemically generated heat addition relaxes to a specified downstream pressure that is somewhat constrained by the flow dynamics itself. The heat addition is considered to be significant relative to the initial enthalpy of the reactive gas. The perfect gas mixture is assumed to have general transport properties and specific heats as well as arbitrary values of the Prandtl and Lewis numbers. The generality of the model should be contrasted with a variety of related efforts [for example, Bush and Fendell 1971; Kapila et al. 1983; Clarke 1983b] that contain unnecessary restrictions. Bush and Fendell (1971) invoke an artificial ignition temperature and consider only Chapman-Jouguet-type processes. Kapila et al. (1982) diminish the importance of their problem to some extent by considering vanishingly small amounts of chemical heat addition. Their solutions describe small perturbations to an inert flow. The results are entirely dependent on an ignition temperature that would be difficult to determine independent of an apparatus and particular experiment.

The mathematical model, described in detail in Kassoy and Clarke (1984), is based on the complete reactive, compressible flow equations including transport properties. Solution development is by asymptotic methods based on the limit $\theta \to \infty$. Rather than presenting a full description of the theoretical effort, the results will be given here along with an interpretation of the physical implications.

The reaction process is examined first in a zone of characteristic thickness $x_R = \mu_1/\rho_1 u_1 = O(10^{-5}$ cm$)$ for the high-speed flows considered here. In the asymptotic limit, the nondimensional temperature variation from the initial state is described by

$$\hat{T} \sim 1 + \beta(\theta)\hat{\phi}(\hat{x};\theta)$$

$$\beta = \frac{\mu_1/p_1}{(1/B)e^\theta \gamma M_1^2} = \frac{x_R/u_1}{(1/B)e^\theta}$$

(4)

where $\hat{x} = x/x_R$; the subscript 1 denotes initial conditions; B is the pre-exponential factor in the Arrhenius rate law; and M_1 is the initial Mach number. The parameter β, which is the ratio of the mean collision time to the initial reaction time when $M_1 = O(1)$ or, alternatively, the ratio of the flow residence time to the reaction time, must be exponentially small with respect to $\theta \to \infty$. The $\hat{\phi}$ equation describes a balance of convective, conductive, reactive, and compressibility effects. The $\hat{\phi}$ solution demonstrates explicitly that when $0 < M_1^2 < \gamma^{-1}$ a temperature rise will occur in this narrow zone. When $\hat{x} = O(1/\theta\beta) \gg 1$ the local analysis fails because the effect of heat addition due to chemistry becomes larger than that due to conductive losses. This asymptotic singularity suggests that there is a much wider downstream region in which a more substantial ignition process develops. There the temperature variation can be written as

$$\hat{T} \sim 1 + \theta^{-1}\overline{\phi}(\overline{x};\theta); \quad \overline{x} = \beta\theta\hat{x} \qquad (5)$$

For physically viable reaction rate parameters, the dimension scale of the ignition region can be as large as 10^3 cm. The $\overline{\phi}$ equation, a basic balance of convection, reaction, and compressibility, has the simple solution

$$\overline{\phi} = \ln[1/(1 - R\overline{x})], \quad R = (1 - \gamma M_1^2)h/(1 - M_1^2)$$
$$0 < M_1^2 < \gamma^{-1} \qquad (6)$$

where $h = O(1)$ is the nondimensional representation of chemical heat addition. One may observe that when $\overline{x} \to R^{-1} = \overline{x}_i$, a singularity occurs that is reminiscent of that in an adiabatic thermal explosion (Kassoy 1975). In the present context, $\overline{x}_i$ is the ignition delay distance, relevant, say, to experiments like those of Nicholls (1963). Other variables, particularly fuel concentration and pressure, behave in a very similar manner. It follows from Eqs. (5) and (6) that when $\theta^{-1}\overline{\phi} = O(1)$ a rapid reaction zone must be examined in which $O(1)$ changes in the dependent variables occur.

The rapid combustion process is described by using the nonlinear asymptotic transformation

$$\overline{x} = \overline{x}_i - H(\hat{T};\theta)\exp[-\theta(1-\hat{T}^{-1})] \qquad (7)$$

to derive basic conservation equations in terms of the independent variable $\hat{T}$ (Kassoy 1977). In the limit $\theta \to \infty$, it is found that

$$H(\hat{T}) = \frac{\hat{T}^2\hat{u}}{h\hat{y}} \frac{d}{d\hat{T}} \left(\int_1^{\hat{T}} C_p d\tau + \frac{\gamma-1}{2} M_1^2 \hat{u}^2 \right) \qquad (8)$$

$$\hat{u}(\hat{T}) = \frac{1+\gamma M_1^2}{2\gamma M_1^2} \left(1 \pm \frac{1-\gamma M_1^2}{1+\gamma M_1^2} [1 - \alpha(\hat{T} - 1)]^{\frac{1}{2}} \right) \quad (9)$$

where

$$\alpha = 4\gamma M_1^2 / (1 - \gamma M_1^2)^2$$

and

$$h\hat{y} = h - \int_1^{\hat{T}} C_p d\tau - \frac{\gamma-1}{2} M_1^2 (\hat{u}^2 - 1) \quad (10)$$

Formally, Eq. (9) and its derivative can be used in Eqs. (8) and (10) to give an explicit $H(\hat{T})$ and $y(\hat{T})$. Then Eq. (7) can be employed to find the spatial distribution $\hat{T}(\bar{x})$.

Continuing with the case $0 < M_1^2 < \gamma^{-1}$, the results in Eqs. (8-10) show that for a given M_1 and h value the temperature will rise by an $O(1)$ amount before the fuel in Eq. (10) vanishes. The maximum possible temperature is given by the critical value

$$\hat{T}_c = 1 + \alpha^{-1} \quad (11)$$

where

$$h\hat{y}_c = h - h_c$$

$$h_c = \hat{T}_c - 1 + (\gamma-1)(M_1^2/2)(u_c^2-1) > 0 \quad (12)$$

Unless M_1 is very small, $\hat{T}_c$ is considerably less than the adiabatic explosion temperature. The limitation is, of course, related to the effects of compressibility, which convert some of the thermal energy of reaction into kinetic energy. Equation (12) shows that if $h > h_c$ there will be fuel left at the critical point. Then Eq. (10) can be used to demonstrate that further fuel consumption is accompanied by a temperature drop. For a given M_1 value, there is a maximum value of heat addition

$$h_{cJ} = (1 - M_1^2)^2 / 2(\gamma+1) M_1^2 \quad (13)$$

for which the fuel concentration in Eq. (10) vanishes when

$$T_{cJ} = [4\gamma/(\gamma+1)^2]T_c \qquad (14)$$

and the local Mach number is

$$M_{cJ} = 1 \qquad (15)$$

The temperature in Eq. (14) is the lowest value obtainable for a given M_1. The speed in the rapid combustion zone increases monotonically. One must change from the upper sign branch to the lower sign branch of Eq. (9) if the critical point in Eqs. (11) and (12) is passed. The behavior of the $H(\hat{T})$ function in Eq. (8) shows that the stagnation temperature increases monotonically in the zone.

Formally, the solution fails when $\hat{y} = O(\theta^{-1})$, which means that one has to develop transitional solutions that describe the approach of the basic variables to their final values far downstream of the origin. The details of these mathematical manipulations can be found in Kassoy and Clarke (1984).

Relative to the ignition delay length $x_i = O(x_R/\beta\theta)$, the thickness of the rapid combustion zone is characterized by

$$x_i - x = O\left(\exp\left(-\theta(1-\hat{T}^{-1})\right)x_i\right) << O(x_i) \qquad (16)$$

As $\hat{T}$ increases from near the initial value, the distance to x_i becomes quite short.

The results of the study for subsonic initial Mach numbers show that a high-speed deflagration will be initiated at the initial temperature of the compressible reactive flow for a very large spectrum of initial values of the strain rate and heat flux. A very small finite chemical reaction rate exists at the origin because a normal Arrhenius rate law is employed without invoking an artificial ignition temperature. As a result, there is a tiny back-diffusion of product at the origin that emphasizes that the initial point is slightly out of equilibrium. In a typical example, the temperature will rise gradually over most of the ignition delay distance. Then very close to x_i, there will be an $O(1)$ increase in temperature accompanied by flow acceleration and Mach number increase. For sufficiently large h values, the temperature may reach a maximum $\hat{T}_c$ and then decline slightly as the flow continues to accelerate.

The solution for a shock-free supersonic initial Mach number can be obtained only when the initial strain rate, heat flux, and species gradient (all exponentially small with respect to $\theta \to \infty$) have very specific values. In practice it would be impossible to control the variable derivatives as required. It is likely (Clarke 1983b) that an embedded shock would appear somewhere downstream of the origin in any experimental situation.

Acknowledgments

Research support was provided by a Senior Visiting Fellowship from the Science and Engineering Research Council of the United Kingdom, by a U.S. Army Research Office contract DAAG 29-83-K-0069, and by a Faculty Fellowship from the University of Colorado.

References

Bush, W. and Fendell, F. (1971) Asymptotic analysis of the structure of a steady planar detonation. Combust. Sci. Technol. 2 (5 and 6), 271-285.

Buckmaster, J. D. and Ludford, G. S. S. (1982) Theory of Laminar Flames. Cambridge University Press, Cambridge, England.

Clarke, J. F. (1983a) On changes in the structure of steady plane flames as their speed increases. Combust. Flame 50 (2) 125-138.

Clarke, J. F. (1983b) Combustion in plane steady compressible flow, I. General considerations and gas-dynamical adjustment regions. J. Fluid Mechanics 136, 139-161.

Hirschfelder, J. O., Curtiss, C. F., and Campbell, D. E. (1953) The theory of flames and detonations. Fourth Symposium (International) on Combustion, pp. 190-211. The Combustion Institute, Pittsburgh, Pa.

Kapila, A. K., Matkowsky, B. J., and van Harten, A. (1983) An asymptotic theory of deflagrations and detonations. SIAM J. Appl. Math. 43 (3), 491-519.

Kassoy, D. R. (1975) Perturbation methods for mathematical models of explosion phenomena. Q. J. Mech. Appl. Math. 28 (1), 63-74.

Kassoy, D. R. (1977) The supercritical spatially homogeneous thermal explosion: initiation to completion. Q. J. Mech. Appl. Math. 30 (1), 71-89.

Kassoy, D. R. and Clarke, J. F. (1984) The propagation and structure of high speed deflagrations. CoA Memo 8309, Aerodynamics, Cranfield Institute of Technology, Bedford, England.

Nicholls, J. A. (1963) Standing detonation waves. <u>Ninth Symposium (International) on Combustion</u>, pp. 488-497. The Combustion Institute, Pittsburgh, Pa.

Williams, F. A. (1965) <u>Combustion Theory</u>. Addison Wesley, Reading, Mass.

Numerical Simulations on the Establishment of Gaseous Detonation

Shiro Taki*
Fukui University, Fukui, Japan
and
Toshitaka Fujiwara[†]
Nagoya University, Nagoya, Japan

Abstract

Numerical simulations are performed on the establishment and failure of one- and two-dimensional unsteady gaseous detonations with a two-step reversible reaction model. Though a one-dimensional Chapman-Jouguet(CJ) detonation is unstable, the numerical solution of such a detonation can be stabilized by the inclusion of artificial diffusion in the finite-difference scheme. The decay of a one-dimensional CJ detonation may be caused by a disturbance whose strength depends on the space mesh size because it affects artificial diffusion. An attenuating one-dimensional detonation is reestablished by explosions in the induction region without any additive energies. In this process, an explosion in shock-heated or fire-heated gas is followed by subsequent explosions, which strengthens the leading shock. The gases are heated successively and, finally, detonative explosion occurs. This mechanism may be important for transition to detonation. A plane CJ detonation in two-dimensional space is also unstable to disturbances. It is confirmed that the triple-shock detonation structure is essential to sustain the gaseous detonation in multidimensional space. The strength of explosion in detonation must be sufficient to generate a blast wave coupled with exothermic reaction, otherwise the triple-shock structure decays.

Presented at the 9th ICODERS, Poitiers, France, July 3-8, 1983. Copyright © American Institute of Aeronautics and Astronautics, Inc., 1984. All rights reserved.
*Associate Professor, Department of Mechanical Engineering.
†Professor, Department of Aeronautical Engineering.

NUMERICAL SIMULATIONS OF DETONATION

Introduction

The initiation and quenching of gaseous detonations have been investigated for many years. It has been understood qualitatively that explosions initiate detonation in front of the accelerating turbulent flame in a tube (Urtiew and Oppenheim 1966), and that for critical conditions transverse shock waves play an important role for the direct initiation of a detonation in a vessel (Bach et al. 1969). On the other hand, Erpenbeck (1966) showed theoretically that the detonation is unstable for normal gas mixtures. However, it is difficult to find the relation between the instability and the initiation of detonation. The goal of this study is to find quantitatively the condition for the onset of detonations. For this purpose, numerical simulations may be most powerful technique, because the phenomena are strongly nonlinear and unsteady. The numerical technique for solving the two-dimensional unsteady gaseous detonation was developed by Taki and Fujiwara (1978, 1981) who found that the spacings of transverse waves of solutions were in good agreement with existing experimental values. The objective of the present paper is to find the criterion for such phenomena as onset or quenching of detonations and the relation to the instability of detonation through use of the finite-difference method.

Model and Method for Numerical Analysis

The model and method for analysis will be briefly described, although the details are already shown in the paper by Taki and Fujiwara (1981).

Two-Step Reversible Reaction Model

Instead of chain elementary reaction, the following two-step reaction model is adopted for simplicity. The progress parameters α and β, for induction reaction and for exothermic recombination reaction, respectively, are both unity at first, followed by decrease of α to zero, then β decreases untill the equilibrium state is reached. The rate equations are given as follows, according to Korobeinikov et al. (1972): Induction reaction,

$$w_\alpha \equiv \frac{d\alpha}{dt} = -\frac{1}{\tau_{ind}} = -k_\alpha \rho \, \exp\left(-\frac{E_a}{\hat{R}T}\right) \tag{1}$$

exothermic reaction,

$$w_\beta \equiv \frac{d\beta}{dt} = 0 \qquad \alpha > 0$$

$$= -k_\beta p^2 [\beta^2 \exp\left(-\frac{E_b}{\hat{R}T}\right) - (1-\beta)^2 \exp\left(-\frac{E_b+Q}{\hat{R}T}\right)] \qquad \alpha \leq 0 \qquad (2)$$

where ρ denotes the mass density, T the temperature, p the pressure, Q the exothermicity, and $\hat{R}$ the universal gas constant; E_a and E_b are the sctivation energies, and k_α and k_β are the constants of reaction rates.

Fundamental Equations

For negligible transport phenomena, Cartesian two-dimensional unsteady reactive fluid dynamic equations can be written in the following quasi conservation form:

$$\frac{\partial f}{\partial t} + \frac{\partial F}{\partial x} + \frac{\partial G}{\partial y} + H = 0 \qquad (3)$$

where

$$f = \begin{bmatrix} \rho \\ \rho u \\ \rho v \\ \rho e \\ \rho \beta \\ \rho \alpha \end{bmatrix} \qquad F = \begin{bmatrix} \rho u \\ \rho u^2 + p \\ \rho u v \\ \rho u (e+p/\rho) \\ \rho u \beta \\ \rho u \alpha \end{bmatrix}$$

$$G = \begin{bmatrix} \rho v \\ \rho v u \\ \rho v^2 + p \\ \rho v (e+p/\rho) \\ \rho v \beta \\ \rho v \alpha \end{bmatrix} \qquad H = \begin{bmatrix} 0 \\ 0 \\ 0 \\ 0 \\ -\rho w_\beta \\ -\rho w_\alpha \end{bmatrix} \qquad (4)$$

where u and v denote the velocities for the x and y directions, respectively; and e is the specific total energy. As perfect gas mixtures are assumed, the following equations hold:

$$p/\rho = RT \qquad (5)$$

$$e = RT/(\gamma - 1) + (u^2 + v^2)/2 + q\beta \qquad (6)$$

where R is the gas constant, q is the specific exothermicity, and γ is the specific heat ratio.

Finite-Difference Scheme

Basically, the MacCormack (1971) explicit finite-difference method with second-order accuracy is used. To reduce the vibrations at the shock front, the flux-corrected transport(FCT) technique (Book et al. 1975) is utilized. When a moving coordinate with CJ detonation velocity is used, fourth-order diffusion is added. For two-dimensional simulations, the time splitting method is adopted, so that the numerical integrations are always done in one-dimensional space. For one-dimensional differential equations,

$$\frac{\partial f}{\partial t} + \frac{\partial F}{\partial x} + \Phi = 0 \tag{7}$$

the finite-difference scheme can be written as follows.
1) MacCormack: i) predictor,

$$f_i^p = f_i^n - \lambda(F_{i+1}^n - F_i^n) - \Delta t\, \Phi_i^n \tag{8a}$$

ii) corrector,

$$f_i^c = \tfrac{1}{2}[f_i^n + f_i^p - \lambda(F_i^p - F_{i-1}^p) - \Delta t\, \Phi_i^p] \tag{8b}$$

2) Fourth-order diffusion:

$$\hat{f}_i^{n+1} = f_i^c - \mu(f_{i-2}^n - 4f_{i-1}^n + 6f_i^n - 4f_{i+1}^n + f_{i+2}^n) \tag{9}$$

3) FCT: i) diffusion,

$$\tilde{f}_i^{n+1} = \hat{f}_i^{n+1} + \eta(f_{i+1}^n - 2f_i^n + f_{i-1}^n) \tag{10a}$$

ii) anti diffusion,

$$f_i^{n+1} = \tilde{f}_i^{n+1} - (\delta_{i+\frac{1}{2}}^c - \delta_{i-\frac{1}{2}}^c) \tag{10b}$$

where

$$\delta_{i+\frac{1}{2}}^c = S\cdot\mathrm{Max}[\,0,\, \mathrm{Min}(S\cdot\Delta_{i-\frac{1}{2}}, |\tilde{\Delta}_{i+\frac{1}{2}}|, S\cdot\Delta_{i+\frac{3}{2}})\,] \tag{10c}$$

$$\Delta_{i+\frac{1}{2}} = \bar{f}_{i+1}^{n+1} - \bar{f}_{i}^{n+1}, \quad \tilde{\Delta}_{i+\frac{1}{2}} = \eta(\hat{f}_{i+1}^{n+1} - \hat{f}_{i}^{n+1}), \quad (10d)$$

$$S = \text{sgn}(\tilde{\Delta}_{i+\frac{1}{2}})$$

The stability criterion of this scheme:

$$\lambda \, (\, a \, + \, |u| \,) \leq 1 \quad (11)$$

must be satisfied at any lattice point, where $\lambda = \Delta t/\Delta x$ is the ratio of time step size to space mesh size; a is the local frozen sound velocity; η and μ are arbitrary constants, in principle; and subscript i and superscript n denote the position of space and time lattice points, respectively.

Normalization and Values of Constants

All the variables and constants are normalized by the reference values, which are the induction distance ℓ^* of CJ detonation, the specific exothermicity q of the mixtures, and the initial mas density ρ_0, so that the reference time $t^* = \ell^*/q^{1/2}$. The dimensioless constants for mixture are selected to fit $2H_2+O_2$ diluted by 70% of inert gas, as follows:

$$\frac{q}{p_0/\rho_0} = \frac{q}{RT_0} = 20.0, \quad E_\alpha/Q = 1.7, \quad E_b/Q = 0.347$$

$$k_\alpha/k_\beta \rho_0 q^2 = 20.0, \quad \gamma = 1.4 \quad (12)$$

The space mesh sizes used are in the range of $\Delta x/\ell^* = \Delta y/\ell^* = 0.2000 - 0.2111$. The channel width for two-dimensional simulations is $w_d/\ell^* = 45 \cdot \Delta y/\ell^* = 9.50$ or 9.26. Other constants are $\eta = 1/8$ and $\mu = 1/48$.

Results and Discussion

Instability of One-Dimensional CJ Detonation

Although one-dimensional Chapman-Jouguet detonation is unstable for the normal gas mixtures, a steady solution for such detonation can be obtained from one-dimensional unsteady finite-difference equations. It is postulated that the detonation is stabilized by the artificial diffusion inherent in the finite-difference equations.

To determine the effect of artificial diffusion on the instability of a steady one-dimensional CJ detonation,

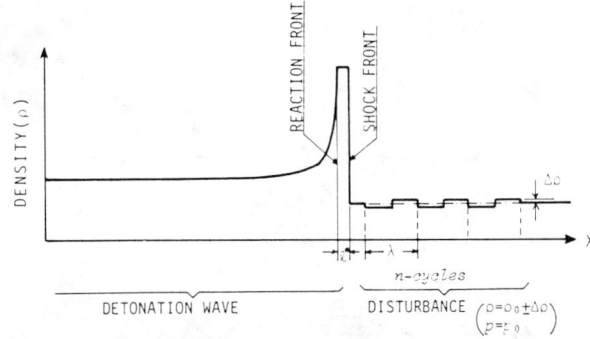

Fig. 1 Initial conditions shown as the density profile around the detonation front. One-dimensional steady CJ detonation is disturbed by n cycles of alternating density variations put in front of the leading shock.

numerical simulations are carried with added density disturbances, as shown in Fig. 1. N cycles of alternating density variations are put in front of the detonation. For space mesh size $\Delta x/\ell^* = 0.2058$, density variation $\Delta\rho/\rho_0 = 0.005$, cycle number n=2, the time development of density profiles and the detonation velocity variation with time are shown in Fig. 2. The effect of the disturbance is noted at time $t/t^* \simeq 13$, after which the detonation decays. Another case, for $\Delta x/\ell^* = 0.2100$ is shown in Fig. 3. Although the added disturbances are stronger, i.e., $\Delta\rho/\rho_0 = 0.03$ and n=3, the detonation does not decay. If $\Delta x/\ell^*$ is reduced to less than 0.200, any steady solution could not be obtained for the present parametric constants [Eq.(12)] with use of the finite-difference scheme [Eqs.(8-10)]. If a first-order accurate finite-difference method such as van Leer's is utilized, steady solutions can be obtained even when $\Delta x/\ell^*$ is less than 0.10.

Reestablishment of ONE-Dimensional Detonation by Explosions

An attenuating one-dimensional detonation, which is obtained in the above section (see Fig. 2), can be reestablished by explosions in the induction region behind the leading shock wave. For the present numerical simulations, instantaneous ignition spots are inserted in the zone between the shock front and the reaction front, when the separation increases to $12 \cdot \ell^*$, as shown in Figs. 4a and 4b, where initial conditions for the two cases are illustrated. The x-t wave diagram for the first example is shown in Fig. 5. An instantaneous ignition spot at time t/t^* equals around 10 generates pressure waves for both directions. The forward

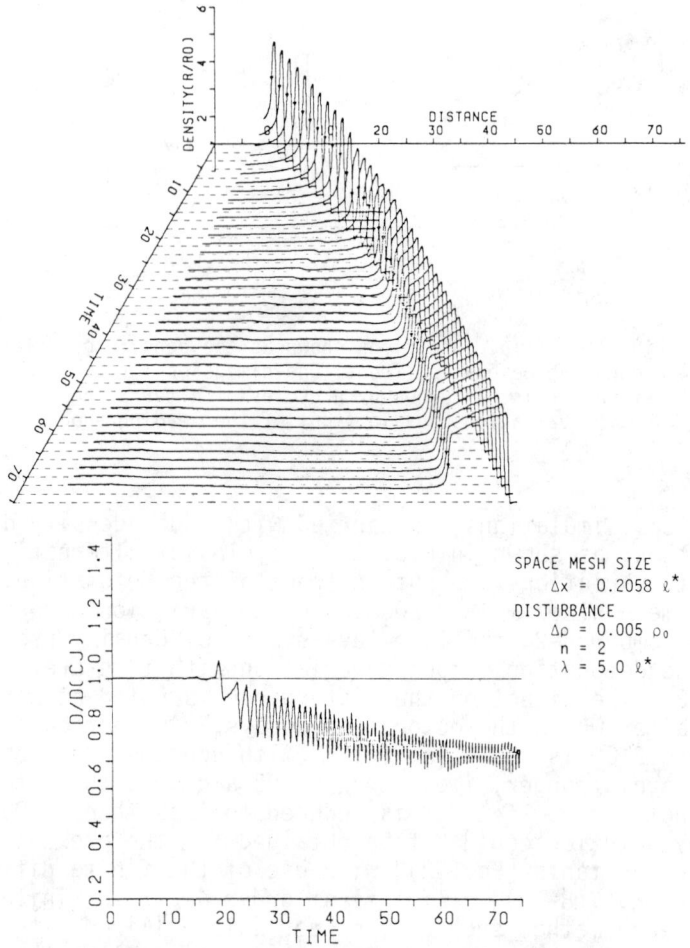

Fig. 2 Time sequences of density profiles and the time variation of leading shock velocity for the case of $\Delta x = 0.2058$, $\Delta \rho = 0.005$, and $n = 2$.

pressure wave overtakes the leading shock and strengthens it. On the contact surface made by this wave interaction, an explosion occurs at $t \simeq 67$. This explosion is not strong enough to initiate a detonation, but makes the leading shock stronger. The next explosion on another contact surface forms overdriven detonation. Figures 6 and 7 show the results for the second example (see Fig. 4b). In this case, two smaller instantaneous ignition spots are inserted between the shock front and the reaction front. Figure 6 shows the x-t wave diagram in the same manner as in Fig. 5,

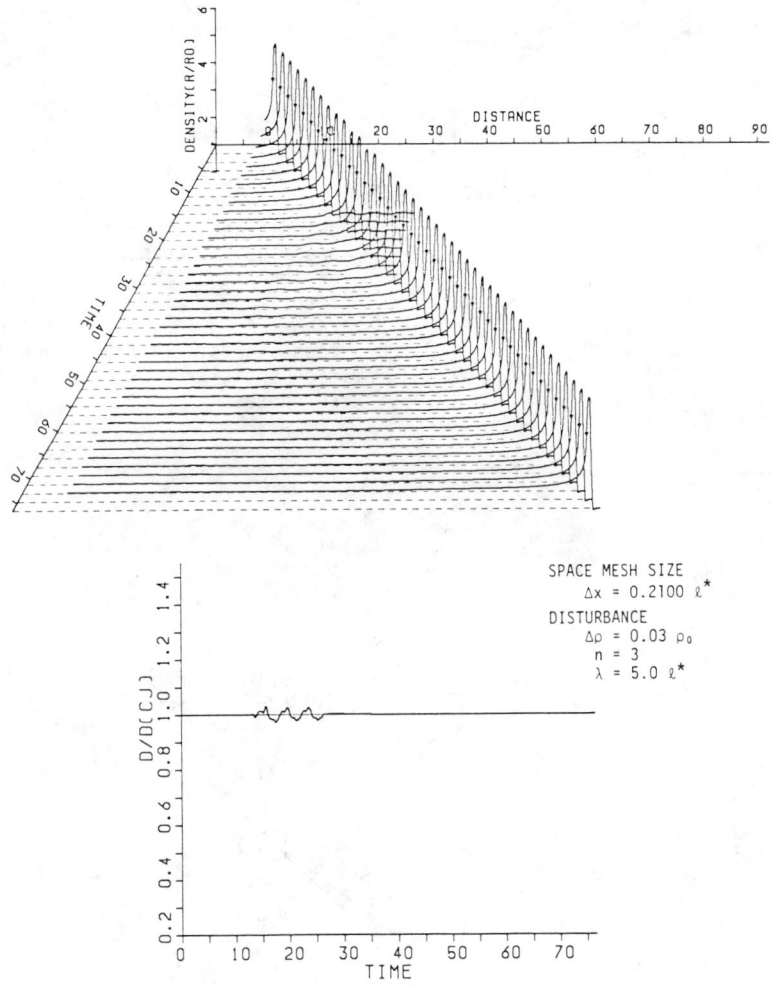

Fig. 3 Time sequences of density profiles and the time variation of leading shock velocity for the case of $\Delta x = 0.2100$, $\Delta \rho = 0.03$, and $n = 3$.

and the time sequences of pressure profiles are shown for the first half-time of the whole simulations in Fig. 7, where dark dots on pressure profiles denote the position of flame fronts. Both of two smaller instantaneous ignition spots generate weaker pressure waves than in the first case. It is clear that the merging of flames also generates pressure waves. Such pressure waves form fairly strong shock waves, as seen in the final pressure profile in Fig. 7. In this case, the second explosion occurs not on the contact surface but on the reaction front.

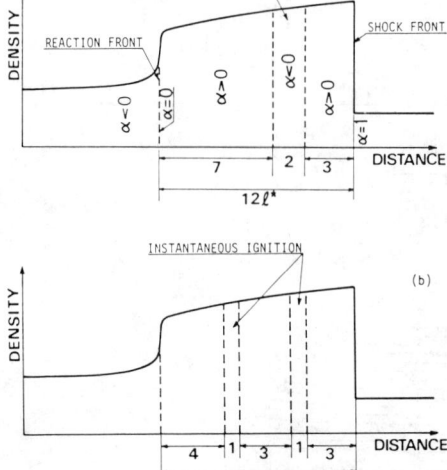

Fig. 4 Initial conditions for the establishment from the decaying detonations, shown as density profiles. α denotes the parameter of induction reaction progress. At first, $\alpha = 1$; then it reduces to zero. At $\alpha = 0$, exothermic reaction starts. Instantaneous ignition spots are inserted as disturbances.

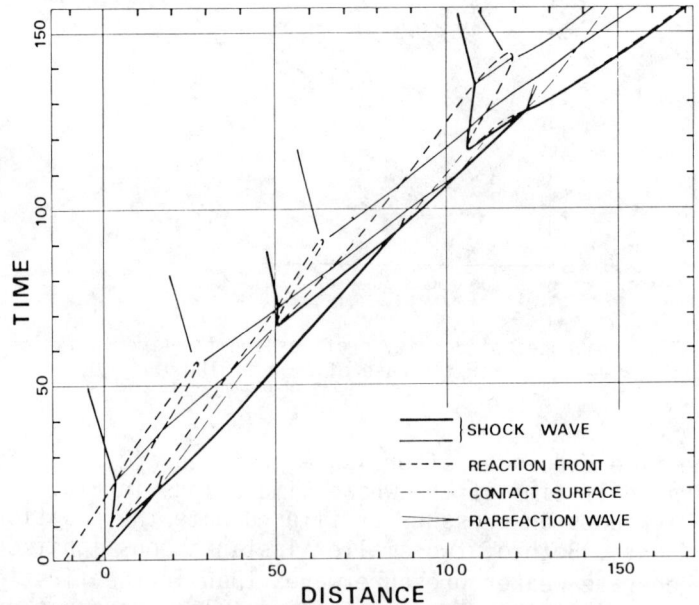

Fig. 5 The x-t wave diagram of the establishing process for the case of Fig. 4a. At time around 10, when distance from shock to reaction front equals 12, an instantaneous ignition spot is inserted as a disturbing source.

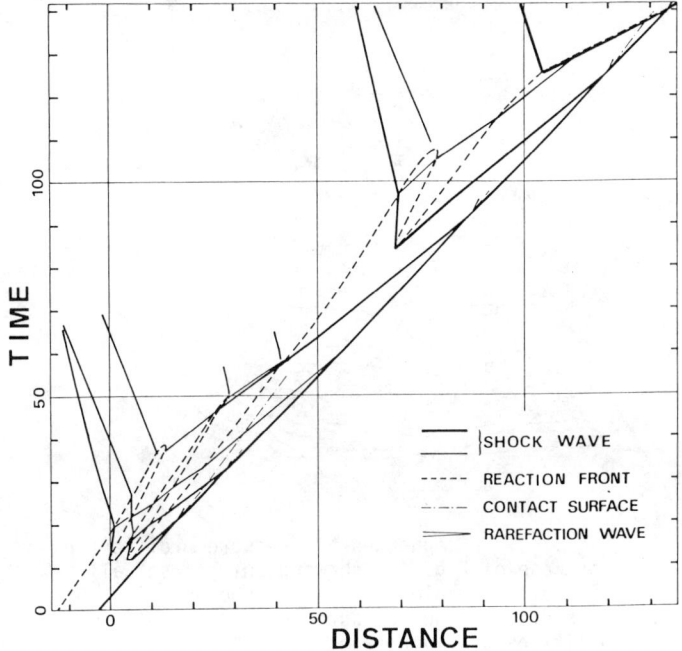

Fig. 6 The x-t wave diagram of the establishing process for the case of Fig. 4b. Two instantaneous ignition spots are inserted to generat more, but weaker, pressure waves than in the case of Fig. 5.

Although these two simulation cases are one dimensional, it is suggested that a series of explosions plays an important role in deflagration to detonation transition. Such phenomena as the successive explosions followed by detonation were observed in the experiments of Urtiew and Oppenheim (1966). It should be noted that the size of the space mesh has little effect on the results of these simulations, since the phenomena considered are not controled mainly by diffusion.

Establishment of Two-Dimensional Detonation

It is clear from above discussions that the plane shock-combustion wave structure is unstable without the effect of stabilization by the artificial diffusion of the finite-difference scheme. When a plane CJ detonation is disturbed by density inhomogeneities in two-dimensional space, as shown in Fig. 8a, the distance from shock front to reaction front increases as time goes on (Fig. 8b) (i.e., detonation is decaying). It is expected that the detonation in multi-

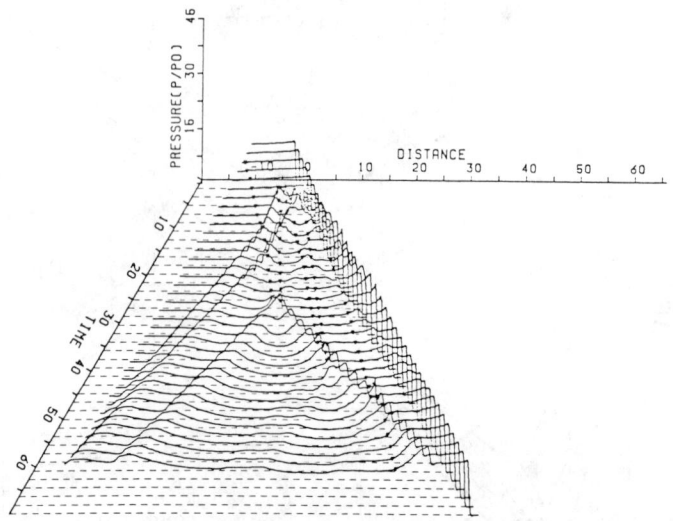

Fig. 7 Time sequences of pressure profile for the case of Fig. 4b, showing only first half time of the whole simulations shown in Fig. 6. Dark dots on the pressure profiles denote the positions of flame fronts.

dimensional space is self-sustained only when the triple-shock structure is formed by strong explosions. In the present simulations, a plane CJ detonation in a two-dimensional channel is disturbed so strongly by a couple of exothermicity spots as to cause an explosion. The arrangement and size of the spots are illustrated in Fig. 9 as the initial conditions. All the parameters are fixed except $\Delta\beta$, so that the parameter on the strength of disturbance is $\Delta\beta$. This arrangement of exothermicity spots generates a disturbance mainly for the lateral direction to the channel. The channel width equals 9.5 ℓ^*, and the space grid size Δx and Δy are even and equal 0.2111 ℓ^*. Figures 10 and 11 show the sequences of isobars after CJ detonations are disturbed for the cases of $\Delta\beta$ = 0.4 and 0.3, respectively, where the intervals of isobars are $5 \cdot p_0$, and the broken lines show the exothermic reaction front. Since the frames of isobars in Figs. 10 and 11 are shifted with CJ propagation velocity D_{CJ}, it is easy to compare the leading shock velocity with D_{CJ}. Immediately after the plane detonations are disturbed, the shock velocities begin to reduce and the induction lengths to increase. In Fig. 10, the transverse pressure wave generated by the initial added disturbance propagates to the wall of y=9.5 at time ≃ 13, followed by the reflection at the wall (time ≃ 15), which causes an explosion at

NUMERICAL SIMULATIONS OF DETONATION

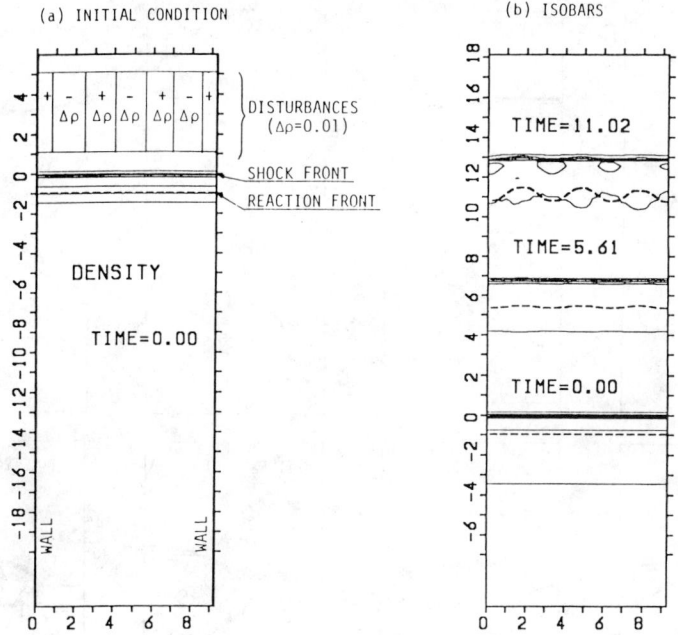

Fig. 8 a) Initial conditions to disturb a plane CJ detonation wave propagating in a two-dimensional channel. Density disturbances are inserted laterally in front of the detonation. b) Results for the initial condition (a), shown as contours of pressure at times 0.0, 5.61, and 11.02. Broken lines denote exothermic reaction fronts.

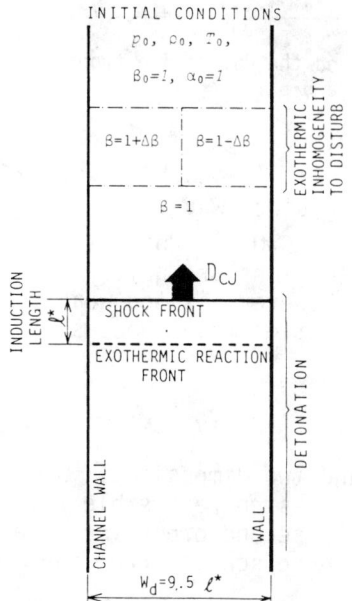

Fig. 9 Initial conditions for the simulations of the transient process of detonation in a two-dimensional channel. A plane CJ detonation is strongly disturbed by a pair of exothermic inhomogeneity, where the parameter of exothermic reaction progress $\beta = 1+\Delta\beta$, and $1-\Delta\beta$. The space mesh size $\Delta x = \Delta y = 0.2111\ \ell^*$.

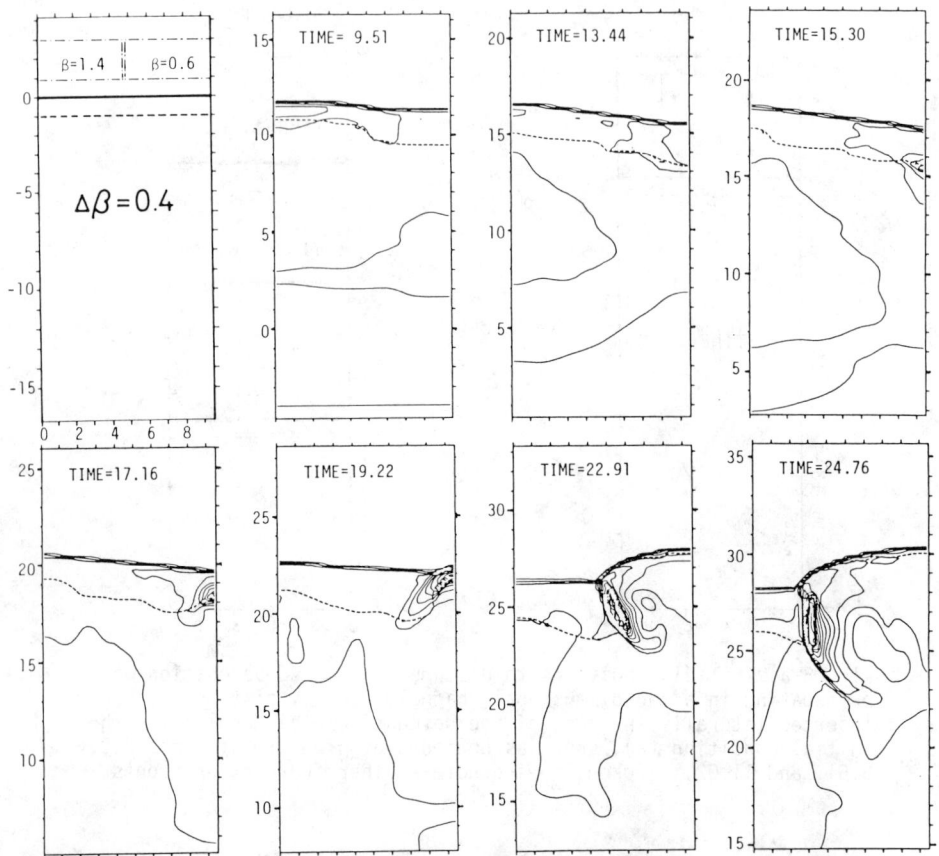

Fig. 10 Time sequences of isobars around the detonation front, when $\Delta\beta = 0.4$. Intervals of isobars are $5 \cdot p_0$.

time $\simeq 17$. The blast wave produces a triple-shock structure of detonation coupled with the leading shock wave. In Fig. 11, the disturbance is a little weaker ($\Delta\beta=0.3$). Until the explosion occurs (time $\simeq 17$), the situations are quite similar to Fig. 10. The blast wave, however, is too weak to couple with the flame, so that the reaction front receds from the leading shock.

Conclusions

Numerical simulations of one- and two-dimensional unsteady gaseous detonations with a two-step reversible reaction model are performed using the second-order accurate MacCormack-FCT explicit finite-difference scheme. From the

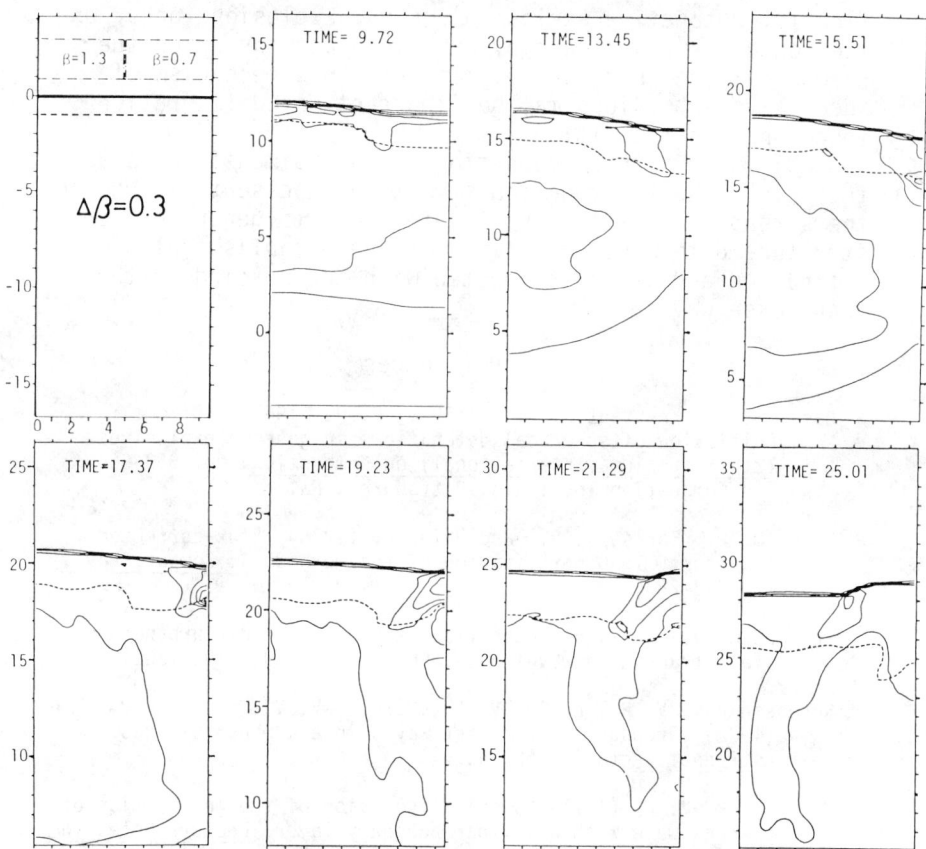

Fig. 11 Time sequences of isobars around the detonation front, when $\Delta\beta$ = 0.3. Broken lines denote the exothermic reaction fronts.

results, it is concluded that:

1) Although one-dimensional Chapman-Jouguet detonation is unstable for normal gas mixtures, the numerical solution of such a detonation can be obtained with one-dimensional unsteady finite-difference equations, if the artificial diffusion included in the scheme is sufficient to stabilize the unstable detonation. The one-dimensional CJ detonation, stabilized by the artificial diffusion of the finite-difference scheme, is decayed by a density disturbance. The strength of the disturbance necessary to decay the detonation depends on the space mesh size Δx, which affects the artificial diffusion.

2) An explosion in shock-heated or flame-heated gas is followed by subsequent explosions, which strengthens the leading shock. The gases are heated successively and,

finally, detonative explosion occurs. Explosions occur on the contact surface made by wave interactions or on the reaction front in these simulations. This mechanism of successive explosions may be important for deflagration to detonation transition.

3) To sustain a detonation in multidimensional space, the triple-shock detonation structure is essential. Whether the strength of explosion is sufficient to generate such structure determines the limit or the establishment of detonation and may be connected with the spacing of the transverse waves.

References

Bach, G. G., Knystautas, R., and Lee, J. H. (1969) Direct initiation of spherical detonations in gaseous explosives. 12th Symposium (International) on Combustion, pp. 853-864. The Combustion Institute, Pittsburgh, Pa.

Book, D. L., Boris, J. P., and Hain, K. (1975) Flux-corrected transport II: Generalization of the method. J. Comput. Phys. 18, 248-283.

Erpenbeck, J. J. (1966) Detonation stability for disturbances of small transverse wavelength. Phys. Fluids 9, 1293-1306.

Korobeinikov, V. P., Levin, V. A., Markov, V. V., and Chernyi, G. G. (1972) Propagation of blast waves in a combustible gas. Astronaut. Acta 17, 529-537.

MacCormack, R. W. (1971) Numerical solution of the interaction of a shock wave with a laminar boundary layer. Lecture Notes in Physics, Vol. 8, pp. 151-163. Springer-Verlag, New York.

Taki, S. and Fujiwara, T. (1978) Numerical analysis of two-dimensional nonsteady detonations. AIAA J. 16, 73-77.

Taki, S. and Fujiwara, T. (1981) Numerical simulation of triple shock behavior of gaseous detonation. 18th Symposium (International) on Combustion, pp. 1671-1681. The Combustion Institute, Pittsburgh, Pa.

Urtiew, P. A. and Oppenheim, A. K. (1966) Experimental observations of the transition to detonation in an explosive gas. Proc. Roy. Soc. Sec. A 295, 13-28.

A Shock Tube Study of the Chlorine Azide Decomposition

C. Paillard* and G. Dupré†
*Centre National de la Recherche Scientifique
et Université d'Orléans, France*
and
N. A. Fomin‡
Academy of Sciences, Minsk, USSR

Abstract

The thermal decomposition of chlorine azide N_3Cl in argon diluant was studied behind incident shock waves within a temperature range of 500 - 1200 K, between 50 and 600 Torr. The reaction was monitored by observation of emitted radiation at several wavelenghts in uv, visible, and i.r. ranges. The N_3Cl consumption was deduced from an emission at 4.89 μm. The effect of reaction exothermicity on shock wave propagation was notable with mixtures containing 5 - and 2- mole % N_3Cl. A coupling between the reaction and the shock wave was observed above a minimum shock strengh. Before the onset of the coupling, an induction period was detected. When the shock strength was increased, the induction period cannot be observed in the 5 - mole % N_3Cl mixture ; and the wave rapidly accelerated but remained at a velocity lower than the theoretical Chapman-Jouguet velocity. With the mixture containing 0.5 - mole % N_3Cl, the exothermicity effect was small. In addition to the

---

Presented at the 9th ICODERS, Poitiers, France, July 3-8, 1983. Copyright © American Institute of Aeronautics and Astronautics, Inc., 1984. All rights reserved.
*Maître Assistant, Laboratoire de Cinétique Chimique de l'U. E.R. Sciences Fondamentales et Appliquées et Centre de Recherches sur la Chimie de la Combustion et des Hautes Températures.
†Chargée de Recherche, Centre de Recherches sur la Chimie de la Combustion et des Hautes Températures.
‡Head of Laboratory, Heat and Mass Transfer Institute.

pyrolysis of N_3Cl, self-photolysis was evident. Before the shock wave transfer, radiation induced partial decomposition in front of the shock wave was observed. For the lowest pressures and the leanest mixture in N_3Cl, self-photolysis was not detected. In this case, N_3Cl half-life results led to an apparent first-order rate constant :

$$k_I \,(s^{-1}) = 3.7 \; 10^6 \exp\left[-4550/T(K)\right]$$

characterized by a very low value of the activation energy ($E_a = 39 \pm 1.5$ kJ mole^{-1}). The validity of the rate constant value is the topic of a further investigation.

Introduction

Since gaseous azides such as N_3X (X = H, Cl, Br) are compounds of great instability and exothermicity (Fair and Walker 1977), they liberate a large quantity of energy on decomposition. Because of their relative molecular simplicity, a small number of final products are produced. As a consequence gaseous azides are of fundamental interest in the verification of models of propagating flames (Dupré et al. 1975) or of heat tranfer behind a detonation wave (Paillard et al. 1979). They are often used as a chemical source of radicals such as N_3, NH, NCl, or NBr for elementary reactions studies (Setser 1979) and also used as an energy source for chemically pumped lasers. §

Previous experimental studies of explosive decomposition of N_3Cl at low pressure led to some interesting peculiarities. The critical diameters for flame quenching are smaller than those observed at a given pressure for any other system, and a detonation propagation velocity through very small diameter tubes is about one half of the predicted Chapman-Jouguet velocity (Paillard et al. 1973; 1974).

A rather qualitative investigation of N_3Cl thermal decomposition (Gleu 1926) showed that decomposition begins at about 400°C with the formation of nitrogen and chlorine and production of an intense emission between 628 and 667 nm. This emission, attributed to (b$^1 \Sigma^+ \rightarrow$ X$^3 \Sigma^-$) NCl transition, was observed during N_3Cl laser photolysis by Coombe et al. (1981) and Piper et al. (1982). Their studies, mainly

---

§ See Basov et al. (1969) ; Helmsin (1972); Rice and Jensen (1972) ; Zaslonko et al. (1972); Piper et al. (1980).

directed towards electronically and vibrationally excited NCl radicals, produced anomalous results.

Therefore, it was considered useful to define precisely the conditions which initiate the explosive reaction so as to obtain a coupling between the reaction zone and the shock wave and thereby some information about the rate and the mechanism of N_3Cl thermal decomposition.

Experimental

Gaseous chlorine azide was generated by acidification of an aqueous solution of sodium hypochloride and sodium azide with boric acid (Rashig 1908). The gas emitted flowed slowly through a $CaCl_2$ drying tube and was stored at pressures less than 30 Torr and maintained in a dark reservoir at 0°C. The purity of N_3Cl was checked by means of mass spectrometry and i.r. absorption. The maximum concentration of hydrogen azide was found to be 1.5 %. The decomposition products were molecular chlorine and nitrogen. The shock tube (previously described by Dupré et al. (1982) and the gas generation apparatus were constructed of Pyrex glass because of the corrosive nature of chlorine azide. As the KRS 5 observation windows for i.r. emission were gradually damaged by chlorated products, CaF_2, suitable for uv, visible emissions, and for i.r. emissions, was used.

Helium was used as a driver gas. Thin diaphragms of terphane (5 - 13 μm thick according to pressure range) were used to avoid limitation of the gas flowing rate during the diaphragm bursting. Gas parameters behind the incident shock were calculated from the jump conditions with the observed shock velocities which were measured over a series of intervals by tiny piezoelectric pressure gages (LEM 20H 48) mounted flush with the inside wall. An interference filter, centered at 4.89 μm ($\Delta\lambda = \pm 0.06$ μm), and a cooled liquid nitrogen InSb detector were used to monitor this emission. Other filters centered at 208.5 nm ($\Delta\lambda$ = 13 nm), 270 nm ($\Delta\lambda$ = 9.7 nm), 326.5 nm ($\Delta\lambda$ = 9.3 nm), 660.8 nm ($\Delta\lambda$ = 11 nm) were chosen to monitor the time dependent evolution of the reaction intermediates and products.

Results

Pyrolysis of chlorine azide in argon diluant at concentrations of 95, 98 and 99.5 mole was studied over a range of pressures (50 < P_2 < 600 Torr) and of temperatures (500 < T_2 < 1200 K) behind the incident shock wave. Figure

1 shows an intense band at 4.89 µm in the infrared absorption spectrum of N_3Cl. Under these conditions, at this specific wavelength, the heated molecule could emit a detectable signal for temperatures equal to or higher than 450 K. Because of the high exothermicity of N_3Cl decomposition ($\Delta H°_{298}$ = -93 kcal. mole^{-1})(Paillard et al. 1967) and its strong sensitivity to light, the mechanism of pyrolysis was difficult to elucidate.

Behavior of $\{N_3Cl-Ar\}$ Mixtures Behind an Incident Shock Wave

The incident shock temperature must be higher than 540K to induce thermal decomposition of N_3Cl behind incident shock at a postshock pressure of 100 Torr. Between 550 and 600 K, an induction period was observed before a rapid decomposition. With the mixtures containing 5- and 2- mole % N_3Cl, the induction period decreases inversely with the distance between the shock origin and the observation station. This effect is a consequence of the high

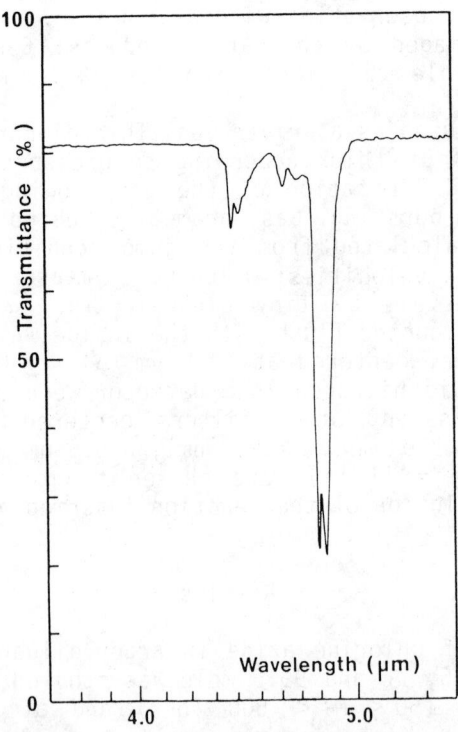

Fig. 1 Infrared absorption spectrum of gaseous N_3Cl.

reaction exothermicity, which induces a signifiant pressure rise as shown on simultaneous recordings, at C_4 station, of pressure and emissions at 208 nm (A) and 326 nm (B) presented in Fig.2 . The third oscillogram (C) is a simultaneous recording of the overall infrared emission between 1 and 6 μm and of pressure behind the incident shock at an observation point C_5 (located at a distance of 100 mm from C_4 station). On the oscillogram (C) it is clear that the initiation of the reaction occurs just after the shock wave and that exothermicity causes a second pressure rise which appears much nearer to the shock front than on oscillograms (A) and (B).

Figure 3 gives the oscillograms obtained at a similar temperature but at a pressure twice as high as just before in Fig. 2. At the observation point C4, an induction

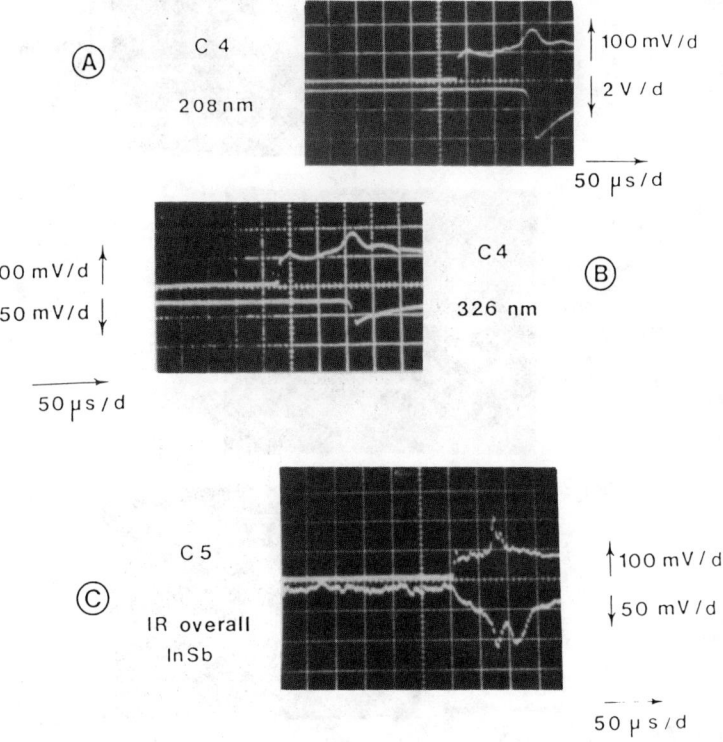

Fig. 2 Simultaneous recording of the pressure and the emission signals at 208 nm (A), 326 nm (B), and the overall infrared emission between 1 and 6 μm (C) behind the incident shock. C_4 and C_5 represent the positions of the fourth and fifth piezoelectric gages. Distance $C_4 - C_5$ = 100 mm. Mixture : {0.05 N_3Cl + 0.95 Ar}. T_2 = 595 K ; P_2 = 76 Torr ; P_2/P_1 = 2.392.

period is still observed [oscillograms (A) and (B)] then strong coupling occurs between the shock wave and the reaction zone [oscillogram (C)]. This coupling is complicated by a self-photolysis phenomenon which induces a partial decomposition before the shock wave. Self-photolysis was observed with mixtures containing 5- and 2-mole % argon when the induction period became negligible, that is to say when temperature was equal to or higher than 600 K. This phenomenon was shown on the infrared emission and visible, uv emissions at 660, 326, and 250 nm. The 208-nm emission was detectable just after the passage of the wave. A different mechanism for pyrolysis and photolysis has to be considered, not only for the initiation step

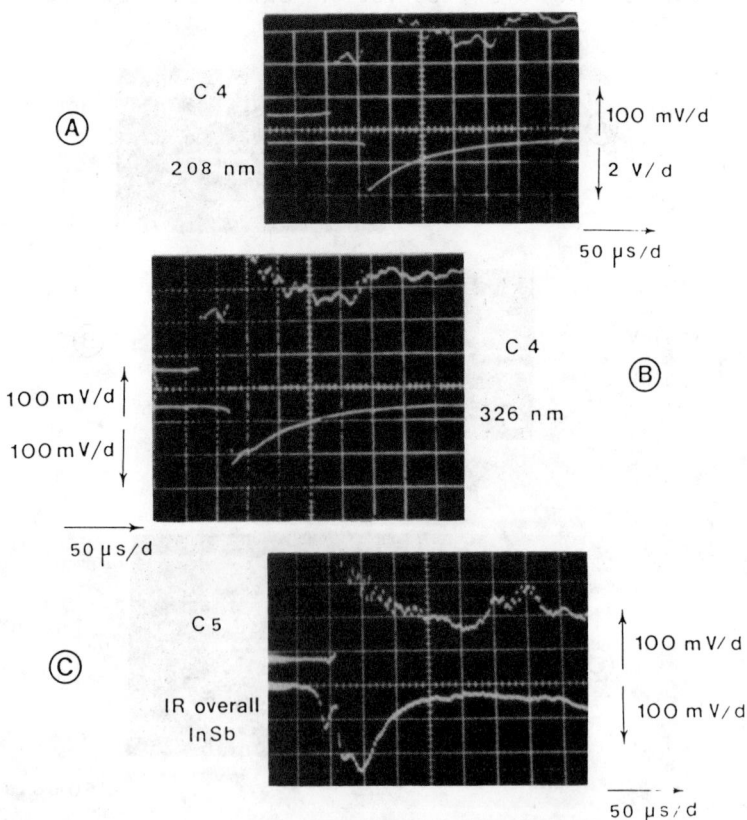

Fig. 3 Simultaneous recording of the pressure and the emission signals at 208 nm (A), 326 nm (B), and the overall infrared emission between 1 and 6 μm (6) behind the incident shock. C_4 and C_5 represent the positions of the fourth and fifth piezoelectric gages. Distance $C_4 - C_5$ = 100 mm. Mixture : {0.05 N_3Cl + 0.95 Ar}. T_2 = 588 K ; P_2 = 149 Torr ; P_2/P_1 = 2.376.

which cannot produce emitting species at such a short wavelength, but also for further reaction steps.

Simultaneous recordings of the shock pressure and the overall infrared emission (between 1 and 6 µm) are shown in Fig. 4 for three different shock pressures and temperatures. Each oscillogram represents one different scheme: on oscillogram (A), a radiation from the heated gas without reaction, which is indicated by a constant pressure and a constant emission signal ; on (B), the reaction preceded by a long induction period, characterized by a pres-

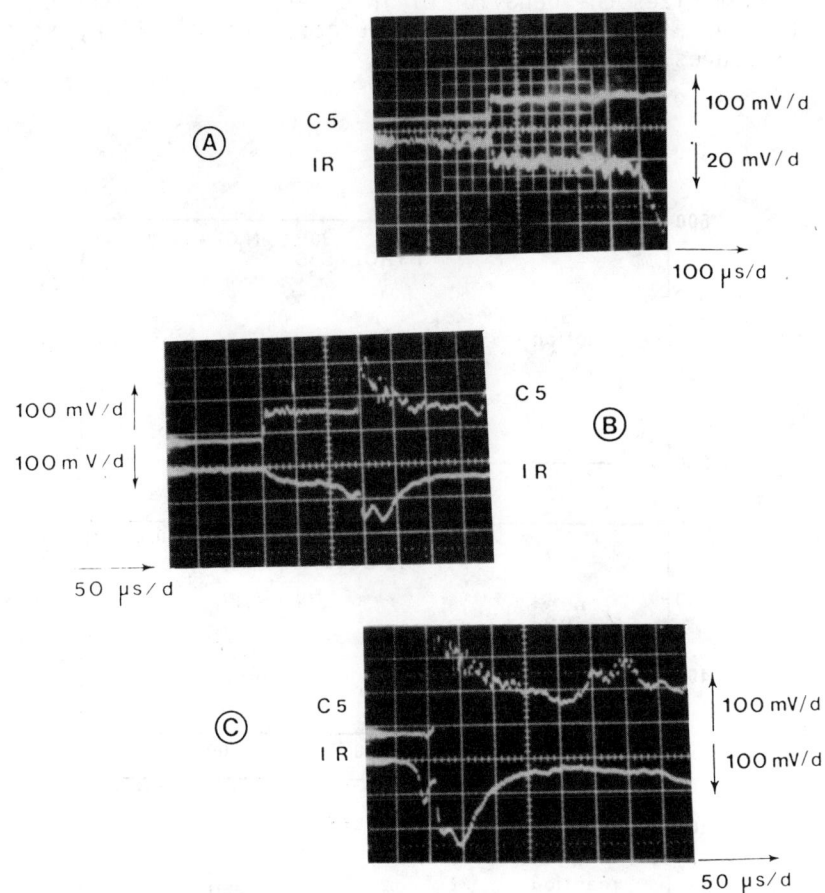

Fig. 4 Simultaneous recording of the pressure and the overall infrared emission between 1 and 6 µm for three different pressures and temperatures behind the incident shock, at the same observation point C_5. Mixture : $\{0.05\ N_3Cl + 0.95\ Ar\}$. A : T_2 = 485 K ; P_2 = 102 Torr ; P_2/P_1 = 1.998 ; B : T_2 = 558 K ; P_2 = 133 Torr ; P_2/P_1 = 2.278 ; C : T_2 = 588 K ; P_2 = 149 Torr ; P_2/P_1 = 2.376.

sure rise and an increase of emission; on (C), the photolysis phenomenon in front of the shock front, followed by the thermal decomposition. These results indicate that N_3Cl is extremely sensitive to visible or uv radiations. Piper et al. (1982) have observed that complete decomposition of N_3Cl occurs after a laser shot of 450 μJ at a 300-nm wavelength while several shots were necessary to induce a partial photolysis of HN_3. Among the numerous experiments carried out behind an incident shock wave in HN_3 mixtures, self-photolysis has never been observed before thermal decomposition (Dupré et al. 1982). Figure 5 summarizes the behavior of {N_3Cl-Ar} mixture with regard to the incident shock as a function of dilution, shock pressures, and temperatures.

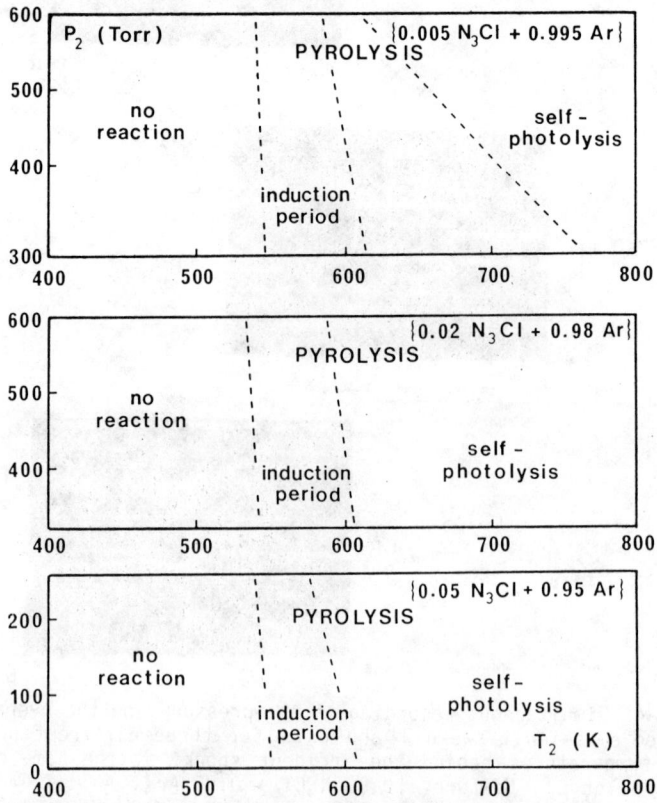

Fig. 5 Behavior of {$N_3Cl - Ar$} mixtures towards an incident shock wave as function of shock pressures and temperatures.

Effect of the Exothermicity of the Reaction

The exothermicity of the reaction has a noticeable influence on the shock propagation and consequently on the gas parameters behind the incident wave. Highly energetic reactions can induce a coupling between the shock wave and the reaction zone and can result in the formation of a detonation wave. However, in the 5- mole % N_3Cl mixture, a self-sustained detonation with a constant velocity was not observed. The existence of some coupling is related to the shock strength (expressed as P_4/P_1, that is the initial applied pressure ratio on either side of the dia-

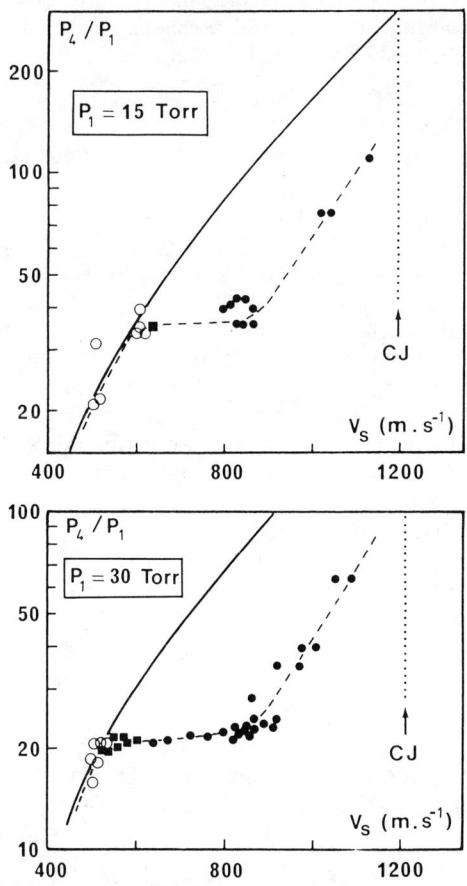

Fig. 6 Incident shock velocity as a function of the logarithm of shock strength, for two different initial pressures. Mixture : $\{0.05\ N_3Cl - 0.95\ Ar\}$.

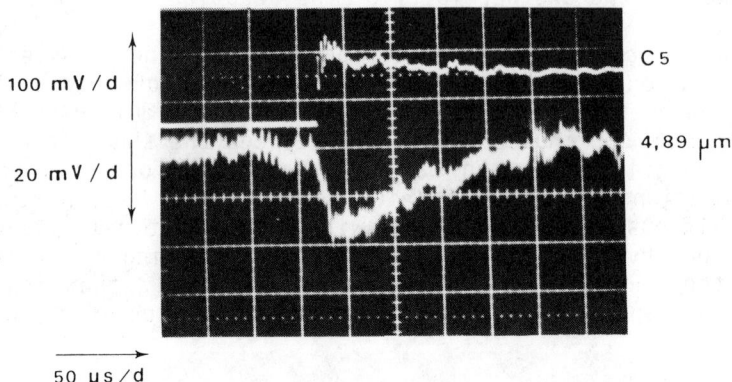

Fig. 7 Evolution of the pressure and the emission signal of N_3Cl at 4.89 μm behind the incident shock. Mixture: $\{0.005\ N_3Cl + 0.995\ Ar\}$; $T_2 = 639$ K; $P_2 = 467$ Torr; $\rho_2/\rho_1 = 2.373$.

phragm). For a given diaphragm thickness, the velocity V_s of the shock without reaction depends on P_4/P_1 and also on the gas nature. The full curve drawn in Fig. 6 represents the relation between $\log(P_4/P_1)$ and V_s. This curve was obtained from experiments carried out in pure argon and in nonreactive mixtures of a composition near to that of the studied mixture. For shock temperatures lower than that at which the pyrolysis can occur in the mixture, the experimental points are fitted with the curve shown in Fig. 6. As soon as pyrolysis has been initiated (that is to say for shock temperatures and pressures higher than the critical values), the experimental points deviate significantly from the expected curve. The speed of the shock front increases rapidly but never reaches the Chapman-Jouguet detonation velocity (Paillard, 1973). This results is in agreement with a previous determination of detonation limits of $\{N_3Cl\text{-diluent}\}$ mixtures : with argon, the composition limit corresponds to 92- mole % Ar.

It was observed that the initiation of a reaction depends on P_4/P_1. The shock must be all the stronger since the initial pressure of the shocked gas is low. These results are in good agreement with those which Bradley et al.(1980) obtained with hydrogen-oxygen mixtures.

As dilution increases, the effect of exothermicity is attenuated. Shock acceleration is still observable for the 98 - mole % Ar mixture) but it is almost reduced to zero for the mixture containing 99.5- mole % Ar.

Pyrolysis of Chlorine Azide

When the exothermicity of the reaction and the self-photolysis downstream the shock front become negligible, some kinetic parameters concerning N_3Cl pyrolysis are brought about at temperatures and pressures calculated from shock speed measurements. The reaction rate can be obtained from half-life measurements deduced from the 4.89-μm emission signal. A typical oscillogram is given in Fig. 7.

In the operating pressure range of 100-520 Torr, the half-life time is found to be nearly independant of shock pressure. If the overall decomposition reaction is assumed to be of first order, the rate constant k_I deduced from experiments (Table I) performed with 99.5-mole % Ar mixture between 580 and 950 K (Fig. 8) can be expressed as

$$k_I \ (s^{-1}) = 3.7 \ 10^6 \ \exp\left[-4550 \ / \ T \ (K)\right]$$

A low activation energy E_a, equal to 39 ± 1.5 kJ mole^{-1} is found to be in good agreement with the values obtained previously from other studies (Paillard 1973) on the

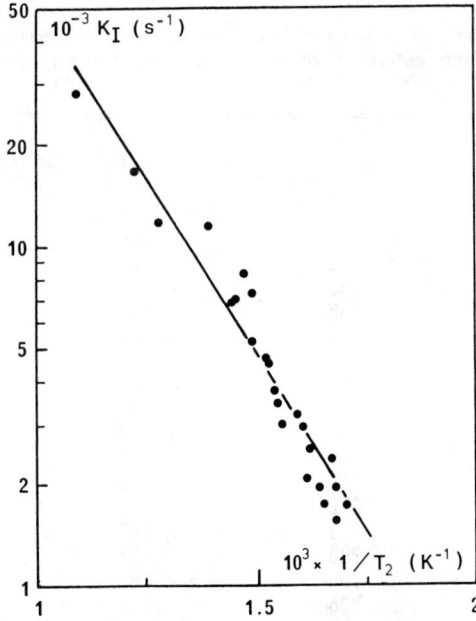

Fig. 8 Logarithmic Arrhenius plot of the rate constants of N_3Cl decomposition vs inverse shock temperatures. Mixture : $\{0.005 \ N_3Cl + 0.995 \ Ar\}$.

critical conditions of the propagation of laminar flames and detonation waves in chlorine azide (E_a = 35 kJ mole^{-1}). A higher value (E_a = 54.4 kJ mole^{-1}) was determined from measurements in a fast flow reactor (Combourieu et al. 1977). The rate constants of this reaction (assumed to be of first order) were determined between 630 and 850 K. They are about 100 times lower than the values found in this work. This disagreement can be explained as follows. The fast-flow reactor is not a well-suited equipment to study thermal decomposition of a single reagent, since the reaction has a strong heterogeneous character ; and the reaction order, equal to one, may correspond to the initial step of a wall catalytic mechanism. Moreover, the partial pressures of N_3Cl mixed with 99.9 - mole % helium were in these experiments 2000 - 5000 times lower than in our experiments. At such pressure conditions it is likely that a second-order reaction occurs in the gaseous phase. Nevertheless the value determined in the present study can be questioned for the following reasons since

1) The exothermicity effect of the reaction is not quite negligible, even at 0.5-mole % N_3Cl. Also, at a highest dilution, the detector noise/i.r. signal ratio becomes so

Table 1 Summary of experimental data for the determination of the first-order rate constant of N_3Cl thermal decomposition k_I. Mixture : {0.005 N_3Cl + 0.995 Ar}

$T_2(K)$	P_2(Torr)	P_2/P_1	$k_I(s^{-1})$
586	398	2.226	1711
595	388	2.252	1952
598	441	2.250	2350
610	452	2.297	1953
617	528	2.307	2520
623	447	2.333	2950
626	497	2.333	3381
635	381	2.357	2988
646	264	2.389	3466
648	419	2.384	3788
655	402	2.406	4673
661	275	2.427	4560
672	423	2.447	5096
673	283	2.451	7110
693	168	2.488	6930
720	306	2.533	11550
785	280	2.650	11950
817	277	2.706	16500
915	100	2.851	28000

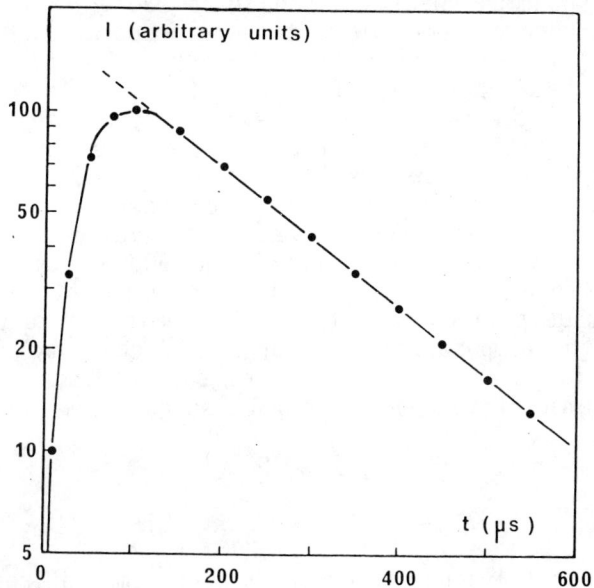

Fig. 9 Emission signal at 660 nm wavelength. Mixture : {0.005 N_3Cl + 0.995 Ar}. T_2 = 795 K ; P_2 = 285 Torr ; P_2/P_1 = 2.667.

important that the determination by i.r. emission of N_3Cl half-life is no longer possible.

2) The i.r. signal feature, the existence of an induction period near the critical conditions of thermal decomposition, as well as the presence of radicals such as NCl or Cl probably formed during the initiation step, lead to the assumption of a chain mechanism.

3) The rate constants measured at the highest temperatures were obtained at lower pressures than those determined at smaller temperatures. If the overall order exceeds the unity, then the activation energy would be underestimated.

4) N_3Cl decomposition may proceed not only through a thermal but also a photolytic mechanism.

These remarks are supported by the fact that the half-life time extrapolated to 273 K should be equal to 5 s. In reality, gaseous chlorine azide stored at this temperature in darkness remains stable for several days. The reaction mechanism has to be carefully established because of the contribution of photodissociation to N_3Cl pyrolysis. Nevertheless one can make the following remarks.

1) The analysis of the decreasing 660- nm emission (Fig. 9) shows that the radiative lifetime τrad of the

corresponding species is equal to 230 ± 20 µs. This value is in good agreement with the one proposed by Piper et al. (1982) for NCl ($b^1 \Sigma^+$) : 250 ± 40 µs but is much smaller than that reported by Coombe et al. (1981) : 630 ± 70 µs. In the present work, the rate constant of the quenching by argon, k_{Ar}, can be roughly assessed from the curve $K = f(P_2/T_2)$ with $K = 1/\tau$ rad + k_{Ar} [Ar], traced in Fig. 10. k_{Ar} is found to be of the order of $0.5\ 10^{-16}$ cm^3 mole^{-1} s^{-1}, which seems a reasonable value.

2) At temperatures higher than 900 K, interference of a different species which also emits at about 660 nm, was made obvious from the broken emission curve. The higher the temperature, the larger the quantity of this species is generated. To initiate the decomposition of N_3Cl two competitive pathways can be suggested :

$$N_3Cl \longrightarrow NCl + N_2 \quad (1)$$
$$ \longrightarrow N_3 + Cl \quad (2)$$

Reaction (2) tends to occur at high temperatures.

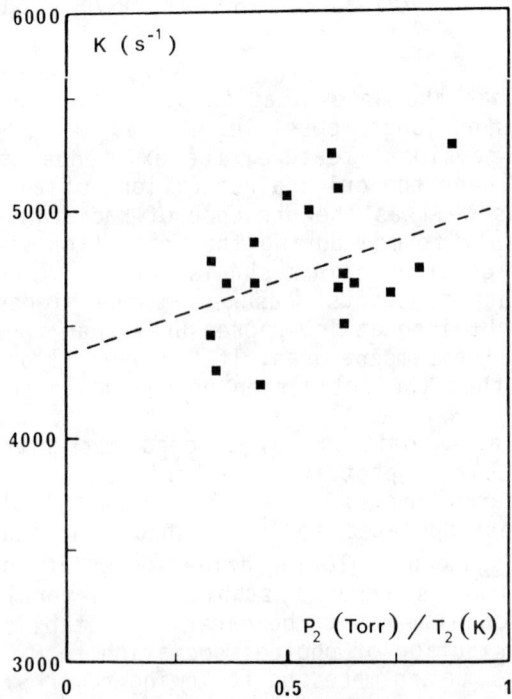

Fig. 10 Effect of shock pressure/shock temperature ratio on the rate constant of the emission decrease at 660-nm wavelength. Mixture : {0.005 N_3Cl + 0.995 Ar}.

3) Other radiation was monitored behind the incident shock front. Emissions at 208- and 250- nm wavelengths are likely due to excited nitrogen, deriving from the highly exothermic and fast reaction (Paur 1973):

$$2 N_3 \longrightarrow 3 N_2 \quad (\Delta H°_{298} = -832 \text{ kJ mole}^{-1})$$

The ultraviolet absorption spectrum of N_3Cl shows a strong band in the vicinity of 205 nm and a smaller one at 250 nm. It follows that a resonant absorption may occur and initiate the photolytic decomposition of N_3Cl, which would contribute to the initiation of the thermal decomposition.

Conclusion

This shock tube study led to a more precise definition of the behavior of N_3Cl heated by an incident shock and to evidence of its extreme sensitivity to light. In a restricted range of pressure (100 - 520 Torr) and temperature (600 - 950 K), N_3Cl decomposition was characterized by an overall reaction of first order with a rate constant equal to :

$$k_I \; (s^{-1}) = 3.7 \; 10^6 \exp \left[- 4550/T \; (K)\right]$$

However, the reaction mechanism remains complex and its complete description requires additional studies on N_3Cl vibrational relaxation times, on the induction period, and also on the evolution of species produced during the reaction.

References

Basov, N.G., Gromov, V.G., Koshelev, E. N., Markin, E.P., and Oraevskii, A. N. (1969) Emission stimulated by explosion of HN_3 in CO. Zh ETF Pis. Red. (Soviet Physics JETP Letters) 10 (1), 5-8.

Bradley, J.N., Capey, W.D., and Farajii, F. (1980) The effect of reaction exothermicity on shock propagation. 12th Symposium (International) on Shock Tubes and Waves, (edited by A. Lifshiftz and J. Rom) pp. 524-532. Jerusalem. The Magnes Press.

Combourieu, J., Le Bras, G., Poulet, G., and Jourdain, J.L. (1977) Reaction $Cl + N_3Cl$ and reactivity of N_3, NCl and NCl_2 radicals in relation to the explosive decomposition of N_3Cl. 17th Symposium (International) on Combustion, pp 863-870 The Combustion Institute, Cambridge, Mass.

Coombe, R.D., Patel, D., Pritt, A.T., and Wodarczyk, F.J. (1981) Photodissociation of ClN_3 at 193 and 249 nm. J. Chem. Phys. 75 (5). 2177-2190.

Dupré, G., Paillard, C., and Combourieu, J. (1975) Etude de la flamme de décomposition de l'azoture de chlore par spectrométrie de masse à temps de vol. 2ème Symposium Européen sur la Combustion, Vol. II, pp. 454-459. Section Française du Combustion Institute, Orléans, France.

Dupré, G., Paillard, C., Combourieu, J., Fomin, N.A., and Soloukhin, R.I. (1982) Decomposition of hydrogen azide in shock waves. 13th Symposium (International) on Shock Tubes and Waves (edited by C.E. Treanor and J.G. Hall), pp.626-634. Niagara Falls, N.Y.

Fair, H.D. and Walker, R.F. (1977) Energetic Materials Vol. 1; Physics and Chemistry of the Inorganic Azides, Vol. 2,503 pp.; Technology of the Inorganic Azides, 296 pp. Plenum Press, New York.

Gleu, K. (1926) Die Lichtemission beim Zerfall von Chlorazid. Z. Phys. Chem. 38, 176-201.

Helmsin E. K. (1972) The $HN_3 + CO_2$ laser: An analytical and experimental investigation. Master's Thesis, U.S. Naval Post-graduate School, Monterey, Calif.

Paillard, C., Moreau, R., and Combourieu, J. (1967) Chaleur de décomposition et température maximale de flamme de l'azoture de chlore. C.R. Acad. Sci. Paris C 264, 1721-1722.

Paillard, C. (1973) Etude de la propagation et de l'extinction des flammes de simple décomposition. Thèse de Doctorat d'Etat, Université d'Orléans, Orléans, France.

Paillard, C., Dupré, G., and Combourieu, J. (1973) Etude de la détonation de composés endothermiques gazeux: I. Célérités de détonation de l'azoture de chlore dans des tubes cylindriques - limites de détonation de l'azoture d'hydrogène. J. Chim. Phys. 70 (5), 811-818.

Paillard, C., Dupré, G., and Combourieu, J. (1974) Etude de la détonation de composés endothermiques gazeux:II. Propagation et conditions critiques d'extinction dans des tubes capillaires de la flamme de décomposition de l'azoture de chlore gazeux pur ou dilué. J. Chim. Phys. 71 (2), 175-181.

Paillard, C., Dupré, G., Lisbet, R., Combourieu, J., Fokeev, V.P., and Gvozdeva, L.G. (1979) A study of hydrogen azide detonation with heat transfer at the wall. Acta Astronomica 6,277-242.

Paur, R.J. (1973) Photolysis of hydrazoic acid. Ph.D Thesis, Indiana University, West Lafayette, Indiana.

Piper, L.G., Krech, R.H., Pugh, E., and Taylor, R.L. (1980) Investigation of concept of efficient short wavelength laser. Technical Report. Physical Sciences, Inc., Woburn, Mass.

Piper, L.G., Krech, R.H., and Taylor, R.L. (1982) The UV photolysis of chlorine azide. J. of Photochemistry 18, 125-136.

Raschig, F., (1908) Uber Chlorazid N_3Cl. Ber dtsch. Chem. Gesell. 41, 4194-4195.

Rice, W.W. and Jensen, R.J. (1972) Hydrogen chloride chemical laser from hydrogen atom-chlorine azide reaction, J. Phys. Chem. 76 (6), 805-810.

Setser, D.W. (1979) Reactive Intermediate in the Gas Phase Generation and Monitoring. 358 pp. Academic Press, New York.

Zaslonko, I. S., Kogarko, S. M., Mozzhukhin, E. V., and Demin, A. I. (1972) Possibility of obtaining an inverse population in exothermic decomposition reactions. Dokl. Akad. Nauk. SSSR 202 (5), 1121-1124.

{ # Chapter II. Detonations in Two-Phase Systems

Dust, Hybrid, and Dusty Detonations

C. W. Kauffman,* P. Wolanski,† A. Arisoy,‡
P. R. Adams,§ B. N. Maker,π and J. A. Nicholls**
The University of Michigan, Ann Arbor, Michigan

Abstract

Dust explosions occur rather frequently in grain elevators and feed mills, coal mines, and other industrial processing plants. In some cases a combustible dust exists in the presence of air. In other cases, a combustible gas and/or an inert dust may be present, either with or without the combustible dust. Undoubtedly, most of these explosions are a manifestation of a deflagrative combustion process. However, in other cases, there are indications that detonations may have occurred. In recognition of the latter, this experimental study was directed at three facets of detonation involving dust. The first was the study of the propagation of detonation in grain dust dispersed in a gaseous oxidizer (dust detonations). The second was detonation in a mixture of coal dust and methane-air (hybrid detonation). The third facet involved the influence of inert particles on the detonability of a gaseous mixture (dusty detonation). Experiments were conducted in a long vertical tube wherein the dust was distributed at the top and allowed to fall through the gaseous mixture. Pressure, velocities, ignition delays, and particle temperatures were determined and streak and instantaneous photographs were made. Supporting experiments were conducted in a horizontal shock tube, relating

---

Presented at the 9th ICODERS, Poitiers, France, July 3-8, 1983. Copyright © 1984 American Institute of Aeronautics and Astronautics, Inc. All rights reserved.
*Associate Research Scientist, Aerospace Engineering Department.
†Visiting Scholar, from Technical University of Warsaw, Poland.
‡Visiting Scholar, from Teknik Universite, Istanbul, Turkey.
§Undergraduate Research Assistant, Aerospace Engineering Dept.
πGraduate Research Assistant, Aerospace Engineering Department.
**Professor, Aerospace Engineering Department.

to the ignition delay period for inert dust and combustible gaseous mixtures. These experiments showed that: 1) it is possible to have a detonation like wave propagating at near Chapman Jouguet velocities through a combustible dust and oxidizer mixture; 2) the structure of such a wave consists of an ignition delay zone and then a reaction zone behind the leading shock wave; 3) the presence of coal dust alters the detonation limits of methane-air mixtures; and 4) the presence of inert particles may serve to quench or enhance the detonation of gaseous mixtures, dependent upon the size and concentration of the particles.

Introduction

While gaseous detonations have been studied for over a century, only in the last two decades were similar studies initiated for heterogeneous detonations; that is, where there may be a mixture of phases (gas, liquid, solid). Such detonations were observed for a gaseous oxidizer and a liquid fuel in the form of a spray or a film (Cramer 1963; Dabora et al.1966). Detonations were also reported for a fuel in the form of a finely divided solid material (dust or powder) and a gaseous oxidizer (Strauss 1968; Cybulski 1971). In these efforts the conclusion as to the existence of a detonation was premised upon large combustion front velocities and large overpressures. In some cases, only a portion of the fuel may be in the condensed phase. Therefore, the possibility of detonations should be examined when a reactive dust is added to a detonable homogeneous mixture which is leaner than its lean limit. It is well known that such hybrid mixtures do support deflagrations (Bartknecht 1981). Furthermore, the possibility has been raised that a nonreactive dust may enhance the propagation of a deflagration through a premixed gaseous fuel and oxidizer (Moore and Weinberg 1981). Two survey articles relating to dust combustion (Nettleton 1975; Griffith 1978) point to a lack of understanding of dust, hybrid, and dusty detonations. Additionally, although numerous tests have been devised in order to evaluate the explosion hazards of dust (Dorsett et al.1960), none of these tests can determine whether a dust and gaseous oxidizer mixture is capable of detonating. Hence, a systematic program was initiated to investigate dust, hybrid, and dusty detonations.

Experimental Apparatus

Data characterizing the dust detonations and the effect of dust on gaseous detonations was collected using a vertical detonation tube (Wolanski et al. 1981) in which the atmosphere could be controlled. The tube had a length of 6.1 m and a square cross section with an inside dimension of 6.35 cm. A dust feeding device was located at the upper end of the tube. Slightly below the dust feeder, a hydrogen-oxygen initiator tube was attached at an oblique angle to the main tube. Pressure transducers and pressure switches were located along the tube. The center of a 33-cm optical test section was located 1.71 m from the downstream end of the tube. In investigating detonation phenomena the tube was filled with the desired gaseous atmosphere; the dust was dispersed into the tube at a rate which would give the desired concentration; and the hydrogen-oxygen driver, which was filled to a desired pressure, was ignited with a glow plug.

In order to investigate the effect of an inert dust on the detonation of a combustible gaseous mixture, a conventional shock tube (Ranger 1968) having a rectangular driven section with internal dimensions of 3.81 x 6.35 cm was employed. At the 30-cm optical test section, located 2.76 m from the diaphragm, it was possible to disperse a small dust cloud across the tube. Operationally, the driven section was filled with the desired mixture of combustible gases and the driver was pressurized to give the appropriate strength shock wave. The dispersion of the dust cloud and the diaphragm breakage occurred in a controlled fashion so that the interaction could be photographed at the test section.

Research Results

Dust Detonations

It is well known that the ignition delay time plays an important role in the establishment of detonation. It has been found from shock tube ignition studies (Ural 1981) that some combustible dust particles, apparently those with a high ballistic-coefficient and a large surface reaction area, have ignition delay times shorter than those reported for combustible gaseous mixtures for a given value of an incident shock wave Mach number. The dust produced by the grains of wheat and oats have these short ignition delay times. Accordingly, a number of experiments were conducted in the vertical detonation tube using grain dust dispersed in air or in oxygen enriched air. A laser Schlieren photograph of a decaying shock wave passing through a corn dust (d < 74 μm)/air mixture at a dust concentration of 140 gms/

m^3 is shown in Fig. 1. The top of the detonation tube is toward the top of the photograph and the shock wave is propagating downward. Toward the bottom, ahead of the shock wave, agglomerations of the fine dust may be seen. The shock wave itself appears to be quite thick, presumably as a result of the dust loading. Behind the shock wave the gas is flowing in the direction of the shock wave propagation and the relatively stationary, larger dust agglomerations are exposed to a supersonic flow with attendant bow shocks. The resultant high pressures and shear forces break up the agglomerations, thereby producing opaque clouds of fine dust. Finally, all the clouds coalesce producing a **uniform dust/air mixture.**

Not all initiating shock waves from the hydrogen-oxygen initiator decay. Figure 2 shows the output of three pressure transducers from a run with oats dust (d < 75 μm) in air at a concentration of 270 gms/m^3 and with an initiator pressure of 100 psia. The transducers were located at the middle of the test section, 1.01-m upstream, and 0.92-m downstream of the middle one. The transducers show a non-

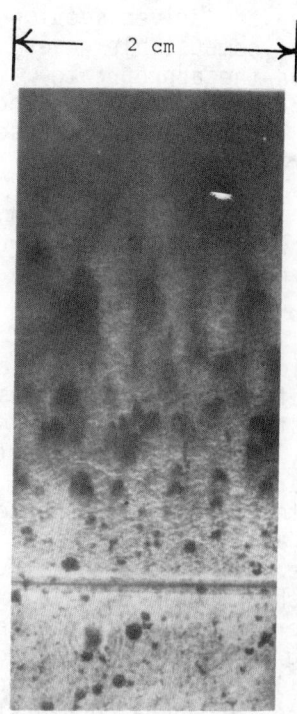

Fig. 1 Decaying blast wave.

Corn Dust (d < 75 μm)/Air, 140 gms/m³

decaying shock wave to be propagating down the tube. The peak pressures were approximately 24 atm and the velocity was 1500 m/s. For oat dust, stoichiometric conditions correspond to 295 gms/m^3, which gives calculated Chapman Jouguet values (Gordon and McBride 1971) for the velocity of 1796 m/s, pressure of 22.36 atm, and temperature of 2835 K. The fourth trace at the bottom shows the response of one of the photodiodes from a four channel radiometer (Bouriannes et al.1977), which was also located at the midpoint of the test section.

A number of factors, such as type of dust, oxidizer, dust size, dust concentration, and initiator energy, are important in determining whether the combustion wave will lead to detonation or simply decay as it progresses down the tube. In Fig. 3 the wave velocity is given as a function of distance along the tube. This data is for oat dust (d < 74 μm) in a 40% oxygen-nitrogen mixture at a concentration of 784 gms/m^3, which corresponds to an equivalence ratio of 1.4. The pressure in the initiator tube is the parameter. For driver pressures of 40 and 60 psia, a constant velocity wave, presumably detonation, in excess of 1500 m/s was obtained. However, for the low driver pressure of 20 psia a steady-state wave was not achieved. It begins at a low velocity, accelerates, then oscillates about the steady-state value, and then begins to decay. For a given higher initiator pressure, similar oscillatory behavior may be seen for mixtures which are very lean or

Oats dust (d < 75 μm)/air, 270 gms/m^3,
1500 m/s, p_{max} = 23.67 atm

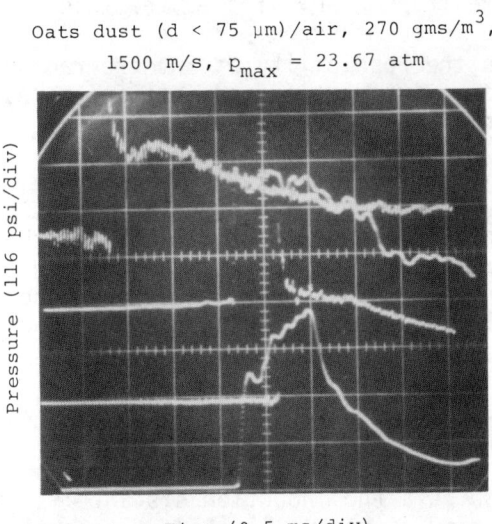

Time (0.5 ms/div)

Fig. 2   Pressure history along tube.

very rich. In some situations there is just a continuous weakening of the blast wave produced by the initiator.

Examination of the velocity and pressure history of the wave for many conditions allows determination of a curve showing the limits of detonation. Figure 4 shows such information for oat dust suspended in air and in a 40% O_2/ 60% N_2 mixture. The ordinate represents the pressure of the hydrogen-oxygen mixture in the initiator required for initiation (and hence is an indicator of initiation energy) and the abscissa is the equivalence ratio. The parameter is the size of the dust particles. For particles in air, the detonation limits are quite narrow. The limits are slightly broadened by an increase in the particle size, but with the very largest particles, $106 < d < 300$ μm, detonation was not achieved for the initiator pressures employed. Increasing the oxygen content of the air to 40% considerably lowers the critical initiator energy and broadens the limits of detonability. Again, a particle size effect may be noted, particularly with the large particles and at the rich limit. The largest particles require a larger initiator energy and the upper limit is increased with increasing particle size. The lower limit is affected significantly by the largest size particles. This size effect is probably attributable to the decreasing surface to volume ratio for the larger particles, thus reducing the fuel availability.

The use of the radiometer, in conjunction with the coincident pressure transducer, allowed the thickness of the induction zone to be measured. An interesting aspect to this phase was the possibility of intense radiation from the reaction zone irradiating and initiating reaction in the dust ahead of the shock wave. An optical fiber was used in a few experiments which allowed the radiometer to sense upstream, ahead of the shock. No radiation ahead of the shock was observed, thus indicating that the mixture acted as an optically thick gas. In most of the measurements the input to the radiometer was confined to a very narrow view angle looking across the tube. The arrival of the pressure front was detected by the transducer and the arrival of the reaction zone was then detected by the radiometer. The difference in arrival time allowed calculation of the induction distance. This data was obtained from channels two and four of Fig. 2. For the specific conditions of oat dust ($d < 75$ μm) in air with an initiator pressure of 100 psia the induction distance is given in Fig. 4. Its similarity with the limits of detonability may be noted.

Although the length of the induction zone or the ignition delay time may be comparable to some gaseous reactants, the length of the reaction zone may be quite long because of the burning time required for each dust particle. This leads to losses from the reaction zone and hence to propagation velocities lower than Chapman Jouguet. A laser Schlieren photograph of a propagating detonation wave is shown in Fig. 5. This is for corn dust (d < 75 μm) at a concentration of 250 gms/m^3 in a 40% oxygen/nitrogen mixture. As in the case of Fig. 1, the shock wave is travelling downward into the undisturbed but agglomerated dust particles. Again, a short time after the shock wave has struck the particles, it is no longer possible to see across the test section nor is it possible to see the initiation of combustion (which is detected by the radiometer). The reaction zone is seen to be opaque for an appreciable distance, until the combustion process consumes some of the light scattering particles. However, the loss of opaqueness should not be construed as the end of the reaction zone. As previously noted, a fast response, four-channel radiometer was located in the middle of the test section and the light emission from the burning particles was recorded at four different wavelengths. Assuming that graybody radiation is emitted by the burning particles, the temperature may be calculated through the reaction zone (Wolanski 1982). Such a temperature history is given in Fig. 6 for oats (d < 75 μm) in two different oxidizer mixtures. In air, the measured maximum temperature is 2800 K while Chapman Jouguet calculations for this mixture give a temperature of 2750 K. In the oxygen enriched mixture the maximum temperature is, of course, higher, as is the rate of temperature rise. If the location of the maximum temperature is taken as the Chapman Jouguet plane, the CJ plane would be taken as being 71 cm and 38 cm behind the induction zone for the air and oxygen enriched cases, respectively. In Fig. 5, for corn dust in an oxygen enriched situation, the reaction zone became transparent approximately 11 cm after the shock wave. For a given gaseous oxidizer, mixtures near stoichiometric gave the highest peak temperatures. The effect of particle size was also examined at a number of equivalence ratios. Except for very lean and very rich mixtures, the largest size particles gave the largest maximum temperature and the smallest size particles gave the lowest maximum temperature. This, of course, fits well with the notion that convective heating is more efficient for the larger particles because of their higher ballistic coefficient.

228   C. W. KAUFFMAN ET AL.

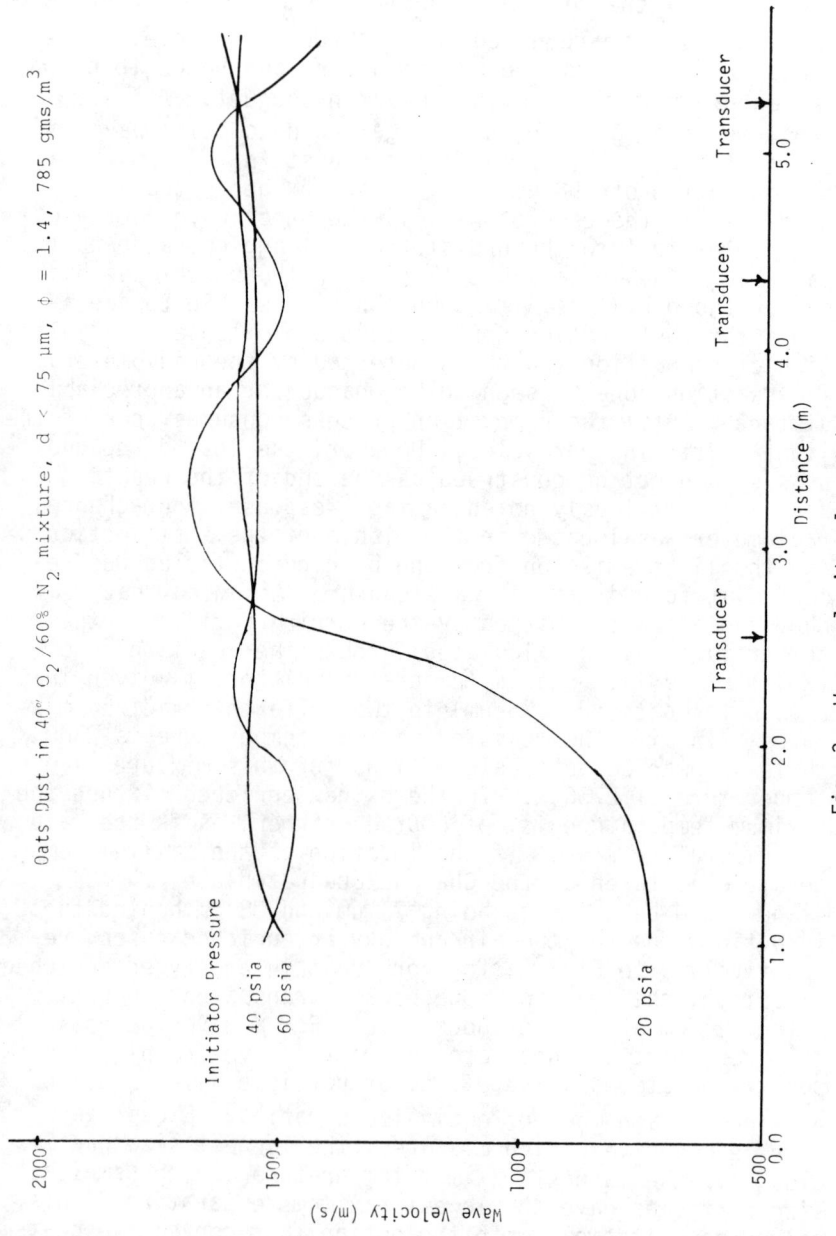

Fig. 3   Wave velocity along tube.

DUST, HYBRID, AND DUSTY DETONATIONS

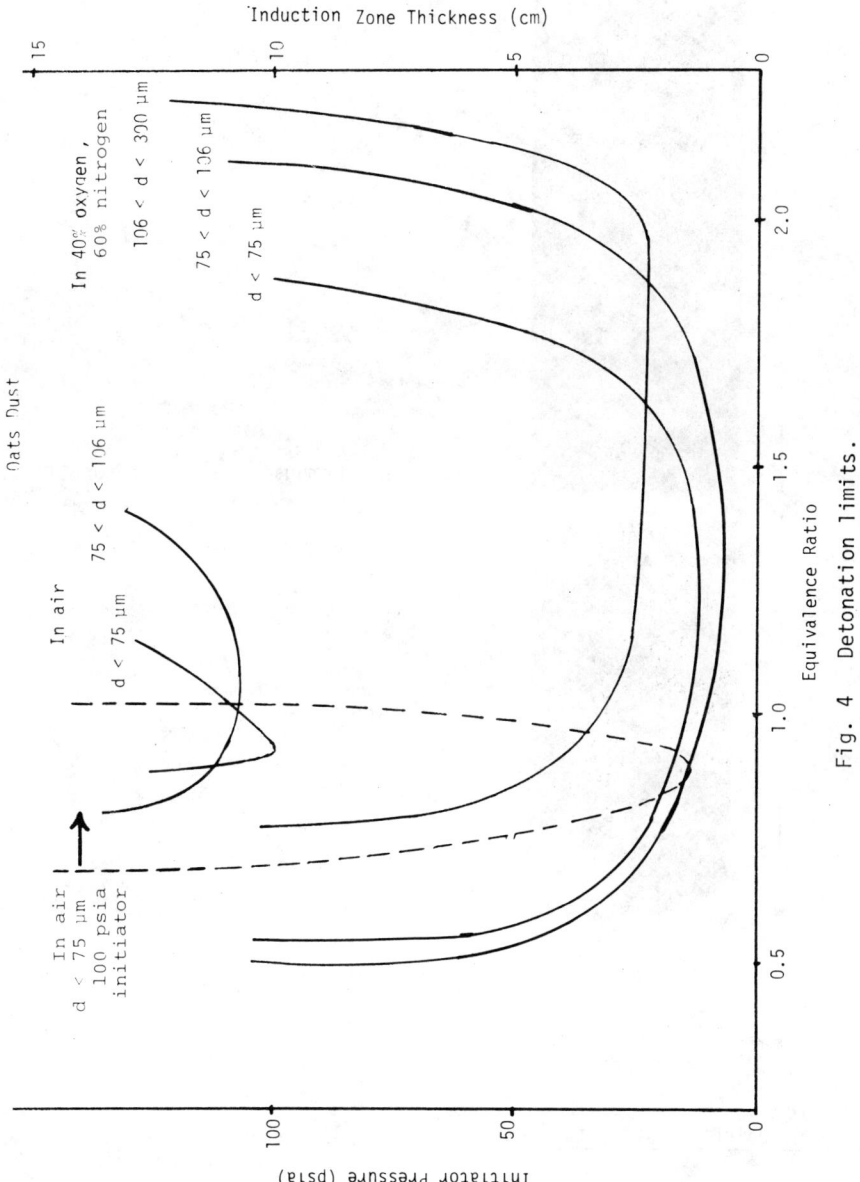

Fig. 4 Detonation limits.

Fig. 5 Combustible dust detonations.

Corn Dust (d < 75 μm)/40% Oxygen, 250 gms/m^3

Hybrid Dust Detonation

Experiments similar to the foregoing, conducted in the vertical detonation tube, were not successful in detonating coal dust/air mixtures. On the other hand, spinning detonations in methane-air were observed (Wolanski et al. 1981). A streak photograph of one of these detonations is shown in Fig. 7 for a stoichiometric mixture. The dust explosion literature reveals that combustible dusts mixed with combustible gases, each leaner than its lower flammability limit, have produced explosive reactions. Accordingly, it was decided to investigate this hybrid case relative to detonation.

The desired concentration of coal dust (d < 74 μm) was dropped into a preselected methane-air mixture. Using the maximum safe initiator pressure of 120 psia, the dust concentrations and methane concentrations required for apparent detonation were determined. The wave structure was also examined. A streak photograph of such a hybrid detonation is shown in Fig. 8. The mixture consists of 7% CH_4 in air and 100 gms/m^3 of coal dust. The spinning structure of the detonation wave is easily recognized and the shat-

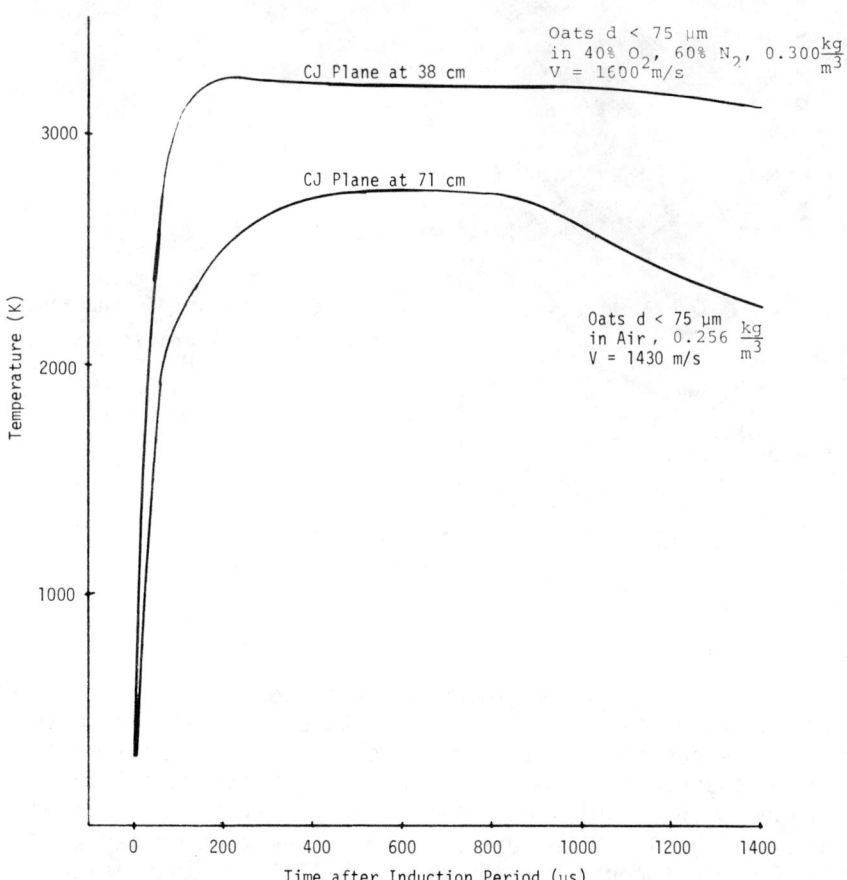

Fig. 6   Reaction zone temperature profile.

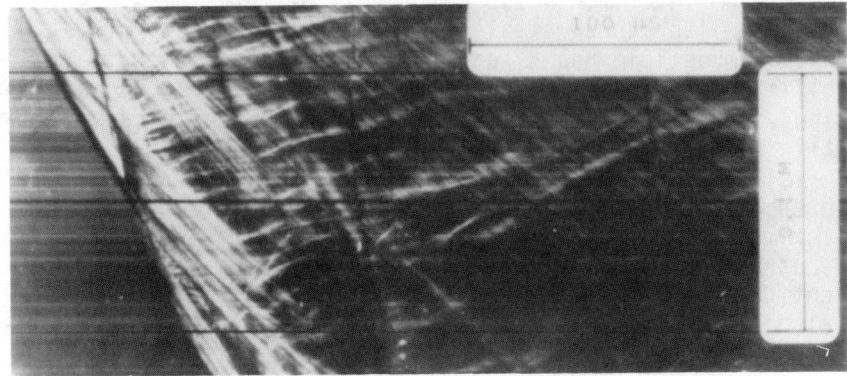

Fig. 7   Streak photography of stoichiometric CH₄/air detonation.

Fig. 8   Hybrid detonation.
Coal Dust (d<74 μm) and 7% Methane/Air
100 gms/m³, 1670 m/s

tering of the small agglomerations of coal can be seen. Without the coal, a 7% CH₄/air mixture would not detonate.

Numerous combinations of concentrations of coal dust and methane were examined to see if detonation could be sustained. The results are given in Fig. 9 for two different types of coal. With zero added coal, detonation may be achieved between 8 and 14% methane. But, as increasing amounts of coal are added, the upper limit is lowered and the lean limit is lowered up to a point. Further addition causes the lean limit to increase, eventually meeting with the rich limit. For coal concentrations above this amount, no detonations are possible. The volatility of the coal has an affect on the amount of broadening of the limits, as well as the maximum amount that may be added. A lesser amount of the high volatility coal will quench the detonation. This effect and the behavior of the lean limit would seem to indicate that the coal and its volatile material are providing fuel to supplement the gaseous methane which is already available. The increase in the lower limit at higher coal concentrations may indicate that the coal is now making available too much fuel to supplement the methane.

Dusty Detonation

Another area of investigation pursued was that of the role of inert particles in a detonable gaseous mixture. Early work indicated that small particles are quite effective in quenching a detonation (Lafitte and Bouchet 1959).

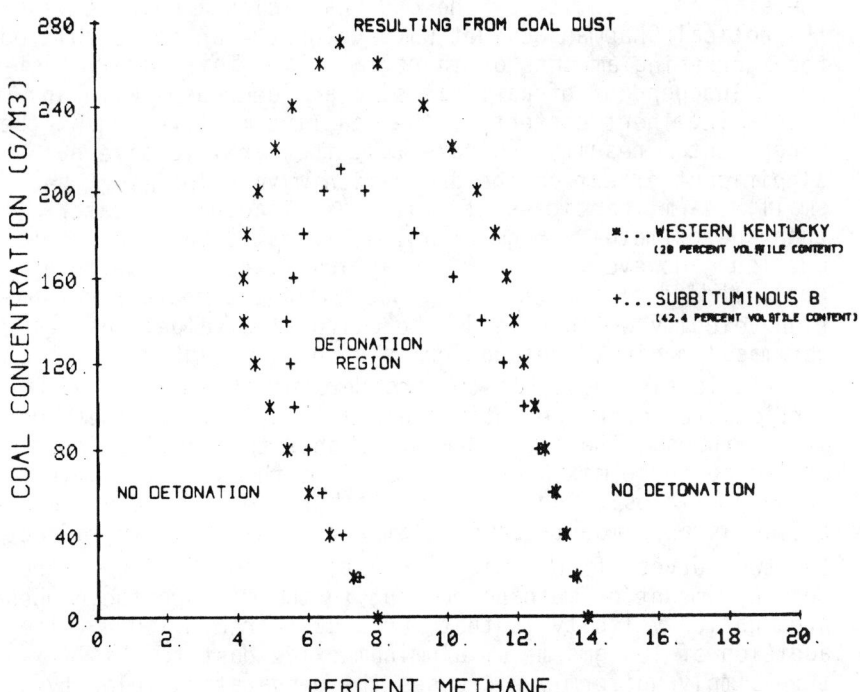

Fig. 9 Effect of pulverized coal on methane-air detonation limits.

In the studies reported here, irregular shaped aluminum oxide particles were dropped into a gaseous mixture of methane and air. For a given hydrogen-oxygen initiator pressure of 120 psia, the percentage of methane and the dust loading for which a detonation would propagate was noted for two different size dust particles. A laser Schlieren streak photograph for a dust loading of 36 gms/m^3 of 14-μm particles in a 12% methane mixture is shown in Fig.10a. The wave is not spinning and is not a detonation; it is a decaying wave. The opaqueness is due to the efficient scattering of light from the particles. Figure 10b shows the propagation of a detonation wave through a mixture of 7.5% methane contaminated with 450-μm aluminum oxide particles at a loading density of 125 gms/m^3. Although the methane concentration is slightly below the previously reported lower limit of 8%, a spinning type of detonation is shown propagating down the tube. For this case, the larger particles are not as effective at scattering the illuminating laser light.

The inert particles definitely have an affect on the detonation wave velocity as is shown in Fig. 11. The results here are for a gaseous mixture of 9.75% methane where the size and quantity of the dust particles are varied. A theoretical Chapman Jouguet wave velocity can be calculated for increasing amounts of inert material. This wave velocity is independent of particle size and decreases with an increasing inert content in the reacting mixture. Now, the experimental results indicate that the particle size has a significant effect on the detonation wave velocity. The small, 14-μm, particles at small mass loadings (greater than approximately 15 gms/m^3) lead to rapid failure of detonation. However, the larger particles, 450 μm, up to a mass loading of approximately 300 gms/m^3, produce a detonation velocity which exceeds the calculated value. At larger mass loadings, detonation is soon quenched.

It is also possible to consider the effect of the inert particulates over the entire detonability range of methane-air mixtures. The data in Fig. 12 show the results from adding to these mixtures 100 gms/m^3 of the 450-μm aluminum oxide particles. There is no data here for the 14-μm particles as any amount above 15 gms/m^3 has a quenching affect. The two curves of analytical results should first be noted. For any amount of methane present in the mixture the highest propagation velocity is that for a dust free mixture. The addition of 100 gms/m^3 of aluminum oxide dust to this mixture simply uniformly decreases the propagation velocity. A particle size need not be specified. For a dust free mixture, the experimentally measured propagation velocities

 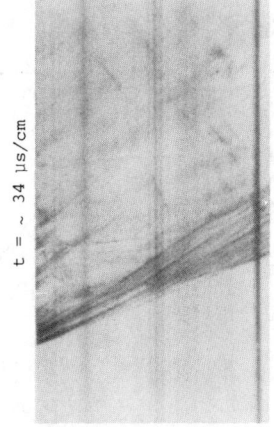

a. Decaying wave
14 μm Particles, 36 gms/m^3
12% CH$_4$/Air

b. Detonation
450 μm Particles, 125 gms/
7.5% CH$_4$/Air

Fig. 10 Influence of inert particles on CH$_4$/air detonation.

lay only slightly below the theoretically computed curve. The addition of the large, inert dust at a modest mass loading has only a very slight affect in decreasing the wave velocity from the measured dust free value. The actual wave velocity remains well above what is calculated for the dust case. If the mass loading of these 450-μm particles were approximately doubled, there would be a significant difference between the measurements for the dust free and dusty gas cases.

As in the case of the coal dust contaminated methane, the effect of the aluminum oxide particles on the detonation limits can also be mapped out. In Fig. 13 the region of detonability is inside of the two curves for the two different size particles. For the small particles there is no effect on the detonation limits and for each methane concentration the relatively small mass loading that does not quench the detonation can be seen. For the larger particles, a larger mass loading is seen to be required to quench the detonation. Also, the rich and lean limit curves are seen to have a slight initial negative slope. They, of course, converge at the higher particle concentrations.

As the inert particles are not adding any fuel, an explanation for their effect on the methane-air mixture must be sought. The inert particles must be effective in absorbing thermal energy, acting as a catalytic or noncatalytic surface, radiating thermal energy, acting as turbulence generators and flame holders, or producing stagna-

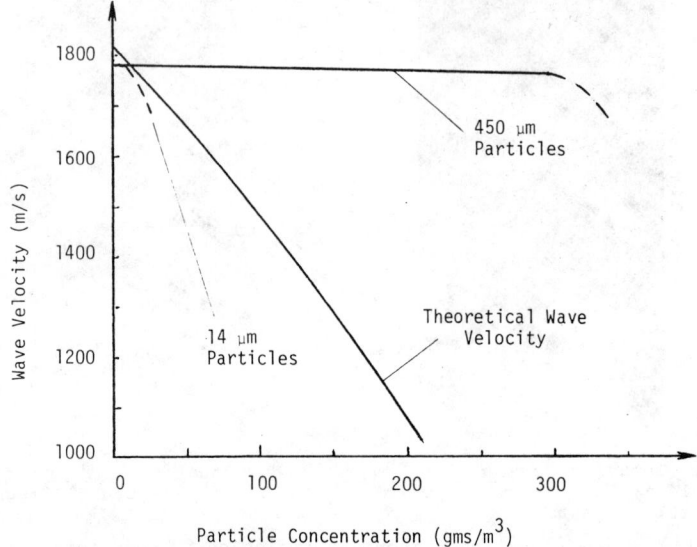

Fig. 11 Effect of particle loading on detonation wave velocity.

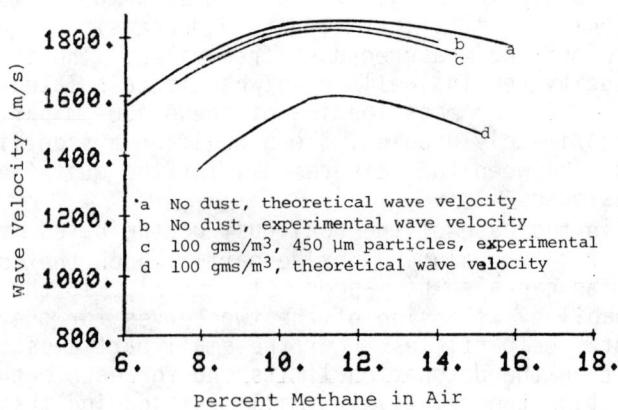

Fig. 12 Effect of particles on detonation wave velocity.

tion temperature hot spots. In order to sort out these possibilities the effect of a small cloud of particles on the shock wave ignition of a methane-air mixture was examined using a horizontal shock tube. A mixture of methane-air was placed in the driven portion of the tube and its ignition delay time was measured for a given strength incident shock wave. The streak photograph in Fig. 14a shows a case with no particles present. Here the incident shock wave is seen near the bottom passing through a 5% methane-air mixture, which is outside the limits of detonability. The ignition delay of approximately 562 μs, which is lower than what would be predicted for this condition, is attributable to premature ignition by the diaphragm. The interaction of an incident shock wave with a small cloud of 450-μm particles in air only is shown in Fig. 14b. After being struck by the shock wave, the initial cloud is increasingly dispersed as it is accelerated downstream by the convective flow. Now, if 5% methane is added to the air mixture and the incident shock wave has a Mach number of 4.07, the inert particles have a marked effect on the initiation of the combustion process. They may be observed in Fig. 14c. Again, the incident shock wave is seen passing through the combustible mixture and striking the cloud of aluminum oxide particulates. The gas mixed with the particulate cloud is seen to be burning almost immediately after the shock wave strikes the cloud. Thus, under the conditions existing here, the presence of the particles leads to a very short ignition delay time. For the same size aluminum oxide particles, 450 μm, a decreasing incident shock wave strength leads to an increase in the ignition delay time. If the smaller 14 μm of 88-μm aluminum oxide particles are used, no effect is seen on the ignition delay time. These smaller particles are very quickly swept downstream by the convective flow. The inertia of the larger

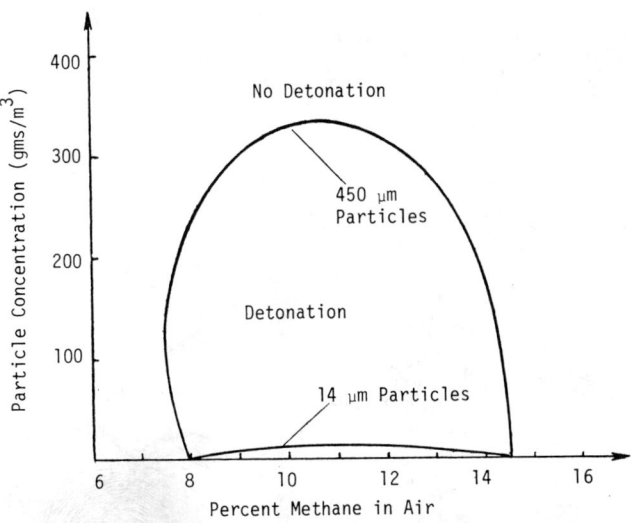

Fig. 13   Effect of particles on detonation limits.

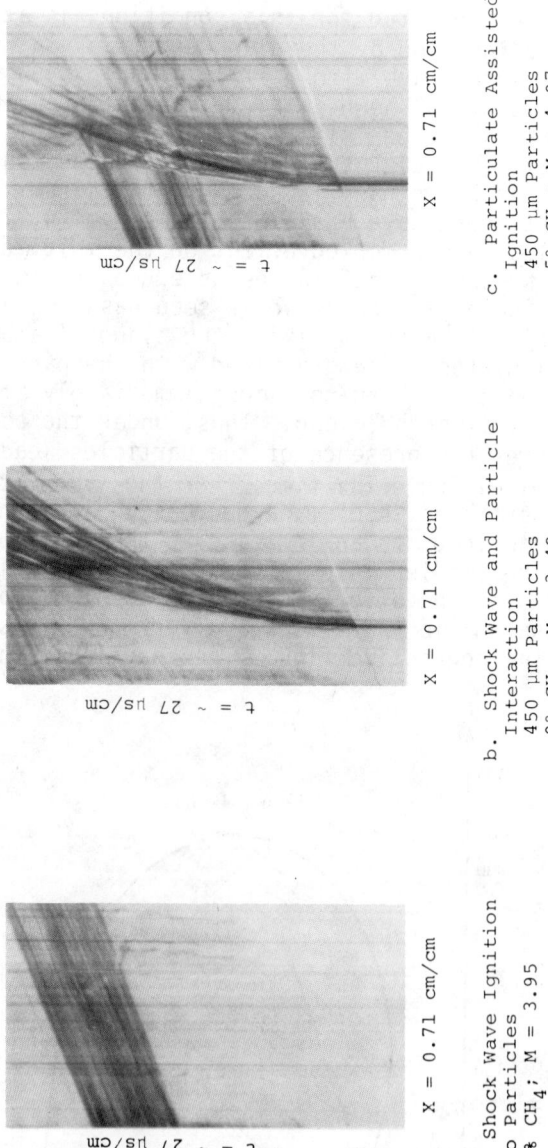

Fig. 14 Influence of inert particles on CH$_4$/air ignition.

particles (ballistic coefficient) leads to higher stagnation temperatures on the particles which leads to more rapid ignition and enhanced combustion.

Conclusions

The following major conclusions may be drawn from this investigation.

1) "Dust detonations" are possible for certain combustible dusts wherein particle size, concentration, the gaseous oxidizer, and initiation energy are important. Velocities, pressures, and temperatures close to Chapman Jouguet conditions have been measured. In general, the lead shock wave is followed by a short induction zone and then an extended reaction zone.

2) "Hybrid detonations" can occur when a combustible dust is added to a combustible gas in a gaseous oxidizer, wherein neither fuel would detonate without the other for the mixture ratios used. Conversely, the addition of the second fuel may put the mixture outside the detonability range.

3) Inert particles added to a combustible gaseous mixture can enhance detonation (dusty detonation), or preclude detonation, dependent upon the particle size and concentration.

Acknowledgment

This work was supported by the National Science Foundation under NSF Grant No. CPE-7622958, Dr. Royal Rostenbach, Grant Monitor.

References

Barthnecht, W., Explosions, Springer-Verlag, New York, 1981.

Bouriannes, R., Moreau, M., and Martinet, J., "Un pyrometre rapide a plosieurs couleurs." Revue de Physique Applique e 12, 5, 1977.

Cramer, F. B., "The onset of detonations in a droplet combustion field." Ninth Symposium (International) on Combustion, Academic Press, New York, 1963.

Cybulski, W. B., "Detonation of coal dusts." Bulletin de L'Academic Pulonaise des Sciences 19, 37, 1971.

Dabora, E. K., Ragland, K. W., and Nicholls, J. A., "A study of heterogeneous detonations." Astronautica Acta 12, 9, 1966.

Gordon, S. and McBride, B. J., Computer program for calculation of complex chemical equilibrium compositions, rocket performance, incident and reflected shocks, and Chapman-Jouguet detonations. NASA SP-273, 1971.

Griffith, W. C., "Dust explosions." Annual Review of Fluid Mechanics, 10, 93, 1978.

Laffitte, P. and Bouchet, R., "Suppression of explosion waves in gaseous mixtures by means of fine powders." Seventh Symposium (International) on Combustion, Butterworths, London, 1959.

Moore, S. R. and Weinberg, F. J., "High propagation rates of explosions in large volumes of gaseous mixtures." Nature, 290, 39, 1981.

Nettleton, M. A., "A review of shock wave chemistry in dusty and foggy gases." Report RD/L/N/97175. Central Electricity Generating Board, 1975.

Ranger, A. A., "The Aerodynamic Shattering of Liquid Drops." Ph.D. Thesis, The University of Michigan, Ann Arbor, Michigan, 1968.

Strauss, W. A., "Investigation of the detonation of aluminum powder-oxygen mixtures." AIAA J. 6, 37, 1953.

Ural, E. A., "Shock Wave Ignition of Pulverized Coal." Ph.D. Thesis, The University of Michigan, Ann Arbor, Michigan, 1981.

Wolanski, P., "Temperature measurement by means of the four wavelengths pyrometer." Internal Report, The Department of Aerospace Engineering, The University of Michigan, Ann Arbor, Michigan, Sept. 1982.

Wolanski, P., Kauffman, C. W., Sichel, M., and Nicholls, J. A., "Detonation in methane/air mixtures." Eighteenth Symposium (International) on Combustion, The Combustion Institute, Pittsburgh, Pa., 1981.

The Structure of Dust Detonations

P. Wolanski,* D. Lee,† M. Sichel‡
C. W. Kauffman,§ and J. A. Nicholls‡
The University of Michigan, Ann Arbor, Michigan

Abstract

Detonation of dusts with different gaseous oxidizers have recently been of interest because of their possible role in dust explosions. It is found that some dusts detonate in air while many more are detonable in pure oxygen or air enriched by oxygen. A model for the reaction zone of a dust detonation taking two-phase flow effects and wall losses into account has been developed. The flow in the reaction zone is assumed to be one-dimensional and steady. The dust particles are assumed to be spherical, monodisperse, and at the same temperature as the surrounding gas. A one fluid model with the gas-particle mixture treated as a single fluid is used in the formulation; however, particle acceleration caused by the difference between gas and particle velocity is taken into account. The model used for the dust combustion rate is the greatest source of uncertainty, especially since experimental data for the wheat dust considered here does not appear to be available. It was assumed that combustion occurs heterogeneously with a surface regression rate which is proportional to the oxygen concentration, an Arrhenius factor, a factor to account for the blocking effect of ash

---

Presented at the 9th ICODERS, Poitiers, France, July 3-8, 1983. Copyright © 1984 American Institute of Aeronautics and Astronautics, Inc. All rights reserved.
  *Associate Professor, Instytut Techniki Ciepinej Politechniki Warszawskiej, Poland.
    †Graduate Student, Department of Aerospace Engineering.
    ‡Professor, Department of Aerospace Engineering.
    §Associate Research Scientist, Department of Aerospace Engineering.

formed during combustion, and a factor to account for increased burning area due to porosity. Combustion starts only after an empirically determined ignition delay time. By comparison of computed and measured reaction zone pressure profiles for a wheat dust-air detonation it was possible to establish values for the significant combustion parameters. Then reaction zone profiles of particle and gas velocity, temperature, and density could be computed. The model also was used to compute detonation velocities and results were in good agreement with measured values for oats dust and wheat dust-air detonations. The combustion parameters developed as described above were used to compute the pressure profile in the reaction zone of an oats dust-oxygen detonation and results were in good agreement with experiment. Radiation and convection losses were relatively small for the 6.45-cm^2 shock tube considered here. While the results must be considered preliminary, they point out some of the main features of dust detonations and also suggest that detonation measurements may provide a means of establishing dust combustion properties.

Nomenclature

A = pre-exponential factor
b = duct perimeter
C_D = particle drag coefficient
C_f = wall skin-friction coefficient
$\bar{C}_p$ = average gas specific heat
C = particle specific heat
D = detonation velocity
D_h = hydraulic diameter
d_p = particle diameter
E = activation energy
f = ash blocking factor
h = gas enthalpy
h_p = particle enthalpy
h_w = gas enthalpy at the wall
K = particle reaction rate
M_p = Mach number relative to the particles
m_p = particle mass
m_f = fuel mass in particle

m_a	=	ash mass in particle
n	=	particle number density
p	=	pressure of the gas
Q_0	=	heat of combustion per unit particle mass
q_w	=	convective heat flux to the wall
q_r	=	radiative heat flux to the wall
q_s	=	viscous work done by wall on fluid per unit area
R	=	gas constant
Re	=	Reynolds number = $\rho u x/\mu$
Re_p	=	relative Reynolds number = $(\rho\|u-u_p\|d_p)/\mu$
T	=	temperature
u	=	gas velocity
u_p	=	particle velocity
u_t	=	surface burning velocity
γ	=	ratio of specific heats of the gas
ε	=	particle volume per unit mixture volume
ε'	=	emissivity
η	=	mass fraction of particles in unburned mixture
μ	=	gas viscosity
ρ	=	gas density
ρ_p	=	density of the particle material
σ_p	=	particle mass per unit mixture volume
σ_{O_2}	=	oxygen mass per unit mixture volume
σ'	=	Stefan-Boltzmann constant
τ_w	=	viscous shear stress at the wall
τ_i	=	ignition delay time
ϕ_a	=	ratio of ash to original fuel mass
ϕ_s	=	internal surface area factor
ϕ_{O_2}	=	stoichiometric factor

Subscripts

0	=	initial value
p	=	particle
f	=	fuel
a	=	ash

Introduction

Detonations of dusts with different gaseous oxidizers have recently been studied by numerous authors (Cybulski, 1971; Bartknecht, 1978; Kauffman et al., 1979; Nettleton and Stirling, 1973; Wojcicki, 1973) because of their possible role in dust explosions. It is found that some dusts detonate in air while many more are detonable in pure oxygen or oxygen enriched air.

Dust detonations both in the field and in the laboratory occur in ducts or chambers of finite size, and since their reaction zones can be quite long wall losses play a significant role. The character of the reaction zone also determines the detonability and detonation limits of fuel-oxidizer mixtures (Lee, 1977). For a better understanding of dust detonations a careful study of the detonation structure is therefore important.

The aim of this paper is to examine the structure of a dust detonation and to find a method of calculating the detonation velocity. A model for the two-phase flow and combustion behind the leading shock of the detonation taking wall losses into account has, therefore, been developed. This model has been used to calculate the detonation velocity. The theoretically computed reaction zone structure and the detonation velocity have been compared with experimental measurements. The model has also been used to establish the influence of ignition delay, dust concentration, and mean particle size upon the detonation structure.

Formulation

The flow in the reaction zone, shown in shock-fixed coordinates in Fig. 1, is assumed to be one-dimensional and steady. The dust particles are taken to be spherical, monodisperse, and at the same temperature as the surrounding gas. Combustion starts after an induction period computed using the results of shock tube measurements. Because of the relatively small dust concentrations the particle volume was neglected, that is the volume fraction $\varepsilon \ll 1$. Losses to the wall due to friction, convection, and radiation from the particles are taken into account. A one fluid model in which the gas particle mixture is treated as a single fluid is used in the formulation.

The mass and momentum conservation equations are then as follows:

$$\frac{d(\rho u)}{dx} + \frac{d(\sigma_p u_p)}{dx} = 0 \qquad (1)$$

THE STRUCTURE OF DUST DETONATIONS

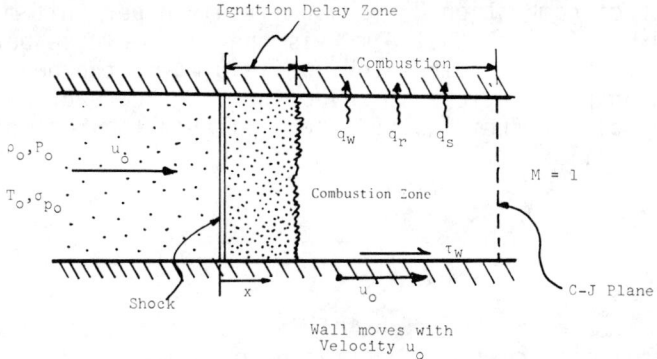

q_w - convective heat flux to wall
q_r - radiative heat flux to wall
q_s - viscous work done on fluid

Fig. 1   Dust detonation shown in shock-fixed coordinates.

$$\frac{d}{dx}(\rho u^2 + \sigma_p u_p^2) = -\frac{dp}{dx} - \frac{4}{D_h}\tau_w \qquad (2)$$

where $D_h = 4A_c/b$ is the hydraulic diameter and τ_w is the viscous shear stress at the wall. All symbols are defined in the nomenclature. The number of particles is conserved so that

$$nu_p = n_o u_{po} = n_o u_o \qquad (3)$$

In formulating the energy conservation equation it is necessary to consider that the combustion products of the dust particles may include solid residue or ash which is a fraction ϕ_a of the original fuel mass. At any given point in the reaction zone the particle mass m_p consists of fuel and ash so that

$$m_p = m_f + m_a = m_f + \phi_a(m_{fo} - m_f) \qquad (4)$$

where the subscript zero refers to the unburned particles and conditions upstream of the detonation. Thus the fuel mass m_f is

$$m_f = \frac{m_p}{1-\phi_a} - \frac{\phi_a m_{po}}{1-\phi_a} \qquad (5)$$

The heat of combustion Q_o is usually given per unit mass of unburned fuel so that $Q_o m_{po}$ is the heat release per particle for complete combustion, while $Q_o m_f$ is the heat of combustion per particle at any point in the reaction zone. It then follows from Eqs. (3) and (5) that the heat of combustion flux is

$$n u_p Q_o m_f = \frac{Q_o}{1-\phi_a} (\sigma_p u_p - \phi_a \sigma_{po} u_o) \qquad (6)$$

The second term on the right of Eq. (6) is constant so that the energy conservation equation becomes

$$\frac{d}{dx}\left[\rho u(h + \frac{1}{2} u^2) + \sigma_p u_p (\frac{1}{2} u_p^2 + h_p + \frac{Q_o}{1-\phi_a})\right]$$

$$= \frac{4}{D_h}\left[-q_w - q_r + q_s\right] \qquad (7)$$

In Eq. (7) q_w and q_r are the convective and radiative heat flux to the wall while q_s is viscous work per unit area done by the wall which moves with velocity u_o in shock-fixed coordinates.

The contribution of the particles to the pressure can be neglected, and assuming the surrounding gas to be ideal, the equation of state is

$$p = \rho RT \qquad (8)$$

The variation of the fuel mass flow density is given by

$$\frac{d}{dx}(\sigma_{pf} u_p) = \frac{1}{1-\phi_a}\frac{d}{dx}(\sigma_p u_p) = -\kappa \qquad (9)$$

where κ is the reaction rate of the particles which will be discussed below. The relation between $\sigma_{pf} u_p$ and $\sigma_p u_p$ in Eq. (9) follows from Eqs. (3) and (5) and accounts for the ash which is assumed to remain on the particles. The density σ_{O_2} of oxygen, an important parameter in the expression for the rate κ, is determined by

$$\frac{d}{dx}(\sigma_{O_2} u) = -\phi_{O_2} \kappa \qquad (10)$$

where ϕ_{O_2} is a stoichiometric factor. From the conservation of particles it follows that the particle diameter d_p and mass flow density $\sigma_p u_p$ are related by

$$\frac{d_p}{d_{po}} = [(\sigma_p u_p)/(\sigma_{po} u_o)]^{1/3} \qquad (11)$$

Following Rudinger (1970), but neglecting the volume fraction ε, the particle equation of motion is

$$u_p \frac{du_p}{dx} = \frac{3}{4} C_D \frac{1}{d_p} \frac{\rho}{\rho_p} (u-u_p)|u-u_p| \qquad (12)$$

The above equations are relatively standard. Evaluation of the particle combustion rate κ is the major source of uncertainty and is discussed in the following section. Empirical results are used to compute the loss terms and the drag coefficient as described below.

Combustion Model

The structure of the detonation is determined mainly by the burning rate κ of the dust particles. At the same time modeling of this rate is, perhaps, the greatest source of uncertainty. The combustion rate depends on many factors such as temperature, oxygen concentration, surface area of the particle, ash formation on the particle, the relative velocity $(u-u_p)$, and the reactive properties of the dust material itself.

The basic process is the burning of individual particles, which is assumed to be independent of adjacent particles except for the effect on the ambient medium. If m_p is the mass of a single particle the reaction rate κ and the single particle burning rate, $-dm_f/dt$ are related by

$$\kappa = -n \frac{dm_f}{dt} = -\frac{n}{1-\phi_a} \frac{dm_p}{dt} \qquad (13)$$

From the conservation of particles Eq. (9) and the relation $n_o = \sigma_{po}/m_{po}$ it is readily shown that

$$n = \frac{6}{\pi} \frac{\sigma_{po}}{\rho_p} \frac{u_o}{u_p} \frac{1}{d_{po}^3} \qquad (14)$$

so that the number density increases as the particles decelerate behind the shock wave.

In the present model the particles are treated like a solid propellant with a regression rate or surface burning belocity u_b. The burning rate will then be

$$-\frac{dm_p}{dt} = \rho_p (\pi d_p^2) \phi_s (1-\phi_a) u_b \qquad (15)$$

where ϕ_s is a factor which accounts for the increased surface area due to the particle porosity. It is assumed that the burning velocity is proportional to an Arrhenius factor, the surrounding oxygen concentration σ_{O_2}, and a blocking factor f which accounts for surface O_2 ash formation. Then

$$u_b = A f [\sigma_{O_2}/(\sigma_{O_2})_0]^{s_1} \exp(-E/RT) \qquad (16)$$

The blocking factor will depend on the ratio of unburned fuel mass m_f to the total mass including ash from material already burned. For generality it is assumed that f varies with this ratio raised to some power s so that

$$f = \left(\frac{m_f}{m_p}\right)^s = \left[\left(1-\phi_a \frac{m_{po}}{m_p}\right)\bigg/(1-\phi_a)\right]^s \qquad (17)$$

From Eqs. (11), (13), and (14-16) it now follows that

$$\kappa = 6(\sigma_p/d_p)\phi_s A f [\sigma_{O_2}/(\sigma_{O_2})_0]^{s_1} \exp(-E/RT) \qquad (18)$$

The reaction rate is coupled to the local conditions in the reaction zone through σ_p, d_p, σ_{O_2}, and T. The parameters A, s_1, s, ϕ_s, and E are combustion parameters specific to the particular type of dust under study, and generally are not well known. In the present study reasonable values of these parameters are established by comparison of computed and experimental reaction zone pressure profiles in a few selected cases.

The onset of particle combustion is preceded by an ignition delay period τ_i during which the particle reaction rate $\kappa = 0$. Ignition delay times of dust particles behind shock waves have been measured by Kauffman et al. (1979) and by Ural et al. (1982) among others, and the data can be correlated by relations of the form:

$$\tau_i = B \exp(E_i/T) \tag{19}$$

The procedure has then been set to $\kappa=0$ for $t < \tau_i$ and to use Eq. (18) for $t > \tau_i$ in the calculation of the detonation structure.

Method of Solution

Equations (1), (2), and (7) for the conservation of mass, momentum, and energy can be integrated to yield the following algebraic equations:

$$\rho u + \sigma_p u_p = A = \tilde{\rho}_0 u_0 \tag{20}$$

$$\rho u^2 + \sigma_p u_p^2 + p + WL = B = \tilde{\rho}_0 u_0^2 + p_0 \tag{21}$$

$$\rho u \left(h + \frac{u^2}{2}\right) + \sigma_p u_p \left(h_p + \frac{u_p^2}{2} + \frac{Q_0}{1-\phi_a}\right) + QW + QR - QS = H_0$$

$$= \tilde{\rho}_0 u_0 \left[(1-\eta)h_0 + \frac{u_0^2}{2} + \eta h_{p_0} + \frac{\eta Q_0}{1-\phi_a}\right] \tag{22}$$

where

$$\tilde{\rho}_0 = \sigma_{po} + \rho_0$$

is the total density of the initial mixture and $\eta = \sigma_{po}/\tilde{\rho}_0$ is the initial particle mass fraction.

Following Ragland et al. (1968), the momentum loss WL and the heat loss and viscous work terms QW and QS are evaluated from

$$WL = (4/D_h) \int_0^x \tau_w dx = 4(x/D_h) C_f [\rho(u_0-u)^2/2] \tag{23}$$

$$QW = (4/D_h) \int_0^x q_w dx = 4(x/D_h)[C_f/2]\rho(u_o-u)\left\{h + [(u_o-u)^2/2]\right.$$
$$\left. - h_w\right\} \quad (24)$$

$$QS = (4/D_h) \int_0^x q_s dx = 4(x/D_h)[C_f/2]\rho u_o(u_o-u)^2 \quad (25)$$

Here C_f is the mean friction coefficient for the flow between the shock and the station x. Since the reaction zone flow in detonations is usually observed to be turbulent it appears reasonable to use (Ragland et al. 1968),

$$C_f = 0.074(Re)^{-1/5} = 0.074[(\rho u x)/\mu]^{-1/5} \quad (26)$$

Reynolds analogy has been used (Ragland et al. 1968) to evaluate the convective heat loss term QW.

The radiation heat loss QS has been approximated by

$$QR = (4/D_h) \int_0^x q_r \, dx = (4x/D_h)\varepsilon'\sigma'T^4 \quad (27)$$

It is assumed that the dust radiates to the tube walls as a grey body with emissivity ε'. Since the local temperature T is used this expression overestimates the radiative flux. However, even with $\varepsilon' \approx 1$ the radiative loss from Eq. (27) was only a small fraction of the heat of combustion so that the main result from Eq. (27) was to show that radiation losses are negligible within the reaction zone.

The algebraic equations (23-25), (8), and (11) together with the differential equations (9), (10), and (12) provide eight equations for determining the eleven unknowns ρ, u, σ_p, u_p, p, T, σ_{O_2}, d_p, σ_{pf}, h, and h_p. The two additional required relations are provided by

$$h_p = CT, \quad h = \bar{C}_p T \quad (28)$$

where C is the specific heat of the solid and $\bar{C}_p$ is the average specific heat of the gas phase. Equations (9) and (10) are readily combined to obtain an algebraic relation between σ_p and σ_{O_2}. This leaves the rate equation (9) and the trajectory equation (12) which have been integrated numerically using a fourth-order Runge-Kutta method.

As is typical of two-phase flow problems, the relative particle Reynolds and Mach numbers vary over a wide range. In the present case $0 \leq Re_p \leq 1000$ and $0 \leq M_p \leq 1.6$ so that it is essential to take the variation in the drag coefficient C_D into account in integrating the trajectory equation (12). The empirical drag coefficient equation developed by Walsh (1975), which is based on an extensive collection of experimental results, was used for this purpose.

The relations Eqs. (23-25) and (8) can be solved for the gas velocity u with the result

$$u = [-F_2 \pm \sqrt{F_2^2 - 4F_1F_3}]/2F_1 \qquad (29)$$

where

$$F_1 = (A - \sigma_p u_p)(\frac{1}{2} - \frac{C_p}{R}) - \frac{C}{R}\sigma_p u_p$$

$$F_2 = (B - WL - \sigma_p u_p^2)(\frac{C_p}{R} + \frac{C\sigma_p u_p}{R(A-\sigma_p u_p)})$$

$$F_3 = \sigma_p u_p \left(\frac{u_p^2}{2} + \frac{Q_o}{1-\phi_a}\right) + QW + QR - QS = H_0 \qquad (30)$$

For lossless gaseous detonations the positive and negative signs in Eq. (29) correspond to strong and weak detonations, respectively, so that only the positive sign is appropriate here.

In a gaseous detonation without losses the Chapman-Jouguet (CJ) detonation corresponds to the case in which the quantity under the radical vanishes, and this condition makes it possible to calculate the propagation velocity D of a CJ detonation. For a strong detonation such that h_o, p_o, and h_{po} can be neglected, without losses, and with $\phi_a = 0$ the conditions $F_2^2 - 4F_1F_3 = 0$ leads to the result

$$D = \sqrt{2(\gamma^2-1)\eta Q_0} = \sqrt{2(\gamma^2-1)\tilde{Q}_{eff}} \qquad (31)$$

which is identical to the Zel'dovich formula (Zel'dovich and Kompaneets 1960) for the CJ velocity. Here $\tilde{Q}_{eff} = \eta Q_0$ is the heat release per unit mass of the dust air mixture.

A similar procedure has been used with losses and ash formation taken into account; now, however, it also is necessary to integrate the differential equations (9) and (10) to compute u_p and σ_p at each step, and the loss terms must also be recalculated at each step. The equations are singular near the CJ plane where the velocity u passes through the sonic value (Zel'dovich and Kompaneets 1960; Fay 1959) and there must be an exact balance between heat release and loss terms. The character of this singularity in the two-phase flow is not treated here. Rather, an iterative procedure has been used to compute the detonation velocity D. A detonation velocity is assumed initially and then calculation of the reaction zone structure is advanced to the point where (u/a) = 0.95. A new value of D then is computed using the Zel'dovich formula Eq. (31). With losses and $\phi_a \neq 0$, $\tilde{Q}_{eff}$, is now to a good degree of approximation

$$\tilde{Q}_{eff} = \theta n Q_0 - [(QW+QR-QS)/\tilde{\rho}_0 u_0] - n[1-\theta(1-\phi_a)](u_c^2/2) \quad (32)$$

The iteration is repeated until using this new value of D until there is no further significant change in the detonation velocity. The factor $\theta < 1$ is a factor which accounts for incomplete combustion at the end of the reaction zone, i.e., for the fact that $\sigma_p u_p > \phi_a \sigma_{po} u_0$ at the end of the reaction zone, and will be

$$\theta = \frac{\sigma_{po} u_0 - (\sigma_p u_p)_{CJ}}{\sigma_{po} u_0 (1-\phi_a)}$$

The value of θ is determined by the iterative procedure described above.

Computational Results

In order to obtain the best comparison between the numerical calculations and the experiments, computations using different combinations of the constants in the particle combustion model Eq. (18) have been performed. For this purpose detonation of a wheat-air mixture propagating with a velocity of 1520 m/s and with an initial dust concentration of σ_{po} = 0.305 kg/m^3 was chosen. The dust used in the experiments had particle sizes which ranged to less than 25 μm; for the calculations the initial diameter d_0 was taken to be 50 μm. The variation of the pressure with time

Table 1 Parameters Used in Wheat Dust-Air Detonation Model[a]

ϕ_a = 0.26	[b]	C_{po} = 1.005 × 10^7 cm^2/s^{2}K		
T_o = 298 K		C = 1.005 × 10^7 cm^2/s^{2}K	[c]	
ρ_p = 0.75 gm/cm^3	[c]	Q_0 = 1.272 × 10^{11} cm^2/s^2	[b]	
ε' = 0.8		ϕ_{O_2} = 0.877	[d]	
ρ_o = 1.184 × 10^{-3} gm/cm^3		P_o = 1 atm		
		D_h = 6.45 cm		

In reaction zone:

$$C_p = [1.005 + 2.56 \times 10^{-5} (T - 295)^{1.296}] \times 10^7 \text{ cm}^2/\text{s}^2\text{K}$$
$$\mu = 1.85 \times 10^{-4} + 1.54 \times (T - 300)^{0.762} \text{ gm/cm-s}$$

[a] See Nomenclature for definition of symbols.
[b] Ref. 14.
[c] Estimated properties of cellulose were used where no others are available.
[d] Assumes wheat is 75% starch by mass.

after shock wave passage was chosen as the matching parameter since this also is the quantity which is recorded by a pressure transducer. Other quantities are also shown as functions of distance behind the leading shock front. The other parameters used in this calculation are shown in Table 1.

Many combinations of the combustion parameters were tested but so far only a partial fit of the calculations to the experimental data has been possible. The best matching has been achieved for the following values of the combustion parameters:

A = 80 cm/s, E/R = 5000 K, S_1 = 1.0, S = 2.5

The fact that the reaction rate is proportional to the first power of the oxygen concentration σ_{O_2} suggests that the reaction may occur at the particle surface. The value of 2.5 for S suggests that there is a relatively large in-

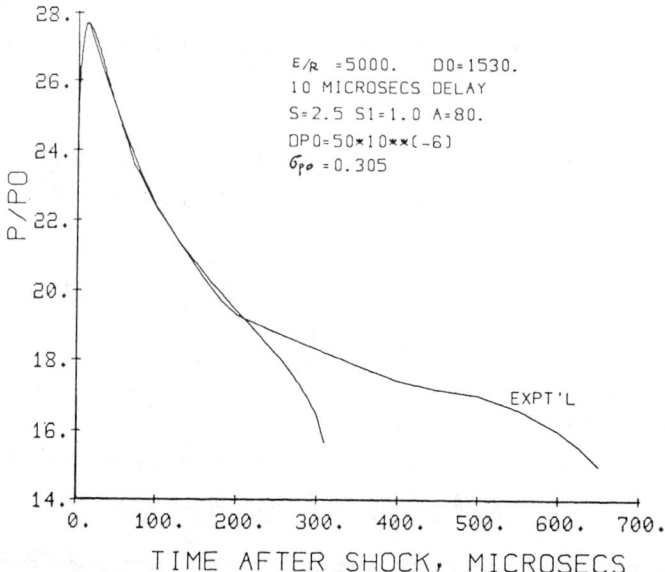

Fig. 2 Comparison of computed and experimental pressure profiles.

fluence of the surface area reduction due to ash formation. In all trials an activation temperature E/R of 5000 K appeared to give the best results. The ignition delay was taken as 10 μs in agreement with experimental observations.

The theoretically computed and observed pressure profiles are shown in Fig. 2. The experimental curve is a replot of an oscilloscope trace of a pressure transducer output. The experimental measurements are described by Kauffman et al. (1979). The agreement between the model and experiment are encouraging for the first 200 μs after shock passage, especially since the pressure is a parameter which is very sensitive to the reaction mechanism. After that in the vicinity of the CJ plane the computed pressure drops much more rapidly than the observed value. Failure to properly treat the CJ condition or to take the unsteady flow behind the CJ plane into account may explain this divergence between theory and experiment. The computed detonation velocity was 1558 m/s in reasonable agreement with the observed value of 1520 m/s.

In addition to the above, the model has also been used for parametric studies to determine the influence of initial conditions on the pressure variation behind a shock propagating through a combustible dust-air mixture. The influence of ignition delay, dust concentration, particle size, and oxygen concentration were examined. Figure 3 shows the

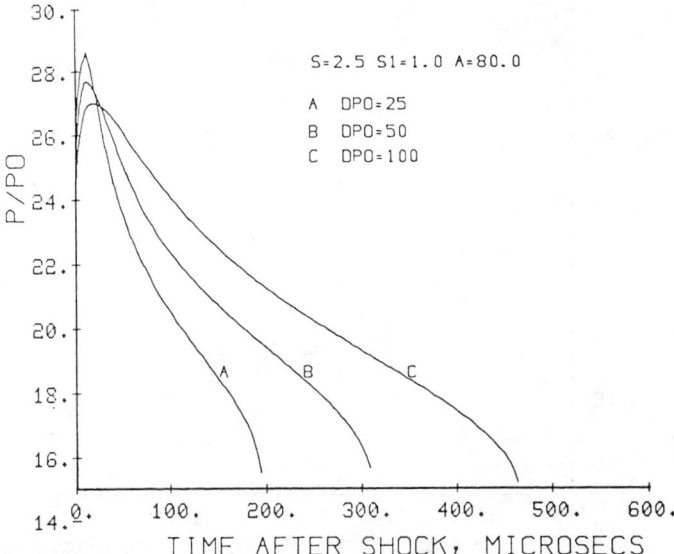

Fig. 3 The influence of particle diameter on the reaction zone pressure profile.

influence of particle diameter on the pressure variation. It can be seen that there is a larger pressure increase just behind the shock for the smaller particles. This is a two-phase flow effect which is caused by the more rapid deceleration of the smaller particles. The biggest influence is however on the combustion time which decreases significantly for the smaller particles because of the larger effective burning surface.

Figure 4 shows that variation of dust concentration σ_{po} has a similar effect on the postshock pressure history; higher dust loading for the same shock velocity results in a higher initial pressure rise. It is interesting that the pressure variation near the CJ plane for the low article cle loading of 0.2 kg/m^3 is much more like that observed experimentally. Combustion in pure oxygen significantly decreases reaction time (Fig. 5). A longer ignition delay will only result in a different initial pressure profile while the total combustion time remains almost unchanged.

The computed property profiles through the reaction zone of the detonation corresponding to Fig. 2 are shown vs time since shock passage in Figs. 6(a-d). Figure 7 shows the variation of T/T_0 and ρ/ρ_0 with distance behind the shock providing an indication of the reaction zone thickness. The variation of the other properties with distance are very similar to the time profiles and so are

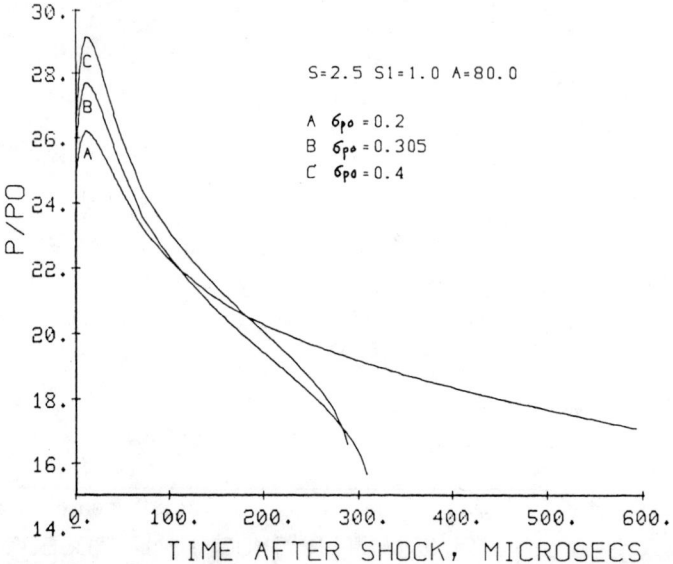

Fig. 4 The influence of initial particle density on the reaction zone pressure profile.

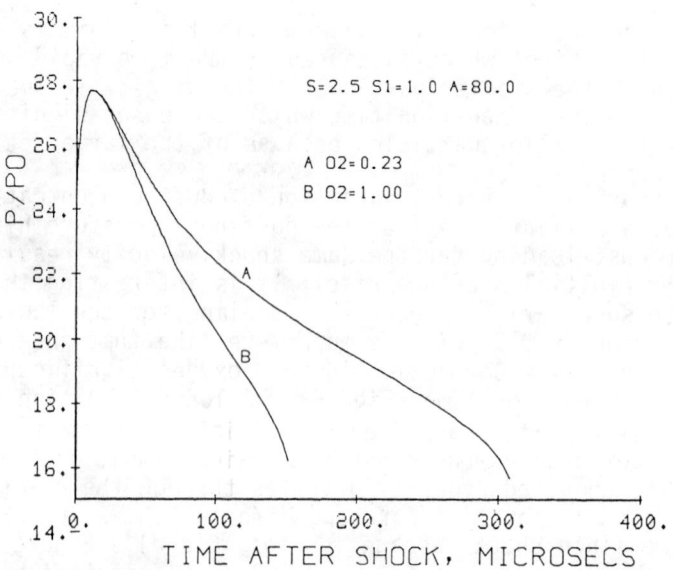

Fig. 5 The influence of oxygen mass fraction on the reaction zone pressure profile.

not shown. It can be seen that the temperature rises
rapidly to a plateau after ignition, but then drops slightly near the end of the reaction zone. The density of the
gas increases insignificantly during the induction period,
then drops rapidly during the first phase of combustion,
and continues to drop slowly to the CJ value.

In shock-fixed coordinates the particles decelerate
very rapidly during the ignition delay and the first phase
of combustion. Then, after the gas has already accelerated to a velocity higher than that of the particles (due
to combustion heat release), the unburned part of the particles follows the gas velocity with a small lag up to the
CJ plane. Because of this velocity history there is a local excess of oxygen which creates a favorable condition
for combustion at the initial phase. Later, when the particle velocity is below that of the gas, the oxygen concentration drops very rapidly.

The detonation velocity has been calculated over a
range of initial particle concentrations σ_{po} following the
procedure described in the preceding section using the combustion model corresponding to Fig. 2. The computed value
of D is compared to that measured for oats and wheat dust-
air mixtures in Fig. 8. The agreement is quite good although the computed value of D appears to be somewhat above
the measured values. Of course it must be recognized that
the detonation velocity is relatively insensitive to the
mechanism of reaction, but depends mainly on the effective
heat release. The calculation of detonation velocity provides no information about detonation limits. Thus Fig. 8
shows values of D for initial dust concentration $\sigma_{po} >$
0.305 kg/m^3 even though such mixtures could not be detonated.

As a further test of the model computed and measured
pressure profiles are compared in Fig. 9 for a detonation
propagating through a mixture of oats dust with a particle
loading of 0.401 kg/m^3 and pure oxygen. Due to a lack of
data the property values listed in Table 1, and combustion
parameters identical to those used for the wheat dust-air
detonation shown in Fig. 2 were used. The agreement, although not as good as in Fig. 2 is quite satisfactory.

Discussion and Conclusions

The model proposed here, which includes the effects of
two-phase flow, produced reaction zone lengths of the
proper magnitude. The computed and experimental pressure
distributions were in excellent agreement immediately behind the shock, but the computed pressure then dropped

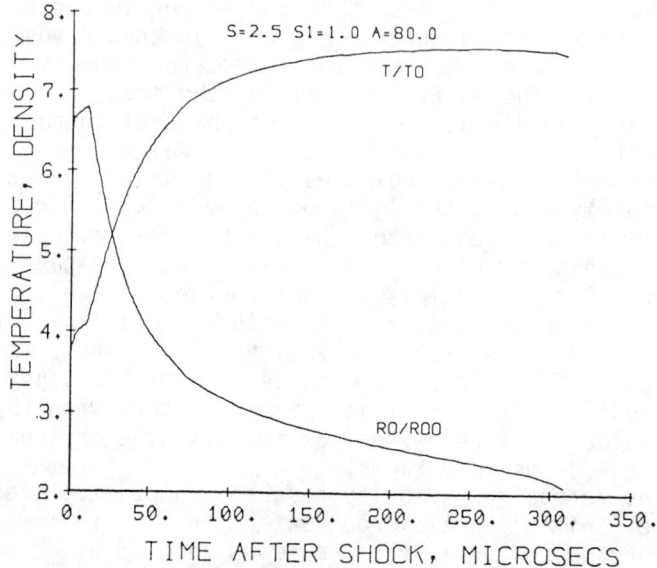

Fig. 6a Reaction zone temperature and density profiles.

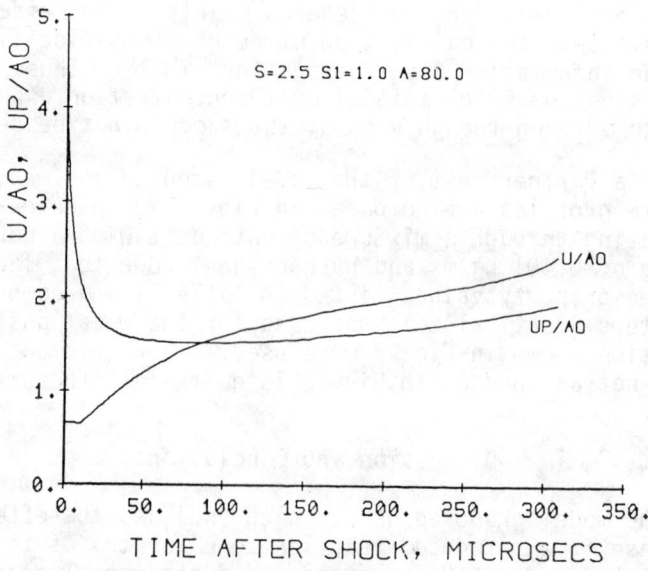

Fig. 6b Particle and gas velocity variation in the reaction zone.

THE STRUCTURE OF DUST DETONATIONS

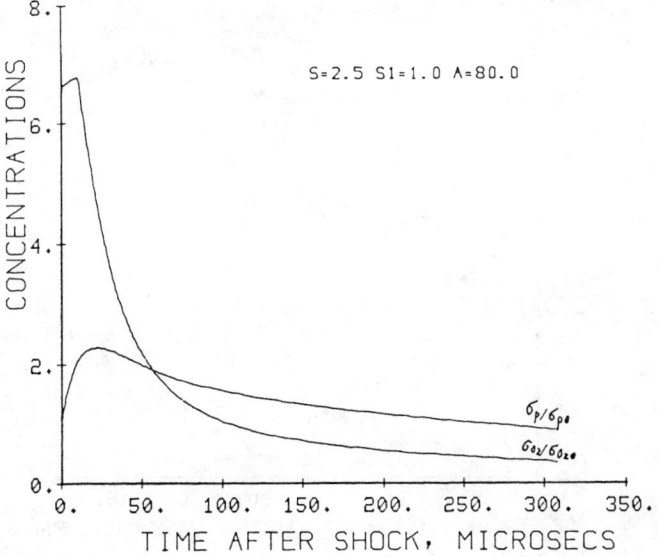

Fig. 6c Particle and oxygen density variation in the reaction zone.

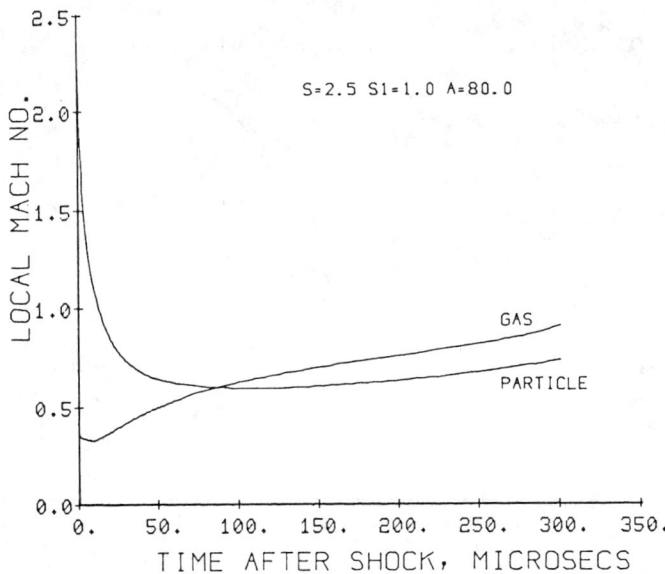

Fig. 6d Variation of the gas and particle Mach number in the reaction zone.

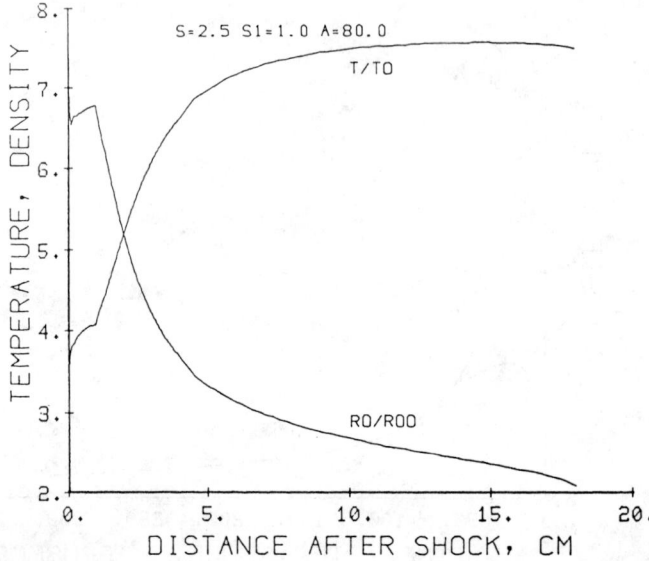

Fig. 7 The variation of reaction zone temperature and pressure with distance behind the shock.

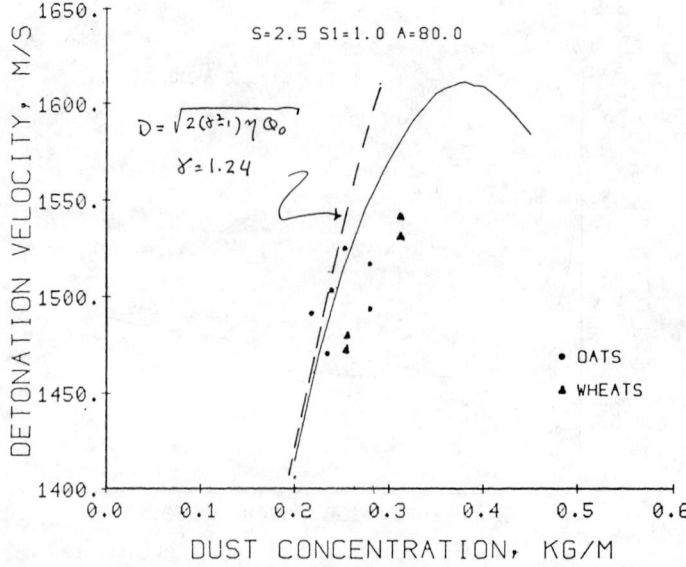

Fig. 8 Measured and computed detonation velocities for wheat-air and oats-air mixtures.

below the measured value, probably because of the many simplifications in the theoretical model, and failure to properly model the flow near the CJ plane.

The good agreement between the computed and measured pressure profiles provides support for the correctness of the computed profiles of temperature, gas and particle density, and gas and particle velocity, quantities which are not readily measured in a two-phase reaction zone.

Surprisingly, the losses due to radiation are negligibly small, even though radiation appears to play an important role in dust flames. In part this may be due to the very short time the dust particles spend in the reaction zone. Radiation in the axial direction of the shock tube has been neglected; however, in preliminary experiments very little radiation flux was observed ahead of the detonation suggesting that longitudinal radiation may not be important in the reaction zone. The maximum values of the convective heat loss QW, and the friction work QS are never more than about 5% of the total heating value Q_o for the 6.5-cm^2 tube considered here and QW and QS tend to cancel each other in the expression Eq. (32) for $\tilde{Q}_{eff}$. It is thus of interest to compute D using the Zel'dovich formula Eq. (31) and assuming complete combustion. The value of D computed in this way is shown as the dotted curve in Fig. 8 and indicates that incomplete combustion becomes important for $\sigma_{po} > 0.25$ kg/m^3. It is interesting that detonations could not be initiated in mixtures with $\sigma_{po} \gtrsim 0.3$ kg/m^3 when it also appears that incomplete combustion becomes a significant factor.

The particle combustion model is the greatest source of uncertainty in the present analysis. The combination of parameters finally used here, i.e., A = 80 cm/s, E/R = 5000 K, S_1 = 1.0, and S = 2.5, is by no means unique. Reasonable agreement also could be obtained using A = 30 cm/s, E/R = 5000 K, S_1 = 2.0, and S = 1.0. Unique determination of combustion parameters will require comparison of computed and measured results over a wide range of experimental conditions.

The key questions which must be considered in particle burning (Glassman 1977) are whether the rate limiting process is surface burning or diffusion of oxidizer to the surface, the effect of surface porosity, and the reduction in reaction rate due to ash formation. All three factors are considered in the present model. The basic assumption is that the surface burning rate is the limiting process; hence, the diffusivity does not enter the rate expression. The particles are taken as nonporous since ϕ_s = 1, and an empirical factor f to account for ash formation has been

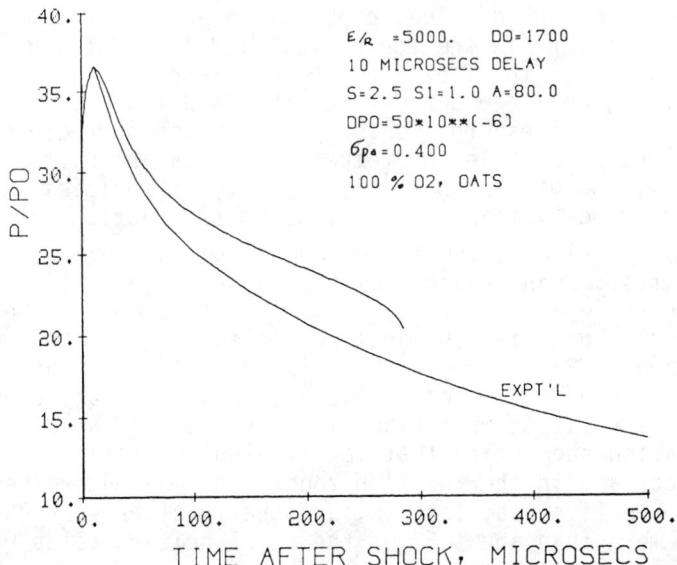

Fig. 9 Computed and measured pressure profiles for a detonation through an oats dust-O_2 mixture.

introduced. The reaction profiles were found to be extremely sensitive to the activation energy. The activation temperature of 5000 K (or 10,000 cal/mole), which provided best agreement with the data, is fairly typical of hydrocarbon fuels.

The authors could not find independent data for grain dust which could be applied to the present model. Actually, the reaction zone profiles are mainly determined by the combustion process because of relatively small magnitude of radiative and convective losses to the shock tube walls. This observation suggests that reaction zone pressure measurements carried out over a wide range of parameters may provide an experimental technique for determining dust combustion properties. This approach may well provide more realiable results than single particle measurements which often involve serious experimental difficulties.

The results presented in this paper must be considered preliminary in nature although they do illuminate the details of dust detonation reaction zone structure. The analysis also points out the great need for wheat dust reaction data.

Acknowledgment

The authors would like to express their appreciation for support for the work reported here by the National

Science Foundation under NSF Grant CPE-8023865, R. Rostenbach, Project Monitor, and by the Air Force Office of Scientific Research, under AFOSR Grant No. 79-0093, B. T. Wolfson, Project Monitor.

References

Bartknecht, W. (1978) Proc. of the Intl. Symp. on Grain Elevator Explosions. National Academy of Sciences, Washington, D.C.

Cybulski, W. (1971) Bulletin del academie polonaise des sciences. 19, 37.

Fay, J. A. (1959) Phys. Fluids 2, 283.

Glassman, I. (1977) Combustion. Academic Press, New York.

Kauffman, C. W., Wolanski, P., Ural, E., Nicholls, J. A. and Van Dyke, R. (1979) Proc. of the Intl. Symp. on Grain Dust, p. 164. Manhattan, Kansas.

Lee, J. H. (1977) Ann. Rev. Phys. Chem. 28, 75.

Martin, C. R. (1978) "Characterization of grain dust." Paper No. 78-3020, presented at meeting of Amer. Soc. of Agricultural Engineers, Logan, Utah, June 27-30.

Nettleton, M. A. and Stirling, R. (1973) Comb. and Flame 21, 307.

Ragland, K. W., Dabora, E. K., and Nicholls, J. A. (1968) Phys. Fluids 11, 2377.

Rudinger, G. (1970) "Relaxation in gas particle flow." Non-Equilibrium Flows, Marcel Dekker, New York.

Ural, E. A., Sichel, M., and Kauffman, C. W. (1982) "Shock wave ignition of pulverized coal." Proc. of 13th Intl. Symp. on Shock Tubes and Waves, p. 809. SUNY Press, New York.

Walsh, M. J. (1975) AIAA J. 13, 1526.

Wojcicki, S. and Zalesinki, M. (1973) Proc. of Ninth Intl. Shock Tube Symp. Stanford University Press, Palo Alto, Calif.

Wolanski, P. (1982) Private communication. The University of Michigan, Ann Arbor, MI.

Zel'dovich, Ya. B. and Kompaneets, S. A. (1960) Theory of Detonation. Academic Press, New York.

"Double-Front" Detonations in Gas-Solid Particles Mixtures

Bernard Veyssière*
Centre National de la Recherche Scientifique
Poitiers, France

Abstract

The detonation of gas-solid particles mixtures has been studied in a vertical detonation tube. Gaseous components were H_2, C_2H_4 or C_2H_2 with O_2/N_2 mixtures with suspended aluminium particles (mean diameter 10 μm). Velocity, pressure, and brightness measurements were made together with streak photographs. For mixtures with H_2, but not with C_2H_2 or C_2H_4, two successive fronts have been observed. Under some experimental conditions, the "double-front" appears to be quasistationary. The first front is due to the reaction of gaseous components and the second to the combustion of aluminium in the gaseous products. The conditions for the appearance of such a structure are discussed.

Introduction

During the past few years, dust explosions have been studied at the Laboratoire d'Energétique et de Détonique of Poitiers. Current research has been concerned with the study of two-phase systems composed of fine reactive particles suspended in explosive gas mixtures. The purpose of this work is to characterize the changes in detonation structure due to the presence of such particles. Veyssière et al. (1980) and Veyssière et al. (1981), described the experimental device designed for the study of detonations in such two-phase systems and reported results obtained in two mixtures $C_2H_4 + XO_2 + ZN_2$, equivalence ratio r = 1.15, with Z/X = 3.76 (called E1) and with Z/X = 3.14 (called E2). Experiments were

---

Presented at the 9th ICODERS, Poitiers, France, July 3-8, 1983. Copyright © American Institute of Aeronautics and Astronautics, Inc., 1984. All rights reserved.
*Chargé de Recherches, Laboratoire d'Energétique et de Détonique, L.A. CNRS 193, E.N.S.M.A.

performed with aluminium particles (mean diameter 10 μm) at concentrations σ ranging between 5-50 g/m³.

The main experimental observations were the following: 1) The detonation front velocities of the two mixtures were about 3% below those measured in the mixtures without suspended aluminium particles. 2) A secondary smooth pressure maximum, which occured about 50-200 μs after the detonation front was observed. 3) About 10-70 μs behind the front the reaction between aluminium and the gaseous mixture became appreciable and led to a rapidly increasing luminous emission. These

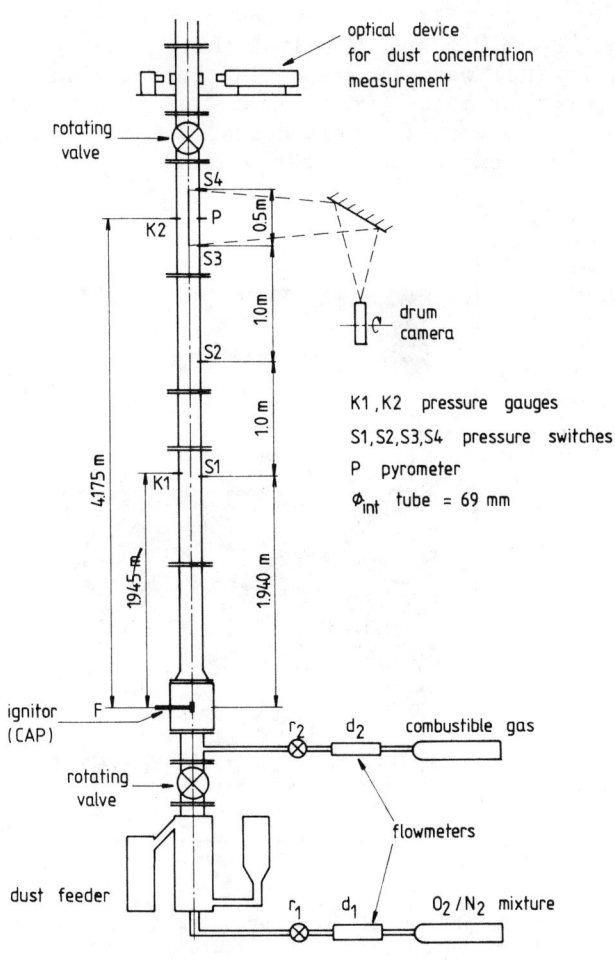

Fig. 1 Schematic of experimental apparatus.

observations suggested that the change in the pressure profile was probably due to the rapid reaction between solid particles and gaseous products. The solid-gas reaction did not appear to influence the propagation of the front. Consequently, it appeared necessary to obtain further observations on the detonation structure in such two-phase mixtures.

Experimental Results

A new series of experiments, with the same apparatus (see Fig. 1) and at the same conditions as previously (Veyssière et al. 1981; Veyssière, and Manson 1982) are reported in this work. Different mixtures with hydrogen: $H_2 + XO_2 + ZN_2$, $Z/X = 3.76$ and $r = 0.78$ (called below H1), $Z/X = 3.76$ and $r = 1.06$ (H2), and $Z/X = 2.2$ and $r = 0.75$ (H3) were experimented. The concentration σ of aluminium particles (mean diameter 10 μm) was varied between 15-80 g/m^3 (in most cases, $\sigma = 65 \pm 15$ g/m^3). As in previous experiments, the velocity of the detonation

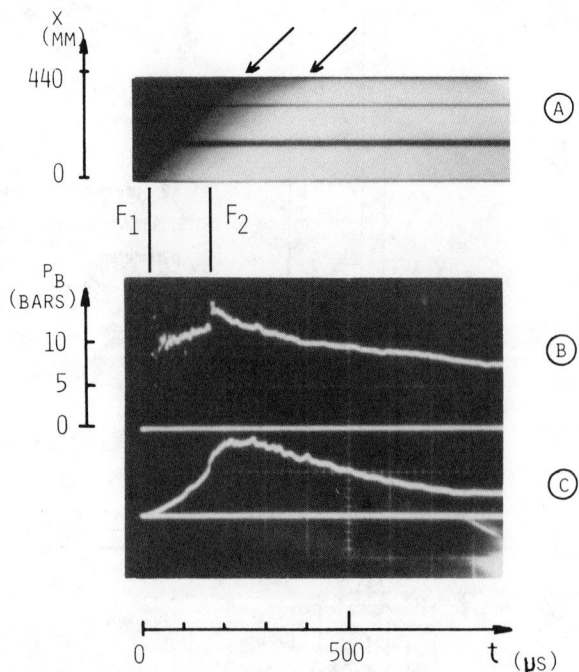

Fig. 2 Streak photograph (A), pressure (B), and brightness (C) records of the detonation in a mixture H_2/air/Al at equivalent ratio $r = 0.78$. ($\bar{d}_{A\ell} = 10$ μm ; $\sigma_{A\ell} = 65$ g/m^3).

front, the pressure profile, and the brightness profile (for more details see Veyssière et al. 1981) were recorded. The pressure was recorded by means of a KISTLER 603 B pressure gage. The pressure gage and the pyrometer were located 4.175 m from the ignition point. A drum camera was used to record simultaneously either self-light or Schlieren deflections to produce streak photographs from which a (x, t) diagram of the phenomenon could be deduced.

When aluminium particles were suspended in mixtures H1 and H2, a sharp discontinuity F_2 in pressure and brightness was observed at about τ = 100-250 μs after the detonation front F_1. A typical experimental result is displayed as Fig. 2 which includes the pressure and brightness profiles and the streak photograph obtained with mixture H1 and aluminium particle concentration of about 65 g/m^3.

On Fig. 3, pressure profiles are shown for the same mixtures H1 with and without particles. The second front F_2 had not been observed in the previous study of the detonation of mixtures E1 and E2 (with ethylene). As those mixtures were slightly rich ones (equivalence ratio r = 1.15), which could adversely affect the exothermic reaction of aluminium particles with gaseous detonation products, observations were made in a mixture Z/X = 2.2, r = 0.8 (E3), which could support sufficiently steady detonations. A second front was not observed in mixture E3 which contained suspended aluminium particles. A typical record of the pressure profile in mixture E3 is given as Fig. 4. In experiments performed in mixtures H3 having the same ratio Z/X = 2.2 and nearly the same equivalent ratio r as E3, the second discontinuity was observed. Additional experiments were performed in two acetylene-air-Aℓ particles mixtures: Z/X = 3.76,

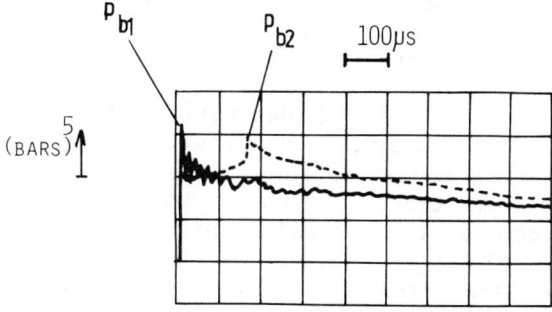

Fig. 3 Pressure profile in the detonation products of mixture H1 (r = 0.78). ($\bar{d}_{Aℓ}$ = 10 μm ; $\sigma_{Aℓ}$ = 65 ± 15 g/m^3).──── without particles, ----- with particles.

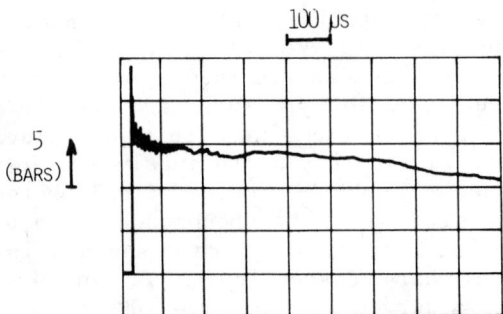

Fig. 4 Typical pressure profile of the detonation in a mixture E3 with aluminium particles.

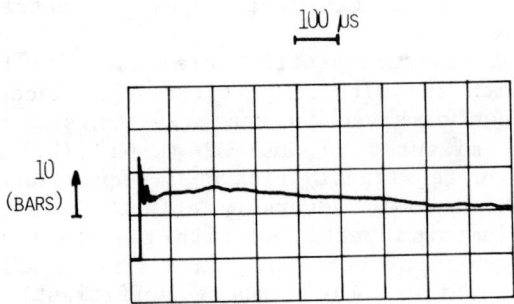

Fig. 5 Typical pressure profile of the detonation in a mixture A2 with aluminium particles.

$r = 1.07$ (A1) and $Z/X = 3.76$, $r = 1.25$ (A2). In those mixtures, a second discontinuity was not observed. Fig. 5, a typical record of pressure profile in mixture A2, indicates that, the pressure profile recorded in mixtures with acetylene is very similar to that recorded in mixtures with ethylene.

The Second Front

The characteristics of detonations (velocity D_{CJ}, pressure p_{CJ} - CJ: Chapman-Jouguet state) in mixtures H1 and H2, and for a concentration of aluminium particles $\sigma = 65$ g/m^3 (mean concentration of our experiments) were calculated from the thermochemical properties reported by Veyssière et al. (1980) and Veyssière et al. (1981). These results are compared in Table 1 with the measured velocity D and the maximum pressures p_{b1} and p_{b2} measured at F_1 and F_2. Whereas in mixtures E the reaction of aluminium particles did not seem to support the

Table 1 Calculated and measured values of velocity and pressure for mixtures H1 ($r = 0.78$) and H2 ($r = 1.06$) with and without aluminium particles ($\bar{d}_{Al} = 10$ μm; $\sigma_{Al} = 65 \pm 15$ g/m^3). τ is the time interval between the two fronts F_1 and F_2

Mixtures	H1 ($r = 0.78$)			H2 ($r = 1.06$)		
	Without Aℓ	With Aℓ		Without Aℓ	With Aℓ	
		Reactive Aℓ	Inert Aℓ		Reactive Aℓ	Inert Aℓ
D (m/s)	1795 ± 40	1965 ± 25		1985 ± 25	2000 ± 15	
D_{CJ} (m/s)	1855	2041	1612	1992	2122	1790
P_{b1} (bar)	17.5 ± 0.5	17.5 ± 1.5		17.5 ± 0.5	17 ± 1	
P_{b2} (bar)	...	16		...	13	
P_{CJ} (bar)	14.8	18.4	11.2	15.9	18.5	13
τ (μs)		135 ± 35			250 ± 50	

Table 2 Comparison of calculated and measured values of velocity and pressure for mixtures E1 and E2 with and without aluminium particles. ($d_{A\ell}$ = 10 μm ; $\sigma_{A\ell}$ = 30 ±10 g/m³). Pressure recorded by means of an ANVAR 11568 pressure gage

Mixtures	E1 (Z/X = 3.76; r = 1.15)			E2 (Z/X = 3.14; r = 1.15)		
	Without Aℓ	With Aℓ		Without Aℓ	With Aℓ	
		Reactive Aℓ	Inert Aℓ		Reactive Aℓ	Inert Aℓ
D (m/s)	1879 ± 20	1811 ± 20		1936 ± 20	1863 ± 20	
D_{CJ} (m/s)	1866	1920	1805	1918	1974	1866
P_{b1} (bar)	15.8 ± 1.5	15.7 ± 1.5		18.1 ± 1.5	17.9 ± 1.5	
P_{CJ} (bar)	19.4	20.7	18.1	20.1	21.8	19.5

propagation of the front F_1 (see Tables 2 and 3; Veyssière et al. 1980; Veyssière et al. 1981), the reaction of aluminium particles must support detonations in mixtures H since the recorded mean velocity is higher for H1, and does not diminish for H2, when aluminium particles are in suspension.

In an attempt to identify the nature of the second front, calculations have been performed in a simplified configuration for which (1) behind the first front F_1, the flow was expanding according to the Taylor-Zel'dovich theory, and (2) the second front F_2 was a steady wave having the same velocity as F_1. This assumption was supported by the fact that on streak photographs, the interval between the two fronts was quasiconstant along the observation length - 440 mm. If F_2 is assumed to be a nonreactive wave, the pressure jump calculated at F_2 is in a good agreement with the measured value for H1 (calculated: 16.7 bar; experimental p_{b2}: 16 bar); but the agreement is not as good for H2 (calculated: 18.9 bar; experimental p_{b2}: 13 bar). If F_2 is assumed to be a detonation wave sustained by the reaction of aluminium with H_2O produced by the reactions of gaseous components, there is no agreement between the calculated velocity and pressure jump and the experimental values. Thus, this second front, cannot be regarded as a detonation wave in a strict sense since the shock front and the reaction zone of aluminium particles (which supports this shock wave) do not appear to be coupled as in a detonation. Besides, as displayed by the simultaneous records of Fig. 2, the maximum of brightness is reached at least 50 μm after F_2.

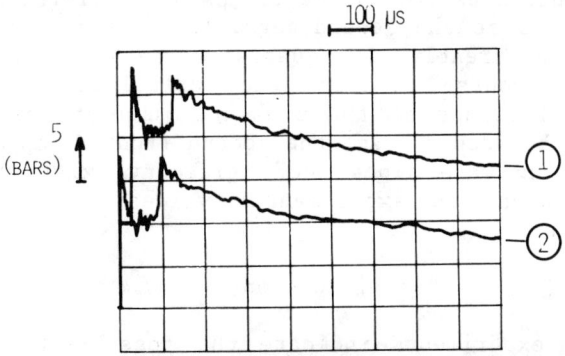

Fig. 6 Pressure profile of the detonation in a mixture H1 with aluminium particles. ① at 1.945 m from the ignition point. ② at 4.175 m from the ignition point.

Table 3 Comparison of calculated and measured values of velocity and pressure for mixture A2 with and without aluminium particles. ($\bar{d}_{Al}$ = 10 μm, σ_{Al} = 65 ± 15 g/m^3). Pressure recorded by means of a Kistler 603 B pressure gage

Mixtures	A2 (Z/X = 3.76; r = 1.25)		
	Without Al	With Al	
D (m/s)	1891 ± 20	1879 ± 20	
D_{CJ} (m/s)	1932	Reactive Al	Inert Al
		2040	1813
P_{b1} (bar)	25	28	
P_{CJ} (bar)	20.7	Reactive Al	Inert Al
		23.2	18.0

To clarify the question of the steady state propagation of the double front, new experiments were undertaken to determine whether the interval between the two fronts remains constant as the detonation progresses along the tube. For this purpose, two pressure gages were located 1.945 and 4.175 m, respectively, from the ignition point. The first results indicate that in mixtures H1 (see Fig. 6), this interval is quasiconstant (about 100-150 μs); but in mixtures H2 (see Fig. 7), the interval increases from about 150 μs at the first gage to almost 300 μs at the second gage. In the first case, the structure apparently is quasistationary, since good agreement occurs between the calculated and the experimental values of the pressure jump at the second front. In the second case, the second front F_2 apparently is a decaying wave since poor agreement exists between the calculated and experimental values of the second pressure jump.

Discussion

These experiments indicate the possibility of the existence of a "double-front" structure in the detonation of two-phase systems of detonable gaseous mixtures and reactive solid particles in suspension. The existence of

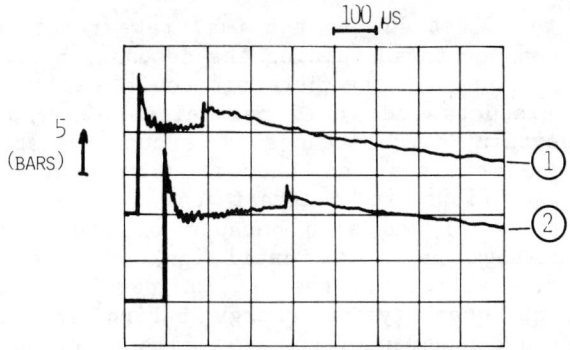

Fig. 7 Pressure profile of the detonation in a mixture H2 with aluminium particles. ① at 1.945 m from the ignition point. ② at 4.175 m from the ignition point.

a secondary shock wave has already been studied in details by Lee et al. (1969). They showed that, for diverging detonations under cylindrical and spherical configuration, a secondary shock wave may be generated in the flowfield behind the detonation front under certain conditions when the detonation is supported by a piston. This situation theoretically should not occur in the one-dimensional case because there is no flow divergence. Edwards et al. (1975) have reported that a shock wave occurred behind a driven planar detonation in a shock tube when a piston motion was provided by the contact surface between the detonated and expanding driver gas. Their experiments were conducted in oxygen-acetylene mixtures at initial pressures of 30-50 Torr. This secondary shock wave was attributed to the flow divergence produced by the wall boundary-layer displacement effect, and had a tendency to move slightly away from the detonation front.

Khasainov et al. (1979) have shown that exothermic reactions downstream of the CJ plane of a detonation might result in the occurrence of a secondary nonstationary shock wave. Afanasieva et al. (1983) have recently reported computational results for a model of a detonation of a two-phase system for which the energy is released in two successive, time-separated stages with heat loss to the confinement walls behind the detonation front included. The model results suggest that the propagation of a "double-detonation" is possible in the plane case. The experimental results of the present work seem to support the findings of prior investigators. As a consequence, the following model is proposed: the two-phase system energy is released in two successive stages, the first due to the fast rate reactions of the gaseous components sustaining the detonation front F_1 and the second due to the heterogeneous reactions between gaseous products and solid particles which may support under certain circumstances a secondary shock wave. Because of momentum and thermal transport effects in particulate flows in contrast to that of gases (see Veyssière, 1983) and also because of the necessity to heat aluminium particles until ignition temperature of 2310 K is achieved, the solid particles absorb a significant quantity of energy behind the detonation front. As a consequence, the particles contribute to flow deceleration and to heat loss from the gas resulting in phenomena analogous to flow divergence.

These conclusions remain restricted to the experimental configuration for which these results have been obtained. The role played by the detonation tube may be of great importance. The influence of the tube diameter on the propagation of detonations in gaseous mixtures has been reported by Manson and Guénoche (1957) and more recently by Desbordes et al. (1983) and Moen et al. (1981). Veyssière et al. (1980), have shown that a detonation, in the gaseous mixtures and conditions studied in this work, are self-sustained near the terminal part of the detonation tube. The addition of particles seems to modify slightly the velocity of the detonation front F_1. This observation suggests that F_1 is mainly sustained by the energy released by the reactions of gaseous components (the combustion of aluminium particles does not appear to influence significantly the propagation of the front F_1). Heat losses from the gaseous combustion products downstream of the CJ-plane are due to the heat absorbed by the particles and to heat losses to the tube wall and play an important role in the occurrence of the second discontinuity. These considerations suggest that the "double-front" structure might not be observed in a tube of larger diameter. While, the existence of the "double-front" structure has been observed within a distance of about 4 m from ignition, there is no experimental proof that the structure will persist in a longer tube. Additional experiments are necessary prior to the extension of the present results to other configurations (larger tube diameters or longer tubes).

Conclusions

The available experimental evidence establishes the existence of "double-front" detonations (quasistationary under certain circumstances). This structure seems due mainly to the intrinsic properties of the mixture (gas-solid particles mixture; heat release in two successive, time-separated stages). The conditions under which "double-front" detonations occur are not yet well defined. The structure may depend on the composition of the gaseous mixture, the concentration of particles, the heterogeneous reaction processes, the ignition energy and device, and the confinement (tube diameter and length). Additional experiments are required to clarify the present results. Reliable experimental information is not easy to obtain, particularly because it is difficult to perform reproducible experiments in heterogeneous systems and also to know the concentration of solid particles with good accuracy. Consequently, one cannot easily draw definitive conclusions on the basis of a few experiments.

Acknowledgment

The author is grateful to Mr. R. Charpentier for his help in the achievement of the experiments.

References

Afanasieva, L. A., Levin, V. A., and Tunik, Yu. V. (1983) Multifront combustion of two-phase medium. Shock Waves, Explosions, and Detonations: AIAA Progress in Astronautics and Aeronautics, (edited by J. R. Bowen, N. Manson, A. K. Oppenheim, and R. I. Soloukhin), pp. 394-413. Vol. 87, AIAA, New York.

Desbordes, D., Manson, N., and Brossard, J. (1983) Influence of walls on pressure behind self-sustained expanding cylindrical and plane detonations in gases. Shock Waves, Explosions, and Detonations: AIAA Progress in Astronautics and Aeronautics, (edited by J. R. Bowen, N. Manson, A. K. Oppenheim, and R. I. Soloukhin), pp. 394-413. Vol. 87, AIAA, New York.

Edwards, D. H., Jones, A. T., and Philips, D. E. (1975) The secondary shock wave in supported detonations with flow divergence. J. Phys. D: Appl. Phys. 8, pp. 891-901.

Khasainov, B. A., Ermolaev, B. S., Borissov, A. A., and Korotkov, A. I. (1979) Effect of exothermic reactions downstream CJ plane on detonation stability. Acta Astronautica 6, pp. 557-568.

Lee, J. H. S., Knystautas, R., and Bach, G. G. (1969) Theory of explosions. MERL Report No.69-10, McGill University, Canada.

Manson, N. and Guénoche, H. (1957) Effect of the charge diameter on the velocity of detonation waves in gas mixtures. 6th Symposium (International) on Combustion, p. 631, The Combustion Institute, Rheinhold.

Moen, I. O., Donato, M., Knystautas, R., and Lee, J. H. (1981) The influence of confinement on the propagation of detonations near the detonability limits. <u>18th Symposium (International) on Combustion</u>, pp. 1615-1622. The Combustion Institute, Pittsburgh, Pa.

Veyssière, B., Bouriannes, R., and Manson, N. (1980) Détonation des mélanges biphasiques éthylène-oxygène-azote-particules d'aluminium. C.R.A.S., t. 290, B, pp. 147-149.

Veyssiere, B., Bouriannes, R., and Manson, N. (1981) Detonation characteristics of two ethylene-oxygen-nitrogen mixtures containing aluminium particles in suspension. <u>Gasdynamics of Detonations and Explosions: AIAA Progress in Astronautics and Aeronautics</u>, (edited by J. R. Bowen, N. Manson, A. K. Oppenheim, and R. I. Soloukhin), pp. 423-438, Vol. 75, AIAA, New York.

Veyssière, B. and Manson, N. (1982) Sur l'existence d'un second front de détonation des mélanges biphasiques hydrogène-oxygène-azote particules d'aluminium. C.R.A.S., t. 295, II, pp.335-338.

Veyssière, B., (1983) Ignition of aluminium particles in a gaseous detonation. <u>Shock Waves, Explosions, and Detonations: AIAA Progress in Astronautics and Aeronautics</u>, (edited by J. R. Bowen, N. Manson, A. K. Oppenheim, and R. I. Soloukhin), pp. 362-375. Vol. 87, AIAA, New York.

Unconfined Aluminum Particle Two-Phase Detonation in Air

Allen J. Tulis* and J. Robert Selman†
Illinois Institute of Technology, Chicago, Illinois

Abstract

Analytical and experimental studies were conducted for the unconfined detonation of aluminum particles dispersed in air. It is shown that 1) the primary requirement is sufficiently large surface-to-mass ratio, e.g., 3-4 m^2/g; 2) initiation requirements are large, e.g., 2.27 kg of high explosive charge; 3) radiant energy associated with high shock strength appears to improve initiation; 4) the convective flow behind very strong initiation shock waves causes homogeneity disproportionation that tends to decouple the incident shock wave from the reaction front; and 5) explosive dissemination, with its attendant inhomogeneous particle-air dispersion in the absence of secondary particle-shattering mechanism such as exists with liquid droplets, makes any solid-fuel particle-air detonation very difficult to achieve. Although incipient detonation was induced under the conditions and constraints of the experiment in the case of flaked-aluminum powder with a surface-to-mass ratio equivalent to that of spherical aluminum particles smaller than 1 μm, stable propagation in large-scale experiments wherein the effects of the initiation charge are completely eliminated has yet to be demonstrated. With the larger-particle sized atomized aluminum of 5-μm mean diameter, detonation was not achievable, although blast enhancement due to exothermic energy release behind the attenuating incident shock wave from the high-explosive initiation charge was observed. Heat-transfer computer calculations based on a propagating aluminum particle-air detonation confirm that spherical

---

Presented at the 9th ICODERS, Poitiers, France, July 3-8, 1983. Copyright© American Institute of Aeronautics and Astronautics, Inc., 1984. All rights reserved.
* Engineering Advisor, IIT Research Institute.
† Associate Professor, Chemical Engineering Department.

particles larger than about 5 μm probably cannot be heated sufficiently fast within the induction times experimentally observed in previous detonation tube studies on the one-dimensional detonation of aluminum particles dispersed in air.

Introduction

The detonation of unconfined two-phase, liquid-fuel-in-air dispersions was achieved many years ago (Robbins and Thuman 1965; Thuman and Robbins 1965; Zabelka and Smith 1969), and the mechanism is now well understood (Kauffman 1971; Fishburn 1974). The overall process is a unique combination of physical and chemical mechanisms and has been postulated to occur as follows (Lu and Slagg 1974; Dabora and Weinberger 1974):

1) The liquid fuel is initially dispersed in air to form an aerosol, generally achieved in a fraction of a second utilizing explosive dissemination techniques. The resultant aerosol is not homogeneous, and the droplet sizes vary over many orders of magnitude. If simply ignited in this state, only a deflagration could result.

2) Hence, use is made of high-explosive charges that are detonated within the aerosol after it is formed. The resultant shock wave and attendant high-velocity convective flow behind the shock wave shatter the droplets by a micro-mist stripping mechanism (Dabora and Fox 1972). This orders-of-magnitude reduction of particle size within the high-pressure and -temperature region behind the incident shock wave allows chemical reaction to proceed sufficiently fast to create secondary shock waves that coalesce with and support the incident shock wave so that a detonation propagates throughout detonable regions of the aerosol.

Thus, the controlling mechanism in the unconfined detonation of liquid fuels dispersed in air appears to be the physical breakup of the relatively large liquid droplets into micron-sized micro-mists, with attendant tremendous increases in surface-to-mass ratios. Such a micro-mist stripping process is not plausible for solid particles, particularly high-melting-point metals. Hence, it is concluded that, in the case of solid particles, they must be sufficiently small to begin with, i.e., of adequate surface-to-mass ratio. Even with liquid fuels it has been estimated that in order to react sufficiently fast for a detonation to propagate, the droplets would have to be smaller than 10 μm (Burgoyne and Cohen 1954). With solid fuels, particularly such as aluminum that has an aluminum-oxide coating of very high melting point, the initial particle size has to be even smaller (Tulis 1982).

This presents a major problem for the initiation of detonation in micron-sized solid particles dispersed in air by very strong shock waves; i.e., they will be accelerated almost instantly by the high-velocity convective flow behind the incident shock wave. Unless they react instantly, they will disproportionate as the incident shock wave attenuates, with a heavy concentration behind the incident shock wave and a particle-depleted inner core (Gerber and Bartos 1974; Suzuki et al. 1976). It should be noted that with very strong shock waves, as can be associated with the detonation of high explosives, the initial outwardly propagating shock wave velocity will be much greater than the inherent detonation velocity of the solid-fuel particle-air aerosol. Both of these effects will tend to inhibit initiating reaction fronts and decouple them from the incident shock wave. Assuming the solid-fuel particle-air aerosol is capable of detonation, when and if this decoupling is stabilized, detonation should propagate as in liquid-fuel droplet-air aerosols.

In early attempts to achieve detonation in unconfined aluminum particle-air dispersions, we circumvented this problem by using a planar initiation source of comparable shock strength to that of the inherent detonation of the aluminum particle-air system (Tulis 1980). Furthermore, a novel shock-dispersal technique was utilized to form a relatively homogeneous linear unconfined aluminum particle-air dispersion. Using this technique with an approximately stoichiometric aluminum particle-air dispersion, a steady propagating detonation was achieved at about 1900 m/s, which compares favorably to the analytically predicted Chapman-Jouguet (CJ) value of 1850 m/s. The length of the linear dispersion was about 4 m, and a periodic variation in detonation velocity was observed, i.e., galloping detonation, which was attributed to concentration variation within the dispersion. This result was achieved only with flaked-aluminum particles of about 3-4 m^2/g, which is equivalent to spherical aluminum particles of under 1 μm. With atomized-aluminum particles of 15-25 μm particle size, attempts to achieve detonation failed.

This work was followed with a comprehensive one-dimensional detonation tube study of the characteristics of detonation in aluminum particle-air dispersions (Tulis 1981). The results confirmed that initiation is difficult; initiation overdrive (even in a one-dimensional geometry) causes decoupling of the reaction and shock fronts; and that particle size, or more precisely the surface-to-mass ratio, is probably the most important factor. Specifically, with the flaked-aluminum particles, detonation velocities as high as 1650 m/s and detonation pressures of about 5 MPa were achieved. With an atomized-aluminum powder of about 5 µm mean particle size, we also achieved detonation, but with more difficulty and with diminished detonation characteristics, i.e., at most, 1350 m/s detonation velocity and about 2.5 MPa detonation pressure. Furthermore, with the flaked-aluminum particles, the induction time between the shock and reaction fronts approached the resolution of our instrumentation, about 1 µs, whereas with the atomized-aluminum particles, these induction times were 14 µs as a minimum and as high as 100 µs or more. Spinning detonation was identified in the case of the flaked-aluminum particles, but it is not known whether spinning detonation was present in the case of the atomized-aluminum particles. Initiation of detonation in all cases, however, could only be achieved by using small charges of high explosive, e.g., several grams.

In the work reported here, our subsequent effort to achieve detonation of unconfined, explosively dispersed aluminum particle-air dispersions, will be discussed. Such an unconfined, three-dimensional detonation of solid particle-air dispersion is believed to be very difficult and has not been achieved, to the best of our knowledge, with any solid fuel particles (solid monopropellant particles are excluded from this statement). The major factors that were investigated were 1) particle size of the aluminum, i.e., flaked- and atomized-aluminum particles as previously described; 2) the amount of high explosive needed for initiating detonation, i.e., up to 2.27 kg; and 3) the type of explosive charge, i.e., highly brisant C4, highly brisant high-radiant output H6, and an IITRI-developed low-brisance high-radiant output explosive composition. These experiments will now be described, along with an analysis of the particle-motion effects that are manifest in the high-explosive charge initiation of solid particle two-phase detonation in the unconfined state.

Analysis

Consideration has to be given to the particle motion induced by the high-velocity convective flow behind the incident shock wave. For analytical purposes, only the two extreme conditions will be considered here; i.e., 1) the particles are sufficiently large so that they will resist the acceleration effects of the convective gas flow within the time frame of the induction zone, and 2) the particles are sufficiently small so that they are instantaneously accelerated to the velocity of the convective gas flow. In the former case, the measurement of the induction time between shock and flame fronts will then yield the actual induction time. In the latter case, however, the actual induction time will be considerably larger, i.e.,

$$t_{act} = \frac{t_{meas}}{1 - u_p/D} \quad (1)$$

where u_p is the convective flow velocity, and D is the detonation velocity. By way of example, for the CJ detonation of aluminum powder in air at 1.85 km/s, the convective flow velocity induced by such associated shock strength would be about 1.53 km/s, neglecting the drag effects of the particles. Hence, the actual induction time would be about 5.76 times as great as the measured induction time. In the case of flaked-aluminum particles, which could be expected to be accelerated almost instantly, such correction appears to be justified.

Particle motion in the induction zone is particularly significant as far as heat transfer is concerned. If the particle is considered large with sufficient inertia to resist the convective gas flow, then it will be heated by convective heat transfer, and viscous heating has to be taken into account. On the other hand, if the particle is swept along instantaneously with the convective gas flow, then there will be negligible velocity gradient between the particle and its surrounding high-temperature and -pressure gas environment. In this case, heating of the particle will be simply due to heat conduction from the gas "stagnant" at the surface of the particle.

In either case, the heating of the particle will also be dependent on whether the particle is "large" or "small" in context of the penetration theory (Bird et al. 1960). If large particles are considered, the penetration thickness can be considered so small that the heat conduction in the particle is equal to that in a semi-infinite solid. Hence, for radiation and convection heat transfer, we have, respectively,

$$\sigma T_g^4 \varepsilon_g = [k/(\pi \alpha t)^{1/2}](T_s - T_0) \quad (2)$$

$$h(T_g - T_s) = [k/(\pi \alpha t)^{1/2}](T_s - T_0) \quad (3)$$

where k is the thermal conductivity, ε the emissivity, α the thermal diffusivity, σ the Stefan-Boltzman constant, h the film heat-transfer coefficient, t the time, and T the temperature. Subscripts g, s, and o are gas, surface, and initial conditions, respectively.

If small particles are considered, then the particle may be considered to acquire a uniform temperature instantaneously. Hence, for radiation and convection heat transfer in the case of small particles, we have, respectively,

$$\sigma T_g^4 \varepsilon A t = \rho V C_p (T_p - T_o) \qquad (4)$$

$$h(T_g - T_p) A t = \rho V C_p (T_p - T_o) \qquad (5)$$

where ρ is the density, A the area, V the volume, C the specific heat, and subscript p stands for particle.

Using these relationships with viscous heat dissipation considered (Eckert and Drake 1959; Kays 1965) and with the aid of a computer program, the heating of aluminum particles, under both extreme conditions of no particle acceleration and instantaneous particle acceleration and for both large and small particles in context to the penetration theory, was evaluated. The conditions assumed were a propagating aluminum particle-air detonation under CJ characteristics. Figure 1 illustrates the results obtained with this program for the case of radiation heating. It is observed that even very small particles are inadequately heated to cause ignition within the induction times measured in our detonation tube studies. Figure 2 illustrates the results obtained for the case of convection heating with viscous dissipation. It is readily observed that 1-µm spherical particles can be heated to above the melting point of the aluminum-oxide coating on the aluminum particles within about 6 mm. Assuming a detonation velocity of 1.85 km/s and particle acceleration, this translates to an induction time of under 1 µs, which was indeed observed in the detonation tube studies with the flaked-aluminum particles. In the case of the 5-µm atomized-aluminum particles, the highest detonation velocity obtained in the detonation tube studies was 1.35 km/s, while the measured induction times varied between 14 and over 100 µs. This translates to induction distances of 19 to over 135 mm with no particle acceleration and up to 700 mm or more with instantaneous particle acceleration. Figure 2 illustrates that 5-µm particles would indeed require induction distances of this order.

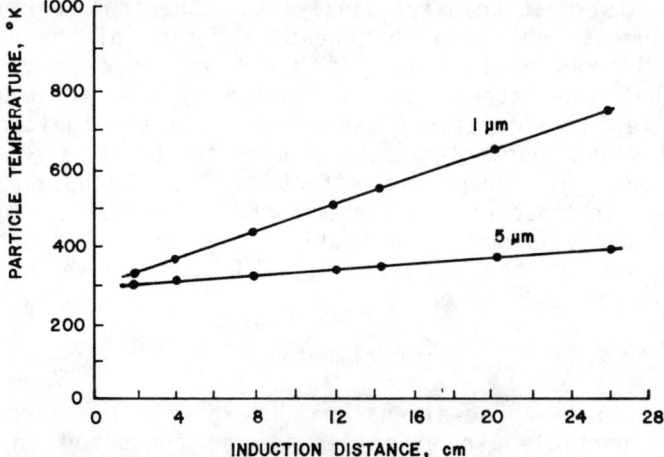

Fig. 1 Temperature of aluminum particles of different particle size heated by radiation from a detonating aluminum particle-air system as a function of the distance of the induction zone.

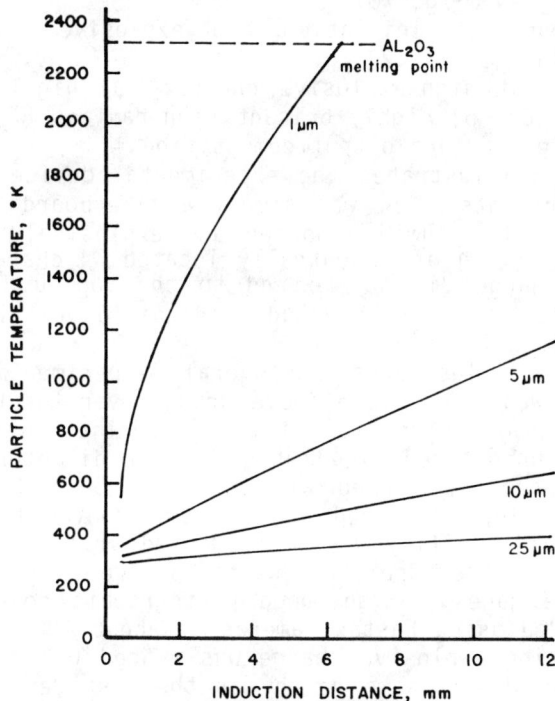

Fig. 2 Temperature of aluminum particles of different particle size heated by convective/viscous energy from the shock-compressed convective flow behind the incident shock wave of a detonating aluminum particle-air system.

In this heat-transfer analysis, spherical particles were assumed, and the latent heat of fusion of the aluminum itself was neglected. It has been shown (Veyssiere 1981) that the latent heat of fusion of the aluminum is negligible in calculations associated with the ignition of fine aluminum particles behind the front of a gaseous detonation. The geometric effect of the flaked-aluminum particles in respect to the assumed spherical particles would probably enhance the heating rate and allow greater acceleration of the particles by the convective flow in the induction zone.

Experimental

Unconfined three-dimensional, explosively dispersed aluminum particle-air experiments were conducted in this study using 4.54 kg of the aluminum powder in all cases. The following major factors were investigated.

1) Particle size or surface-to-mass ratio (S/M): a) 5-μm atomized aluminum of about 0.34 m^2/g S/M; b) flaked aluminum of 3-4 m^2/g S/M.

2) Amount of initiation high-explosive charge: a) 100 g; b) 459 g; c) 2.27 kg.

3) Type of high-explosive charge: a) highly brisant composition C4; b) highly brisant high-radiant H6; c) low-brisance high-radiant output composition.‡

Figure 3 illustrates the experimental device used for these experiments. It was simply a fiberboard container filled with the aluminum powder and explosively disseminated using 100 g of a centrally located C4 charge. This explosive charge did not extend to the top or bottom of the container so as to preclude preignition of the aluminum upon dissemination.

Figure 4 illustrates the overall experimental setup. The device was placed 1 m above ground over the blast pad facility. Upon detonation of the dissemination explosive charge, an unconfined cloud about 6 m in diameter and 1 m in height readily formed within 25 to 50 ms. In all cases, the cloud remained about 0.5 m above the ground. The blast pad facility was instrumented with two lines of piezoelectric pressure tranducers as well as numerous static fuze gages. High-speed photographic coverage was also provided using Fastax cameras at about 4000 frames/s. The initiation explosive charge was placed 0.5 m from the test device at the same height as the test device. This

---

‡ Mixture of aluminum powder and ammonium perchlorate sensitized with nitroguanidine (Tulis 1976).

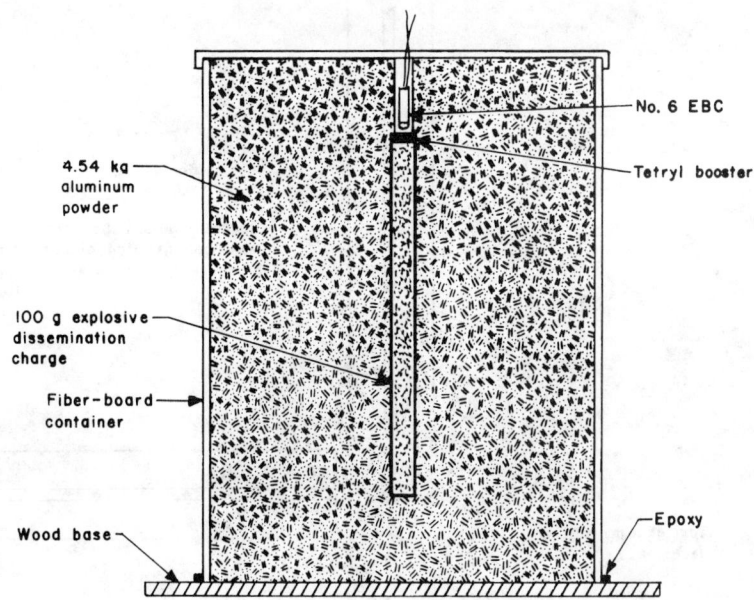

Fig. 3 Explosive-dissemination device for generating unconfined three-dimensional aluminum particle-air clouds.

charge, generally referred to as a second-event charge, was detonated 25 ms after the dissemination charge detonated. The aluminum particle-air concentration was estimated based on total weight of the aluminum powder and the volume of the cloud as determined from the film records. It was, as intended, about the same for all tests and averaged out at 0.16 kg/m^3, or about half of stoichiometric based on the formation of aluminum oxide. The piezoelectric pressure transducers, located flush with the ground, monitored blast output but also provided information on the arrival time of the blast wave from which incremental blast-wave velocities could be obtained. Hence, in the case of initiation of detonation of the aluminum particle-air cloud, although not necessarily stable, supported shock waves in the cloud could be identified.

In the experiments using the largest amount of initiation-explosive charge, i.e., 2.27 kg, the output from such charge became appreciable in comparison to that anticipated from the detonation of the aluminum particle-air cloud. Hence, in order to differentiate between output from the initiation charge and subsequent aluminum particle-air detonation, a calibration test was conducted using a 2.27-kg charge of C4 alone at the same location as in

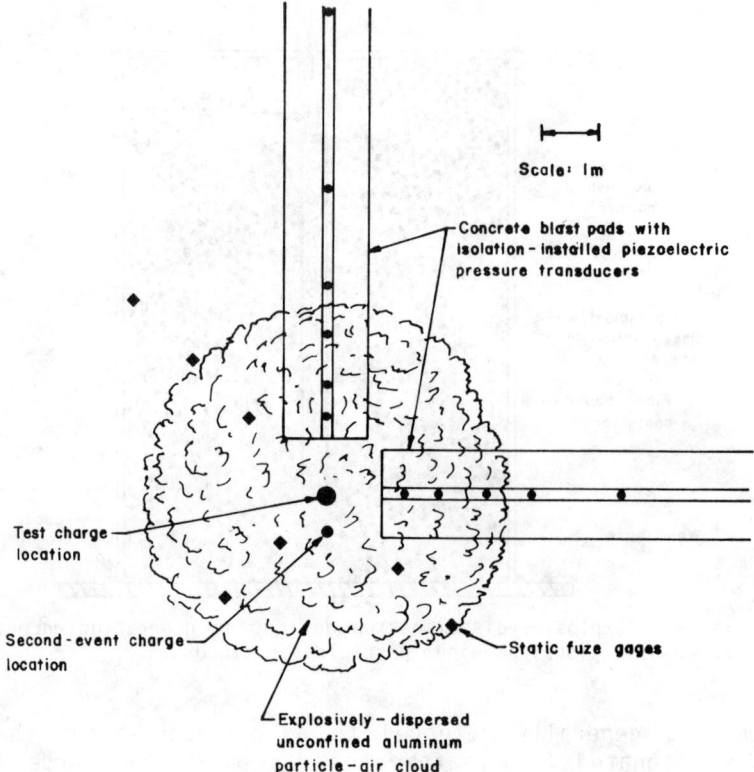

Fig. 4 Schematic layout of the IITRI blast pad facility as viewed from above, showing an explosively generated aluminum particle-air cloud.

the experiments with 4.54-kg aluminum particle-air clouds. Thus, measured blast pressures that were greater than those obtained with the 2.27-kg C4 calibration charge had to be attributed to detonation, or at least blast enhancement, from the aluminum particle-air cloud. The Fastax film records also provided information on the results of these experiments; i.e., the velocity of the flame front could be detected. Because of the extreme amount of radiant output from either a detonation or deflagration of an aluminum particle-air cloud, another Fastax camera was used that was shuttered down to the extent that only the highly radiant fireball was visible.

Results

In all experiments using either 100 or 454 g of initiation-explosive charge, the aluminum particle-air cloud simply deflagrated. Results of these experiments will not

Table 1 Ground blast pressures measured at various distances
from the 2.27-kg initiation-explosive charge
in three-dimensional unconfined atomized- and flaked-aluminum
powders explosively dispersed in air

Distance m	Calibration C4	Atomized aluminum[a] Initiation-explosive type			Flaked aluminum[a] Initiation-explosive type		
		C4	EMO[b]	H6	C4	EMO[b]	H6
1.3	1048	1159	1586	1193	793	2166	2331
1.9	634	455	676	800	1303	1021	1324
2.7	359	290	441	469	1076	607	993
3.5	166	152	241	214	297	345	352
5.1	83	103	172	110	159	152	186
8.2	28	41	55	41	62	55	76

[a] Aluminum powder weight in all tests was 4.54 kg.
[b] Explosive-metal-oxidizer composite, i.e., nitroguanidine sensitized ammonium perchlorate mixed with aluminum powder.

be further described here. Table 1 presents the results of the six experiments conducted with 2.27 kg of initiation-explosive charge. The pressures reported are the average of the two piezoelectric pressure transducers located at equal distances from the ground-zero point of the experiment in the two blast pads located 90 deg apart. It should be pointed out that in these experiments the aluminum particle-air cloud upon initiation of the second-event explosive charge had a radius of about 3 m. Hence, distances recorded that were greater than 3 m were beyond the edge of the cloud.

Table 2 presents these same data except that the pressure values from the three types of initiation-explosive charges were averaged. Thus, the pressures for the C4 calibration charge are the average of two points, whereas the pressures for the atomized- and flaked-aluminum data are the average of six points. In this way, the overall results of the two types of aluminum powders investigated can be better compared. Figure 5 illustrates these results as a function of the radial distance from ground zero. The cloud boundary is also included. It can be readily observed that the atomized-aluminum particle-air clouds did not detonate, although some blast enhancement is indicated. With the flaked-aluminum particle-air clouds, incipient detonation was induced, providing an average blast pressure within the cloud 610 kPa greater than that of the C4 alone.

Table 2 Averaged data for all 4.54-kg explosively dispersed aluminum particle-air experiments using a 2.27-kg initiation-explosive charge [a]

Distance m	Scaled distance[b] m/kg$^{1/3}$	C4 kPa	Atomized kPa	Flaked kPa
1.3	1.0/0.6	1048	1313	1763
1.9	1.4/0.9	634	644	1216
2.7	2.1/1.2	359	400	892
3.5	2.7/1.6	166	202	331
5.1	3.9/2.3	83	128	166
8.2	6.2/3.8	28	46	64

[a] Note: Each value for the aluminum powders is the average of three tests with two values per test from the blast pads 90 deg apart.
[b] First value based on weight of initiation-explosive charge only, second value based on total weight of initiation-explosive charge and the weight of the aluminum powder.

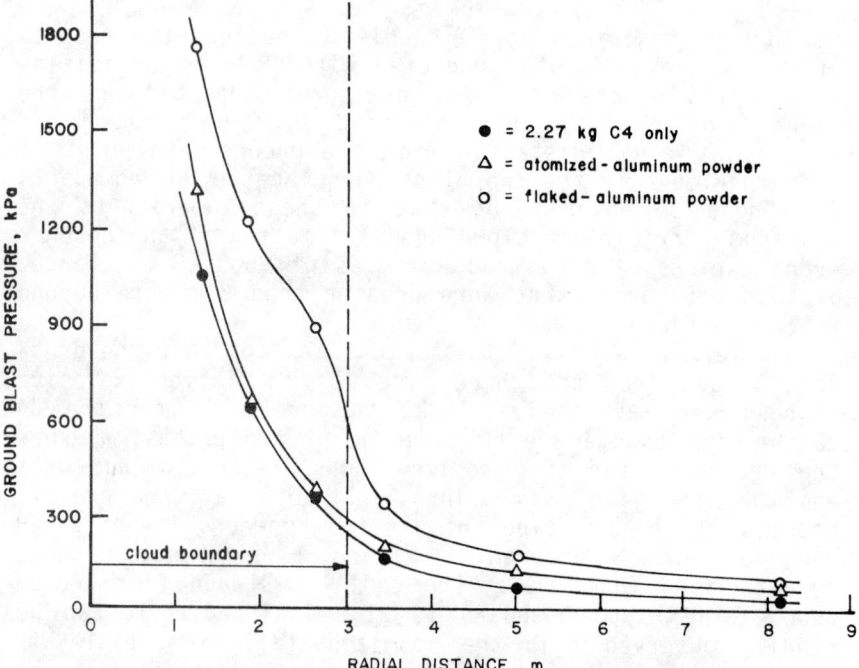

Fig. 5 Ground blast pressures at various distances from ground zero for attempted detonations of atomized- and flaked-aluminum particle-air clouds compared to the detonation of the 2.27-kg initiation-explosive charge by itself.

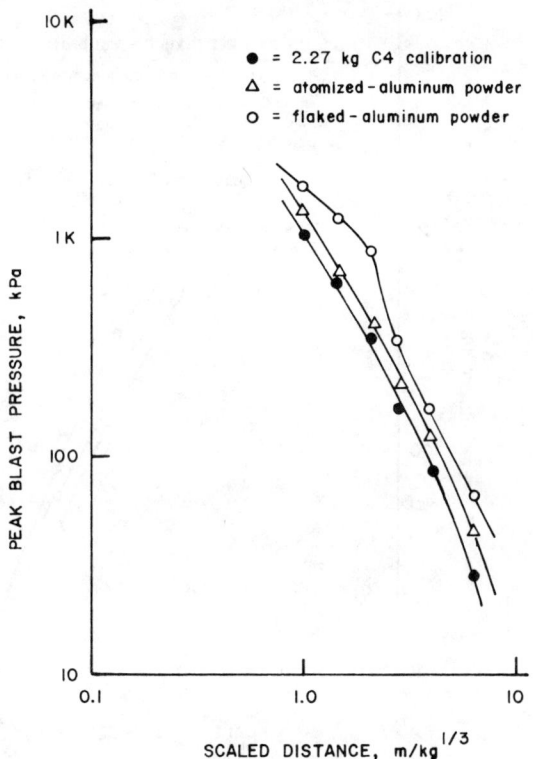

Fig. 6 Log-log plot of peak pressure vs scaled distance based on the weight of the initiation-explosive charge only.

In order to assess these data from an energetic-output point of view, we have included Figs. 6 and 7 here. In Figure 6, the data of Table 2 are plotted as a function of scaled distance based on the weight of the initiation-explosive charge only. Note that the output from the atomized-aluminum particle-air clouds exceeds that of the C4 alone by a very small amount, whereas the output from the flaked-aluminum particle-air clouds definitely shows the typical spatial-output effect anticipated for the detonation of a fuel-air system, with pronounced fall-off beyond the edge of the cloud. Figure 7 is similar, except that the scaled distance is based on the total weight of initiation-explosive charge as well as that of the aluminum-powder charge. Again, it is apparent that in the case of the atomized-aluminum particle-air clouds there is no indication of energetic output from the aluminum, whereas in the case of the flaked-aluminum particle-air clouds the aluminum evidently provides energetic output comparable to

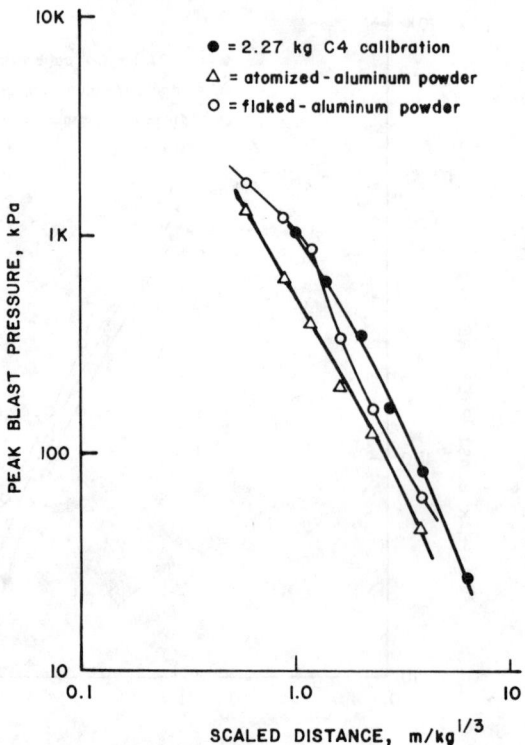

Fig. 7 Log-log plot of peak pressure vs scaled distance based on the total weight of the initiation-explosive charge and the aluminum powder.

that of an equal amount of explosive within the cloud boundaries. Chemical equilibrium calculations confirm that this degree of energetic output can be anticipated and would manifest itself appreciably as overpressure under the conditions of an aluminum particle-air detonation (Tulis 1981).

Conclusions

Based on the large-scale experiments conducted in this investigation, it is concluded that the detonation of totally unconfined, explosively generated aluminum particle-air clouds will require, as a minimum, 1) aluminum particles with S/M in the range of 3-4 m^2/g (or greater), and 2) high-explosive initiation charges on the order of 2.27 kg (or greater). Because it was necessary to use initiation charges of this size, it cannot be ascertained whether or not stable detonation propagation might have resulted from the incipient initiations that were obtained in the experiments conducted in this investigation, at least in the case of the flaked-aluminum powder tests. It will be necessary to conduct much larger experiments in order to demonstrate propagation of detonation rather than blast enhancement due to chemical-energy release behind a decaying blast wave.

Regarding the influence of radiant energy for initiation and propagation, analytical calculations indicate it to be negligible, whereas experimental observations indicate considerable influence. More definitive experiments will have to be conducted in order to isolate and identify the influence of radiation.

Acknowledgment

This work was conducted as part of an Internal Research and Development program sponsored by IIT Research Institute.

References

Bird, R. B., Steward, W. E., and Lightfoot, E. N. (1960) Transport Phenomena, p. 354. John Wiley and Sons, New York.

Burgoyne, J. H. and Cohen, L. (1954) The effect of drop size on flame propagation in liquid aerosols. Proc. R. Soc., London Ser. A, 225, 375-392.

Dabora, E. K. and Fox, E. (1972) The breakup of liquid droplet columns by shock waves. Astronaut. Acta 17(4 and 5), 669-674.

Dabora, E. K. and Weinberger, L. P. (1974) Present status of detonations in two-phase systems. Acta Astronaut. 1(3 and 4), 361-372.

Eckert, E. R. G. and Drake, W. (1959) Analysis of Heat and Mass Transfer. McGraw-Hill Co., New York.

Fishburn, B. D. (1974) Boundary layer stripping of liquid drops fragmented by Taylor instability. Acta Astronaut. 1(9 and 10), 1267-1284.

Gerber, N. and Bartos, J. M. (1974) Strong spherical blast waves in a dust-laden gas. AIAA J. 12(1), 120-122.

Kauffman, C. W. (1971) Shock wave ignition of liquid drops. Ph.D. Thesis, The University of Michigan, Ann Arbor, Mich.

Kays, W. M. (1965) Convective Heat and Mass Transfer. McGraw-Hill Co., New York.

Lu, P. L. and Slagg, N. (1974) Chemical aspects in shock ignition of fuel drops. Acta Astronaut. 1(9 and 10), 1219-1226.

Robbins, R. C. and Thuman, W. C. (1965) Explosive dispersion of volatile liquids. SRI Report NOTS-TP-3624, p. 341.

Suzuki, T., Ohyagi, S., Higashino, F., and Takano, A. (1976) The propagation of reacting blast waves through inert particle clouds. Acta Astronaut. 3(7 and 8), 517-529.

Thuman, W. C. and Robbins, R. C. (1965) The particle size distribution of an explosively dispersed liquid. SRI Final Report (AD-367-152).

Tulis, A. J. (1976) Sympathetic detonation of ammonium perchlorate by small amounts of nitroguanidine. ACR-184, Office of Naval Research, Arlington, Va.

Tulis, A. J. (1980) On the detonation of unconfined aluminum particles dispersed in air. Shock Tubes and Waves, p. 539. The Magnus Press, The Hebrew University, Jerusalem, Israel.

Tulis, A. J. (1981) Criteria for the detonation of aluminum particles dispersed in air. Ph.D. Thesis, Illinois Institute of Technology, Chicago, Ill.

Tulis, A. J. and Selman, J. R. (1982) Detonation tube studies of aluminum particles dispersed in air. 19th International Symposium on Combustion, p. 655. The combustion Institute, Pittsburgh, Pa.

Veyssiere, B. (1981) Ignition of fine aluminum particles behind the front of a gaseous detonation. Progress in Astronautics and Aeronautics, 87, 362-375.

Zabelka, R. J. and Smith, L. H. (1969) Explosively dispersed liquids. NWC TP 4702 (AD 863 268).

Dynamics of Dispersion and Ignition of Dust Layers by a Shock Wave

V. M. Boiko* and A. N. Papyrin†
Institute of Theoretical and Applied Mechanics
Novosibirsk, USSR
and
M. Woliński‡ and P. Wolański§
Technical University of Warsaw, Warsaw, Poland

Abstract

Studies of dust layer dispersion were carried out in a shock tube with a cross section of 50×50 mm^2 at the Institute of Theoretical and Applied Mechanics. The phenomenon was observed with a Schlieren system with a stroboscopic laser light source and a streak mirror camera. In all the experiments, dust was placed either at the bottom of the tube in a layer 2 mm thick or in a small cavity. Dust ignition experiments were carried out in a shock tube with a 35×35 mm^2 cross section at the Technical University of Warsaw. The shock tube was driven with helium heated by detonation of a $2H_2 + O_2$ mixture. Pressure switches were employed for velocity measurements, while ignition delay was recorded with a photodiode. The Novosibirsk tests showed that dust can be effectively dispersed by a shock wave. The dispersion is more effective when the Mach number is higher, and the dust particles smaller and lighter. The Warsaw tests showed that fine organic dust particles lying on the floor can be ignited with a sufficiently strong shock wave. For

---

Presented at the 9th ICODERS, Poitiers, France, July 3-8, 1983. Copyright American Institute of Aeronautics and Astronautics, Inc., 1984. All rights reserved.
*Graduate Research Assistant.
†Associate Research Scientist.
‡Graduate Research Assistant.
§Associate Professor.

a Mach number close to 4, the dust is first lifted up into the air and then ignited. For a very strong shock wave (a Mach number higher than 6), the ignition occurs nearly instantaneously, upon shock impact with the dust particles remaining virtually at rest during the ignition period. The addition of oxygen decreases the ignition delay, especially for lower shock Mach numbers. The ignition delay for a dust layer and a dust prism in air are similar to the ignition delay in a detonation of dust-air mixtures.

Introduction

While there is an increasing number of publications dealing with the problem of ignition of dust particles by a shock wave (see Nettleton and Stirling 1967; Fox et al. 1978; Kauffman et al. 1979; Kauffman et al. 1982), the problem of a dust layer ignition has only been mentioned in the past decade (Wolański 1975); and the study of a dust layer dispersion by a shock wave has been reported only recently (Boiko et al. 1982). It is essential to understand the mechanism of ignition of a dust layer, since there is a possibility of a dust layer detonation analogous to a detonation of the liquid fuel layer in a gaseoux oxidizer (Borisov et al. 1968; Sichel et al. 1971). The main objective of this paper is the study of the initial phase of a dust layer dispersion and ignition.

Dust Dispersion Experimental Studies

Studies of the dust layer dispersion were conducted in a shock tube with a cross section of 50×50 mm^2 at the Institute of Theoretical and Applied Mechanics. A Schlieren system with a stroboscopic laser light source and a streak mirror camera were used to observe dispersion phenomena. In all the experiments, dust was placed either at the bottom of the tube in a layer 2 mm thick or in a small cavity.

Typical laser stroboscopic pictures, showing the process of the dust dispersion from the layer or from the cavity, are presented in Fig. 1. As the shock wave velocity is the same for both cases, differences between the initial phases of dust dispersion from a layer or from a cavity can be noted. At the beginning, more dust is dispersed from a layer than from a cavity. After 300 μs, approximately half of the tube's cross section is filled by a dust cloud from a layer, while at the same time, only a quarter of the tube's cross section is filled by a dust cloud from a cavity. The differences diminish at larger times.

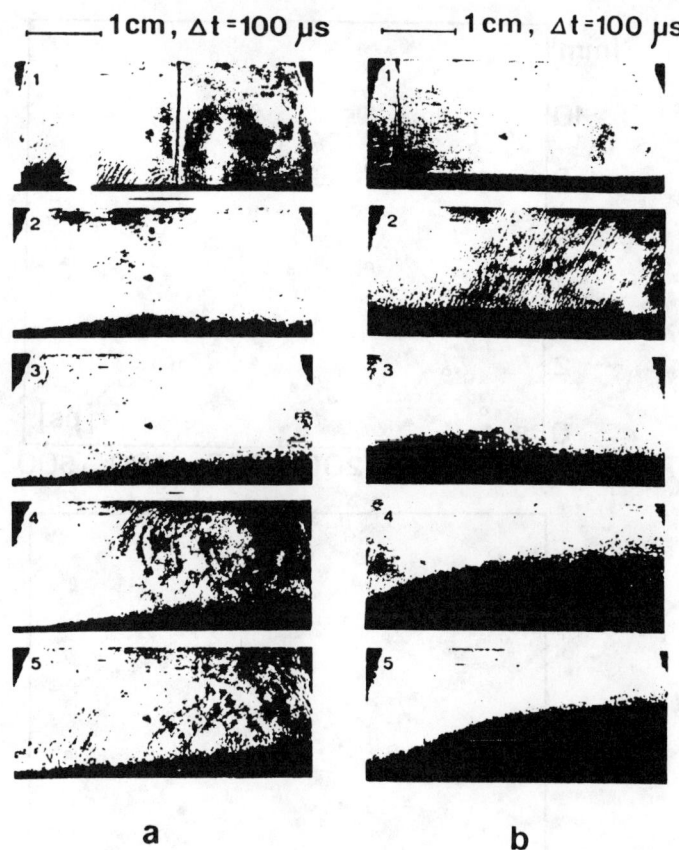

Fig. 1   Typical laser stroboscopic pictures of dust dispersion in air from the cavity (a) and from the layer (b). Spherical particles of organic glass; $d_m = 80$ μm, $\rho = 1200$ kg/m^3, initial shock tube pressure = 1 atm, $M_s = 2.6$.

In Fig. 2, the relation between the height of the lifted dust and the time after the shock wave for two different dusts is shown. The metallic dust particles are lifted more slowly than the plexiglass dust particles of the same size. The experiments indicate that, all other things being equal, dispersion ability increases as the shock Mach number increases and as the particle diameter and specific density decrease. The best dispersion of dust can be obtained for small, light dust particles and a relatively strong shock wave.

For fine particle dusts, monitoring of the dispersion process is much more difficult, because individual fine

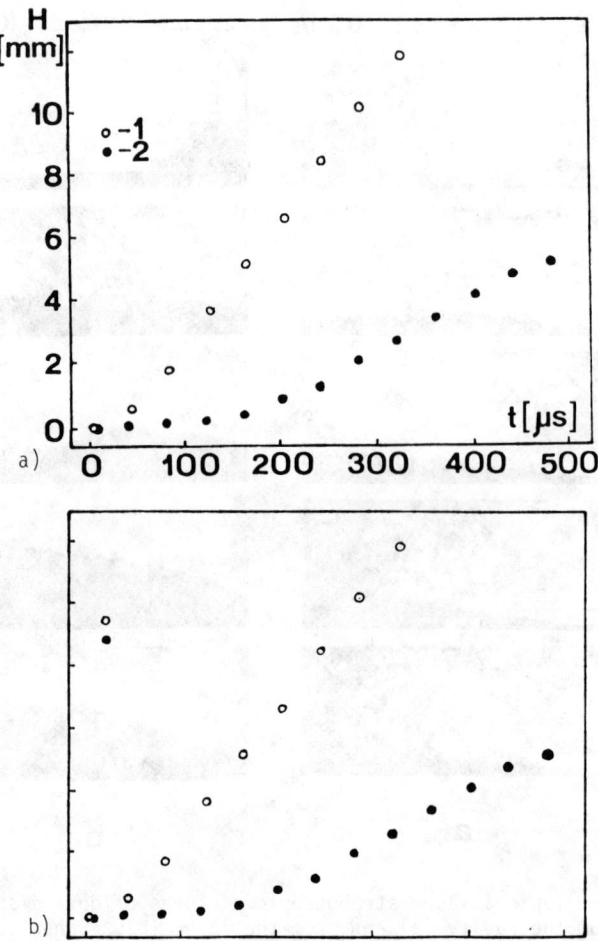

Fig. 2 Relation between the altitude of the dust lifted up from the layer and the time after the shock wave for two different dusts: (a) Organic glass; spherical particles, d_m = 200 μm, ρ = 1200 kg/m^3. (b) Bronze; spherical particles, d_m = 200 μm, ρ = 8700 kg/m^3, initial shock tube pressure = 1 atm, M_s = 2.6 for both dusts.

dust particles or clouds with low particulate densities fall below the resolution of the optical system. As such, dusts can be observed only for relatively high particulate concentrations, only qualitative data can be obtained. Even so, the information necessary to describe the dust dispersion in the period prior to ignition can be obtained from the studies.

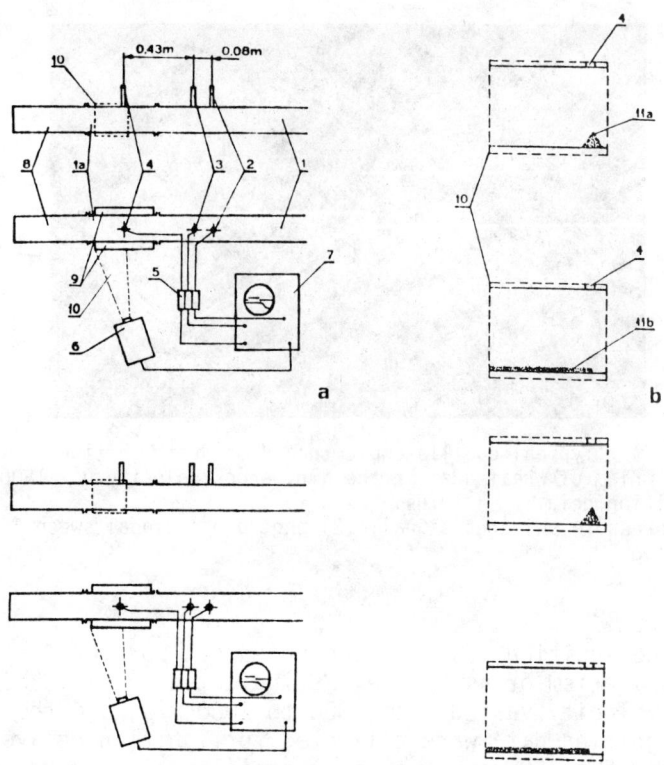

Fig. 3 a) Schematic of the test stand:
1) shock tube; 1a) test region; 2-4) pressure switches;
5) electronic circuit and power supply of the pressure switches;
6) photodiode; 7) oscilloscope; 8) dump tank; 9) windows;
10) area "observed" with the photodiode.
b) Forms of the dust samples:
11a) a prism; height h = 3 mm; 11b) a layer, thickness h = 1 mm.

The Ignition Study

The dust ignition experiments were conducted at the Technical University of Warsaw. A shock tube of 35×35 mm^2 cross section, driven with helium heated by detonation of a $2H_2 + O_2$ mixture, was used for this purpose. A schematic of the test section is shown in Fig. 3a. Pressure switches were used for velocity measurements, while a photodiode was employed to observe ignition delay. All times were recorded with the dual beam storage oscilloscope. The sensitivity of the photodiode signal was set on such a level that it did not measure light emitted by shocked air (or oxygen).

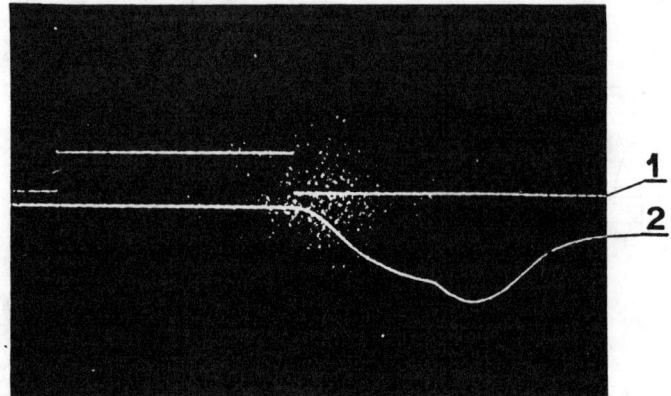

Fig. 4   Typical oscilloscopic record of dust ignition. The prism of wheat dust in the air, shock velocity u = 1800 m/s, ignition delay t_i = 10 μs:
1) pressure switches signal;   2) photodiode signal sweep 50 μs/division.

The ignition tests were carried out for forms of dust set as a prism or as a layer as indicated in Fig. 3b. The shock velocity varied from 1300 to 2000 m/s, and the reliable ignition data were collected for ignition delays ranging from 5 to 50 μs. A typical oscilloscope record of dust ignition is shown in Fig. 4. The ignition delay is assumed to be the time difference between the passage of the shock wave at pressure switch 4 (set at the beginning of a dust layer or a prism (see Fig. 3a) and the time when the deflected oscilloscopic signal due to light emission from the ignition area exceeds the thickness of the recorded signal.

The ignition delay measurements for wheat dust are shown in Fig. 5. The dots represent individual measurements, while the lines are the result of the least-squares fitting of the data to the exponential expression coupling the ignition delay with the static postshock temperature. In this graph, the results for the ignition of the wheat dust in a prism and in a layer for both air and oxygen atmosphere are presented. The additional line shows the results for the ignition delay of the same dust, obtained at The University of Michigan, but for the case of a small amount of dust dispersed in air prior to the shock interaction. The ignition delay in the wheat-air detonation is shown as a point. The data were evaluated from the streak picture of the wheat dust detonation in air presented by Kauffman et al. (1982).

These results indicate that the ignition delay for the same dust is significantly influenced by the mixture condition during the induction period. Some of the differences are due to different and arbitrarily chosen levels of the light emission interpreted as ignition, but some are due to the differences of dust consistency in the induction period.

Dust particles dispersed in air prior to shock interaction have the smallest ignition delay, while the ignition delay due to shock wave passage over a dust layer or a prism is much longer. The ignition delay measurements at The University of Michigan were performed with a very sensitive photomultiplier, so the measurements show the very initial period of the ignition, while the ignition evaluated on the basis of streak pictures and photodiode signal reveals a relatively longer period of the ignition delay characteristic for the onset of combustion. The ignition delay of dust in a layer or prism should be longer than the ignition delay of identical dust properly dispersed in air prior to shock interaction; but the differences should be evaluated under identical experimental conditions and with the same instruments.

Figure 5 indicates that there are relatively small differences in the ignition delay between dust in a prism and in a layer. With a high Mach number of the incident shock, the ignition proceeds more rapidly for a prism than for a layer. This is probably due to the higher level of the prism and the higher stagnation temperature. The same behavior was also observed for the oxygen environment, but the addition of oxygen always resulted in a smaller ignition delay. Since there were no big differences between a prism and a layer, the latter experiments were carried out for a prism only.

Figure 6 shows the measured ignition delay for wheat, lycopodium and charcoal dusts obtained in air and in oxygen. All the results show that the ignition delay is smallest for the lycopodium and longest for the charcoal. In oxygen, the ignition delay is much smaller for a smaller postshock temperature, while for the higher temperature, ignition delay seems to be less dependent on oxygen concentration, especially for lycopodium and wheat.

Discussion and Conclusions

The experiments conducted show that dust can be effectively dispersed with a shock wave. This dispersion is more effective for the higher Mach number, smaller and lighter dust particles. Fine organic dust particles lying on the floor can be ignited by a strong shock wave (a Mach

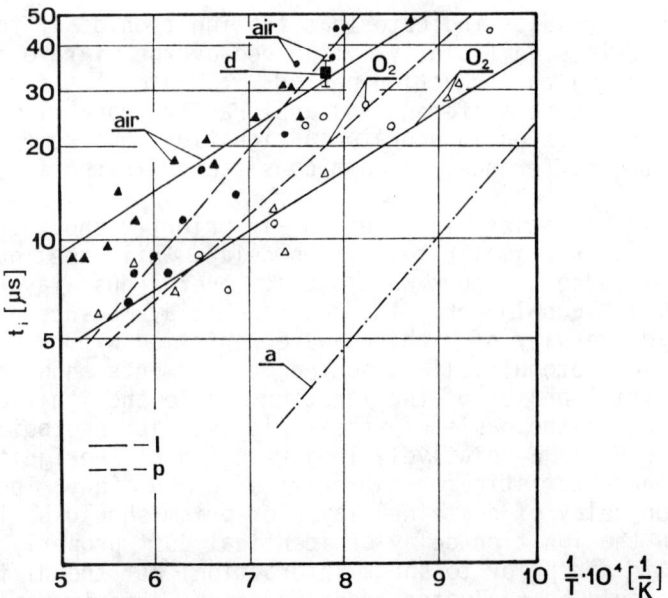

Fig. 5 Ignition delay measurements for wheat dust (particles ≤ 75 μm): (1) layer, (p) prism; initial shock tube pressure = 0.1 atm, (a) Kauffman et al. (1982), (d) wheat air detonation, Kauffman et al. (1982).

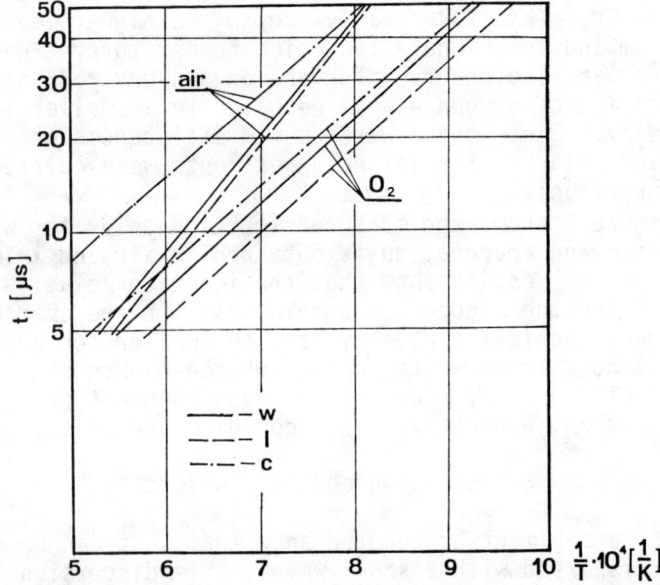

Fig. 6 Ignition delay measurements for different dusts (particles ≤ 75 μm): (w) wheat, (l) lycopodium, (c) charcoal; initial shock tube pressure = 0.1 atm.

number higher than 4). For a shock Mach number close to 4, the dust is first lifted up into the air and then ignited. For a very strong shock wave (a Mach number higher than 6), ignition will occur nearly instantaneously with the shock interaction without appreciable dispersion of dust particles during the ignition period. The addition of oxygen decreases the ignition delay, especially for a lower Mach number of the shock wave, while for a high Mach number, this effect diminishes. As the ignition delay for a dust layer and a dust prism in air are similar to that of dust-air detonation, detonative combustion of a dust layer appears to be possible, at least in an oxygen environment.

References

Boiko, V. M., Papyrin, A. N., Poplavski, S. V., Wolański, P. and Zalesiński, M. (1982) High speed laser photography of shock wave with liquid layer and dust layer. 3rd International School of Explosibility of Industrial Dusts, pp. 53-64, Turawa, Poland.

Borisov, A. A., Kogarko, S. M., and Lyubimov, A. V. (1968) Ignition of fuel films behind shock waves in air and oxygen. Combust. Flame 12:5, 465-468.

Fox, T. W., Rockett, C. W., and Nicholls, J. A. (1978) Shock wave ignition of magnesium powders. 11th International Symposium on Shock Tubes and Waves, University of Washington.

Kauffman, C. W., Wolański, P., Ural, E., Nicholls, J. A., and Van Dyk, R. (1979) Shock wave initiated combustion of grain dust. Symposium on Grain Dust. USDA, Manhattan, Kansas.

Kauffman, C. W., Ural, E., Nicholls, J. A., and Wolański, P. (1982) Detonation waves in confined dust clouds. 3rd International School on Explosibility of Industrial Dusts, pp. 75-88, Turawa, Poland.

Nettleton, M. A. and Stirling, R. (1967) The ignition of clouds of particles in shock heated oxygen. Proc. R. Soc. London 300:1460, 62-77.

Sichel, M., Rao, C. S., and Nicholls, J. A. (1971) A simple theory for the propagation of film detonations. 13th Symposium (International) on Combustion, pp. 1141-1147. The Combustion Institute, Pittsburgh, PA.

Wolański, P. (1975) The ignition of coal dust in layer. Internal Report, Technical University of Warsaw, Poland.

Detonations in Explosive Foams

J. P. Saint-Cloud* and O. Peraldi†
Université de Poitiers, Poitiers, France

Abstract

Previous work has shown that two different regimes of detonation may appear for a foam density range of 5 to 25 kg/m^3 in an aqueous foam filled with a stoichiometric propane-oxygen mixture. When the majority of the bubbles have a diameter greater than 2-3 mm, the phenomenon observed is very similar to that of a premixed gaseous detonation ; this regime is called the Normal Regime of Detonation (NRD). When the predominant diameter is less than 2 mm, the shock and the combustion fronts are clearly separated and their velocities are markedly lower than that of the detonation in the gaseous mixture alone. This regime is called the Double Front Regime (DFR). In the present work, under the same experimental conditions, the influence of the gaseous mixture composition is established. Experiments were performed with stoichiometric mixtures ($2H_2 + O_2 + ZN_2$, $CH_4 + 2 O_2 + ZN_2$, $C_2H_4 + 3 O_2 + ZN_2$ and $C_3H_8 + 5 O_2 + ZN_2$). At a given mean bubble diameter and foam density, a maximum dilution limit - Z_{lim} - is observed beyond which the NRD is not observed. The dependence of the critical predominant bubble diameter on the dilution of nitrogen (Z) at the transition from NRD to DFR was determined.

Introduction

In a previous study on the propagation of detonations in aqueous foams filled with a stoichiometric

---

Presented at the 9th ICODERS, Poitiers, France, July 3-8, 1983. Copyright © American Institute of Aeronautics and Astronautics, Inc. 1984. All rights reserved.
*Assistant Professor, Laboratoire d'Energétique et de Détonique, E.N.S.M.A.
†Postgraduate Student, Laboratoire d'Energétique et de Détonique, E.N.S.M.A.

propane-oxygen mixture (at ordinary pressure and temperature), Saint-Cloud et al. (1976) observed that for a foam density ρ_m ($5 < \rho_m < 25$ kg/m^3), two regimes of propagation could be observed, according to the foam structure. The foam is characterized by the Sauter diameter d_{32} ($d_{32} = \Sigma\, n_i\, d_i^3/\Sigma\, n_i\, d_i^2$, where n_i represents the number of bubbles whose diameter is d_i.). When d_{32} is greater than 2-3 mm, the observed velocity of propagation is $\simeq$ 2300 m/s and is similar to that observed in the premixed gaseous detonation. This regime is called the Normal Regime of Detonation or NRD. If d_{32} is less than 2-3 mm, the structure is that of a dissociated detonation (Saint-Cloud et al. 1972) and the velocity (D $\simeq$ 550 m/s) and pressure ($P_D \cong$ 5-10 bars) of the detonation products are notably less than those observed for premixed gaseous detonations. This regime is called the Double Fronts Regime or DFR.

While these experiments have demonstrated the dominant role of the foam structure, in this work the objective has been to determine, with the same experimental apparatus, the influence of the gaseous reactive system upon the appearance of one of the two regimes of detonation.

In this work, experiments were conducted for the following parameters: 1) the fuel used (H_2, CH_4, C_2H_4 and C_3H_8) ; and 2) the dilution of the mixture defined by the number Z of nitrogen moles in mixtures of:

$$C_nH_m + (n + m/4)\, O_2 + Z\, N_2 \qquad (1)$$

for nearly constant foam diameter d_{32} and density ρ_m.

Moreover, for a given composition of the gaseous explosive mixture, foam diameter d_{32} was varied to determine the limiting foam diameter at which transition from NRD and DFR occurs.

Experimental

Detonation Tube

The tube used had a length of 5 m and an internal diameter of 28 mm (see Fig.1). It was made of 7 plexiglass sections, 500 mm in length ; and 2 plexiglass sections (B_1 and B_2) 720 mm in length, each equipped with piezoelectric gages constructed in the laboratory (Brochet et al. 1969) and located at C_1, C_2, C_3 and C_4. The separation between (C_1-C_2) and also (C_3-C_4) is 500 mm ± 1 mm. On section B_2 and 50 mm away from gage C_3, a pressure gage K was

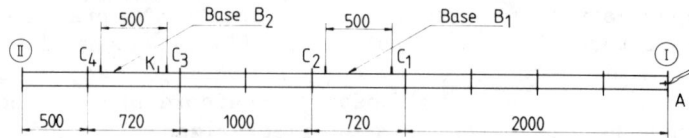

Fig. 1 Detonation pipe.

installed (Kistler 603 B). Lastly, sections B_1 and B_2 were placed, respectively, 2 m and 3.72 m away from the end I of the detonation tube.

Velocity and Pressure Measures

The detonation velocity was measured to the nearest 1% with the aid of gages (C_1 and C_2) and (C_3 and C_4), while the pressure of the detonation products was measured by the gages C_1, C_2, C_3, C_4, and K. Laboratory calibrations indicate that the precision of measurement was around 5% with gage K and 10% with gages C_1, C_2, C_3, and C_4 (Brochet et al. 1969).

Foams

The foams were produced from a solution of 5% in volume of commercial detergent in water with the generator described in detail by Saint-Cloud et al. (1976). The Sauter diameter d_{32} was determined with the help of macrophotographs of a foam sample (Saint-Cloud et al. 1976). This method was preferred to that based upon the diffusion of light of a foam surface illuminated by a ray of light (Saint-Cloud et al. 1981). When the diameter d_{32} is large (> 3-4 mm), the surface of the foam sample is unstable. This instability causes discrepancies between the measurements made with the two methods of up to 30%. For the same reason, the foam density was monitored by weight measurements, as electric measurements (Saint-Cloud et al. 1981) do not give satisfactory results when the foam density $\rho_m < 5$ kg/m^3.

Generation of the Detonation

Detonation initiation was achieved with a Gevelot P53 A electric ignitor, located at the end I of the tube. To obtain transition from deflagration to detonation prior to base B_1 for all mixtures, the ignitor A was placed 2 m away from this basis, and a foam column composed of fine bubbles (3 mm) was created in the volume next to the ignitor. Saint-Cloud et al. (1981) have shown that the fine bubble foam enhances early transition, probably because it causes the preliminary movement of the mixture to be more turbulent (Shchelkine 1945). The mixtures were prepared in advance and used at an initial temperature T_f = 292 ± 5 K and an initial pressure p_f = 1 atm. Mixture composition was determined from partial pressure observations.

Experimental Results

The Effect of the Gaseous Explosive Mixture for a Fixed Diameter D_{32}

For the different mixtures (H_2 + 1/2 O_2 + ZN_2 (A) ; CH_4 + 2 O_2 + ZN_2 (B) ; C_2H_4 + 3 O_2 + ZN_2 (C) ; and C_3H_8 + 5 O_2 + ZN_2 (D), a foam diameter of 5-6 mm was used and Z was varied from 0 to 9. For the mixture (D), the foam density was approximately 4-5 kg/m^3, while for the mixtures (A), (B), and (C), it varied between 2 - 3 kg/m^3. Figure 2 shows the dependence of the detonation velocity on Z for each mixture tested.

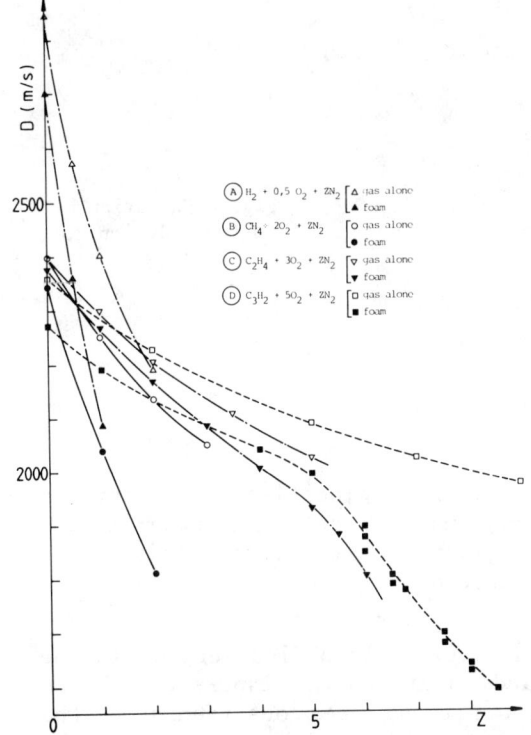

Fig. 2 Variation of detonation velocity vs the dilution Z.

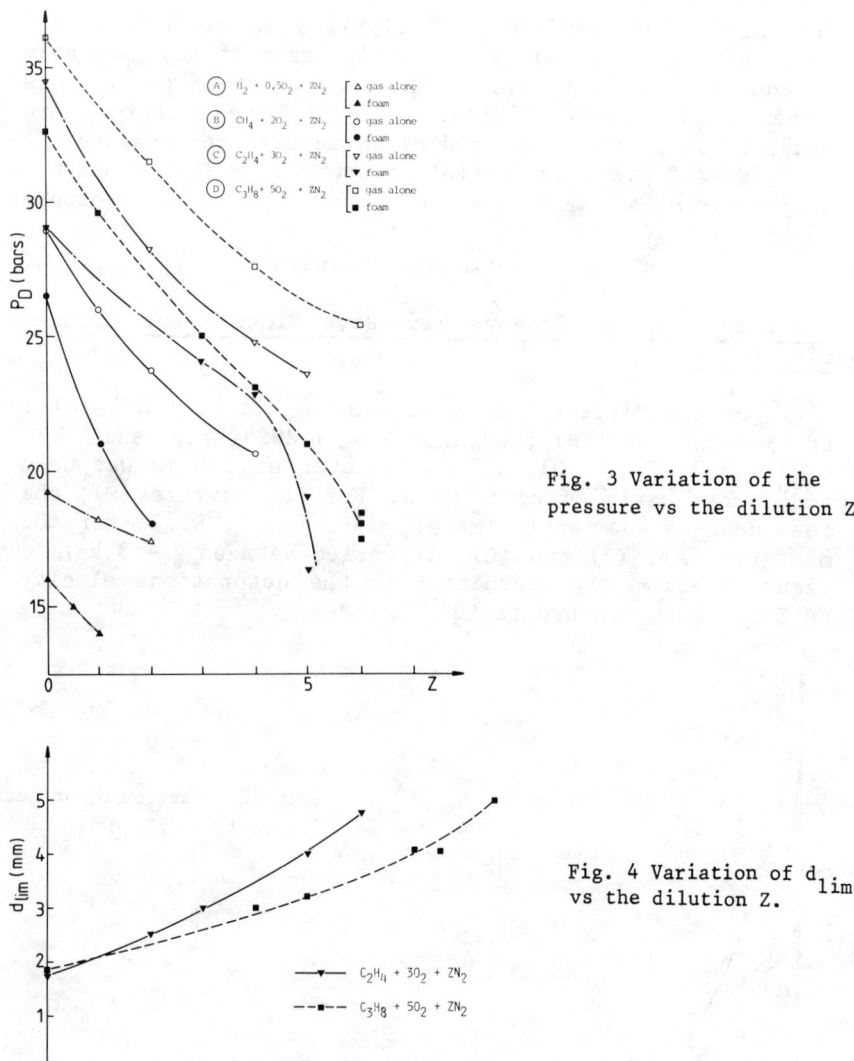

Fig. 3 Variation of the pressure vs the dilution Z.

Fig. 4 Variation of d_{lim} vs the dilution Z.

For $Z < Z_{lim}$, the measured velocities are generally 5-7% less than those measured at the same experimental condition for premixed gaseous detonations. This value Z_{lim} depends upon the nature of the fuel and approaches $Z = 5$ for C_3H_8 and C_2H_4, $Z = 2-3$ for CH_4, and $Z = 1-2$ for H_2. For $Z > Z_{lim}$, one can observe an obvious change in the variation of velocity and an increased dispersion of the values measured. As had already been observed with the stoichiometric propane-oxygen mixture, the pressure of the

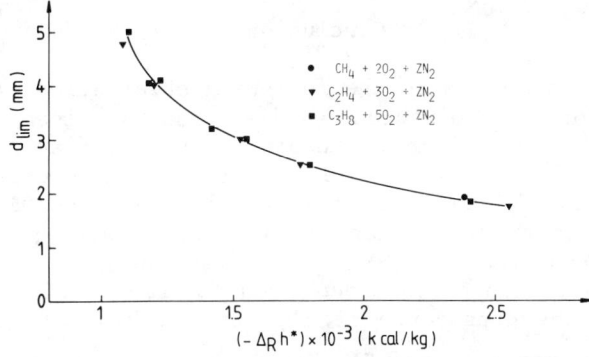

Fig. 5 Variation of d_{lim} vs the heat of reaction of the explosive gaseous mixture.

detonation products is lower than that observed in gaseous mixtures. The differences exceed those for the velocity. The curves representing the variations of pressure with Z (Fig. 3) show that for $Z > Z_{lim}$, the pressure decreases sharply, as does the velocity. For the propane (C) or ethylene (D) and for $Z > Z_{lim}$, unstable and oscillating regimes of propagation similar to those observed in galloping detonations (Saint-Cloud et al. 1972) have been observed. In the cases the regimes of propagation oscillate between the NRD and the DFR regimes.

Transition from the NRD to the DFR

The results of experiments for each mixture of given ρ_m and for Z with d_{32} in the range of 1.5 to 6 mm, indicated that DFR occurs when the diameter d_{32} of the foam bubbles falls below a value of d_{lim}. The variation d_{lim} with Z for propane and ethylene fuels is shown in Figure 4. The curves $d_{lim} = f(Z)$ form the limits of the range of existence of the DFR and NRD regimes. As Z increases, d_{lim} increases. For the mixtures containing either hydrogen or methane, the variation of d_{lim} with Z could not be established. A very slight variation of the nitrogen dilution leads to a strong variation of d_{lim}.

The Effect of the Heat of Reaction

The variation of the diameter d_{lim} with the heat of reaction of the gaseous phase for mixtures containing propane, ethylene, or methane, is shown in Figure 5. This curve shows that if the components of the gaseous explosive mixture and d_{32} are known, one can tell in advance which detonation regime, NDR or DFR, will occur.

Conclusions

This work leads to the following observations :

1) For the mixtures studied and a given bubble diameter ($d_{32} \simeq$ 5-6 mm), one can observe, for an nitrogen dilution $Z > Z_{lim}$, a rapid decrease of the detonation velocity, which leads to an instability and the DFR. This change is particularly marked in mixtures with a propane or ethylene base.

2) Variation of foam density by a factor of two does not alter the dependence of d_{lim} on the heat of reaction of the gaseous mixture. This would tend to show that the mechanical losses (breaking up of the edges) are much greater than the heat losses (heating up or vaporization, droplets of water) in the setting up of the NRD regime. This reinforces the findings of Saint-Cloud et al. (1976) who showed that the detonation velocity and pressure did not depend on foam density. However, when the shock wave dissociates from the flame front, in the case of the DFR, heat exchange with the droplets becomes dominant and the influence of the density is no longer negligible.

References

Brochet, C., Guerraud, C., Manson, N., and Veyssière, M. (1969) Variation de la pression des gas derrière le front des détonations dans les mélanges stricts propane-oxygène, influence du diamètre du tube et de la pression initiale. C. R. Acad. Sci. Paris 268B, pp. 361-364.

Saint-Cloud, J. P., Guerraud, C., Brochet, C., and Manson, N. (1972) Quelques particularités des détonations très instables dans les mélanges gazeux. Acta Astron. 17, pp. 487-498.

Saint-Cloud, J. P., Guerraud, C., Moreau, M., and Manson, N. (1976) Expériences sur la propagation des détonations dans un milieu biphasique. Acta Astron. 3, pp. 781-794.

Saint-Cloud, J. P., Fessou, Ph., Guerraud, C., and Manson, N. (1981) Influence d'interfaces liquides sur la propagation des détonations dans des mélanges stricts propane-oxygène. Colloque International Berthelot, Vieille, Mallard et Le Chatelier, Bordeaux, pp. 511-516, édité par la Société Francaise du Combustion Institute, CNRS, CRCCHT, 45045 Orléans Cédex.

Saint-Cloud, J. P., Moreau, M., and Manson, N. (1981) Détermination de l'aire interfaciale liquide-gaz et de la célérité des ondes de pression faibles dans des mousses aqueuses. C. R. Acad. Sci. Paris 292C, pp. 825-829.

Shchelkine, K. I. (1945) Combustion rapide et détonation hélicoïdale des gaz. Acta Phys. Chem. USSR 20, 1945, pp. 303-306.

Propagation Velocity and Mechanism of Bubble Detonation

Tatsuya Hasegawa*
Nagoya Institute of Technology, Showa-ku, Nagoya, Japan
and
Toshitaka Fujiwara†
Nagoya University, Chikusa-ku, Nagoya, Japan

Abstract

In a previous paper (Hasegawa and Fujiwara 1982), observations of shock wave propagation in glycerin with 70%Ar+30%($2H_2+O_2$) bubbles and associated chain explosion of the bubbles in a vertical shock tube were reported. This phenomenon was called "bubble detonation," because the blast wave generated by a bubble explosion compressed the nearest bubble to explode and this sequence sustained the propagation of explosion. In this work, the bubble detonation was further investigated from the viewpoint of steady propagation by measurements of the decay of the pressure amplitudes at three points, and by observations of the interaction between the two adjacent bubbles using a high-speed framing camera. The results obtained were as follows: 1) The propagation of bubble detonations was observed to be non-steady, and the velocity and pressure amplitudes decreased conceivably due to the insufficient length of the shock tube. However, observations revealed that the pressure amplitude approached a constant value. 2) The pressure wave accompanying bubble detonations had two typical features, a weak precursor wave followed by a strong shock wave. 3) The reciprocal compression-explosion process of bubbles provided the propagation mechanism of bubble detonation. 4) A model of the propagation velocity, which was reduced from the observed behavior of the bubbles, explained the measured detonation velocity and suggested a relation on the velocity dependency upon the pressure amplitude.

Presented at the 9th ICODERS, Poitiers, France, July 3-8, 1983. Copyright © 1984 by the American Institute of Aeronautics and Astronautics, Inc. All rights reserved.
*Research Assistant, Department of Mechanical Engineering.
†Professor, Department of Aeronautical Engineering.

Nomenclature

a_1 = sound velocity of liquid
D = detonation or shock velocity
d_0 = initial bubble diameter
d_{l_0} = initial longitudinal bubble diameter
d_{t_0} = initial transversal bubble diameter
L = interbubble distance
p_s = shock pressure amplitude
p_4 = driver pressure
ρ_1 = density of liquid
τ_c = bubble compression time

Introduction

Shock wave propagation and associated chain explosion in oxyhydrogen-bubbled liquids in a vertical shock tube were observed for the first time in a previous work (Hasegawa and Fujiwara 1982). Their experimental observations indicated that once a bubble started exploding the explosion sustained the wave propagation. The amplitude and propagation velocity of the shock wave were significantly different from a nonexploding shock wave, and as a result, the pressure showed a considerable change. A limit phenomenon, which depends on the initial bubble radius and interbubble distance, was observed in such a chain explosion. These results clearly indicated the existence of bubble detonation because the blast wave generated by a bubble explosion was strong enough to compress and explode the adjacent bubble and because the propagation of bubble explosion was self-sustaining.

A strong motivation to study bubble detonations lies because in nuclear power plant accidents oxyhydrogen bubbles, which can be produced in abnormally high-temperature water, can constitute an explosion hazard.

Thus, in this work, the bubble detonation phenomenon was further studied from the viewpoint of its steadiness and propagation mechanism by measurement of the pressure amplitudes at three different locations and observation of interactions between adjacent bubbles. A phenomenological model of the propagation velocity of bubble detonation was developed for comparison with the measured velocities.

Experimental

As illustrated in Fig. 1, the present shock tube consists of a 445-mm-long driver separated by a plastic diaphragm from a 1840-mm-long test section having 50×50 mm^2

cross section. The test section is filled with pure glycerin up to 700 mm below the diaphragm, while the remaining volume is filled with air at atmospheric pressure. The driver section is pressurized to 1.5 MPa with helium. A syringe at the lower end of the test tube was mounted to produce a series of bubbles using a vessel pressurized up to 0.3 MPa with an argon diluted stoichiometric oxyhydrogen mixture, $70\%Ar + 30\%(2H_2+O_2)$. Because of the high viscosity of glycerin (1500 times as large as that of water), the bubble diameter becomes 10-15 mm with an interbubble distance of 50-60 mm. The shock tube discharge is synchronized with the steady bubble formation. The pressure histories were measured by the three Kistler 601A pressure transducers, mounted at intervals 459 and 240 mm and connected to the low-pass filters with a cutoff frequency of 25 KHz to avoid the noise from the tube wall vibration. Glass windows are located at the section where the second transducer is mounted. Through these windows the simultaneous motion of

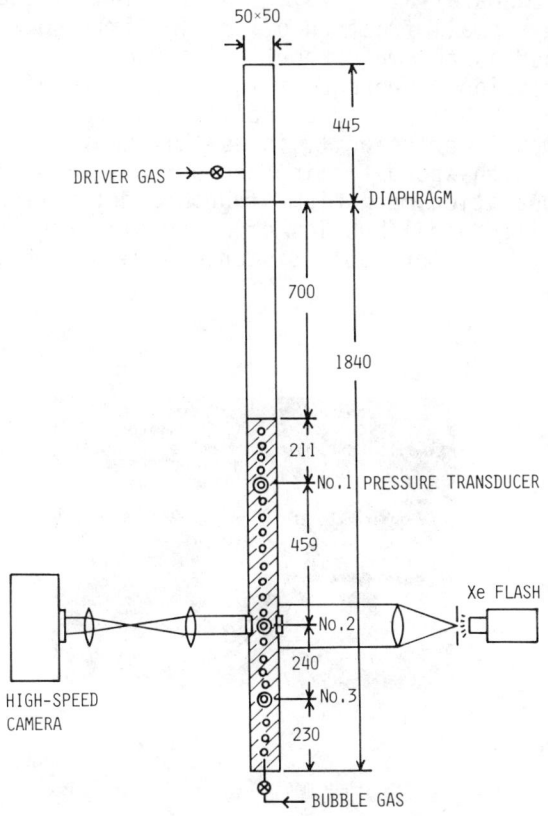

Fig. 1 Sketch of the experimental apparatus.

multiple bubbles is observed by a high-speed framing camera (Nikon-Uemura type UHF-250B with a speed of 30,000 frames per second) with backlighting by flash lamp (with a duration of 8 ms).

Typical characteristics of bubble explosions are evident in the oscillatory pressure profiles (Fig. 2). The propagation is still nonsteady mode due to the insufficient length of the shock tube. The propagation velocity decreased from 444 m/s between the pressure transducers 1 and 2 to 300 m/s between the transducers 2 and 3, while the pressure amplitude decreased from 3.17 MPa at the transducer 1 to 1.82 MPa at the transducer 3.

Examination of the pressure profile in more detail reveals a small-amplitude wave which precedes the main jump (at t=500 μs for the pressure oscillogram 1, t=1000 μs for the pressure oscillogram 2). Since the precursor has a velocity of 1000 m/s, it is a sound wave propagating in pure glycerin under the influence of the elastic deformation of tube wall (Kawada et al. 1973). The small-amplitude wave is sustained by a small fraction of the initial shock wave strength, whereas the major portion is used to generate bubble compression. Subsequent to the passage of the shock wave through a bubble a large-amplitude shock wave propagating with much lower velocity is established.

Typical high-speed framing photographs (Fig. 3) indicate the propagation mechanism of bubble detonation. Here, two oblate ellipsoidal bubbles of 15-mm diam are initially positioned at an interbubble distance 59.4 mm. After being

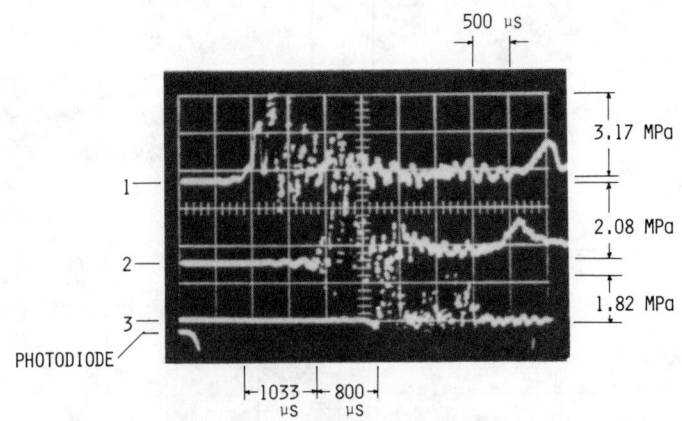

D_{12}=444 m/s , D_{23}=300 m/s

Fig. 2 Pressure oscillograms of bubble detonation.

compressed by a shock wave, the upper bubble eventually attains its minimum size and at this instant it explodes, while the lower bubble remains undeformed at the moment shown in photograph (5). When the upper bubble starts expansion and deforms into a parachute shape (Fujiwara and Hasegawa 1981), the other bubble begins to shrink toward its minimum size and finally explodes in a very short time as shown in photograph (10). The bubble explosion is confirmed from the following observations: (1) in a previous work (Hasegawa and Fujiwara 1982), the size and shape of the light emission from a reacting bubble recorded by open shutter photography agreed with the observed minimum bubble diameter recorded by high-speed shadowgraph; (2) the contrast of shadowgraph picture is decreased by the radiation emission from the explosion only at the instant of minimum size in Fig. 3; and (3) the bubble size becomes larger than the initial size after explosion, as shown in the temporal behavior of the longitudinal bubble diameter (Fig. 4).

The above phenomenon indicates that when a shock wave arrives at a bubble, compression of the bubble results in shock wave decay and leads to disappearance, except for a very weak pressure wave leaking through this bubble dynamic process and propagating downstream with the sound velocity.

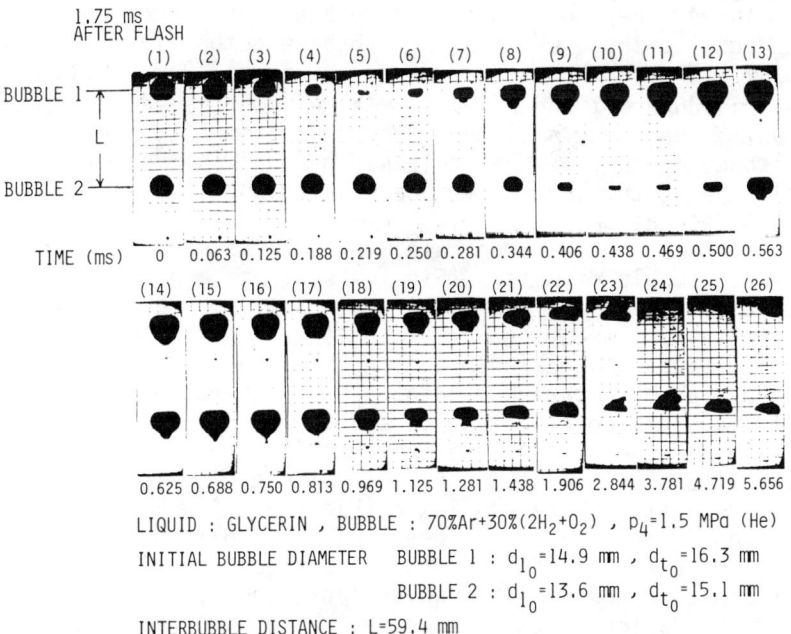

LIQUID : GLYCERIN , BUBBLE : 70%Ar+30%($2H_2+O_2$) , p_4=1.5 MPa (He)
INITIAL BUBBLE DIAMETER   BUBBLE 1 : d_{l_0}=14.9 mm , d_{t_0}=16.3 mm
                           BUBBLE 2 : d_{l_0}=13.6 mm , d_{t_0}=15.1 mm
INTERBUBBLE DISTANCE : L=59.4 mm

Fig. 3 High-speed framing shadowgraphs of bubble behaviors.

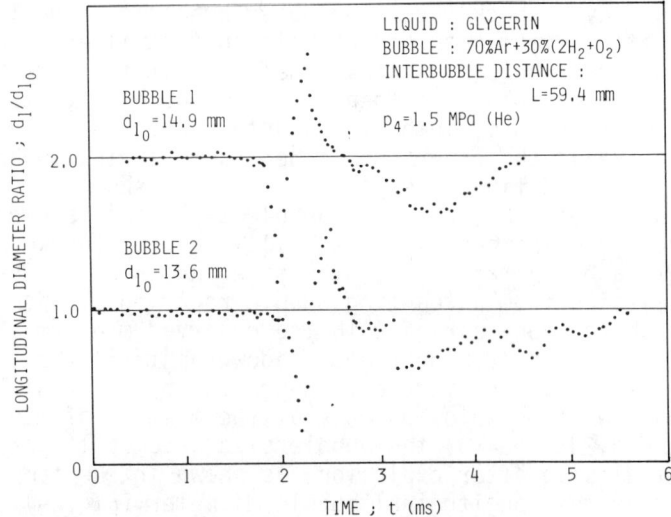

Fig. 4 Temporal behavior of longitudinal bubble diameters.

In contrast, Fig. 5 shows typical pressure profiles for which no explosion occurs. The bubble detonation cannot be sustained because of the smaller bubble diameter (~12 mm), even though the bubble gas mixture and driver pressure are identical to those which sustain bubble detonations as in Fig. 2. The shock velocity decreased from 312 m/s between the transducers 1 and 2 to 247 m/s between transducers 2 and 3, while the shock amplitude decreased from 1.12 MPa at transducer 1 to 0.58 MPa at transducer 3. The velocity and amplitude of the shock wave are smaller than those of the bubble detonation.

In this nonexploding case, the temporal behavior of observed bubble diameters (Fig. 6) suggests a propagation mechanism slightly different from that of bubble detonation. At the instant when the upper bubble is reduced to its minimum size by shock compression, the lower bubble has already begun to shrink. Both bubbles subsequently experience sinusoidal oscillation in size.

The attenuation of the pressure amplitude and its asymptotic approach to a constant value during the detonation propagation were confirmed (Fig. 7). A bubble detonation (open circles) has larger pressure amplitudes than an inert shock wave (open triangles and squares). The pressure difference between the two waves approaches 1.3 MPa at 910 mm below the glycerin surface, while attenuation of the detonation wave decreases with depth.

To determine the influence of the initial shock
strength on the transition to a steady detonation, lower
driver pressures p_4 were used to produce different shock
strength. When $1.5 > p_4 > 1.3$ MPa, a bubble detonation occurred
but were not reproducible, although there was no essential
difference in the behavior of the bubble detonation. As a

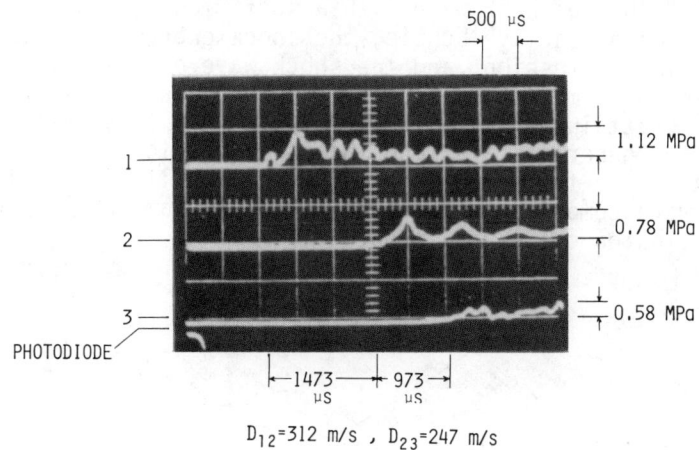

$D_{12} = 312$ m/s , $D_{23} = 247$ m/s

Fig. 5 Pressure oscillograms of inert shock waves.

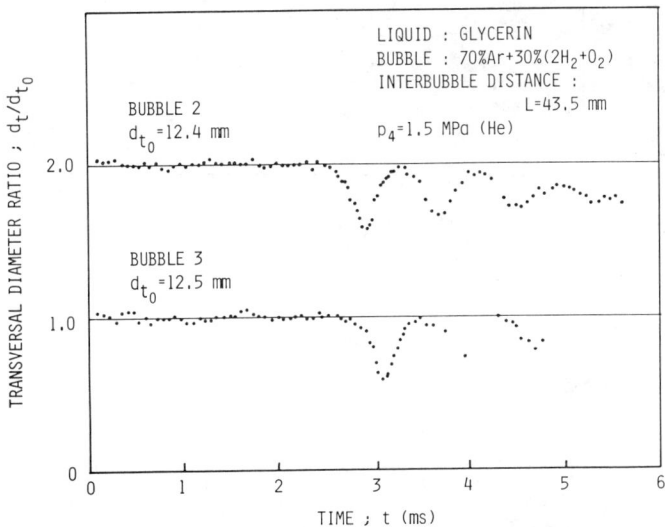

Fig. 6 Temporal behavior of transversal diameters (bubble
detonation does not occur).

bubble detonation could not be generated when p_4=1 MPa, this initial shock strength is below the initiation limit.

A Model of the Propagation Velocity

The bubble motions observed in Fig. 3 suggest a model for the velocity of bubble detonations, as illustrated in Fig. 8. When a shock wave with a sufficient strength arrives at a reactive bubble, the shock energy is absorbed by bubble compression; and the shock wave decays. The compression time is denoted as τ_c. When a critical pressure level is attained, bubble explosion occurs. The bubble explosion generates a strong blast wave which propagates at approximately the sound velocity a_1 of the liquid. This new shock wave compresses the adjacent downstream bubble, and the explosion chain continues.

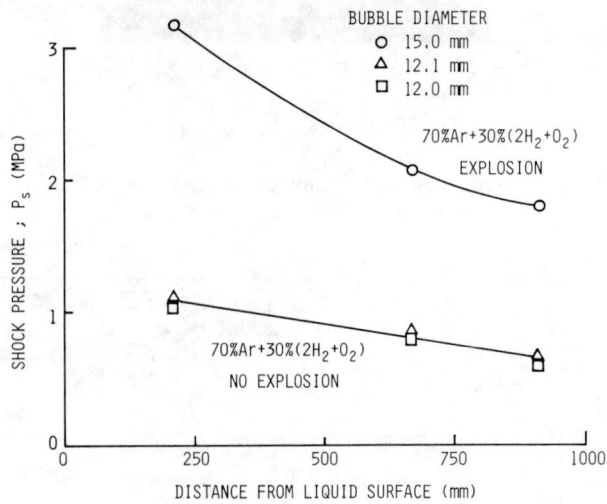

Fig. 7 Decay of the shock pressure amplitude.

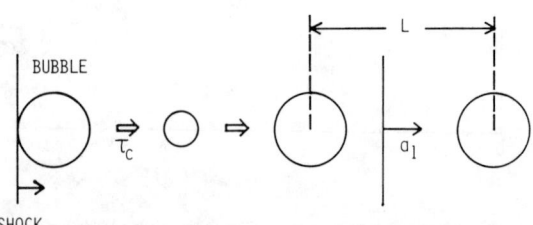

Fig. 8 Proposed model for the propagation of bubble detonation.

The propagation velocity D of bubble detonation is written as

$$D = L/(\tau_c + L/a_1) \qquad (1)$$

where L is the interbubble distance. With the empirical values τ_c =190 μs and L =59.4 mm observed in Fig. 3 and a_1= 1986 m/s for glycerin, Eq.(1) predicts a velocity D of 270 m/s. This prediction is in reasonable agreement with the measured velocity 300 m/s between the transducers 2 and 3 in Fig. 2.

As τ_c>>L/a_1 for the present experimental conditions, Eq.(1) can be approximated as

$$D \simeq L/\tau_c \qquad (2)$$

Rayleigh's solution for the collapse time of a cavitation bubble can be used to estimate τ_c (Knapp et al. 1970):

$$\tau_c \propto d_0 \sqrt{\rho_1/p_s} \qquad (3)$$

where τ_c is the time of collapse, d_0 the initial bubble diameter, ρ_1 the liquid density, and p_s the pressure at infinity (shock pressure).

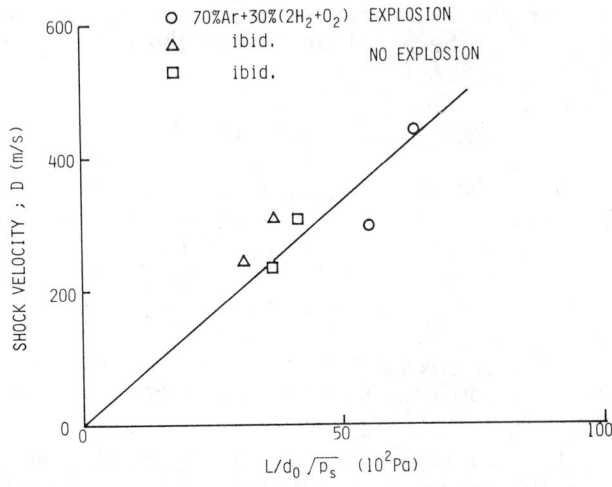

Fig. 9 Proportional relation between D and $L/d_0 \sqrt{p_s}$.

Table 1 Uniformity of the longitudinal diameter and interbubble distance of rising bubbles

Height of measuring point			230 mm	470 mm	929 mm
Case 1	d_{10}	(mm)	15.8	15.7	16.5
	L_{10}	(mm)	55.6	57.2	57.8
Case 2	d_{10}	(mm)	12.5	12.8	13.3
	L_{10}	(mm)	55.2	56.0	55.9
Case 3	d_{10}	(mm)	14.7	15.4	15.9
	L_{10}	(mm)	54.4	55.2	56.4

If Eq.(3) is substituted into Eq.(2), the propagation velocity of a bubble detonation is

$$D \propto L/d_0 \sqrt{p_s/\rho_1} \qquad (4)$$

The properties of the detonable gas mixture implicitly appear in the form of the shock pressure p_s ; p_s is the amplitude of the rectangular-profile shock wave in Rayleigh's analysis. Eq.(4) is verified by the comparison with the experimental values for both bubble detonations and inert shock waves (Fig. 9).

Unifomity of the initial bubble diameter d_0 and the interbubble distance L was checked by a separate experiment. The result shows that the bubble diameter and interbubble distance increase at most 8 and 4%, respectively along the liquid column, as shown in Table 1. Thus, the nonsteady decrease of the wave propagation velocity for bubble detonation appears to be attributable to the decrease of shock amplitude p_s in Eq.(4).

Conclusions

1) Observed bubble detonations are still nonsteady, with decreasing velocity and pressure amplitudes, although the pressure tends to approach a constant.
2) The pressure profile of a bubble detonation has a double discontinuity structure; a weak precursor followed by a strong shock wave.
3) The reciprocal compression-explosion process of bubbles is a dominant mechanism for self-sustained bubble detonations.
4) An empirical model of the detonation velocity explains the measured values and the velocity dependency on the shock pressure amplitude.

References

Fujiwara, T. and Hasegawa, T. (1981) Shock propagation in liquid-gas media. Shock Tubes and Waves. pp.724-732. State University of New York Press, Albany, N.Y.

Kawada, H., Hotta, M., and Makiguchi, M. (1973) An experimental study with a hydrodynamic shock tube. Trans. Japan Soc. Aero. Space Sci. 16, pp.195-206.

Knapp, R. T., Daily, J. W., and Hammit, F. G. (1970) Cavitation. McGraw-Hill Book Co., New York, 99 pp.

Hasegawa, T. and Fujiwara, T. (1982) Detonation in oxyhydrogen bubbled liquids. Nineteenth Symposium (International) on Combustion. pp.675-683. The Combustion Institute, Pittsburgh, Pa.

Nonsteady Shock Wave Propagating in a Bubble-Liquid System

Tadayoshi Sugimura*
Meijo University, Tenpaku-ku, Nagoya, Japan
Kazushi Tokita†
Nippon Oil & Fats Co. Ltd., Taketoyo, Chita, Aichi, Japan
and
Toshitaka Fujiwara‡
Nagoya University, Chikusa-ku, Nagoya, Japan

Abstract

This investigation is concerned with the shock wave/gas bubble interactions which occur in cavitation phenomena and slurry explosives. These interactions are numerically simulated with a model system: steam bubbles in water, and attention is focused on the behavior of disturbances near a gas bubble. The two-phase flowfield is treated as one dimensional, and is solved by both a second-order finite-difference method with the FCT (flux-corrected transport) scheme and the random choice method. Tait and Tammann equations are used as the equation of state for water. A shock wave propagates with a velocity of 2950 m/s (M = 2) in water (ρ = 1.0 g/cc), and interacts with a stationary gas bubble (ρ = 7.92 x 10^{-4} g/cc) initially at 1 atm, where the Rankine-Hugoniot pressure behind the shock wave is 15,800 atm. The numerical results obtained by the random choice method indicated that a transmitted shock wave in the one-dimensional bubble (6-mm width) propagates with a mean velocity 1159 m/s, which is in good agreement with the results obtained by the impedance matching method.

Presented at the 9th ICODERS, Poitiers, France, July 3-8, 1983. Copyright © 1984 American Institute of Aeronautics and Astronautics, Inc. All rights reserved.

*Associate Professor, Department of Transport Machine Engineering.
†Research Scientist, Taketoyo Plant.
‡Professor, Department of Aeronautical Engineering.

Introduction

The behavior of a shock wave in a two-phase medium, i.e., in gas bubble existing as inhomogeneities in liquid, has been investigated in connection with cavitation problems and detonation characteristics of explosives. Experimental evidence indicates that the presence of gas bubbles in slurry or emulsion explosives strongly influences the propagation of a detonation. The hot spot formed by the interaction between a shock wave and a gas bubble produces a temperature which is sufficient to maintain exothermic chemical reactions required for detonation propagation. While microscopic processes in a bubble leading to the formation of hot spots have been conjectured (Evans et al. 1962; Mader 1965), they have not been observed so far. Recently, Hasegawa and Fujiwara (1982) introduced gas bubbles filled with an oxyhydrogen mixture into glycerin, observed bubble explosions caused by shock compression, and modeled the process. In this study, whose aim is the clarification of gas bubble motion in a liquid, a bubble is modeled as a density inhomogeneity. The interaction between a one-dimensional nonsteady shock wave and a bubble was numerically analyzed for a steam bubble in water.

Numerical Analysis

The schematic of the flowfield analyzed in the numerical simulation is shown as Fig. 1(a). A one-dimensional gas bubble is positioned in front of a steady shock wave propagating with Mach number 2 (D = 2950 m/s) in water. The initial pressure and density profiles prior to the interaction are shown as Fig. 1(b).

Fundamental Equations

It is assumed that water vapor is a perfect gas with constant specific heat, water is compressible, and both the gas and liquid are inviscid, nonheat-conducting and nonreactive. Thus, one-dimensional nonsteady fluid dynamic equations can be expressed as

$$\frac{\partial f}{\partial t} + \frac{\partial F}{\partial x} = 0 \tag{1}$$

$$f = \begin{bmatrix} \rho \\ \rho u \\ \rho e \\ \rho Y \end{bmatrix} \quad F = \begin{bmatrix} \rho u \\ \rho u^2 + p \\ \rho u(e + p/\rho) \\ \rho u Y \end{bmatrix} \tag{2}$$

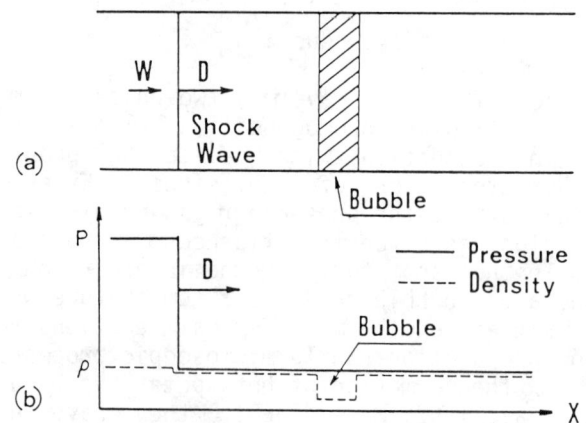

Fig. 1 The schematic of the flowfield used in numerical simulation. a) Arrangement of one-dimensional gas bubble. b) Initial pressure and density profiles.

The equations of state for the gas region are

$$p = \rho R T \qquad (3)$$

$$e = \frac{1}{2} u^2 + \frac{1}{\gamma - 1} \frac{p}{\rho} \qquad (4)$$

while for the water region

$$p + p_c = \rho K T \quad \text{(Tammann equation)} \qquad (5)$$

or

$$p = A\left[\left(\frac{\rho}{\rho_0}\right)^n - 1\right] + p_0 \quad \text{(Tait equation)} \qquad (6)$$

$$e = \frac{1}{2} u^2 + \frac{1}{\eta - 1} \frac{p + p_c}{\rho} + \frac{p_c}{\rho} \qquad (7)$$

where u, p, ρ, T, and Y denote the velocity, pressure, mass density, temperature, and mass fraction (Y=1 for water), respectively. In addition, R, K, γ, and η are the gas constant, the liquid constant equivalent to the gas constant, the specific heat ratio of the gas, and the adiabatic index of water, respectively, and p_c and A are the physical constants for water. The following constant values were used in the calculation:
R = 4.56 atm cm^3 g^{-1} K^{-1}, K = 10.84 atm cm^3 g^{-1} K^{-1}, γ = 1.4,

$\eta = 7.15$, $p_c = 3000$ atm, $n = 7$, and $A = 3268$ atm. Equation (5) was used for the MacCormack method and Eq. (6) for the random choice method.

Dimensionless Forms

The characteristic quantities used to nondimensionalize the present problem are the initial pressure $p_0 = p_{\ell 0} = p_{g0}$, the initial density $\rho_0 = \rho_{g0}$, and arbitrary choice of the length x^*, where a combination of $x^* = 1$ mm and the dimensionless mesh size $\Delta X = 0.2$ was assumed for the MacCormack method and 3.5 mm and 0.1 for the random choice method. The fundamental equations (1 - 6) can be written in dimensionless forms with the introduction of:

$$p^* = p_0, \quad \rho^* = \rho_0, \quad v^* = (p^*/\rho^*)^{1/2}, \quad t^* = x^*/v^* \qquad (8)$$

Finite-Difference Equations

1) The differential equation (1) can be approximated by the second-order explicit difference method originated by MacCormack (1971) and the FCT scheme (Boris and Book 1976):

$$f_i^0 = f_i^n - \frac{\Delta t}{\Delta x}(F_{i+1}^n - F_i^n) \qquad (9)$$

$$\tilde{f}_i^{n+1} = \frac{1}{2}\left[f_i^n + f_i^0 - \frac{\Delta t}{\Delta x}(F_i^0 - F_{i-1}^0) \right] \qquad (10)$$

$$\bar{f}_i^{n+1} = \tilde{f}_i^{n+1} + \zeta(f_{i+1}^n - 2f_i^n + f_{i-1}^n) \qquad (11)$$

$$f_i^{n+1} = \bar{f}_i^{n+1} - (\delta_{i+1/2}^c - \delta_{i-1/2}^c) \qquad (12)$$

$$\delta_{i+1/2}^c = S \cdot \max[0, \min(S \cdot \Delta_{i-1/2}, |\tilde{\Delta}_{i+1/2}|, S \cdot \Delta_{i+3/2})] \qquad (13)$$

$$\Delta_{i+1/2} = \bar{f}_{i+1}^{n+1} - \bar{f}_i^{n+1} \qquad (14)$$

$$\tilde{\Delta}_{i+1/2} = \zeta(\tilde{f}_{i+1}^{n+1} - \tilde{f}_i^{n+1}) \qquad (15)$$

$$S = \text{sgn}(\tilde{\Delta}_{i+1/2}) \qquad (16)$$

where (i,n) denotes a lattice point in x-t space, and Δx and Δt are the difference steps in space and time. ζ is an arbitrary constant set at 1/6 at present.

The n+1-th time step Δt^{n+1} is determined by the stability condition:

$$\Delta t^{n+1} = C_f / \max(a_s/\Delta x) \qquad (17)$$

$$a_s = u + a$$

$$a = \begin{cases} \sqrt{\eta \dfrac{p + p_c}{\rho}} & \text{(for the liquid region)} \\ \sqrt{\gamma \dfrac{p}{\rho}} & \text{(for the gas region)} \end{cases} \qquad (18)$$

where C_f and a are the Courant number (0.85) and the local speed of sound.

2) Numerical analysis of the flowfield by the random choice method (Sod 1978) is based on the Riemann problem:

$$f_{i+1/2}^{n+1/2} = g_{i+\xi_n}^{n+1/2} \qquad (\xi_n : \text{random variable}) \qquad (19)$$

$$p_* = (u_1 - u_r + \frac{p_r}{M_r} + \frac{p_1}{M_1})/(\frac{1}{M_1} + \frac{1}{M_r})$$

$$u_* = (p_1 - p_r + M_1 u_1 + M_r u_r)/(M_1 + M_r) \qquad (20)$$

The solution of the Riemann problem g_i^n is determined from Eq. (20) for which the subscript $*$, 1, and r are a state of contact surface, left- and right-hand-side boundary conditions, respectively. Momenta M_i's are expressed as follows for the gas region:

$$M_r = (\rho_r\, p_r)^{1/2}\, \Phi(x) \qquad x = p_*/p_r$$

$$M_1 = (\rho_1\, p_1)^{1/2}\, \Phi(x) \qquad x = p_*/p_1 \qquad (21)$$

$$\Phi(x) = (\frac{\gamma + 1}{2} x + \frac{\gamma - 1}{2})^{1/2} \qquad x \geq 1$$

$$= \frac{\gamma - 1}{2\gamma^{1/2}} \frac{1 - x}{1 - x^{(\gamma-1)/2\gamma}} \qquad x \leq 1 \qquad (22)$$

and for the liquid region:

$$M_r = (\rho_r \bar{p}_r)^{1/2} \Phi'(\bar{x}) \quad \bar{x}=\bar{p}_*/\bar{p}_r$$

$$M_l = (\rho_l \bar{p}_l)^{1/2} \Phi'(\bar{x}) \quad \bar{x}=\bar{p}_*/\bar{p}_l \tag{23}$$

$$\Phi'(\bar{x}) = \{(\bar{x} - 1)/[1 - (1/\bar{x})^{1/n}]\}^{1/2} \quad \bar{x} \geq 1$$

$$= \frac{n-1}{2n^{1/2}} \frac{1-\bar{x}}{1-\bar{x}^{(n-1)/2n}} \quad \bar{x} \leq 1 \tag{24}$$

where $\bar{p} = p + A$.

Initial and Boundary Conditions

It was confirmed that the isentropic relation derived from the Tammann equation (Chen and Collins 1971) agreed within 1 % error with the Tait equation (Flores and Holt 1981) for water.

A one-dimensional steady shock wave propagating at Mach number 2 (D = 2950 m/s) in water is produced numerically from an initially discontinuous step profile. In this case, Rankine-Hugoniot pressure behind the shock wave is 15,800 atm and the particle velocity is 543 m/s. Figure 1 and Table 1 show the initial conditions and the constant values of the water and gas bubble.

During numerical computations with the MacCormack method, diffused interface, i.e., artificial diffusion phenomenon, was encountered. In such a region, the following expressions are utilized to avoid computational difficulty:

$$p = \frac{KR}{(1-\alpha)R + \alpha K} \rho T - \frac{(1-\alpha)R}{(1-\alpha)R + \alpha K} p_c \tag{25}$$

$$\rho e = \frac{\alpha}{\gamma - 1} p + \frac{1-\alpha}{n-1}(p + p_c) + (1-\alpha)p_c + \frac{1}{2}\rho u^2 \tag{26}$$

where α denotes the volume fraction. The density ρ in Eq. (25) is expressed as $\rho = (1-\alpha)\rho_\ell + \alpha\rho_g$. ρ_ℓ and ρ_g are determined from the following expression, which is derived from the conditions of $p = p_\ell = p_g$ and $T = T_\ell = T_g$, by the iterative procedure with known ρ, e, and Y.

$$(K\rho_\ell - R\rho_g)\left[\frac{K}{\eta-1}Y + \frac{R}{\gamma-1}(1-Y)\right]^{-1}\{e - [Ye_{h\ell} + (1-Y)e_{hg}]$$

$$+ \left[\frac{(p_{h\ell}+p_c)/\rho_\ell}{\eta-1}Y + \frac{p_{hg}/\rho_g}{\gamma-1}(1-Y)\right]\} - p_c = 0 \quad (27)$$

where the subscripts $h\ell$ and hg denote quantities obtained from the Rankine-Hugoniot relations for liquid and gas, respectively.

Results and Discussions

The solution of Eq. (1) by MacCormack's method was checked by modeling the interaction between a shock wave and a helium (molecular weight 4) bubble in SF_6 gas (molecular weight 146). Figure 2 shows typical numerical results in a

Table 1 Initial and characteristic properties of water and gas bubble at T_0 = 277 K, utilized in the computation

	Water	Gas bubble
Initial pressure, P_0	1 atm	1 atm
Initial density, ρ_0	1.0 g/cc	7.92 × 10^{-4} g/cc
Speed of sound, a_0	1475 m/s	423 m/s
Adiabatic index	7.15	1.4
Shock velocity, D	2950 m/s	...
Size	...	6.0 mm

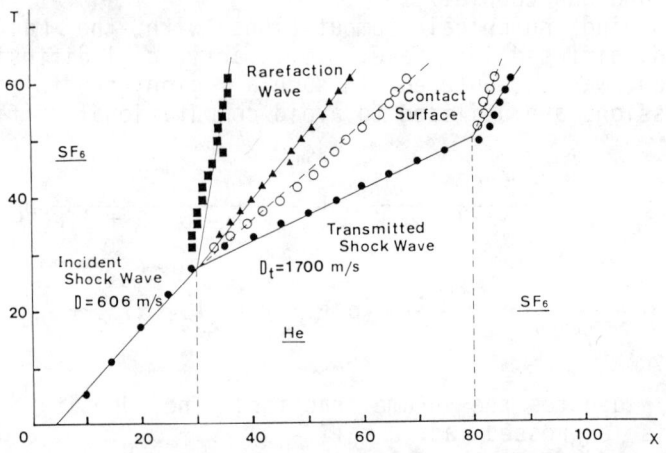

Fig. 2 X-T diagram of disturbances propagating from SF_6 gas region to He bubble region. Velocities of incident and transmitted shock waves: D = 606 m/s (M = 4) and D_t = 1700 m/s (M_t = 1.7).

nondimensional x-t diagram. Solid and open circles are the traces of computed locations of shock waves and contact surfaces, respectively. Penetration of an incident shock wave propagating at Mach number M = 4 into the He-gas bubble produces a transmitted shock wave of Mach number M_t = 1.7, a contact surface with the velocity U_c = 835 m/s, and a rarefaction wave. The behavior of disturbances are in good agreement with the results, obtained with the characteristic method, as shown by the solid and broken lines. The following approximate formula, derived from the impedance matching method, was satisfied: U_c (velocity of interface) = 835 m/s ≃ 2 U_b (particle velocity of gas behind the incident shock wave)=2 x 430 m/s. This method, which is more accurate

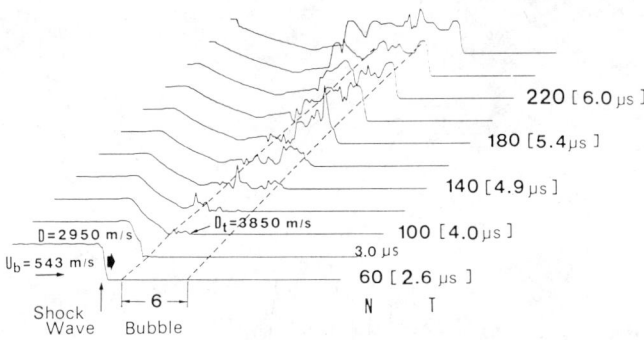

Fig. 3 Pressure profiles under interaction between shock and bubble. Mach number of shock wave in liquid: M = 2 (D = 2950 m/s). Rankine-Hugoniot pressure behind incident shock wave: P_b = 15,800 atm. Liquid = water and bubble = H$_2$O gas.

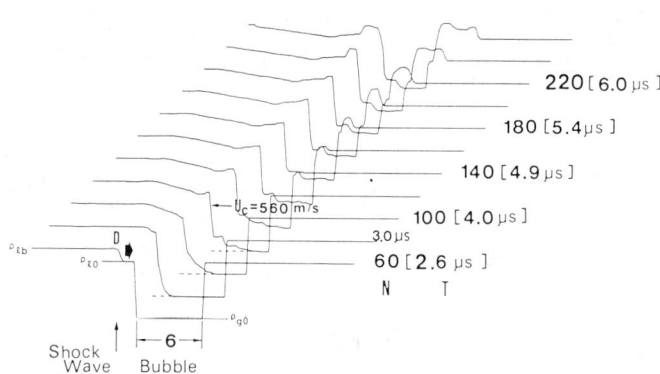

Fig. 4 Density profiles under interaction between shock and bubble. Initial densities of liquid and bubble: $\rho_{\ell 0}$ = 1.0 g/cc and ρ_{g0} = 7.92 x 10^{-4} g/cc. $\rho_{\ell 0}/\rho_{g0}$ = 1260, $\rho_{\ell b}/\rho_{\ell 0}$ = 1.225.

than the free surface approximation, is used to determine a compression state behind a transmitted shock wave, and an interface velocity, from a Hugoniot curve for main medium and the shock impedance of bubble medium.

As a consequence, MacCormack's difference scheme is applicable to the interaction problem between a shock wave and a mass discontinuity.

Typical numerical results of the interaction applied to a gas-liquid system are shown in Figs. 3 and 4. Figure 3 shows a sequence of pressure profile from the time 2.6 µs, just prior to the initiation of the interaction at 3.0 µs. Coincident with the penetration of the incident shock wave into the bubble, a rarefaction wave propagates upstream in the liquid; and a transmitted shock wave, with a mean velocity of 3850 m/s (M_t = 9), propagates in the gas bubble.

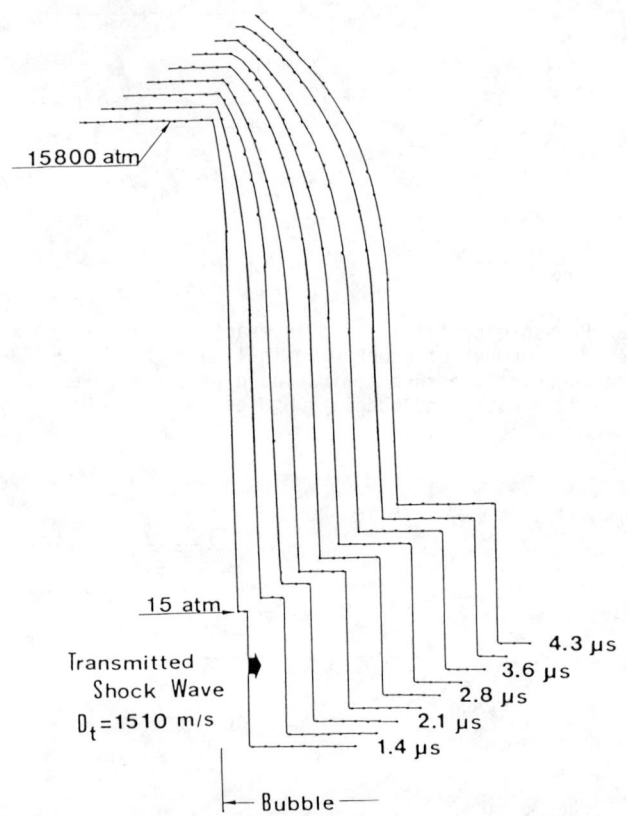

Fig. 5 Pressure profiles at the initial stage of interaction between shock and gas bubble (random choice method). Velocity and Rankine-Hugoniot pressure of incident shock wave: D = 2950 m/s and P = 15,800 atm.

SHOCK PROPAGATION IN A BUBBLE-LIQUID SYSTEM 329

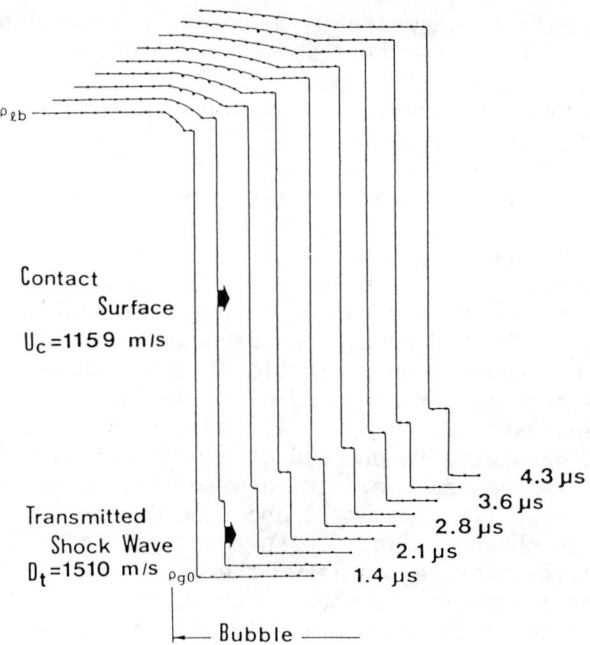

Fig. 6 Density profiles at the initial stage of interaction between shock and gas bubble (random choice method). Initial densities of liquid and bubble: $\rho_{\ell 0}$ = 1.0 g/cc and ρ_{g0} = 7.92 x 10^{-4} g/cc. Density behind incident shock wave: $\rho_{\ell b}$ = 1.23 g/cc.

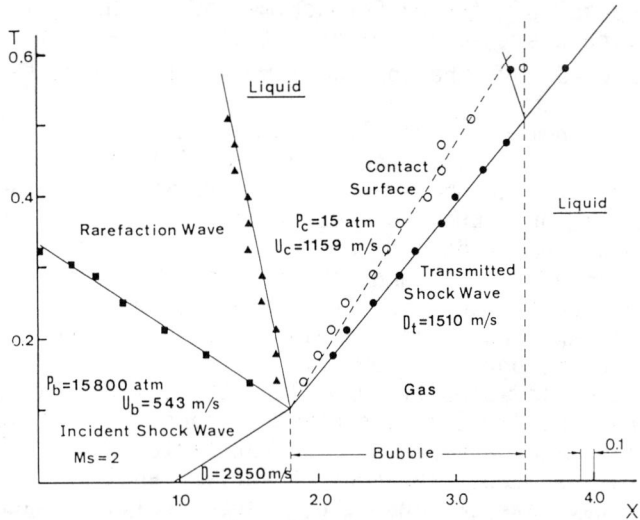

Fig. 7 X-T diagram of disturbances at the initial stage of interaction. Velocities of incident and transmitted shock waves: D = 2950 m/s and D_t = 1510 m/s.

Colliding with the downstream boundary between the bubble and water, the transmitted shock wave produces a reflected and a new transmitted shock. Although such qualitative behaviors are reasonable, a serious discrepancy occurs. During the several collisions between the two interfaces, the pressure in the bubble becomes the same order of magnitude of the Rankine-Hugoniot pressure for the incident shock wave (6.0 µs), even when the bubble size is reduced to merely half of its original size.

Figure 4 shows the behavior of the flowfield in terms of its density distribution. The "well" profile at 2.6 µs representing the gas bubble is compressed by the incident shock. The initial density ratio of water to gas ($\rho_{\ell 0}/\rho_{g0}$ =1260) is much larger compared with the density ratio across the incident shock ($\rho_{\ell b}/\rho_{\ell 0}$ = 1.225). Even if the transmitted shock wave were to proceed at a high velocity (D_t = 3850 m/s) in the gas bubble, the density increase in the gas would be naturally very small and could not fill the "well" profile, as shown in Fig. 4. Furthermore, ideally, a step-like density profile, unlike the one in Fig. 4, should proceed as a transmitted shock wave in the gas bubble.

In spite of the high velocity of the transmitted shock wave (D_t = 3850 m/s), supported by the piston effect of the contact surface, the mean velocity of the contact surface (U_c = 560 m/s, shown in Fig. 4) is almost as low as the particle velocity behind the incident shock wave (U_b = 543 m/s, shown in Fig. 3). At present, the flowfield predicted by application of the MacCormack method to the analysis of the gas-liquid system is not in satisfactory agreement with that predicted by the application of the characteristic method.

Typical numerical results based on the random choice method calculation for the same initial and boundary conditions are shown in Figs. 5 - 7. Figure 5 shows a sequence of pressure profiles from the time 1.4 µs. The "pressure plateau" behind the transmitted shock wave can be confirmed more clearly than in the previous case. In this case, the propagation velocity of the transmitted shock wave is 1510 m/s and lower than the incident shock wave 2950 m/s.

The corresponding density distribution is shown in Fig. 6. In this density profile, there are two density discontinuities: One is a contact surface with the velocity of 1159 m/s and another is a transmitted shock wave. The density ratios across the contact surface and the transmitted shock wave are 281 and 4.26, respectively. In addition, the results computed with the random choice method are summarized in the form of the x-t diagram in Fig. 7. The solid and broken lines in Fig. 7 are the solutions obtained by the characteristic method and numerical results fit these lines well. The generated contact surface travels with the velocity of 1159 m/s, about twice that of the particle velocity behind the incident shock wave (U_b = 543 m/s). This result is in good agreement with the result derived from the impedance matching method.

An effort to overcome the above-mentioned difficulties with the application of the MacCormack method is in progress. Extension of the model to a two-dimensional problem with solutions by the random choice method promises to give interesting results.

Conclusions

1) In numerical analysis based on the MacCormack method, the Tammann equation is preferable to the Tait equation for two-phase shock propagation.

2) Use of the volume fraction α is valid to avoid the computational difficulty arising from the artificial diffusion in the vicinity of the liquid-gas boundary.

3) Numerical results obtained with the random choice method show good agreement with the flowfield derived from the characteristic method.

References

Boris, J. P. and Book, D. L. (1976) Flux-corrected transport. III. Minimal-error FCT algorithms. J. Comp. Phys. 20, 397-431.

Chen, H. T. and Collins, R. (1971) Shock wave propagation past on ocean surface. J. Comp. Phys. 7, pp. 89-101.

Evans, M. W., Harlow, F. H., and Meixner, B. D. (1962) Interaction of shock or rarefaction with a bubble. Phys. Fluids 5, 651-656.

Flores, J. and Holt, M. (1981) Glimm's method applied to underwater explosions. J. Comp. Phys. 44, 377-387.

Hasegawa, T. and Fujiwara, T. (1982) Detonation in oxyhydrogen bubbled liquids. Proceedings of the 19th (International) Symposium on Combustion, pp. 675-683. The Combustion Institute, Pittsburgh, Pa.

MacCormack, R. W. (1971) Numerical solution of the interaction of a shock wave with a laminar boundary layer. Lecture Notes in Physics Vol. 8, pp. 151-163. Splinger-Verlag, New York.

Mader, C. L. (1965) Initiation of detonation by the interaction of shocks with density discontinuities. Phys. Fluids 8, 1811-1816.

Sod, G. A. (1978) A survey of several finite difference methods for systems of nonlinear hyperbolic conservation laws. J. Comp. Phys. 27, 1-31.

Ignition of Dust Suspensions Behind Shock Waves

A. A. Borisov,* B. E. Gel'fand,* E. I. Timofeev,*
S. A. Tsyganov,*and S. V. Khomic*
Academy of Sciences, Moscow, USSR

Abstract

Existing data on ignition and detonation of dust suspensions in air and oxygen are reviewed. Relations between the kinetic characteristics of ignition (ignition delays, burning times), initiation of detonation and its limits are pointed out. Absence of secondary shattering in solid particles and the substantially lower heat release rates in heterogeneous reactions (when compared with homogeneous mixtures) results in dust clouds that are less detonable than liquid-fuel sprays and gaseous mixtures. It is noted that data available in the literature on the detonability of dust suspensions in air are not reproducible being sometimes contradictory. Ignition delays of particles of solid materials in air were measured in reflected shock waves in the range of $10^{-5} - 10^{-3}$ s. Noted is the strong effect on the temperature dependence of ignition delays, local particle concentration, and the way by which a high-temperature pulse in the gas is created. Increase in particle concentration as well as experiments carried out using incident shock waves lead to an apparent devaluation of ignition delays. These two factors increase the discrepancy between the calculatated and real temperatures behind the shock front as a consequence of reflections due to density inhomogeneities. The experiments, conducted with different powder materials (TNT, wheat, naphtalene, etc.) showed that ignition delays of particles in

---

Presented at the 9th ICODERS, Poitiers, France, July 3-8, 1983. Copyright © American Institute of Aeronautics and Astronautics, Inc., 1984. All rights reserved.
*Senior Scientist, Institute of Chemical Physics.

reflected shock waves are close to those for kerosene drops -- i.e., the detonability of particle suspensions does not differ from that of kerosene sprays.

Introduction

A direct connection between self-ignition characteristics of explosive mixtures of gases and their detonability has been shown by previous studies of gaseous detonations. Critical energies of detonation initiation and detonation limits have been related directly to ignition delays that are functions of temperature and mixture content. Utilization of the data on ignition delays for predicting the mixture detonability or assessing the explosion effect is hindered by the fact that most of the data have been obtained at pressures considerably below those likely to prevail in accidents and for mixtures that are highly diluted with inert gases. Only a few investigations (Kogarko and Borisov, 1960) have dealt with mixtures and pressures that can be considered realistic for detonation processes. Nevertheless, the ignition delays and reaction zone thicknesses for a number of gaseous mixtures can be calculated from the available data.

Ignition Delays

Calculation of ignition delays in sprays and suspensions of solid particles in air is more complex than that for gaseous mixtures, since heat transfer, evaporation, and other physical phenomena must be incorporated together with chemical kinetics in the correlation (or prediction) of ignition delay for a heterogeneous mixture. In gaseous mixtures, the burning time τ_b is very much smaller than the ignition delay τ_i. In the two-phase systems, τ_b cannot be neglected in comparison with the ignition delay τ_i. The burning rate of spray particles is defined by their size and by the gaseous flow around the particles and may approach τ_i if the external flow field produces significant drop shattering. However, always $\tau_b > \tau_i$. Based on these observations, one can predict the likelihood of detonability for both the gaseous and the two-phase system from measured or calculated τ_i.

For fuel-air detonations initially at normal conditions, the temperature and pressure ranges that may be of interest are 1100-1550 K and 2-4 MPa and, for fuel-oxygen mixtures, 2000-2500 K and 4-6 MPa. Reaction zone thickness in dust detonations is significantly larger than in gaseous mixtures, while the detonability of particulate systems is lower. This has been directly demonstrated for unconfined explosions by Bull et al. (1980). Other studies of the possibility of establishing a self-sustaining dust detonation in tubes have not yet produced conclusive evidence. Self-ignition of monopropellants (RDX and nitrocellulose) in a nitrogen atmosphere at temperatures of 750-1200 K was studied by Cohen and Decker (1979a, 1979b). Kaufman et al. (1982) and Eidelman and Sichel (1981) found that RDX pariticle suspensions in nitrogen did not detonate as a consequence of their poor ignitability. For their investigations, the former used 3 μm RDX particles and the latter 10 μm RDX particles. Detonations of 17 μm magnesium particles in air has been observed at particle concentrations of 0.6-0.8 kg/m^3 by Ryzhik et al. (1980). However, Austing et al. (1982) and Heberlein (1982) reported no detonation at approximately the same conditions. Strauss (1968) noted that aluminium powder detonated in oxygen. The velocity and pressure of the detonation-like regimes in aluminum powder suspensions in air exceeded those reported for an oxygen atmosphere (Austing et al. 1982; Heberlein 1982). Analysis of their data leads to the conclusion that a decaying

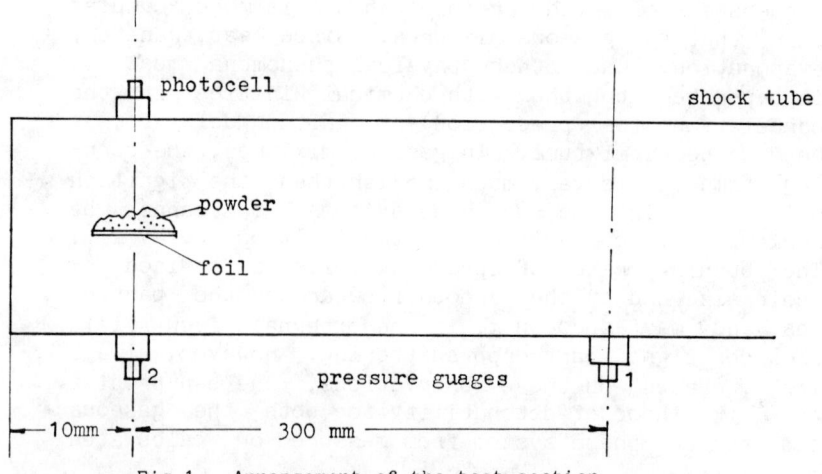

Fig.1 Arrangement of the test section.

explosion process was observed in their experiments. The process was monitored over too small distances to establish whether a self-sustaining detonation was achieved.

These studies suggest that a detonation is unlikely to occur for suspensions of particles with diameters greater than 3 µm at concentrations in air below 1 kg/m^3, even when powerful initiators are used. They demonstrate also how difficult it is to observe self-sustaining detonations in both liquid fuel sprays and dust suspensions. Large HE charges and very large two-phase mixture volumes are required to initiate unconfined detonations. Similar circumstances prevail for experiments using tubular confinement. The use of large HE charge makes experimental data difficult to interpret. The errors that may arise are illustrated by Nicholls et al. (1979).

Detonability of Dust Suspensions

The detonability of dust suspensions is assessed here by adopting mixtures of heptane, butane, and methane with air as a standard. Methane-air mixtures exemplify systems that are extremely difficult to initiate by a HE charge. Heptane-air and butane-air mixtures are examples of those having moderate detonability (Westbrook 1982). To rank various dust-air mixtures with respect to their detonability, ignition delays of the powdered materials have been measured behind reflected shock waves by applying a technique described elsewhere (Nettleton and Stirling 1967; Seeker et al. 1980). Arrangement of the test section is shown in Fig. 1.

The experiments were carried out with powders of aluminium (15-20 µm particle size), TNT (100 µm particles), napthalene, and wheat flower. Ignition delays were compared with those for kerosene drops in air and for gaseous mixtures of heptane and methane. In some of the runs with aluminium particles use was made of O_2+ZN_2 mixtures, Z varied from 0 to 0.3.

Figure 2 presents measured ignition delays of various combustible powders. The main feature of all the dust suspensions is the reduction of ignition delays when the measurement is made behind the reflected shock wave (RS) rather than behind the incident shock wave (IS) and when the oxidizer is changed from an air mixture to oxygen. Ignition delays of dust

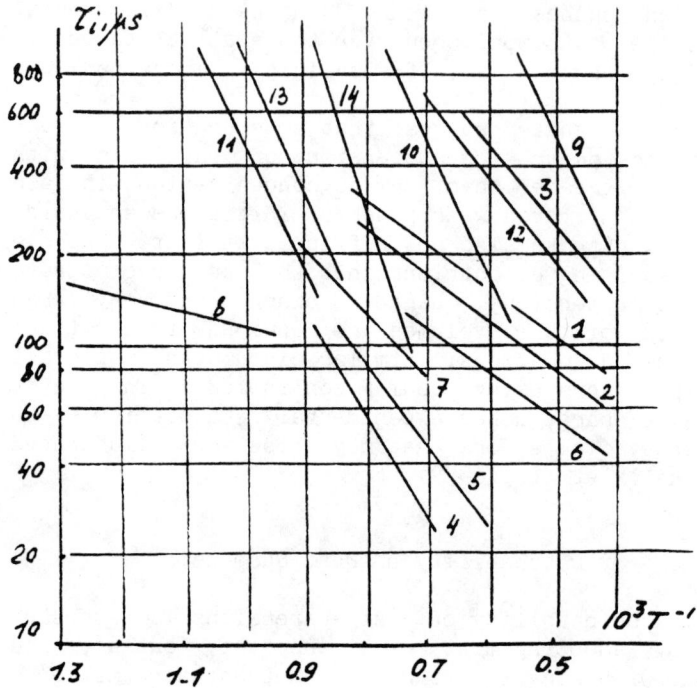

Fig.2 Arrhenius plots of ignition delays of dust suspensions:

1. TNT-air d_p=100 μm
   (reflected shock waves-RS)
2. Wheat Flower (RS) d_p=10-30 μm
3. Naphtalene (RS) d_p=30-60 μm
4. RDX-O_2 d_p=10 μm
   (incident shock waves-IS) /5/
5. Coal dust-O_2 (IS) /5/ d_p=37-55 μm
6. Mg-O_2 (IS) /5/ d_p=17 μm
7. Coal dust-O_2 (RS) /13/ d_p=17-35 μm
8. RDX-N_2 (RS) /3,4/ d_p=3 μm
9. $C_{17}H_{34}$-air (RS)   on the end plate
10. Kerosene-air (IS) d_p=10-50 μm
11. Kerosene-O_2 (RS) d_p=10-50 μm
12. Aluminium (RS) d_p=15-20 μm
13. Heptane-air (RS) /1/   gas
14. Methane-air (RS) /1/   gas

behind the reflected shock waves were found to be close to those of liquid fuel drops and much larger than for gases. The data reported by Cohen and Decker (1979a, 1979b) diverge substantially from the main body of the data and need therefore additional verification.

Detonabilities of some of the combustible mixtures are compared in Fig. 3 on the basis of measured

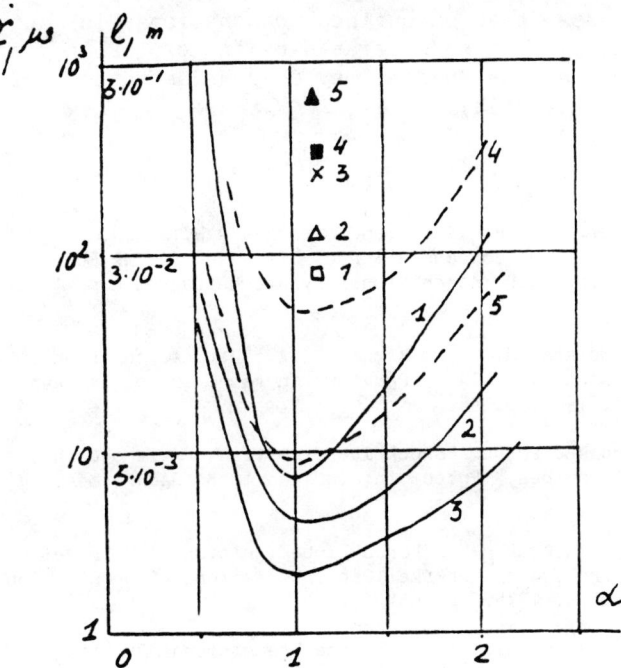

Fig.3 Ignition delays and induction zone lengths in detonation waves of different combustible mixtures:

Curves
1 - CH_4 /1/
2 - C_7H_{16} /1/
3 - C_4H_{10} /1/
4 - C_7H_{16} /12/
5 - CH_4 /12/

Points
1,2 and 3 are for decalene drops in O_2 (120, 160 and 300 μm)
4 - wheat flower /6/
5 - RDX /6/
ℓ - is air or oxygen equivalence ratio

ignition delays, following the method proposed by Westbrook (1982). Curves 1-3 of ignition delays and induction zone lengths for methane, butane, and heptane are based on experimental results of Kogarko et al. (1961) and curves 4 and 5 are based on calculations of Westbrook (1982). Points 1-3 represent the calculated induction zone lengths (Eidelman and Sichel 1981) for the kerosene-drop/oxygen system with drop diameters of 120, 160, and 300 μm. Points 4 and 5 are for 80 μm wheat particles and 80 μm RDX particles, respectively. All the dust suspensions have ignition delays that are higher than the methane-air mixtures.

This suggests that unconfined detonations in dust suspensions cannot be initiated by an explosion, so that unconfined suspensions may be considered as practically nondetonable.

References

Austing, J.S., Tulis, A.J. and Sumida, N.K., "The Direct Measurement of Concentration of Dispersed Powder Fuels in Air," Proc. 8th International Pyrotechnics Seminar, 1982, p. 45.

Bull, D.C., Mcleod, M.A. and Mizner, I.A. "Detonation of Unconfined Aerosols," AIAA: Progress in Astronautics and Aeronautics, 1980, p. 48.

Cohen, A. and Decker, L. "Shock Tube Ignition of Nitrocellulose and Nitramines," Proceedings of the 10th JANAF Meeting, 1978

Cohen, A. and Decker L. "Shock Tube Ignition of Nitrocellulose," Proceedings of the 12th International Symposium on Shock Tubes, 1979, p. 514

Eidelman, S. and Sichel, M. "The Transitional Structure of Detonation Waves in Multi-Phase Media," Comb. Sci. and Tech. 26(3-4), 1981, p. 215

Heberlein, D.C. "Measurements of Detonation Parameters in Two-Phase Heterogeneous Systems," Proceedings of the 8th International Pyrotechnics Seminar, 1982, p. 1028

Kaufman, C.W., Wolanski, P., Ural, E., and Nicholls, J.A. "Detonation Waves in Confined Dust Clouds," Proceedings of the 19th International Symposium on Combustion, The Combustion Institute, 1960, p. 535.

Kogarko, S.M., Borisov, A.A., "Ob Izmerenii Zaderzhek Vosplamenenia pri Visokikh Termperaturakh," Izv. Akademii Nauk SSR, Ser. Khim., 1960, p. 1348

Nettleton, M.A., and Stirling, R., "The Ignition of Clouds of Particles in Shock Heated Oxygen," Proc. Royal Soc. A-300, 1967, p. 62.

Nicholls, J.A., Bar-or, R., and Gabriel, Z., "Recent Experiments on Heterogeneous Detonation Waves," AIAA Paper 89-288, 1979.

Ryzhik, A.P., Makhin, V.S., and Kititsa, V.N., "Detonatsia Aerozolei Dispersnogo Magnia," Fiz. Goren. i Vzryva, No. 2, 1980, p. 78.

Seeker, W.R., Lester, T.W., and Merklin, J.F., "Shock Tube Techniques in the Study of Pulverized Coal Ignition and Burnout," Rev. Sci. Instr. 51(11), 1980, p. 1523.

Strauss, W.A., "Investigation of Detonation of Al-O_2 Mixtures," AIAA Journal 6(12), 1968, p. 1753

Westbrook, C.K., "Chemical Kinetics of Hydrocarbon Oxidation in Gaseous Detonation," Comb. Flame 46 (2), 1982, p. 191.

Chapter III. Condensed Explosives

Characterization of an Overdriven Detonation State in Nitromethane

L. Hamada,* H. N. Presles,† C. Brochet,‡ and R. Bouriannes§
Université de Poitiers, Poitiers, France
and
R. Cheret¶
Centre de l'Energie Atomique, Vaujours, Sevran, France

Abstract

For particular values of the angle of incidence, the interaction of two oblique CJ detonation waves in a condensed explosive can lead to an irregular reflection. In such a situation the Mach stem becomes an overdriven detonation wave, and it is possible to study detonation products for higher values of pressure and temperature than in the CJ state. A simple coaxial geometry has been used to generate an overdriven detonation state in nitromethane. This detonation becomes quasisteady and propagates at a velocity of 8 mm/μs in the last part of the detonation tube so that it is possible to determine its detonation characteristics by the same techniques used to determine the CJ state. The overdriven detonation pressure was observed to be 35.5 GPa, and the brightness temperature of that state is 900 K higher than the CJ state.

Introduction

Many experimental and theoretical studies have been conducted on the CJ detonation state. To check the

---

Paper presented at the 9th ICODERS, Poitiers, France, July 3-8, 1983. Copyright American Institute of Aeronautics and Astronautics, Inc. 1984. All rights reserved.

*Etudiant, Laboratoire d'Energétique et de Détonique, E.N.S.M.A.

†Chargé de Recherche, Laboratoire d'Energétique et de Détonique, E.N.S.M.A.

‡Maître de Recherche, Laboratoire d'Energétique et de Détonique, E.N.S.M.A.

§Professeur, Laboratoire d'Energétique et de Détonique, E.N.S.M.A.

¶Chef du Service Mécanique des Milieux Continus.

theoretical predictions of the Hugoniot away from the CJ state, experimental data for overdriven detonations would be useful. A few studies have been performed to determine the pressure of overdriven detonations by an impedance match method (Skidmore and Hart 1965 ; Argous et al. 1965). In the experiments reported here, a steady overdriven detonation in nitromethane was produced, and its pressure with a magnetic probe (Fisson 1976) and its brightness temperature with a fast pyrometer (Bouriannes et al. 1977) were determined. The experimental data are compared with calculated values.

Experimental Setup

The interaction of two CJ detonation waves in the interior of an explosive can lead to an irregular reflection (Skidmore and Hart 1965; Lambourn and Wright 1965; Gardner and Wackerle 1965; Argous et al. 1965).

Argous et al. (1965) have shown that the overdriven detonation wave becomes steady when the interacting waves are plane and steady. A similar result was obtained with an axisymetric incident detonation wave (Brun et al. 1978; Krishnan et al. 1981). The setup used (Fig. 1) produced such an incident wave. The internal cylindrical chamber contains nitromethane while the outer coaxial chamber contains a high-speed explosive Astrolite. Their detonation velocities are 6.3 and 8 mm/µs, respectively. The expansion of Astrolite detonation products in the nitromethane induces a steady oblique detonation wave which interacts along the axis of the setup.

Waves System Inside The Nitromethane

Nitromethane, being transparent to visible light, allows the study of interacting waves by optical methods

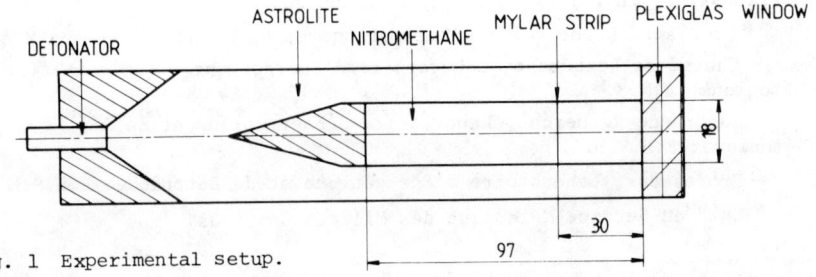

Fig. 1 Experimental setup.

(camera and pyrometer). A transparent "mylar" strip 10-mm wide and 0.1-mm thick is placed perpendicularly to the axis of the cylinder (Fig. 1) at 30 mm from the plexiglas window. As detonation waves interact with the "mylar" strip or the plexiglas window, there is a fast decrease in the light emitted by the detonation fronts which produces a dark zone on a camera record. Thus it is possible to obtain the configuration of the waves system propagating inside the nitromethane.

Figure 2 shows a butt-end streak camera record (sweeping speed is 3.2 mm/μs). A complete analysis of such a record is given by Hamada (1983). In this analysis it is assumed that the central zone (①, in Fig. 3), the most luminous portion of the record, is due to the overdriven detonation wave in the nitromethane. This wave has a varying curvature during its propagation but becomes nearly plane in its central part when impinging upon the plexiglas window. The velocity of its leading edge is about 8 mm/μs during the last 30 mm of propagation. This means that the velocity of the overdriven detonation wave is about 8 mm/μs near the plexiglas window.

Zone ② (Fig. 3) is less luminous than the central one (zone ①) and corresponds to the propagation of the incident detonation wave. The streak camera record of its interaction with the "mylar" strip and the plexiglas window shows that its slope is constant at least on the last 30 mm of the detonation tube. This wave propagates at 8 mm/μs and the normal component of its velocity is 6.3 mm/μs, which is about the CJ detonation velocity of nitromethane.

Zone ③ (Fig. 3) which spreads between zone ② and the wall of the inner tube is dimly luminous. A complete study (Hamada 1983) shows that an oblique

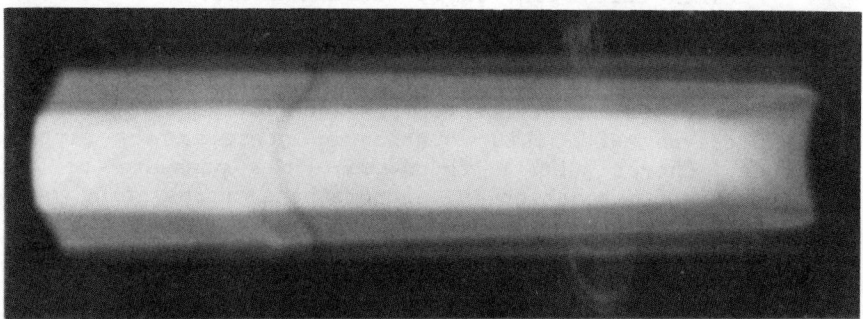

Fig. 2 Streak camera record of the light emitted by the detonation waves propagating in nitromethane.

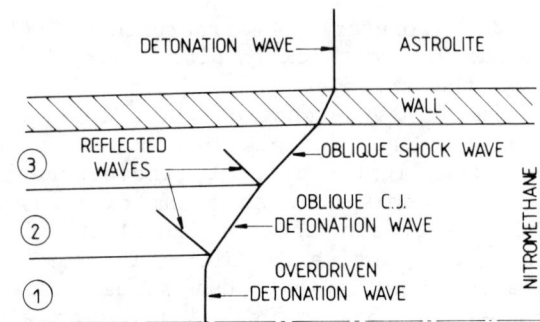

Fig. 3 Scheme of the waves system inside nitromethane.

incident shock wave followed by an oblique reflected detonation wave propagates in zone ③.

Characterization of the Overdriven Detonation State in Nitromethane Propagating at 8 mm/μs

Calculated Data

Using the equation of state of a perfect gas:

$$E = Pv/\gamma-1$$

and assuming the adiabatic exponent γ is constant along the detonation products Hugoniot, Skidmore and Hart (1965) derived a relation between the velocity D and the pressure P of an overdriven detonation:

$$(D/D_{CJ})^2 = Z^2/2Z - 1$$

where $Z = P/P_{CJ}$.

With this relation it was shown that the pressure of an overdriven detonation in nitromethane propagating at 8 mm/μs is 32.5 GPa.

The BKW equation of state has been extensively used to calculate the detonation characteristics of condensed explosives. With this equation of state in the Arpege code, Cheret (1971) found that the pressure of an overdriven detonation in nitromethane propagating at 8 mm/μs is 36 GPa and that its temperature is 1750 K higher than the CJ temperature.

Experimental Data

The particle velocity of the overdriven detonation products has been determined with the electromagnetic gage technique. It is based on the measurement of the voltage generated by a thin metallic foil moving in the flow of the detonation products through a magnetic field. The active part of the foil was 5-mm length and was placed perpendicularly to the axis of the setup at 5 mm from the plexiglas window. A particle velocity of 3.89 ± 0.2 mm/µs leading to an overdriven detonation pressure of 35.0 ± 1.9 GPa has been found. This value is in good agreement with the corresponding data calculated with the BKW equation of state (36 GPa) but is 8% higher than the value deduced from the Skidmore and Hart relation (32.5 GPa).

In the past few years, methods have been developed to measure the brightness temperature of the detonation front in condensed explosives (Urtiew 1976; Kato 1978). It has been found that the CJ detonation front in nitromethane radiates like a blackbody in the visible and near infrared range. Since the density of the products of an overdriven detonation is higher than that of the CJ state, it was assumed that the results about the emissivity of the CJ detonation front are also valid for an overdriven detonation.

The pyrometer operated in the near infrared (λ = 1062.5 nm) and was calibrated with a carbon arc (T = 3800 ± 30 K). Measurements of the brightness temperature of the CJ and the overdriven detonation fronts give 3940 ± 100 K and 4860 ± 150 K, respectively. The brightness temperature of an overdriven detonation wave propagating at 8 mm/µs in nitromethane is 900 K higher than the brightness temperature of the CJ detonation. Calculations based on the BKW equation of state give a difference of 1750 K for the equilibrium temperature between these same states.

If comparison between the calculated equilibrium temperature of the detonation products and the brightness temperature data is valid, the equation-of-state BKW gives a very large variation of the temperature with regard to the detonation velocity. From recent results about nitromethane detonation temperature (Kato et al. 1981), it seems that brightness temperature is really representative of detonation equilibrium temperature. But it must be remembered that the thermal radiation emitted from the detonation products passes through the shocked and partially decomposed explosive before being recorded by the pyrometer. Because the density of the leading shock front in the overdriven detonation is much higher than that of a CJ detonation, absorption of the radiation is likely to be higher for an overdriven detonation wave than for a CJ detonation. That could be checked using a multiwavelength pyrometer.

Conclusion

An overdriven detonation has been studied by optical and electromagnetic techniques. The measurements show that the pressure of an overdriven detonation propagating at 8 mm/µs in nitromethane is 35.5 GPa, which agrees with the calculated data based on the BKW equation of state. In spite of the lack of agreement on the variation of the calculated and measured temperature between the CJ state and the overdriven state, the results show that it is possible to obtain precise information on the detonation products Hugoniot. It is hoped that a more extensive study on overdriven detonation will lead to a better fit of the equation of state of the detonation products.

Acknowledgments

This work was partially supported by the French Atomic Energy Commission. The authors are very grateful to Y. Sarrazin for his experimental assistance and to MM. Cobat and Lemoine from the Centre d'Etudes du Ripault (CEA) for the streak camera records.

References

Argous, J. P., Peyre, C., and Thouvenin, J. (1965) Observations et études de conditions de formation d'une détonation de Mach. 4th Symposium (International) on Detonation pp. 135-141, U.S. Government Printing Office, Washington, D.C.

Bouriannes, R., Moreau, M., and Martinet, J. (1977) Un pyromètre rapide à plusieurs couleurs. Rev. Phys. App. 12, pp. 893-899.

Brun, L., Chéret, R., and Vacellier, J. (1978) Considérations sur les détonations fortes. Actes du Symposium H.D.P., pp. 1-15. Paris, France.

Chéret, R. (1971) Contribution à l'étude numerique des produits de détonation d'une substance explosive. Rapport C.E.A., R.4122.

Fisson, F. (1976) Etude des grandeurs caractéristiques de la détonation d'explosifs liquides. Thèse de Docteur ès Sciences, Université de Poitiers, France.

Gardner, S. D. and Wackerle, J. (1965) Interactions of detonation waves in condensed explosives. 4th Symposium (International) on Detonation pp. 154-155, U.S. Government Printing Office, Washington, D.C.

Hamada, L. (1983) Contribution à l'étude de la réflexion de Mach en milieu condensé inerte et réactif. Thèse de 3eme Cycle, Université de Poitiers, France.

Kato, Y. (1978) Contribution à l'étude des détonations des mélanges hétérogènes de nitrométhane et d'aluminium. Thèse de Docteur Ingénieur, Université de Poitiers, France.

Kato, Y., Bauer, P., Brochet, C., and Bouriannes, R. (1981) Brightness temperature of detonation wave in nitromethane-tetranitromethane mixtures and gaseous mixtures at a high initial pressure. 7th Symposium (International) on Detonation pp. 403-408, U.S. Governement Printing Office, Washington, D.C.

Krishnan, S., Brochet, C., and Chéret, R. (1981) Mach reflection in condensed explosives. Propellants and Explosives 6, pp. 170-172.

Lambourn, B. D. and Wright, P. W. (1965) Mach interaction of two plane detonation waves. 4th Symposium (International) on Detonation pp. 142-152, U.S. Government Printing Office, Washington, D.C.

Skidmore, I. C. and Hart, S. (1965) The equation of state of detonation products behind overdriven detonation waves in composition B. 4th Symposium (International) on Detonation pp. 47-51, U.S. Government Printing Office, Washington, D.C.

Urtiew, P. (1976) Brightness temperature of detonation wave in liquid explosives. Acta Astronautica 3, pp. 555-566.

The Effects of Grain Size on Shock Initiation Mechanisms in Hexanitrostilbene (HNS) Explosive

Robert E. Setchell* and Paul A. Taylor*
Sandia National Laboratories, Albuquerque, New Mexico

Abstract

Hexanitrostilbene (HNS) is a binderless, granular explosive prepared over wide ranges of density and grain size. In the present study, the relationship between initial grain size and shock sensitivity in this explosive was investigated experimentally and analytically. The experiments utilized planar impact techniques and velocity interferometry (VISAR) instrumentation to examine the evolution of a variety of compressive waves, including ramp waves, sustained shock waves, and short-duration shock waves. For each initial loading condition, the waveforms resulting from propagation through a fixed thickness of explosive were measured for both HNS-I (nominal grain size 5-10 μm) pressed to 92% of the theoretical maximum density and HNS-FP (nominal grain size 1-2 μm) pressed to the same density. The observed waveforms show pronounced grain-size effects for most of the loading conditions considered, with the coarse-grained material appearing far more sensitive. When compared with the results of earlier studies, the present experiments establish that a complex relationship exists between grain size and shock sensitivity in HNS. The shock amplitude, shock duration, and particular grain sizes under consideration enter into this relationship. In the analysis, the process of pore collapse is assumed to be the most important mechanism for the formation of hot spots. The corresponding microstructural parameters related to the shock sensitivity are identified. Numerical simulations of pore collapse give results that are consistent with a suggested explanation for the sensitivity behavior observed in the present experiments.

Presented at the 9th ICODERS, Poitiers, France, July 3-8, 1983. Copyright © American Institute of Aeronautics and Astronautics, Inc. 1984. All rights reserved.
*Technical Staff, Shock Wave and Explosives Physics Division.

Introduction

Numerous studies have examined the initiation processes resulting from the propagation of a plane shock wave through an explosive material. In solid heterogeneous explosives, thermal energy is localized into "hot spots" having temperatures much higher than the bulk temperature expected from plane shock heating. Ignition occurs at hot spots that retain sufficient thermal energy. The subsequent release of chemical energy results in shock amplification until detonation conditions are achieved. A number of possible mechanisms for the generation of hot spots and the coupling of released chemical energy into the shock motion have been identified. Recent summaries of possible initiation mechanisms can be found in Lee and Tarver (1980) and Setchell (1981). However, the processes that are dominant during the initiation conditions of interest have not been well established. Direct experimental investigation of the fundamental mechanisms is quite difficult, and most of the studies have been restricted to either examining the effects of material parameters on initiation sensitivity or characterizing the initiation properties of specific explosives.

Sensitivity studies on granular explosives have examined the effects of explosive composition, density, and grain size. For a given composition, the initiation processes depend critically on the shock amplitude and duration, and the heterogeneity of the explosive material. For relatively low initial shock stresses, growth at the shock front is inhibited by mechanical dissipation (Kennedy and Nunziato 1976). Reactions some distance behind the front generate compressive disturbances that overtake and slowly accelerate the shock. Longer shock durations are necessary to complete these processes. When shock stresses are relatively high the wave growth is dominated by the chemical energy release near the front, and shorter shock durations are sufficient. Decreasing the initial density of the explosive results in a greater relative energy release near the shock front. Conversely, increasing the density results in a more pronounced growth due to pressure pulses generated well behind the front, as in a homogeneous explosive (Fauquignon and Cheret 1969; Stirpe et al. 1970). Increasing the density has also been shown to increase the run distance to detonation for a given initial shock pressure (Campbell et al. 1961; Howe et al. 1976).

The effects of grain size distribution on shock sensitivity have been examined for a number of explosives. The early experiments of Campbell et al. (1961) on pressed TNT charges showed that the time required for a given initial shock to reach detonation increased with increasing grain size, for a fixed initial density. A higher density of smaller voids was apparently more efficient in producing chemical reaction, and it was suggested that a surface reaction governs the rate of energy release. In order to establish the likelihood of a grain-burning process governing the rate of energy release following ignition at hot spots, Howe et al. (1976) conducted a series of shock-initiation experiments on nitromethane with controlled inclusions and pressed TNT charges having different densities and grain sizes. Measured run distances to detonation were found to decrease with increased inclusion surface area for a fixed inclusion volume fraction in nitromethane experiments and with decreased grain size for a fixed density in TNT experiments. Critical impact energies for achieving detonation were also measured for pressed TNT samples, and lower values were found for smaller particle sizes if the shock duration was fixed. In addition, reaction thresholds for sustained-shock loading of pressed TNT charges were examined. By monitoring the free-surface velocity of explosive samples impacted by flat gas-gun projectiles, it was found that coarse-grained charges began to react at lower initial shock pressures than fine-grained charges. Recovery experiments were used by Andreev et al. (1976) to examine the effects of grain size on the degree of reaction in cast TNT. Following shock compression to approximately 0.5 GPa for $\sim$ 10 μs, the recovered sample showed far greater decomposition within a central fine-grained zone than within the surrounding coarse-grained material. Von Holle (1980) used infrared radiometry to directly examine hot-spot formation and reaction growth at the back surface of HMX samples having different characteristic grain sizes but a common density. Reaction thresholds in sustained-shock loading were found to be lower for samples with larger grains. As the initial shock pressure was increased, however, samples with smaller grains showed higher reaction rates.

If there were no additional information on grain-size effects to consider, a relatively simple description of these effects could be concluded. The sensitivity of an explosive, in terms of reaching detonation as a result of a given shock wave input, would simply increase with decreasing grain size, as evidenced in both the shorter run distances to detonation for sustained shocks and the lower critical energies achieving detonation by short-duration shocks. This description would be based on the argument that the rate of chemical energy release is governed by a surface reaction and that the smaller grains provide a larger specific surface area for this reaction. Larger grains may have reaction thresholds at lower shock pressures, but the reduced rate of energy release inhibits growth to detonation. However, several studies using different explosives have found a more complicated relation between the critical impact energy needed to achieve detonation and the initial grain size. de Longueville et al. (1976) examined the sensitivity of pressed RDX samples having a common density but two different grain-size distributions. For shock durations greater than 0.8 μs, corresponding to shock pressures below 1.6 GPa, the coarse-grained samples showed lower critical energies than the fine-grained samples. For shock durations less than 0.5 μs, corresponding to shock pressures above 2.0 GPa, the fine-grained samples showed the lower critical energies. Similarly, in a study of several TATB formulations, Honodel et al. (1982) found that samples with the smallest grains were the least sensitive when subjected to long-duration, low-amplitude shocks, whereas the same samples were the most sensitive when subjected to short-duration, high-amplitude shocks. These additional results clearly indicate that a more complete understanding of grain-size effects requires a greater knowledge of hot-spot formation and growth processes.

Hexanitrostilbene

Hexanitrostilbene (HNS) is a binderless, granular explosive prepared over wide ranges of particle size and density. Sheffield et al. (1976) established shock Hugoniot properties and formulated an equation of state for HNS at densities of 1.0-1.7 g/cm^3 (crystal density is 1.74 g/cm^3). Hayes and Mitchell (1978) and Hayes (1983) developed a model for the shock compaction and subsequent chemical decomposition of HNS. These studies utilized the results of planar impact experiments, in which pressure excursions in shocked HNS due to chemical reaction were obtained from velocity interferometry measurements at the impact interface. The more recent model (Hayes 1983) predicts that hot-spot temperatures and hot-spot decomposition times will be similar in HNS samples having different initial grain sizes if the samples are shocked to similar pressures. Local burn front velocities during subsequent grain burning are predicted to depend primarily on the initial shock pressure. Therefore, a fine-grained sample having a large specific surface area is expected to have a faster overall rate of chemical energy release than a coarse-grained sample experiencing the same initial shock pressure. This prediction is consistent with the experimental records shown for HNS samples having two different characteristic grain sizes. It is also consistent with detonation threshold measurements in HNS obtained by Schwarz (1982). Using small-diameter, thin flyers accelerated by an electrically exploding metallic foil, detonation thresholds were established using HNS samples having a common density but three different grain size distributions. For a given shock duration, the largest grain size required the highest impact pressure to ensure detonation within the sample length. Since the experimental approach required very thin flyers, however, all shock durations were less than 0.04 µs.

In the present study, grain-size effects are examined experimentally by using velocity interferometry to observe the evolution of a variety of compressive waves generated in HNS samples using planar impact techniques. By carefully repeating the impact conditions, waves of equal amplitude and duration are generated in explosive samples having a common density but different characteristic grain sizes. By observing the decay or growth of these waves after a fixed propagation distance, the relative effects of mechanical dissipation or chemical energy release due to grain size differences can be examined. A brief description of the experimental approach is given in the next section, followed by a summary of the measurements and a discussion of their implications. An analysis of grain-size effects, based on a pore-collapse mechanism for the formation of hot spots, is presented in a subsequent section. In this analysis, microstructural parameters related to shock sensitivity are identified. A possible explanation for the observed insensitivity of fine-grained HNS is then suggested. Finally, results are given from numerical simulations of pore collapse in HNS that support this explanation.

Experiments

Experimental Approach

The experiments were conducted on a compressed-gas gun capable of producing repeatable, planar impacts of projectiles 6.4 cm in diameter at velocities up to 1.2 km/s. A schematic of the experimental configuration used to generate short-pulse shock waves in HNS samples is shown in Fig. 1. At the front of the projectile a 0.5-mm-thick disk of fused silica was backed by a disk of carbon foam. The target assembly consisted of a disk of HNS surrounded by an aluminum ring, followed by two disks of fused silica having thicknesses of 1.6 and 12.7 mm. The interface between these fused silica disks included a diffuse-reflecting layer of vacuum-deposited aluminum. A double-delay laser interferometer (VISAR) system was used to measure particle velocity histories at this interface. The 1.6-mm-thick "buffer" disk of fused silica was placed between the HNS sample and the VISAR measurement interface in order to smooth spatial irregularities in the wave transmitted from the granular explosive. The elastic properties of fused silica were used in a method-of-characteristics calculation to find the corresponding one-dimensional waveform at the explosive/buffer interface. The measured velocities were also corrected for wave-generated refractive index changes in the 12.7-mm-thick fused silica window. Further details on obtaining VISAR data during impact experiments using the configuration shown in Fig. 1 can be found in Setchell (1981).

The HNS samples used in this study were pressed to a common density of 1.60 ± 0.01 g/cm^3 from two well-characterized lots of powder. One lot consisted of HNS-I grains having a specific surface area of 2.1 m^2/g. Zeiss analysis of this powder gives an average grain length of 8.8 μm and an average width of 4.0 μm. The second lot consisted of HNS-FP grains having a specific surface area of 8.2 m^2/g. Zeiss analysis of this powder gives an average grain length and width of 1.5 and 1.4 μm, respectively. The present HNS-I material is very similar to the material used

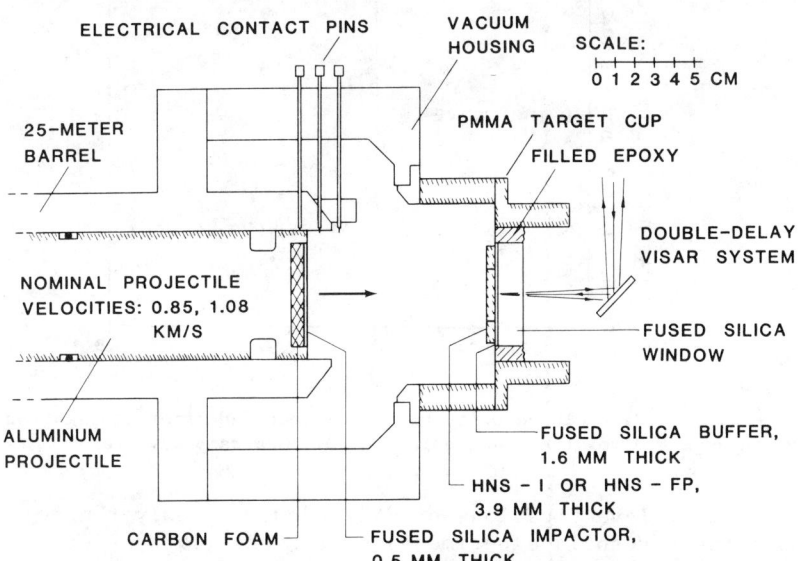

Fig. 1 Experimental configuration for short-duration shock experiments conducted on a compressed-gas gun facility.

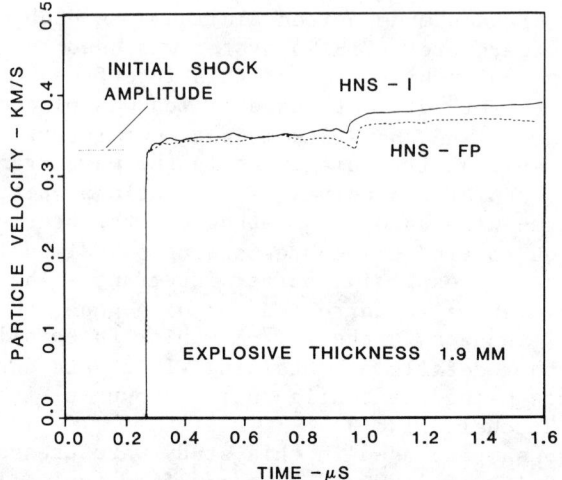

Fig. 2 Particle velocity histories recorded after propagation through 1.9 mm of explosive for the 2.5-GPa sustained-shock case (the profiles represent the waves in the fused-silica buffer at the explosive/buffer interface and have been shifted in time to facilitate comparisons).

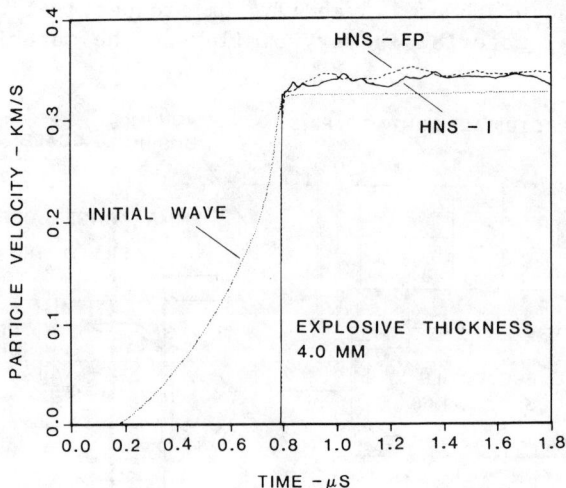

Fig. 3 Particle velocity histories recorded after propagation through 4.0 mm of explosive for the 2.4-GPa ramp wave case.

in earlier investigations of HNS. However, only the detonation threshold experiments of Schwarz (1982) used finer grains in trying to establish grain-size effects. The studies of Hayes and co-workers (1978,1983) used samples pressed from both HNS-I and HNS-II, a powder having grains approximately twice as large as those found in HNS-I.

Measured Waveforms

Experiments were conducted to examine the evolution of a variety of compressive waves, including 2.4-GPa ramp waves having an initial rise time of 0.6 μs, sustained shocks having initial shock pressures of 2.5 and 3.0 GPa, and short-pulse shocks having an initial shock duration of 0.19 μs and initial shock amplitudes of 3.0 and 4.0 GPa. Figure 2 shows the waveforms recorded for the 2.5-GPa sustained-shock case after propagation through explosive samples 1.9-mm thick. These profiles are the transmitted waves in fused silica at the explosive/buffer interface and have been shifted in time to facilitate comparisons. The "initial shock amplitude" shown in this figure is found from the impact velocity using impedance-matching calculations. The measured shock amplitudes are essentially equal for the two grain sizes and neither profile shows evidence of significant chemical energy release. The small rise approximately 0.7 μs after the wave front is probably a weak rereflected wave due to higher impedance materials at both the impact and buffer interfaces of the HNS samples. Figure 3 shows the waveforms recorded for the 2.4-GPa ramp wave case after propagation through explosive samples 4.0-mm thick. The ramp wave was generated by

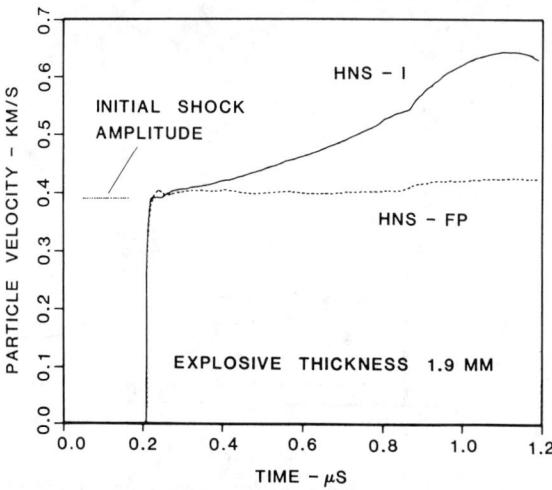

Fig. 4   Particle velocity histories recorded after propagation through 1.9 mm of explosive for the 3.0-GPa sustained-shock case.

impacting a 19-mm-thick disk of fused silica placed in front of the HNS sample (Setchell 1983). The "initial wave" shown in this figure was found from a waveform measured in an experiment conducted without an HNS sample, together with impedance-matching calculations. The measured waveforms in Fig. 3 show that the initial ramp waves have steepened to form shocks. The two shock waves have similar amplitudes and neither profile shows evidence of significant chemical reaction.

Figure 4 shows waveforms recorded after a propagation distance of 1.9 mm for the 3.0-GPa sustained-shock case. The fine-grained material again shows no evidence of chemical energy release, whereas the coarse-grained material shows a progressively rising particle velocity following the shock front. This corresponds to compressive disturbances generated by chemical energy release throughout the shock-loaded explosive material (Setchell 1981). Figure 5 shows velocity histories recorded for the 3.0-GPa short-pulse case after propagation through 3.9-mm-thick explosive samples. Also shown on this figure is a calculation of the initial wave in the explosive if this wave were observed as transmitted into the fused-silica buffer. Clearly, the amplitudes of the two waves have fallen substantially as rarefactions generated by the unloading of the thin fused-silica impactor (Fig. 1) have overtaken the shock fronts. The coarse-grained sample shows signifi-

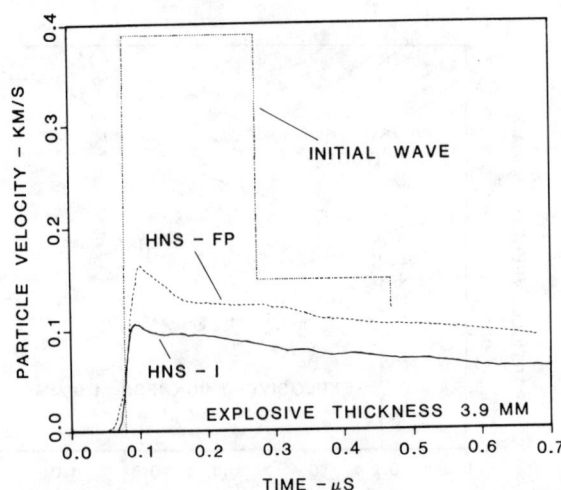

Fig. 5 Particle velocity histories recorded after propagation through 3.9 mm of explosive for the 3.0-GPa short-duration shock case.

cantly greater wave decay, suggesting a grain-size dependence in unloading from the shock-compressed state.

The most striking evidence of grain-size effects found in the present study is shown in Fig. 6. This figure shows the waveforms recorded for the 4.0-GPa short-pulse case after propagation through 3.9 mm of explosive. The fine-grained material shows a behavior similar to that observed in the previous case, with only an attenuated wave front resulting from rarefactions due to the thin impactor. The coarse-grained material shows a shock wave followed by a rapidly increasing particle velocity, with peak values greater than the amplitude of the initial wave. Substantial chemical energy release has obviously occurred, with some of this energy coupled into the wave motion. The attenuating effects of rarefactions due to the thin impactor have been overcome, and the observed wave would probably achieve detonation conditions within a few more millimeters of propagation.

Discussion

The experimental results show a pronounced grain-size effect for most of the initial loading cases considered. The 2.4-GPa ramp wave case, the 2.5-GPa sustained-shock case, and the 3.0-GPa short-duration case illustrate the response of HNS under shock compaction and unloading when hot-spot formation and growth processes are inadequate to result in sustained chemical energy release. Ramp wave

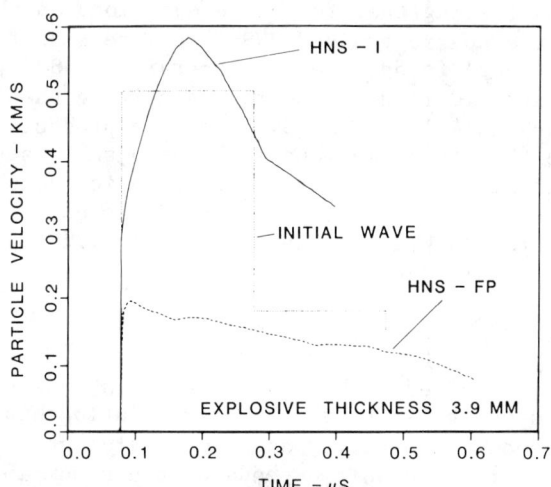

Fig. 6 Particle velocity histories recorded after propagation through 3.9 mm of explosive for the 4.0-GPa short-duration shock case.

loading has previously been found to inhibit hot-spot formation processes (Setchell 1981). In the 2.5-GPa sustained-shock case, hot-spot temperatures are possibly insufficient to produce decomposition. A 3.0-GPa shock is sufficient to produce decomposition in the coarse-grained material, as shown in Fig. 4, but rarefactions overtaking this shock will inhibit further decomposition. In fact, the wave in the coarse-grained material shows greater attenuation following these rarefactions (Fig. 5). The 3.0-GPa sustained-shock waveforms clearly show that the coarse-grained material will experience progressive decomposition at lower shock pressures than the fine-grained material. This is similar to the results found by Howe et al. (1976) for reaction thresholds in pressed TNT. Since the waves are not followed through to detonation, it is not possible to say conclusively that the run distance to detonation will be shorter for the coarse-grained material. Hayes (1983), however, showed that sustained-shock loading of HNS-I at pressures near 3 GPa resulted in more rapid chemical energy release than similar loading of HNS-II material, which has larger grains. Consequently, the relationship between sensitivity and the initial grain size under sustained-shock loading may depend on the particular grain sizes considered.

The results from the 4.0-GPa short-pulse case strongly indicate that the coarse-grained material will achieve detonation more readily than the fine-grained material. The impact energy in this case is the same for both grain sizes, indicating that for pulse durations as short as 0.19 μs the coarse-grained HNS-I is more sensitive. Since Schwarz (1982) found a very fine-grained HNS (specific surface area 10 m^2/g) to be more sensitive than HNS-I at shock durations less than 0.04 μs, the present results indicate that the relationship between shock sensitivity and grain size in HNS under short-duration shock loading depends on the shock amplitude and duration, as has been observed for RDX (de Longueville et al. 1976) and TATB (Honodel et al. 1982).

Analysis

When compared with previous studies, the present experiments establish that a complex relationship exists between shock sensitivity and the initial grain size in HNS. This relationship depends on the shock amplitude, shock duration, and particular grain sizes under consideration. An analysis examining all of the observed behavior could be quite difficult. In the present study, therefore, we restrict our consideration to the regime of relatively long-duration, low-amplitude shock loading. The sensitivity behavior observed for these conditions in other explosives suggests that if grain sizes become sufficiently small, the generation of hot spots that ignite and result in sustained reaction becomes very difficult. Winter and Field (1975) state generally that for initiation to occur, many explosives require hot spots that are at least 0.1-10 μm in diameter, the specific value depending on the hot-spot temperature and duration. Honodel et al. (1982) suggest that if hot-spot dimensions are very small, losses due to thermal conduction and hydrodynamic flow will result in lower hot-spot temperatures and slower reaction. However, they also suggest that these losses may not be important if hot-spot temperatures are sufficiently high, as in high-amplitude shock loading. The dominant mechanism for hot-spot generation may also be important in establishing grain-size effects. Hot spots formed by a spherical void collapse during compression by a planar shock should have temperatures that are insensitive to the initial void diameter (Hayes 1983), whereas hot spots formed by frictional processes could have temperatures that are proportional to the void dimensions (Howe et al. 1976).

Microstructural Parameters

In this analysis, we wish to relate the observed sensitivity behavior to differences in microstructural features found within pellets pressed from different HNS powders. Other factors that can influence sensitivity, such as pellet density, are assumed to be fixed. The earlier results of Hayes (1983), together with the current experiments, provide a particular set of observations that serve to motivate and direct the analysis. In the earlier study, samples of coarse-grained HNS-II and medium-grained HNS-I were subjected to sustained shocks having initial pressures near 3 GPa, with all samples initially at a density of 1.60 g/cm^3. The HNS-I samples, having a higher specific surface area, were found to have faster rates of chemical energy release. In the present study, samples of HNS-I and fine-grained HNS-FP at a common density of 1.60 g/cm^3 were also subjected to 3-GPa sustained shocks. Measured waveforms showed significant chemical reaction occurring in the HNS-I material, but no reaction occurring in the fine-grained material. These combined results clearly show how the different initial grain sizes affect shock initiation processes for the case of sustained, 3-GPa shock loading in HNS. Sensitivity generally increases with decreasing grain size, due to the larger area

Table 1 Microstructural dimensions of the three HNS types before and after pressing

Type	Pressing powder grain dimensions, μm x μm	Average pore size d, μm	Average pore spacing s, μm
HNS-II	167 x 51	5.3	43.0
HNS-I	8.8 x 4.0	0.6	4.9
HNS-FP	1.5 x 1.4	0.15	1.3

Table 2 Values of the microstructural parameters for the three HNS types, assuming $w = 0.2$ μm

Type	Pore size parameter δ, d/w	Pore spacing parameter λ, s/d
HNS-II	27.0	8.1
HNS-I	3.0	8.2
HNS-FP	0.8	8.7

available for a surface reaction following ignition. However, this dependence "reverses" when the grain size becomes sufficiently small. A useful initial goal, therefore, is to provide an explanation for this particular behavior. Eventually, a more complete analysis must examine the influences of shock amplitude and shock duration.

SEM photographs were taken of fragments from the interior of pellets of each HNS type pressed to a density of 1.60 g/cm^3. Careful examination of these photographs reveals a porous microstructure without well-defined grain boundaries, where the average pore size and pore spacing increase with an increase in the characteristic grain size of the original powder. Typical dimensions for these features are listed in Table 1. The listed dimensions, however, reflect only the most visible features in the SEM photographs. A distribution of much smaller voids must also be present to account for the actual porosity of the samples. In view of these observations, average values for the pore size and pore spacing are assumed to be the important dimensions characterizing the microstructure. Furthermore, hot-spot formation is assumed to occur by the mechanism of pore collapse under shock compression. Previous numerical studies of the collapse of isolated spherical pores (Mader 1965; Hayes 1981), however, found hot-spot temperatures that did not depend significantly on the initial pore size. In order to introduce such a dependence, it is necessary either to identify an additional length scale associated with the hydrodynamics of pore collapse or to consider the effects of thermal conduction from the hot-spot to the surrounding material (Hayes and Mitchell 1978). In the present analysis, we consider only the effects of an additional length scale associated with pore collapse. The relative value of the average pore diameter with respect to this length scale should reflect the anticipated behavior of hot-spot temperatures.

Studies of granular explosives experiencing ramp-wave loading (Setchell 1981) have shown that reducing the rate of compression can strongly inhibit the generation and growth of hot spots. If voids within HNS have dimensions

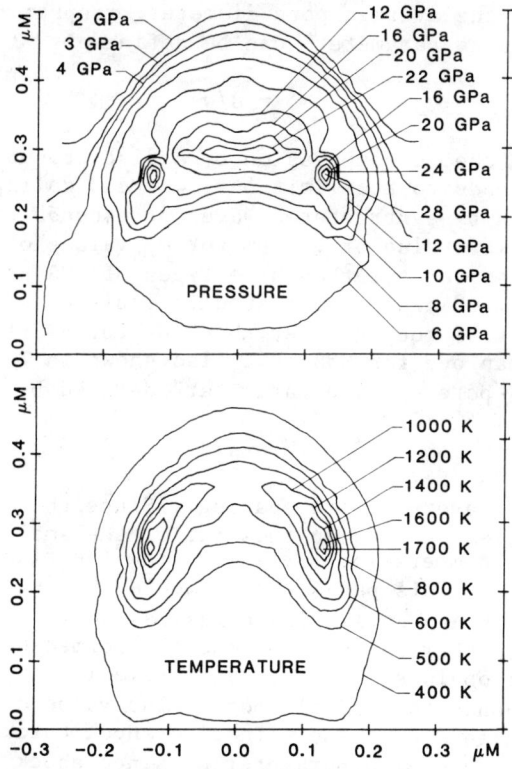

Fig. 7 Calculated pressure and temperature contours at the instant of complete pore collapse for a pore size parameter $\delta = 3.3$ (initial shock pressure is 3.0 GPa; prior to shock arrival the pore is centered at (0,0), with a radius of 0.3 μm).

that are sufficiently small, a similar effect may occur. In the present experiments, the rise times of spatially averaged shock waves were found to be less than the 4-ns response time of the VISAR instrumentation. (The VISAR instrumentation measures particle velocities that are averaged over the cross-sectional area of the reflected laser beam.) An actual rise time for a wave segment approaching an isolated pore in HNS is not known, but this time could be extended as a result of shock diffraction processes at previous pore sites. If we assume an effective rise time of 0.1 ns for a wave approaching an isolated pore, for example, then measured shock velocities predict an effective shock thickness of 0.3 μm. This value is comparable to the characteristic pore diameters listed in Table 1. Consequently, we introduce the effective shock front thickness as the additional length scale associated with the hydrodynamics of pore collapse.

If w and d are defined to be the effective shock front thickness and the average pore diameter, respectively, then a "pore size parameter" can be defined by

$$\delta = d/w$$

Void collapse due to compression by a discontinuous shock front corresponds to the limit $\delta \to \infty$, whereas collapse due to compression by a structured wave corresponds to $\delta \leqslant 1$. Using an assumed value of 0.2 μm for w, values of this pore size parameter for the three types of HNS pressed to a density of 1.60 g/cm^3 are listed in Table 2. The numbers vary from much greater than one for HNS-II to a value less than one for HNS-FP. Also shown in Table 2 are values for a "pore spacing parameter" defined by

$$\lambda = s/d$$

where s is the average pore spacing. Since the porosity of a pressed pellet depends on this ratio, the small variation in the listed numbers for λ results from the fixed density of the samples. This parameter reflects the degree of interaction between collapsing pores, as well as the degree of interaction between the subsequently formed hot spots. These effects obviously become negligible in the limit $\lambda \to \infty$, which corresponds to a single pore. The value of this parameter can influence the effective shock thickness w and therefore the pore size parameter δ, since shock diffraction effects will increase with decreasing values of λ. For present purposes, we restrict our consideration to cases where $\lambda \gg 1$ and consider only the influence of the pore size parameter δ.

Numerical Simulations

The two-dimensional Eularian wave code CSQ (Thompson 1979) was used to simulate the collapse of a symmetric, isolated pore in a cylinder of nonreactive HNS. The explosive was modeled as an isotropic, elastic/perfectly plastic material having a yield stress of 0.7 MPa. The thermal and Hugoniot properties of HNS were taken from Hayes and Mitchell (1978). The interior of the pore was assumed to be a vacuum, as in the present experimental conditions, although the numerical calculation can account for an initial atmosphere. The incident wave was assumed to be a one-dimensional, 3.0-GPa sustained shock, with the shock front thickness determined by the degree of artificial viscosity introduced into the calculations. The viscosity coefficients that were used resulted in a wave with an effective thickness of 0.2 μm, corresponding to an effective rise time of 0.06 ns.

Figure 7 shows calculated pressure and temperature contours at the instant of complete pore collapse for the case $\delta = 3.3$. The shock wave motion is from the bottom to the top of the contour plots. The volume of preferentially heated hot-spot material is approximately equal to the original pore volume[†] as found in previous pore-collapse calculations (Mader 1965; Hayes 1983). The majority of the thermal energy within the hot spot is due to irreversible plastic work occurring near the pore equator, resulting in the axisymmetric region of elevated pressure and temperature shown in the figure. Another mechanism for generating thermal energy is the impact of the shock-accelerated lower pore surface upon the stationary upper pore surface. The contribution from this mechanism becomes more significant as the pore size parameter increases.

Table 3 lists calculated values for the average thermal energy per unit volume within the hot spot, E_T, for several values of the pore size parameter δ. Unfortunately, spatial zoning limitations in the calculations have prevented a complete study of the influence of this parameter. By choosing suitable values for artificial viscosity coefficients, thereby fixing the shock thickness, the initial pore size could not be reduced to values of interest without losing spatial resolution and numerical accuracy. However, the results listed are consistent with the expectation that the hot-spot thermal energy will

---

[†]The hot-spot volume is taken to be the volume enclosed by the outermost contour whose temperature exceeds the bulk temperature expected from plane shock heating.

Table 3 Calculated values
for average hot-spot thermal energy/
volume as a function of pore size parameter

Pore size parameter δ, d/w	Average hot-spot thermal energy/volume E_T, kJ/cm^3
3.3	12.0
2.5	11.6
1.7	11.2

decrease as the value of δ decreases. This reduction in thermal energy is expected to be more rapid for values of δ less than one. At some value of $\delta \ll 1$ the hot-spot thermal energy will be insufficient for ignition to occur, preventing any subsequent surface burning. Despite the larger surface area, the explosive will have become insensitive to this shock loading condition. Such a situation appears to have been reached in the present study during experiments with the fine-grained HNS-FP.

Summary

Experimental investigations have established that a complex relationship exists between the initial grain size and shock sensitivity in HNS. An analysis of microstructural features within pressed pellets has identified the value of the average pore diameter relative to the effective shock front thickness as an important parameter related to sensitivity. The results of numerical simulations of pore collapse are consistent with the expectation that the thermal energy within the hot-spot material will decrease when the pore diameter becomes comparable to the shock thickness. This effect can account for the insensitivity of very fine-grained HNS observed in the present study.

Acknowledgment

This work was performed at Sandia National Laboratories supported by the U.S. Dept. of Energy under Contract # DE-AC04-76DP00789. The authors would like to thank Ms. Merri Lewis for her skillful assistance in preparing and conducting the experiments.

References

Andreev, S. G., Boiko, M. M., and Solov'ev, V. S. (1976) Detonation initiation in step loading. Combust., Explos., Shock Waves 12, 102.

Campbell, A. W., Davis, W. C., Ramsay, J. B., and Travis, J. R. (1961) Shock initiation of solid explosives. Phys. Fluids 4, 551.

de Longueville, Y., Fauquignon, C., and Moulard, H. (1976) Initiation of several condensed explosives by a given duration shock wave. Proceedings of the 6th Symposium (International) on Detonation, p. 105. Office of Naval Research, Arlington, Va., ACR-221.

Fauquignon, C. and Cheret, R. (1969) Generation of detonations in solid explosives. 12th Symposium (International) on Combustion, p. 745. The Combustion Institute, Pittsburgh, Pa.

Hayes, D. B. and Mitchell, D. E. (1978) A constitutive equation for the shock response of porous hexanitrostilbene (HNS) explosive. Symposium (International) on High Dynamic Pressures, p. 161. Commissariat a l'Energie Atomique, Paris.

Hayes, D. B. (1983) Shock induced hot-spot formation and subsequent decomposition in granular, porous hexanitrostilbene explosive. Shock Waves, Explosives, and Detonations, AIAA Progress in Astronautics and Aeronautics, (edited by J. R. Bowen, N. Manson, A. K. Oppenheim, and R. I. Soloukin), Vol. 87, pp. 445-467. AIAA, New York.

Honodel, C. A., Humphrey, J. R., and Weingart, R. C. (1982) Shock initiation of TATB formulations. Proceedings of the 7th Symposium (International) on Detonation, p. 425. Naval Surface Weapons Center, White Oak, Md., NSWC MP 82-334.

Howe, P., Frey, R., Taylor, B., and Boyle, V. (1976) Shock initiation and the critical energy concept. Proceedings of the 6th Symposium (International) on Detonation, p. 11. Office of Naval Research, Arlington, Va., ACR-221.

Kennedy, J. E. and Nunziato, J. W. (1976) Shock-wave evolution in a chemically reacting solid. J. Mech. Phys. Solids 24, 107.

Lee, E. L. and Tarver, C. M. (1980) Phenomenological model of shock initiation in heterogeneous explosives. Phys. Fluids 23, 2362.

Mader, C. L. (1965) Initiation of detonation by the interaction of shocks with density discontinuities. Phys. Fluids 8, 1811.

Schwarz, A. C. (1982) Shock-initiation sensitivity of hexanitrostilbene (HNS). Proceedings of the 7th Symposium (International) on Detonation, p. 1024. Naval Surface Weapons Center, White Oak, Md., NSWC MP 82-334.

Setchell, R. E. (1981) Ramp-wave initiation of granular explosives. Combust. Flame 43, 255.

Setchell, R. E.(1983) Effects of precursor waves in shock initiation of granular explosives. Combust. Flame 54, 171.

Sheffield, S. A., Mitchell, D. E., and Hayes, D. B. (1976) The equation of state and chemical kinetics for hexanitrostilbene (HNS) explosive. Proceedings of the 6th Symposium (International) on Detonation, p. 748. Office of Naval Research, Arlington, Va., ACR-221.

Stirpe, D., Johnson, J. O., and Wackerle, J. (1970) Shock initiation of XTX-8003 and pressed PETN. J. Appl. Phys. 41, 3884.

Thompson, S. L.(1979) CSQ-II - An eularian finite difference program for two-dimensional material response, Part 1: Material Sections. Report SAND77-1339, Sandia National Laboratories, Albuquerque, N. Mex.

Von Holle, W. G. (1980) Temperature measurements of shocked energetic materials by time-resolved infrared radiometry. Fast Reactions in Energetic Systems, (edited by C. Capellos and R. F. Walker), p. 485. D. Reidel Publishing Co., Boston.

Winter, R. E. and Field, J. E. (1975) The role of localized plastic flow in the impact initiation of explosives. Proc. Royal Soc. of London, Ser. A 343, 399.

Theoretical Modeling of Converging and Diverging Detonation Waves in Solid and Gaseous Explosives

C. M. Tarver* and P. A. Urtiew†
Lawrence Livermore National Laboratory
University of California, Livermore, California

Abstract

It is shown that the ignition and growth reactive flow model of shock initiation and detonation accurately calculates detonation velocity-radius data on spherically diverging and converging detonation waves in the solid explosive PBX-9404. Comparison of calculations with the ignition and growth Zeldovich-von Neumann-Doring (ZND)-type model and a simple Chapman-Jouguet (CJ) model for converging detonation waves in the solid explosive LX-17 shows that the additional momentum associated with a finite thickness reaction zone in the ZND model causes a slightly more rapid acceleration of the detonation wave front. The theory of converging detonation waves is briefly reviewed. The effects of the reaction zone structure, the variability of the heat of reaction, and the equations of state in gaseous and solid explosives on converging detonation waves are discussed. It is postulated that the reaction product states are best described as a locus of sonic states as the detonation velocity and pressure increase. In converging gaseous detonations, the heat of reaction decreases and the adiabatic exponent increases as the detonation velocity increases. In converging detonations in solid explosives, equation of state uncertainties dominate the calculations, and more experimental data on overdriven detonation waves is required for improved modeling.

Presented at the 9th ICODERS, Poitiers, France, July 3-8, 1983. This paper is declared a work of the U.S. Government and therefore is in the public domain.
*Chemist, Chemistry and Materials Science Department.
†Engineer, Chemistry and Materials Science Department.

Introduction

The ignition and growth model of reactive flow in heterogeneous solid explosives (Lee and Tarver 1980) has been applied to a great deal of experimental data on one- and two-dimensional shock initiation (Tarver and Hallquist 1981), detonation wave propagation (Hayes and Tarver, 1981), and metal acceleration (Tarver et al. 1983). Most of this experimental data has been generated using the solid explosive PBX-9404, a HMX-based explosive, and LX-17, a TATB-based explosive, whose properties are summarized by Dobratz (1981). An interesting problem which has not been previously addressed is the propagation of detonation waves in converging and diverging geometries. Although converging and diverging shock and detonation waves have long been studied theoretically by self-similar flow analyses (Stanyukovich 1960; Whitham 1974), very little quantitative experimental data is available on the detonation velocity-radius relationships for diverging or converging detonations in solid explosives. In the case of spherically diverging geometries, Green and James (1965) measured arrival times at various radii for several HMX-based explosives, including PBX-9404. Cheret and Verdes (1970) reported similar measurements at large radii for another solid explosive. In the case of spherically converging detonation waves, Cheret et al. (1981) recently published detailed arrival time data for PBX-9404 detonation waves imploding from outer radii of 14 and 15 cm to an inner radius of 1.5 cm. These data provide a means of testing the validity of the PBX-9404 detonation wave model at pressures and velocities much greater than the steady-state Chapman-Jouguet (CJ) values. Detailed comparisons between the existing spherically diverging and converging experimental data and the PBX-9404 model calculations are presented in the next section.

The ignition and growth model of a detonation wave is based on the Zel'dovich-von Neumann-Doring (ZND) model (Zel'dovich and Kompaneets 1960), which consists of a shock wave front followed by a finite width chemical reaction zone whose energy release terminates at the CJ state. This model is contrasted to a simple CJ model in which the detonation wave has an infinitely thin reaction zone and, thus, is described only by the CJ state and the subsequent expansion of the reaction products. It has been demonstrated that the momentum associated with the reaction zone must be included to calculate the fine time resolved experimental data on detonation waves currently being obtained by embedded gages (Tarver et al. 1983) and laser interferometric techniques (Sheffield et al. 1983). Reported in this paper are calculations for LX-17 which compare the velocity-radius relationship for spherically converging detonation predicted by the ignition and growth (ZND) model to that of the CJ model. This comparison shows that the finite thickness reaction zone has a definite effect on the rate of convergence of the detonation wave.

Numerical calculations of converging and diverging detonation waves in solid explosives can be accurate if numerical diffusion is kept negligible. Since the conservation equations are continuously solved, changes in equations of state and in the heat release of the explosive and its reaction products can be modeled as the pressure and propagation velocity change. This is difficult to accomplish analytically, although many self-similar solutions have been developed. The current status of self-similar solutions of converging and diverging detonations in condensed and gaseous explosives is briefly reviewed, and some observations on the chemical and physical processes involved in the acceleration of converging waves and the propagation of diverging waves are discussed in a later section of this paper.

Converging and Diverging Detonations in PBX-9404

Detailed comparisions between the ignition and growth model calculations and the available experimental data on diverging and converging PBX-9404 detonation waves are presented in this section. In the case of a spherically diverging detonation in PBX-9404, Green and James (1965) reported arrival times at three distances in PBX-9404 initiated by a 3.8-mm-diam pellet of PBX-9407 (see Dobratz 1981). Their experimental configuration was modeled exactly in the two-dimensional Lagrangian hydrodynamic code DYNA2D (Hallquist 1982), using the PBX-9404 parameters listed by Tarver and Hallquist (1981). The calculated pressure contours of the spherically diverging PBX-9404 detonation wave are shown in Fig. 1 at four different times. Green and James (1965) reported transit times of 1.115, 1.857, and 2.933 μs at thicknesses of 0.952, 1.588, and 2.540 cm, respectively. Figures 1a, 1b, and 1c show that the calculated positions of the detonation wave front at those three times are very close to the measured positions. The detonation wave has not quite reached the end of the 2.540-cm charge at time =

2.937 μs in Fig. 1c, but has completed its propagation on the charge axis in Fig. 1d, plotted at time = 2.957 μs. Therefore the propagation velocity of a spherically diverging detonation wave in PBX-9404 is very accurately calculated by the ignition and growth model.

In this and other diverging detonation wave calculations, a smooth, gradual increase in detonation

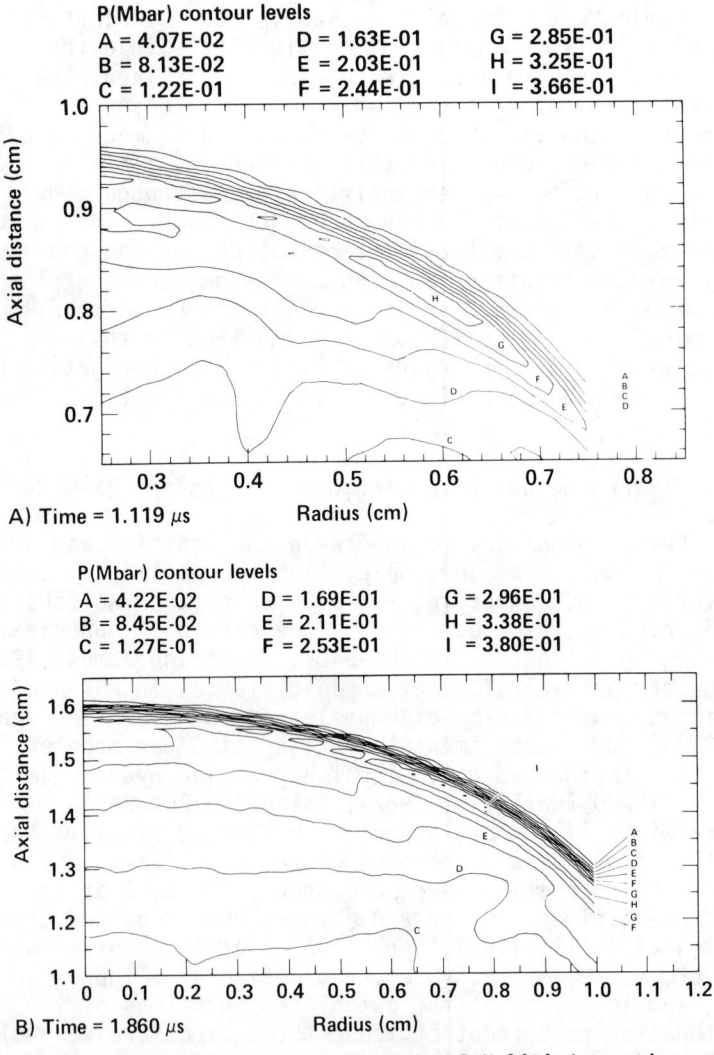

Fig. 1 Pressure contours for a diverging PBX-9404 detonation wave at various times; a) time = 1.119 μs; b) time = 1.860 μs; c) time = 2.937 μs; d) time = 2.957 μs.

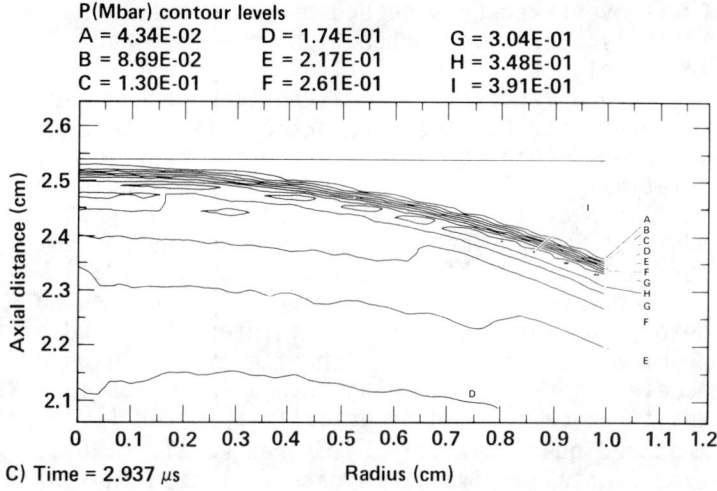

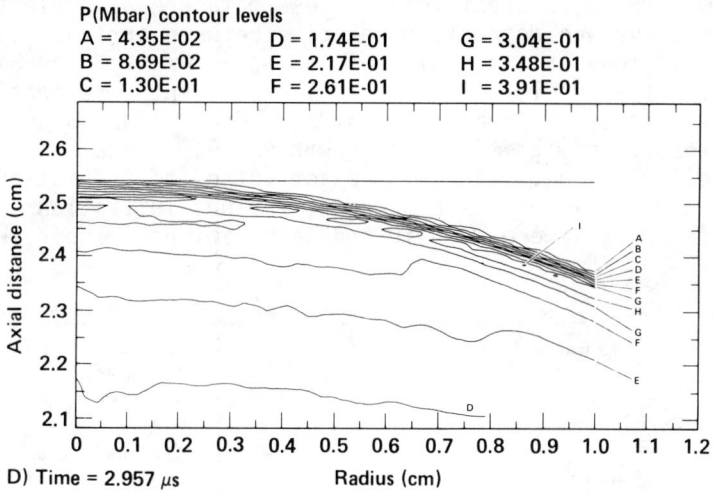

Fig. 1 cont'd Pressure contours for a diverging PBX-9404 detonation wave at various times; a) time = 1.119 µs; b) time = 1.860 µs; c) time = 2.937 µs; d) time = 2.957 µs.

velocity as the radius increases is always observed when the initiation is initially underdriven. Cheret and Verdes (1970) reported a detonation velocity overshoot in a solid explosive. It is not clear from the data whether the velocity is gradually decreasing to CJ value at large radius. The calculations demonstrated that such an overshoot is possible only when an initiating reactive flow propagation whose velocity is considerably less than

CJ is overtaken by a second pressure pulse resulting in a slightly overdriven detonation wave that rapidly decays to CJ velocity. The relatively long-lasting overshoot of Cheret and Verdes (1970) has not been predicted by the calculations, but their reported difference between the measured detonation velocity and the CJ value is relatively small. Thus, spherically diverging detonation waves appear to be adequately calculated by this model. Additional experimental data on other explosives is currently being obtained by Bahl et al. (1983).

The acceleration of a spherically converging detonation wave is a much more interesting and difficult problem than the propagation of a diverging wave. The acceleration of converging detonation waves in PBX-9404 was first calculated numerically by Mader (1979) and measured quantitatively by Cheret et al. (1981). The experimental arrival time data is listed in Tables 1 and 2 for PBX-9404 charges initiated at outer radii of 14- and 15-cm, respectively. These data are also plotted as the average detonation velocity between arrival time pin positions as a function of radius in Figs. 2 and 3 for the 14- and 15-cm outer radii experiments, respectively. Also shown in Tables 1 and 2 and Figs. 2 and 3 are the results of ignition and growth PBX-9404 model calculations. Since the exact method of initiation in the experiments is not known, various initiation techniques were used in the calculations. The results of

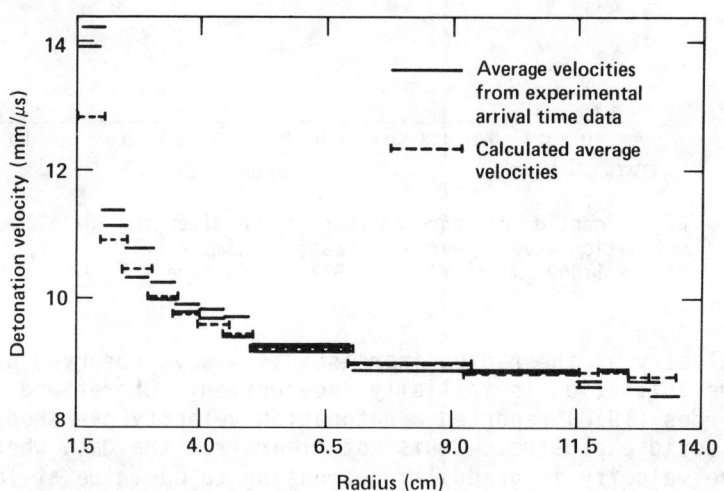

Fig. 2 Detonation velocity vs. radius for a converging PBX-9404 detonation wave from an outer radius of 14 cm.

the calculations proved to be independent of the initiation technique as long as CJ detonation was achieved before the wave reached the first arrival time pin 0.5 cm into the PBX-9404. The early experimental arrival times in Table 2 for the 15-cm outer radius charge have an excess transit time for initiation, while the 14-cm charge does not. In any case, the calculated arrival times and average detonation velocities agree very closely with the experimental data, except perhaps for the last centimeter of convergence (from 2.5 to 1.5 cm). The measured detonation velocity over the last centimeter of propagation seems to be greater than that predicted by the reactive flow model. Since the calculated pressures in this region exceed 100 GPa, one must use a large extrapolation of the unreacted explosive and reaction product equations of state of detonating PBX-9404, which are normalized to steady state CJ values of 40 GPa for the von Neumann spike and 37 GPa for the CJ state. Equation of state experiments on planar overdriven detonation waves, such as those by Kineke and West (1970), are clearly needed to check the calculated equations of state, and Green et al. (1983) have begun such an effort on PBX-9404 and other explosives. Nevertheless, the overall agreement between the experimental arrival times and the numerical results is

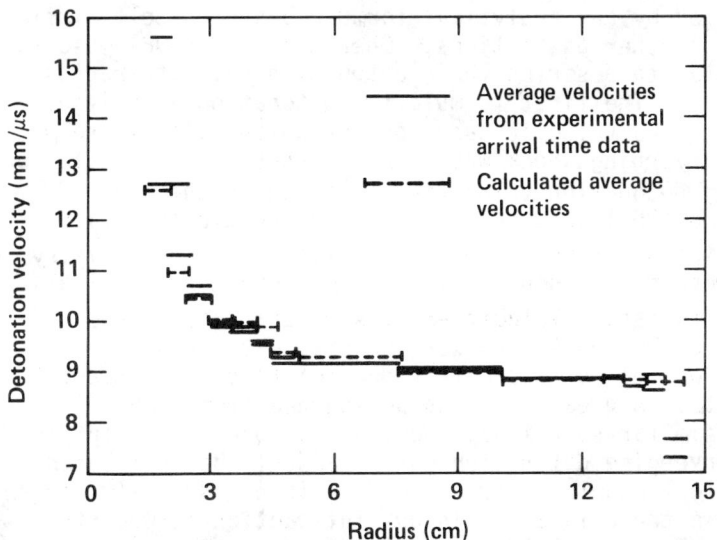

Fig. 3 Detonation velocity vs. radius for a converging PBX-9404 detonation wave from an outer radius of 15 cm.

Table 1 Experimental and calculated arrival times for a converging PBX-9404 detonation wave from an outer radius of 14 cm

Radius R', cm	Experiment time, μs	Calculated time, μs	Radius R', cm	Experiment time, μs	Calculated time, μs
13.483	0	0	13.545	a	----
13.020	0.545	0.525	13.093	0	0
12.598	1.035	1.010	12.640	0.520	0.515
12.092	1.600	1.580	12.150	1.070	1.072
11.630	2.130	2.110	11.700	1.590	1.583
9.284	4.765	4.730	9.359	4.235	4.230
7.003	7.300	7.270	7.060	6.825	6.810
5.010	9.455	9.440	5.070	8.975	8.960
4.506	9.975	9.970	4.576	9.500	9.490
3.998	10.500	10.500	4.065	10.020	10.020
3.501	11.005	11.010	3.573	10.515	10.520
3.005	11.490	11.500	3.063	11.025	11.040
2.504	11.975	11.990	2.562	11.490	11.510
2.004	12.415	12.450	2.073	11.930	11.960
1.502	12.755	12.850	1.575	12.280	12.350

a No experimental result, pin failed.

encouraging, with the model results averaging approximately 0.07-μs longer than the total experimental transit times of 13-14 μs when excess transit times for initiation are subtracted.

The agreement between the numerical calculations and the experimental data is better than that which can be obtained by any analytical formula based on self-similar flow or other assumptions. Cheret et al. (1981) use two formulas to describe their detonation velocity-radius results. The first formula for detonation velocities near the CJ value is based on the Whitham (1974) solution for converging shock waves, while the second formula for higher detonation velocities is based on the work of Damamme (1981) and depends on an adiabatic expansion coefficient, a Gruneisen coefficient, and the geometry. Parameters are then obtained from Cheret et al. (1981) velocity data. Velocity-radius relationships from spherically converging weak or strong shock wave solutions do not accurately predict these experimental results. A great deal can be learned from such self-similar solutions. However, accurate descriptions of converging detonation waves can only be formulated by numerical models because the flow is not self-similar and because there is a continuous interaction between the chemical energy release and the accelerating shock front. This point is discussed at length in a later section of this paper.

Table 2 Experimental and calculated arrival times for a converging PBX-9404 detonation wave from an outer radius of 15 cm

Radius R' cm	Experiment time, μs	Calculated time, μs	Radius R'' cm	Experiment time, μs	Calculated time, μs
14.532	0	0	14.517	0	0
14.024	0.695	0.577	14.023	0.645	0.561
13.532	1.265	1.140	13.532	1.195	1.120
13.037	1.835	1.700	13.035	1.765	1.685
12.537	2.395	2.270	12.535	2.330	2.250
10.032	5.245	5.110	10.034	5.160	5.080
7.552	7.970	7.870	7.535	7.930	7.870
5.048	10.685	10.570	5.035	10.640	10.560
4.556	11.215	11.100	4.553	11.165	11.070
4.010	11.735	11.600	4.050	11.690	11.580
3.569	12.235	12.090	3.551	a	12.090
3.068	12.735	12.580	3.051	12.700	12.590
2.571	13.200	13.060	2.547	13.180	13.070
2.0711	a	13.520	2.044	13.625	13.530
1.568	13.990	13.920	1.544	13.945	13.930

[a]No experimental result, pin failed.

Comparison of ZND and CJ Models of Converging Detonation Waves in LX-17

To determine whether the ignition and growth ZND model is necessary to describe spherically converging detonation waves, calculations were also performed with a CJ model based on an infinitely thin reaction zone followed expansion determined by the reaction product equation of state. In the case of PBX-9404, the calculated reaction zone length at CJ detonation is approximately 0.18 mm or 0.02 μs with the pressure decreasing from a von Neumann spike value of 40 GPa to a CJ value of 37 GPa. The effect of this relatively narrow reaction zone is observable in spherically converging detonation wave calculations because CJ model convergence calculations take 0.12-μs longer than the ignition and growth (ZND) model calculations to converge from 14 or 15 cm to 1.5 cm. To illustrate this effect more dramatically, the results of spherically converging detonation wave calculations using the ZND and CJ models for LX-17, which has a much thicker reaction zone than PBX-9404, are shown in Figs. 4 and 5 for the 14- and 15-cm outer radii geometries, respectively. The LX-17 model reaction zone length for a CJ detonation is approximately 2 mm or 0.25 μs with the pressure decreasing from the spike value of 33.7 GPa to the CJ value of 27.5 GPa, as discussed by Tarver and Hallquist

(1981). The detonation velocity-radius curves for the ignition and growth ZND model in Figs. 4 and 5 predict higher detonation velocities and thus greater wave acceleration than the CJ model due to the extra momentum associated with the reaction zone. The calculated arrival times at 1.5 cm predicted by the ZND model are 15.50 and 16.72 μs for the 14- and 15-cm outer radii, respectively, while the calculated arrival times predicted by the CJ model are 15.72 and 16.94 μs for the 14- and 15-cm cases, respectively. Therefore the CJ model predicts an additional 0.22 μs for detonation wave transit compared to the ZND model for LX-17 in these calculations. Unfortunately, no converging detonation wave velocity data exists for LX-17 to check the overall accuracy of these calculations. The effect of the reaction zone width on the calculated acceleration of a spherically converging detonation wave depends in a rather complex manner on all of the assumptions made about the equations of state and the reaction rates for the particular explosive.

Theory of Converging Detonation Waves

As an offshoot of self-similar blast wave theory, self-similar detonation waves have been studied by Taylor (1950), Oppenheim et al. (1972), and Barenblatt et al.

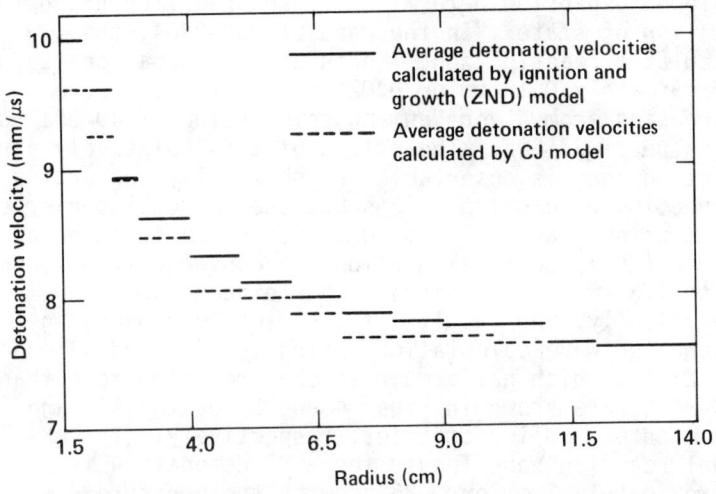

Fig. 4 Detonation velocity vs. radius for a converging LX-17 detonation wave from an outer radius of 14 cm.

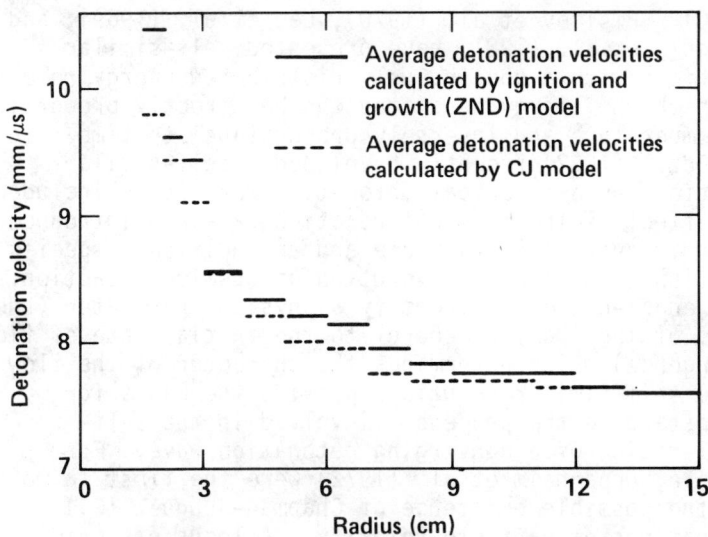

Fig. 5 Detonation velocity vs. radius for a converging LX-17 detonation wave from an outer radius of 15 cm.

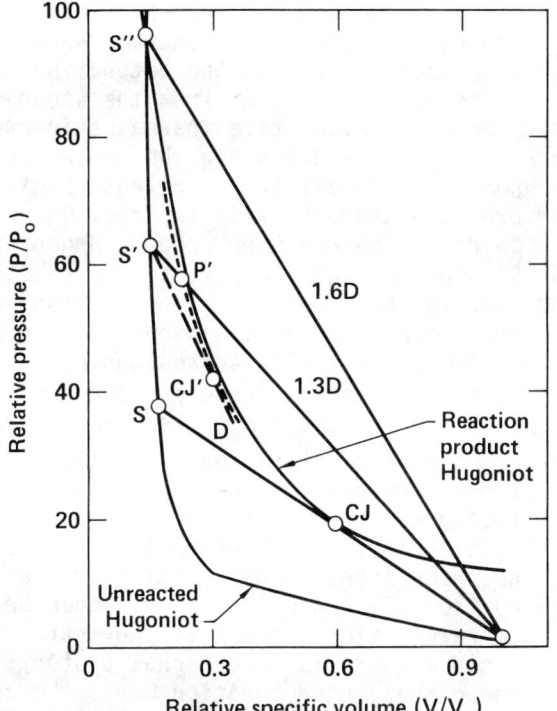

Fig. 6 Pressure-specific volume states attained in a converging gaseous detonation wave.

(1980). Bisimov et al. (1970), Lee et al. (1969), and Guirguis et al. (1981) have presented self-similar solutions with a continuously distributed energy release in which the energy deposited can be directly proportional to temperature and inversely proportional to time. Logan and Bdzil (1982) recently developed a self-similar solution for a spherical detonation wave which included the effect of the chemical reaction rate and introduced an additional state variable and an additional species equation to govern the evolution of chemical reaction. This enabled them to identify a physical parameter (the ratio of the chemical energy to the initial rate of flow divergence) which determines the character of the flow. These self-similar solutions provide the basis for understanding the processes involved in the self-acceleration of a converging detonation wave. For example, Oppenheim et al. (1972) were the first to point out the possible existence of Chapman-Jouguet (CJ) detonations of variable velocity. A locus of sonic, CJ-like states may closely approximate the states attained at the rear of the chemical reaction zone in a converging detonation wave. Consider the hypothetical pressure-specific volume and pressure-particle velocity curves for a converging gaseous detonation wave shown in Figs. 6 and 7, respectively. As the detonation wave converges, its velocity increases from the steady-state CJ velocity D and its spike state increases in pressure from state S in Figs. 6 and 7 along the unreacted Hugoniot. When the chemical energy release begins from this higher pressure state (say S' in Figs. 6 and 7), the reaction proceeds to the reaction product Hugoniot curve centered on S' (the dotted line next to the reaction product Hugoniot for steady-state CJ detonation in Figs. 6 and 7). For a steady detonation overdriven by an external piston with a velocity corresponding to P' in Fig. 7, the chemical reaction proceeds down the 1.3D Rayleigh line and intersects the dashed product Hugoniot at P'. For a converging detonation wave, the reaction product gases previously formed behind the shock front where the velocity was less than 1.3D have a lower particle velocity than point P' which is subsonic with respect to the shock front. The reaction products created behind the 1.3D shock front would then be subjected to a rarefaction process if they attained point P'. However, since there is no external piston overdriving the converging detonation wave, the chemical energy release beginning at S' does not have to intersect the product Hugoniot at P'. The product state can be

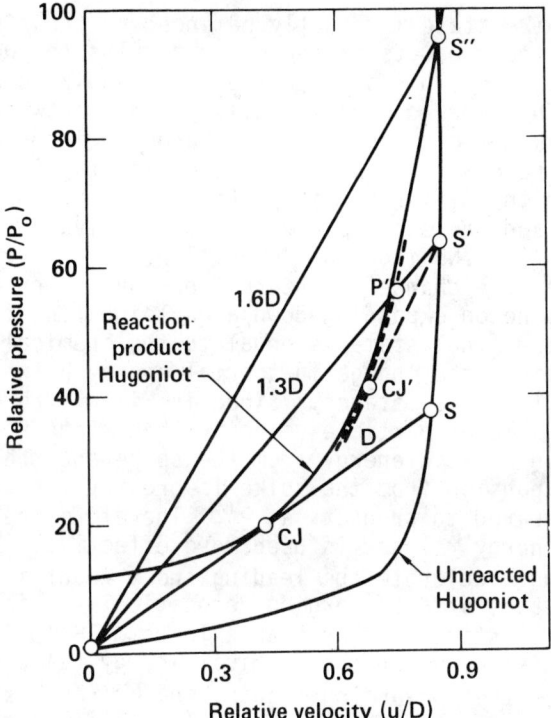

Fig. 7 Pressure-particle velocity states attained in a converging gaseous detonation wave.

represented by any point on the dotted Hugoniot curve in Figs 6 and 7 between P' and the sonic point CJ'. Only reaction products at sonic point CJ' would not be subject to rarefaction processes from the rear. Therefore a locus of CJ' states seems to be the most likely path for the reaction product states in a converging detonation wave. Although, strictly speaking, the CJ state and ZND model are steady-state detonation concepts, the concept of sonic CJ' states describing the end of reaction at each point in the converging flow is a useful model for understanding the mechanism by which the chemical energy release accelerates the shock wave front.

Not only would the attainment of a locus of CJ' states isolate the converging detonation wave from the following product expansion process, it would be a more efficient use of the chemical energy release in accelerating the leading shock wave. Cowperthwaite and Adams (1967) showed that, for a ZND model detonation wave, the work done by the shock front to attain the von

Neumann spike state is exactly balanced by the kinetic energy of the products at the CJ state plus the work done by the reacting fluid between the spike and product states. This work done is equally divided between the change in kinetic energy and the change in internal energy. The change in kinetic energy is of course related to the difference in particle velocity between the spike and the sonic state. In the derivation of the nonequilibrium ZND model of detonation, Tarver (1982) showed that the change in internal energy or equivalently, the work done on expanding down a Rayleigh line from the spike to the final state is equal to the chemical energy released minus the change in thermal energy between the spike and the final states. Since in Fig. 7 the CJ' state has a lower particle velocity than state P', the decrease in kinetic energy from the spike and the work done in expansion from the spike is greater for products at CJ' compared to products at P'. Therefore the chemical energy release is used more effectively to support and accelerate the leading shock front if the product states of a converging detonation wave follow a locus of CJ' states rather than a locus of P' points. The mechanism by which the chemical energy release accelerates the converging leading shock front is most likely the same as that which sustains planar detonation i.e., the amplification of pressure wavelets in the reaction zone by the vibrationally excited reaction products rapidly produced by chain reactions, as discussed by Tarver (1982).

Although the proper description of the chemical reaction zone is essential to model converging detonation, several other factors influence the process. These factors also limit the usefulness of the existing self-similar solutions. In the case of gaseous explosives, the perfect gas law can be used for the equations of state, but, as the detonation velocity increases, the chemical equilibrium changes. The heat of reaction decreases and γ, the ratio of heat capacities, increases as more atoms and radicals are formed at temperatures above the CJ temperature. At velocities of 1.6-1.8D for most gaseous explosives at an initial pressure of 1 atm, the chemical reaction becomes endothermic rather than exothermic. This situation is denoted by point S'' at 1.6D in Figs. 6 and 7 at which the Hugoniot curves cross. Therefore the converging gaseous detonation wave would actually be accelerated by a variable (decreasing) amount of energy release and a variable (increasing) value of γ. Eventually the converging detonation becomes endothermic rather than exothermic and the rate of acceleration of the detonation wave decreases accordingly. Experimental demonstration of these effects is quite difficult, because a very large volume of gaseous explosive would have to be simultaneously initiated at its outer radius to create the converging wave.

In the case of condensed explosives, the overall heat of reaction may also change as the pressure and temperature increase above the CJ values, but to a lesser extent than in gases since stable products such as CO, N_2, CO_2, H_2O, and C are still being formed from metastable large molecules. Furthermore, the equation-of-state problem becomes quite complex at pressures and temperatures in excess of the CJ values. However, there are experimental measurements that can be made to shed light on the problem. The shock Hugoniot curves for many of the individual product species have been measured and theoretically modeled at high pressures (see, for example, Mitchell and Nellis 1982 and Ree 1982 on the water equation of state). As previously mentioned, planar overdriven detonation wave experiments could yield mixture equation of state data for the high pressure reaction products. Calculations using reactive flow models based on improved equations of state could then be compared to experimental converging detonation wave propagation data as an extension of the process to understand converging and diverging detonation wave propagation.

Conclusions

The results presented in this paper suggest that spherically diverging and converging detonation waves in the solid explosive PBX-9404 can be accurately modeled by the ignition and growth (ZND-type) model of detonation. The extrapolation of the equations of state and reaction rates to pressures far above the steady-state CJ values induces uncertainties into the calculations that can perhaps be reduced by models based on more experimental data on planar overdriven detonation waves. Results suggest that a ZND-type model yields a better prediction for a converging detonation than a CJ-type model because the additional momentum associated with the finite thickness reaction zone does effect the acceleration of the shock front. The acceleration of the shock front is also effected by the details of the chemical energy release process such as the overall heat of reaction and the equilibrium state of the reaction products, particularly for converging detonation waves in gaseous explosives. Converging detonation waves in gaseous explosives are shown to actually be variable energy release waves since their overall heat of reaction decreases as the convergence continues. Additional experimental and theoretical research is suggested to discover more of the details of the interesting processes of converging and diverging detonation waves. The initiation and propagation of gaseous diverging detonation waves, as reviewed by Lee et al. (1977), is another fascinating area, particularly the mechanism of the continuous generation of additional cells in the cellular detonation wave front as it diverges into a larger volume.

Acknowledgments

The authors would like to thank G. L. Nutt of Lawrence Livermore National Laboratory and M. Cowperthwaite of SRI International for many interesting discussions of this subject.

This work was performed under the auspices of the U.S. Department of Energy by the Lawrence Livermore National Laboratory under Contract No. W-7405-ENG-48.

References

Bahl, K., Lee, R. S., and Weingart, R. C. (1983) Velocity of spherically diverging detonation waves in RX-26-AF, LX-17, and LX-10. (To be published in the Proceedings of the American Physical Society Conference on Shock Waves in Condensed Matter, Santa Fe, N. Mex, July 1983.)

Barenblatt, G. I., Guirguis, R. H., Kamel, M. M., Kuhl, A. L., Oppenheim, A. K., and Zel'dovich, Y. B. (1980) Self-similar explosion waves of variable energy at the front. J. Fluid Mech. 99, 841-858.

Bisimov, E., Korobeinikov, V. P., and Levin, V. A. (1970) Strong explosion in combustible gaseous mixture. Astronautica Acta 15, 267-274.

Cheret, R., Chasisse, F., and Zoe, J. (1981) Some results on the converging spherical detonation in a solid explosive. Seventh Symposium (International) on Detonation, pp. 602-607 Naval Surface Weapons Center, Annapolis, MD, NSWC MP 82-334.

Cheret, R. and Verdes, G. (1970) Divergent spherical detonation waves in a solid explosive. Fifth Symposium (International) on Detonation, ACR-184, pp. 31-39. Office of Naval Research, Pasadena, CA.

Cowperthwaite, M. and Adams, G. K. (1967) Explicit solutions for steady-and unsteady-state propagation of reactive shocks at constant velocity. Eleventh Symposium (International) on Combustion, pp. 703-711. The Combustion Institute, Pittsburgh, PA.

Damamme, G. (1981) The Chapman-Jouguet detonation and its accelerationfor a perfect fluid without conduction. Seventh Symposium (International) on Detonation, pp. 634-640. NSWC-MP 82-334, Naval Surface Weapons Center, Annapolis, MD.

Dobratz, B. M. (1981) LLNL explosives handbook. Lawrence Livermore National Laboratory Report UCRL-52997.

Green, L. G. and James, E. Jr. (1965) Radius of curvature effect on detonation velocity. Fourth Symposium (International) on Detonation, ACR-126, pp. 86-91. Office of Naval Research White Oak, MD.

Green, L. G., Lee, E. L., Mitchell, A. and Van Thiel, M. (1983) Detonation product EOS: The pressure region above Chapman Jouguet. (To be published in the Proceedings of the American Physical Society Conference on Shock Waves in Condensed Matter, Santa Fe, N. Mex. July 1983.

Guirguis, R. H., Oppenheim, A. K., and Kamel, M. M. (1981) Self-similar blast waves supported by variable energy deposition in the flow field. Progress in Astronautics and Aeronautics Vol. 75, pp. 178-192. American Institute of Aeronautic and Astronautics, New York, NY.

Hallquist, J. O. (1982) User's Manual for DYNA2D. Lawrence Livermore National Laboratory Report UCID-18756 Rev. 1.

Hayes, B. and Tarver, C. M. (1981) Interpolation of detonation parameters from experimental particle velocity records. Seventh Symposium (International) on Detonation, pp. 1029-1039. NSWC MP 82-334, Naval Surface Weapons Center, Annapolis, MD.

Kineke, J. H. Jr. and West, C. E. Jr. (1970) Shocked states of four overdriven explosives. Fifth Symposium (International) on Detonation, pp. 533-543. ACR-184, Office of Naval Research, Pasadena, CA.

Lee, E. L. and Tarver, C. M. (1980) Phenomenological model of shock initiation and heterogeneous explosives. Phys. Fluids 23, pp. 2362-2372.

Lee, J. H. (1977) Initiation of gaseous detonation. Am. Rev. Phys. Chem. 25, pp. 75-104.

Lee, J. H., Soloukhin, R. I., and Oppenheim, A. K. (1969) Current views on gaseous detonation. Astronautica Acta 14, pp. 565-584.

Logan, J. D. and Bdzil, J. B. (1982) Self-similar solution of the spherical detonation problems. Comb. Flame 46, pp. 253-269.

Mader, C. L. (1979) Numerical Modeling of Detonations. p. 122. University of California Press, Berkeley, CA.

Mitchell, A. C. and Nellis, W. J. (1982) Equation of state and electrical conductivity of water and ammonia shocked to the 100 GPa (1 Mbar) pressure range. J. Chem. Phys. 76, pp. 6273-6281.

Oppenheim, A. K., Kuhl, A. L. and Kamel, M. M. (1972). On Self-similar blast waves headed by the Chapman-Jouguet detonation. J. Fluid Mech. 55,pp. 257-270.

Ree, F. H. (1982) Molecular interaction of dense water at high temperatures. J. Chem. Phys. 76, pp. 6287-6302.

Sheffield, S. A., Bloomquist, D. D. and Tarver, C. M. (1984) Subnanosecond measurements of detonation fronts in solid high explosives. (Paper to appear in April 15, 1984 issue of J. Chem. Phys.)

Stanyukovich, K. P. (1960) Unsteady Motion in Continuous Media. Pergamon Press, New York.

Tarver, C. M. (1982) Chemical energy release in detonation waves I, II, and III. Combust. Flame 46, pps. 111-133, 135-156, 157-176.

Tarver, C. M. and Hallquist, J. O. (1981) Modeling two-dimensional shock initiation and detonation wave phenomena in PBX-9404 and LX-17. Seventh Symposium (International) on Detonation, pp. 488-497. NSWC MP 82-334, Naval Surface Weapons Center, Annapolis, MD.

Tarver, C. M., Parker, N. L., Palmer, H. G., Hayes, B. and Erickson, L. M. (1983) Reactive flow modeling of recent embedded gauge and metal acceleration experiments on detonating PBX-9404 and LX-17. J. Energetic Materials, 1, 213-250 (1983).

Taylor, G. I. (1950) The dynamics of the combustion products behind plane and sperical detonation fronts in explosives. Proc. Royal Society 200, 235-247.

Zel'dovich, Y. B. and Kompaneets, A. S. (1960) Theory of Detonation, p. 284. Academic Press, New York.

Model Similarity Solutions for Shock Initiation Containing a Realistic Constitutive Relationship for Condensed Explosive

Michael Cowperthwaite*
SRI International, Menlo Park, California

Abstract

A procedure is formulated for incorporating a realistic description of condensed explosives into similarity solutions for unsteady Zel'dovich-von Neumann-Doering waves. Differential geometry methods are used to find the infinitesimal generator A of the Lie group admitted by the governing equations. The similarity solution is constructed in terms of first integrals of A. In contrast to classical treatments, all the flow equations are written as conservation laws and the Rankine-Hugoniot shock boundary conditions are written as first integrals of the flow equations so that the Lie group is not restricted by the material properties of the explosive. A realistic constitutive relationship for reacting explosive can therefore be incorporated into the similarity solution and used to calculate the global decomposition rate of the explosive as it undergoes the invariant hydrodynamic flow admitted by the group. The construction of a model similarity solution for a type of shock initiation observed in condensed explosives is based on properties of Lagrange particle velocity histories recorded in shocked PBX9404. Procedures used to assign values of group, flow, and constitutive parameters to the solution in a realistic and thermodynamically consistent manner are given, and Lagrange plots for a particular solution are presented to exemplify properties of the initiating flow.

Presented at the 9th ICODERS, Poitiers, France, July 3-8, 1983. Copyright © 1984 by the American Institute of Aeronautics and Astronautics, Inc. All rights reserved.
*Senior Chemical Physicist, Poulter Laboratory.

Introduction

The assumption that an explosive and its reaction products are governed by polytropic equations of state with the same polytropic index has been used widely in mathematical treatments of Zel'dovich-von Neumann-Doering (ZND) waves for the sake of tractibility (Zel'dovich and Kompaneets 1960; Fickett and Davis 1979). This polytropic assumption for the explosive also makes possible the construction of similarity solutions for unsteady ZND waves when the initial pressure in the explosive is ignored (Cowperthwaite and Adams 1967; Sternberg 1970; Cowperthwaite 1978; Logan and Bdzil 1982). These similarity solutions are important for understanding certain aspects of shock-induced reactive flow, but their use in testing hydrocodes and reactive flow-Lagrange analyses (RFLA) (Cowperthwaite 1981) for condensed explosives is limited because the polytropic assumption for the explosive is only valid for gases. It is therefore important to formulate a procedure for incorporating a more realistic description of condensed explosives into similarity solutions for ZND waves. Such a procedure and a particular solution modeling a type of initiation observed in PBX9404 and cast TNT are presented in this paper.

Construction of Similarity Solution

The basic procedure detailed here is the same as the one used previously to construct model solutions for polytropic explosive (Cowperthwaite 1978). Methods of differential geometry are used to find the infinitesimal generator A of the Lie group admitted by the governing equations and the similarity solution is constructed in terms of first integrals of A. This construction differs from the previous one in that it excludes the constitutive relationship of the explosive from the governing equations. All the flow equations are written as conservation laws and the Rankine-Hugoniot (RH) shock boundary conditions are written as their first integrals so that the Lie group and its associated similarity solution are not constrained by the material properties of the explosive. A realistic constitutive relationship for reacting explosive can thus be incorporated into the similarity solution and methods used in RFLA can be used to calculate the global decomposition rate of the explosive as it undergoes the invariant hydrodynamic flow admitted by the group.

ZND waves are based on the premise that the shock is a nonreactive discontinuity and that the one-dimensional flow induced by the shock is adiabatic and inviscid. We let t =

time, h = Lagrange distance, D = shock velocity, u = particle velocity, v = specific volume, p = pressure, and e = specific internal energy; o is used as a superscript to denote the initial unshocked condition, i as a subscript to denote the initial shocked condition, and H as a subscript or superscript to denote the shocked state.

In a geometric approach, all the variables are considered to be local coordinates on a manifold M, and the flow equations and the shock boundary conditions on M, respectively, are expressed by a set of exterior differential two-forms Ω_i (i = 1, 2, 3), and by a one-form w and a set of functions F_i (i = 1, 2, 3). A solution to a flow problem is thought of as a pair of differentiable maps (ψ_F, ψ_H). The solution map ψ_F maps M_2, a two-dimensional integral submanifold of the set of forms Ω_i, into M. This means that the set of two-forms $\psi_F^* \Omega_i$ mapped onto M_2 by the induced map ψ_F^* vanishes ($\psi_F^* \Omega_i = 0$) to give the differential flow equations on M_2. The map ψ_H maps M_1, a one-dimensional submanifold into M_2. The form $\psi_H^* \circ \psi_F^* w$ mapped into M_1 by the induced map ψ_H^* vanishes ($\psi_H^* \circ \psi_F^* w = 0$) to give a differential equation for the shock velocity, and the set of functions $\psi_H^* \circ \psi_F^* F_i$ mapped onto M_1 by the induced map ψ_H^* vanishes ($\psi_H^* \circ \psi_F^* F_i = 0$) to give the Rankine-Hugoniot jump conditions.

To exclude the constitutive relationship of the explosive from the governing equations expressed in (h, t, D, u, v, p, e) coordinates, the set of differential two-forms Ω_i is written as

$$\Omega_1 = dv \wedge dh + v^o du \wedge dt \qquad (1a)$$

$$\Omega_2 = du \wedge dh - v^o dp \wedge dt \qquad (1b)$$

$$\Omega_3 = de \wedge dh + p dv \wedge dh \qquad (1c)$$

the one form w as

$$w = dh - D dt \qquad (2)$$

and the functions F_i as

$$F_1 = v^o D - v(D-u) \qquad (3a)$$

$$F_2 = u^2 - p(v^o - v) \qquad (3b)$$

$$F_3 = 2(e-e^o) - p(v^o - v) \qquad (3c)$$

In this case the integral submanifold M_2 is written as [h, t, $\psi_F^* u = \tilde{u}$ (h, t), $\psi_F^* v = \tilde{v}$ (h,t), $\psi_F^* p = \tilde{p}$ (h,t),

$\psi_H^* e = \tilde{e}(h,t)]$, and the equations $\psi_f^* \Omega_i = 0$ ($i = 1, 2, 3$) reduce to the differential flow equations in Lagrange (t, h) coordinates. When h is considered as the coordinate of M_1, $\psi_H^* t = \tilde{t}_H(h)$ is the equation of the shock path, $\psi_{H}^*{}_0 \psi_f^* D = \tilde{D}(h)$ is the equation for the shock velocity, and $\psi_H^* \tilde{u} = \tilde{u}_H(h)$, $\psi_H^* \tilde{v} = \tilde{v}_H(h)$, $\psi_H^* \tilde{p} = \tilde{p}_H(h)$, and $\psi_H^* \tilde{e} = \tilde{e}_H(h)$ are the equations for the particle velocity, specific volume, pressure, and specific internal energy along the shock path, respectively, that satisfy the Rankine-Hugoniot relations $\psi_{H}^*{}_0 \psi_f^* F_i = 0$ ($i = 1, 2, 3$).

The tangent vector A is written as

$$A = \bar{H}\frac{\partial}{\partial h} + \bar{T}\frac{\partial}{\partial t} + \bar{D}\frac{\partial}{\partial D} + \bar{U}\frac{\partial}{\partial u} + \bar{V}\frac{\partial}{\partial v} + \bar{P}\frac{\partial}{\partial p} + \bar{E}\frac{\partial}{\partial e} \quad (4)$$

and, the Lie derivative with respect to A is written as L_A. The invariant conditions

$$L_A \Omega_j = A_{ji}\Omega_i \quad i = 1,2,3 \quad (5a)$$

$$L_A w = Bw \quad (5b)$$

$$L_A F_j = B_{ji}F_i \quad i = 1,2,3 \quad (5c)$$

are then solved to find $\bar{H}$, $\bar{T}$, $\bar{D}$, $\bar{U}$, $\bar{V}$, $\bar{P}$, and $\bar{E}$ and thereby determine the infinitesimal generator of the Lie group admitted by the flow equations and the shock-boundary conditions. Solving Eq. (5) leads to the following set of generators:

$$\bar{H} = bh+b_1, \quad \bar{T} = at+a_1, \quad \bar{D} = (b-a)D, \quad \bar{U} = cu$$
$$\bar{V} = \delta(v-v^o), \quad \bar{P} = \varepsilon p, \quad \bar{E} = f(e-e^o) \quad (6)$$

where the group parameters are related by the equations, $2(b-a) = \varepsilon - \delta$, and $2c = f = \varepsilon + \delta$.

A similarity solution is constructed in terms of first integrals of A and is thus invariant under the group generated by integrating A. First integrals of A satisfy the equation $L_A f = \langle A/df \rangle = 0$, and are obtained by integrating the system S(A) [$dt/dr = \bar{T}$, ..., $de/dr = \bar{E}$]. If S(A) is formally integrated so that [t = $\tilde{T}$ (r, t^i..., e^i), ..., e = $\tilde{E}$ (r, t^i, ..., e^i)] and [t = t^i, ..., e = e^i] when r = 0, then [t^i, ..., e^i] are first integrals of A and any function $f(t^i, ..., e^i)$ is also a first integral of A. The functions obtained by integrating S(A) are used to construct a set of first integrals f_i(i = 1, ... 6) and a mapping

$\phi: [h,t, \ldots, e) \to (h, f_1, \ldots, f_6)]$ upon which the construction of the similarity solution is based. Under the map ϕ, A is transformed into the tangent vector $\phi_* A = \overline{H} \partial/\partial h$ on $(h, f_1, \ldots, f_6)$ space, and the two-forms Ω_1 given by Eqs. (1a-c) are transformed into the two-forms $(\phi^{-1})^* \Omega_i$ on $(h, f_1, \ldots, f_6)$ space.

In the present case with A defined by Eq. (6), the mapping ϕ is constructed as follows

$$h = h \tag{7a}$$

$$f_1 = (at^i + a_1)/(bh^i + b_1)^{a/b} = (at + a_1)/(bh + b_1)^{a/b} \tag{7b}$$

$$f_2 = D^i/(bh^i + b_1)^{(b-a)/b} = D/(bh + b_1)^{(b-a)/b} \tag{7c}$$

$$f_3 = u^i/(bh^i + b_1)^{c/b} = u/(bh + b_1)^{c/b} \tag{7d}$$

$$f_4 = (v^i - v^o)/(bh^i + b_1)^{d/b} = (v - v^o)/(bh + b_1)^{d/b} \tag{7e}$$

$$f_5 = p^i/(bh^i + b_1)^{\varepsilon b} = p/(bh + b_1)^{\varepsilon b} \tag{7f}$$

$$f_6 = (e^i - e^o)/(bh^i + b_1)^{f/b} = (e - e^o)/(bh + b_1)^{f/b} \tag{7g}$$

by integrating S(A) and using the integral $bh + b_1 = (bh^i + b_1) e^{br}$ to identify r with h and eliminate r from the other equations obtained from the integration. The separation of variables induced by the inverse map ϕ^{-1} can be formally expressed as follows:

$$(\phi^{-1})^* t = f_1 \tilde{t}(h), \quad (\phi^{-1})^* D = f_2 \tilde{D}(h),$$

$$(\phi^{-1})^* u = f_3 \tilde{u}(h), \quad (\phi^{-1})^* v = f_4 \tilde{v}(h),$$

$$(\phi^{-1})^* p = f_5 \tilde{p}(h), \quad (\phi^{-1})^* e = f_6 \tilde{e}(h)$$

with the functions of h defined by Eq. (7). A similarity solution can now be constructed as an invariant of the group generated by A by choosing (h, f_1) as the coordinates of an integral submanifold of the set of forms $(\phi^{-1})^* \Omega_i$ and constructing a solution map from (h, f_1) space to $(h, f_1, \ldots, f_6)$ space. The solution map ψ_f imposes the condition that the first integrals $f_2, \ldots, f_6$ are functions of f_1, $\psi_f^* f_i = \tilde{f}_1(f_1)$ $(i = 2, \ldots, 6)$, and must satisfy the equation $\psi_f^* \circ (\phi^{-1})^* \Omega_i = 0$ to make (h, f_1) an integral submanifold of the forms $(\phi^{-1})^* \Omega_i$. The equations $\psi_f^* \circ (\phi^{-1})^* \Omega_i = 0$ $(i = 1, 2, 3)$ are the differential flow equations on (h, f_1) space. The first integral f_1 is called the similarity variable and from now is set to equal ξ to follow conventional notation. The similarity curves defined by Eq. (7b) form a family of curves of constant ξ in the (t, h) plane, and a particular value of ξ is determined by the values of t^i and h^i at the initial point (t^i, h^i).

At this stage, it is convenient to set $h^i = 0$ and non-dimensionalize the equations by introducing the characteristic distance $\beta = b_1/b$ and the characteristic time $\alpha = a_1/a$. The similarity variable can then be written as

$$\xi = (1 + t^i/\alpha) = (1 + t/\alpha)/(1 + h/\beta)^{(a/b)} \qquad (8)$$

and the value of ξ on a particular similarily curve in the (t, h) plane is determined by the time it passes through the h = 0 axis. Because the shock path is a similarity curve by construction and passes through the origin ($t^i = 0$, $h^i = 0$), the equation for the shock path is obtained by setting $\xi = 1$ in Eq. (8). The solution map ψ_H thus satisfies the equation $\psi_H^*(\xi) = 1$, and the equation for the shock velocity is obtained as

$$D = D_i (h/\beta + 1)^{(b-a)/b} \qquad (9)$$

by setting $\psi_H^* o(\phi^{-1})^* D^i = D_i$. Differentiation of Eq. (8) with $\xi = 1$ then shows that the initial shock velocity D_i is defined by the group parameters b_1 and a_1 through the equation $D_i = b_1/a_1$.

The solution map ψ_f from (h, ξ) to (h, ξ, u^i, $v^i - v^o$, p^i, $e^i - e^o$) space can now be written as

$$u = u_i (1 + h/\beta)^{c/b} U(\xi) \qquad (10a)$$

$$v - v^o = (v_i - v^o)(1 + h/\beta)^{\delta/b} V(\xi) \qquad (10b)$$

$$p = p_i (1 + h/\beta)^{\varepsilon b} P(\xi) \qquad (10c)$$

$$e - e^o = (e_i - e^o)(1 + h/\beta)^{f/b} E(\xi) \qquad (10d)$$

by writing the first integrals u^i, $(v^i - v^o)$, p^i, and $(e^i - e^o)$ as $u^i = u_i U(\xi)$, $v^i - v^o = (v_i - v^o) V(\xi)$, $p^i = p_i P(\xi)$, and $(e^i - e^o) = (e_i - e^o) E(\xi)$, so that $\psi_H^* U(\xi) = \psi_H^* V(\xi) = \psi_H^* P(\xi) = \psi_H^* E(\xi) = 1$ along the shock path where $\psi_H^* \xi = 1$, and:

$$\psi_H^* \tilde{u}(h, \xi) = \tilde{u}_H(h), \quad \psi_H^* \tilde{v}(h, \xi) = \tilde{v}_H(h),$$

$$\psi_H^* \tilde{p}(h, \xi) = \tilde{p}_H(h), \quad \psi_H^* \tilde{e}(h, \xi) = \tilde{e}_H(h).$$

Because the forms Ω_i are preserved by A, the equations $\psi_f^*\, o(\phi^{-1})^*\, \Omega_i = 0$ (i = 1, 2, 3), expressing the condition that (h, ξ) is an integral submanifold, reduce to a set of ordinary differential equations relating $U(\xi)$, $V(\xi)$, $P(\xi)$, and $E(\xi)$.

Model Similarity Solution for Initiation of Condensed Explosive

Construction of Solution

In the type of initiation under consideration, detonation occurs after the lead shock is overtaken and strengthened by a reactive compression wave from the rear. Negative values of a and b must be chosen to model this type of converging flow. In this case, we write our similarity parameter as

$$\eta = (1 - t/\alpha)/(1 - h/\beta)^{a/b} \qquad (11)$$

and our solution map ψ_f from (h, η) to (h, η, u^i, v^i, p^i, e^i) space as

$$u = u_i U(\eta)/(1 - h/\beta)^{c/b} \qquad (12a)$$

$$v - v^o = (v_i - v^o)V(\eta)/(1 - h/\beta)^{\delta/b} \qquad (12b)$$

$$p = p_i P(\eta)/(1 - h/\beta)^{\varepsilon/b} \qquad (12c)$$

$$e - e^o = (e_i - e^o)E(\eta)/(1 - h/\beta)^{f/b} \qquad (12d)$$

with $\psi_H^* U(\eta) = \psi_H^* V(\eta) = \psi_H^* P(\eta) = \psi_H^* E(\eta) = 1$ along the shock path where $\psi_H^*(\eta) = 1$. The conditions for (h, η) to be an integral submanifold, $\psi_f^*\, o(\phi^{-1})\, \Omega_i = 0$, give the following differential equations relating $U(\eta)$, $V(\eta)$, $P(\eta)$, and $E(\eta)$

$$\frac{dV}{d\eta} = \eta \frac{dU}{d\eta} + \frac{c}{a} U \qquad (13a)$$

$$\frac{dU}{d\eta} = \eta \frac{dP}{d\eta} + \frac{\varepsilon}{a} P \qquad (13b)$$

$$\frac{dE}{d\eta} = 2P \frac{dV}{d\eta} \qquad (13c)$$

The equation for the shock velocity is

$$D = D_i (1 - h/\beta)^{1-a/b} \qquad (14)$$

and eliminating h among the shock variables leads to the following equations representing the shock in the (D-u) and (p-v) planes

$$D = D_i (u_H/u_i)^{(\varepsilon/\delta-1)/(\varepsilon/\delta+1)} \quad (15)$$

and

$$P_H = \bar{B} (1-v_H/v^o)^{\varepsilon/\delta} \quad (16)$$

where $D_i = b\beta/a\alpha$ and $\bar{B} = p_i/(1-v_i/v^o)^{\varepsilon/\delta}$.

The model solution presented here is constructed to simulate significant features of particle velocity histories measured in multiple Lagrange gage experiments on PBX9404. It is based on the equation

$$\frac{d^2U}{d\eta^2} = m(\eta - \bar{\eta}) \quad (17)$$

because measured Lagrange particle velocity histories in shocked PBX9404 exhibit a maximum and a point of inflection. Equations for $dU/d\eta$ and U obtained by integrating Eq. (17) are written as

$$\frac{dU}{d\eta} = \frac{m}{2}(\eta^2 - 2\eta\bar{\eta} + U_1) \quad (18)$$

and

$$U = 1 + \frac{m}{2}(1/3\,\eta^3 - \bar{\eta}\,\eta^2 + U_1\eta - U_2) \quad (19)$$

where $U_1 = 2\bar{\eta}\hat{\eta} - \hat{\eta}^2$, $U_2 = (1/3 + U_1 - \bar{\eta})$, and $\hat{\eta}$ locates the maximum in the Lagrange particle velocity. The corresponding expression for V, P, and E are obtained by integrating Eqs. (13) as

$$V = \eta U - (1 - c/a) \int_1^\eta U\, d\eta \quad (20)$$

$$P\eta^{\varepsilon/a} = 1 + \int_1^\eta \eta^{(\varepsilon/a - 1)} \frac{dU}{d\eta} d\eta \quad (21)$$

$$E = 1 + 2 \int_1^\eta P \left(\frac{c}{a} U + \eta \frac{dU}{d\eta}\right) d\eta \quad (22)$$

Equations (18) and (19) were used to evaluate the integrals in Eqs. (20-22) and determine the functional dependence of V, U, and E on η.

With U, P, V, and E known, the initial shocked state must be specified and values of the group parameters and the flow parameters m, $\bar{\eta}$, and $\hat{\eta}$ must be assigned before flowfields can be calculated for a particular solution. A constitutive relationship for shocked reacting explosive is also needed to calculate the global decomposition rate of the explosive from values of e, p, and v along a particle path specified by the similarity solution.

Constitutive Relationship for the Explosive

The constitutive model used to describe the shocked reacting explosive is based on the assumption that the explosive and its reaction products attain mechanical but not thermal equilibrium (Cowperthwaite 1983). The explosive decomposes because it receives heat from the reaction products but it is assumed to undergo isentropic compression or release as the pressure increases or decreases along a particle path. Let λ denote the fraction of unreacted explosive, assume that the reaction products are polytropic, use the subscripts x and p to denote explosive and reaction products, and write their equations of state as

$$e_x = e_x^o + \tilde{e}_x(p, v_x) \tag{23}$$

$$e_p = \Sigma_1^c \alpha_i (e_p^o)_i + [pv_p/(k-1)] \tag{24}$$

where $\alpha_1 .. \alpha_c$ denote the mass fractions of the reaction products and k denotes the polytropic index. Combining Eqs. (23) and (24) with the mixture rules $e = (1-\lambda)e_x + \lambda e_p$ and $v = (1-\lambda)v_x + \lambda v_p$ to eliminate v_p, allows the $e = \tilde{e}(p, v, v_x, \lambda)$ relationship for reacting explosive to be written as

$$e - e_x^o = -\lambda \left[q(\alpha_1, ..., \alpha_c) + \tilde{e}_x(p, v_x) - \frac{pv_x}{(k-1)} \right]$$

$$+ \tilde{e}_x(p, v_x) + \frac{p(v - v_x)}{k-1} \tag{25}$$

where $q(\alpha_1, ... \alpha_c) = - [\Sigma_1^c \alpha_i (e_p^o)_i - e_x^o]$. The mixture rules reduce to $e_H = e_x^H$ and $v_H = v_x^H$ at the shock front because $\lambda^H = 0$ in a ZND wave. In applying Eq. (25) to a particle path, $\tilde{e}_x = \tilde{e}_x^h$ and v_x^h are evaluated on the isentrope passing through the shocked state defining the particle path. It is thus convenient to know v_x^h as a function of p along an isentrope when the $e = \tilde{e}(p, v, v_x, \lambda)$ relationship is used to calculate λ^h from the flow variables, e^h, p, and v^h. Equations for $e_x^h(p)$ and $\tilde{p}^h(v_x)$ were derived (Cowperthwaite 1982) using the form for $\tilde{e}_x(p, v_x)$ describing a

Mie-Gruneisen (MG) solid, but in this case an iterative procedure is generally needed to calculate v_x^h as a function of p. For this reason, an isentropic ($p-v_x$) relationship was chosen to model the isentropes given by the Mie-Gruneisen equation, and the first law of thermodynamics was integrated using the Hugoniot curve as a boundary condition to obtain the corresponding $\tilde{e}_x(p, v_x)$ relationship for the explosive. This isentropic relationship was written as

$$p = p_H (v_x^f - v_x)^r / (v_x^f - v_x^H)^r \tag{26}$$

where the superscript f denotes quantities evaluated on the p = 0 isobar, and r is a parameter to be evaluated. The volume v_x^f is a function of p_H and is therefore constant along an isentrope, but it must be calculated for each isentrope. Integrating $de_x = -p\,dv_x$ from the Hugoniot with Eq. (26) gives the equation for the energy along an isentrope as

$$e_x - e_x^H = p \frac{(v_x^f - v_x)}{(r+1)} - p_H \frac{(v_x^f - v_x^H)}{(r+1)} \tag{27}$$

The equation for calculating v_x^f was derived by equating values of e_x^f computed for two different processes. One of these was the dynamic, adiabatic process of shock compression followed by isentropic release, and the other was the static addition of heat at p = 0. The change in energy in the dynamic process was written as

$$e_x^f - e_x^o = e_x^f - e_x^H + e_x^H - e_x^o \tag{28}$$

with $e_x^H - e_x^o = 1/2\, p_H (v_x^o - v_x^H)$ given by the Hugoniot equation, and

$$e_x^f - e_x^H = - p_H [(v_x^f - v_x^H)/(r+1)] \tag{29}$$

given by Eq. (27). Combining the equation for $e_x^f - e_x^o$ in the heating process

$$e_x^f - e_x^o = [(c_x^o)^2 / \Gamma]\, \rho_x^o\, (v_x^f - v_x^o) \tag{30}$$

and Eq. (29), after some rearrangement, gives the following equation for calculating v_x^f along an isentrope passing through p_H

$$(v_x^f - v_x^o) \left[\frac{(c_x^o)^2 \rho_x^o (r+1)}{\Gamma\, p_H} + 1 \right] = (v_x^o - v_x^H) \frac{(r-1)}{2} \tag{31}$$

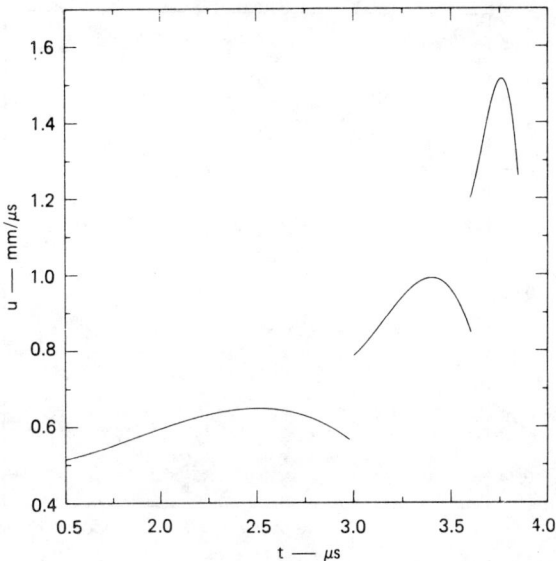

Fig. 1 Particle velocity vs time profiles for a reactive shock at 4.71, 10.12, and 12.75 mm. (p_i = 20 kbar, D_i = 3 mm/μs, e/δ = 2.7, ε/b = 6/7, α = 4μs, $\bar{\eta}$ = 0.85, $\hat{\eta}$ = 0.6, m = 28).

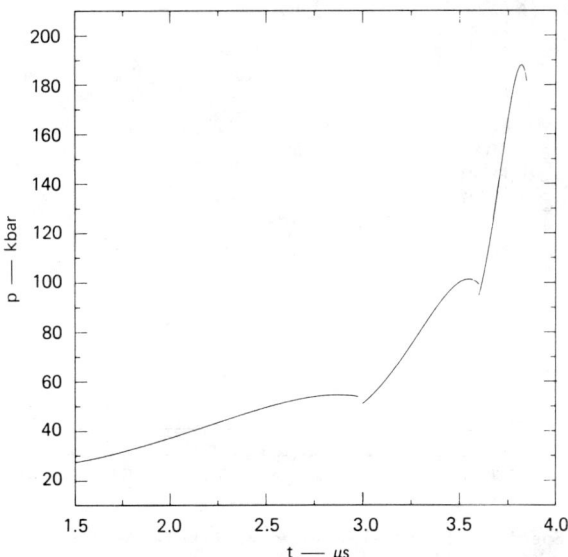

Fig. 2 Pressure vs time profiles for a reactive shock at 4.71, 10.12, and 12.75 mm. (p_i = 20 kbar, D_i = 3 mm/μs, ε/b = 6/7, α = 4μs, $\bar{\eta}$ = 0.85, $\hat{\eta}$ = 0.6, m = 28).

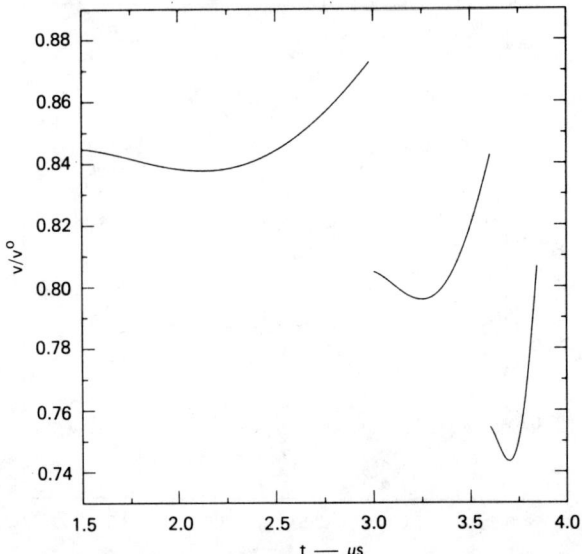

Fig. 3 Relative specific volume vs time profiles for a reactive shock at 4.71, 10.12, and 12.75 mm. (p_i = 20 kbar, D_i = 3 mm/µs, ε/δ = 2.7, ε/b = 6/7, α = 4µs, $\bar{\eta}$ = 0.85, $\hat{\eta}$ = 0.6, m = 28).

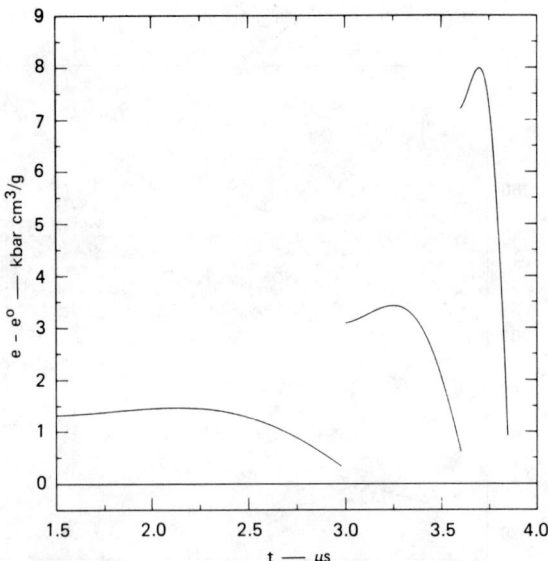

Fig. 4 Specific internal energy vs time profiles for a reactive shock at 4.71, 10.12, and 12.75 mm. (p_i = 20 kbar, D_i = 3 mm/µs, ε/δ = 2.7, ε/b = 6/7, α = 4 µs, $\bar{\eta}$ = 0.85, $\hat{\eta}$ = 0.6, m = 28).

where c_x^0 denotes the sound speed in the unshocked explosive and ρ_x^0 denotes its initial density. Equations (26), (27), and (31) provide the values of $v_x = v_x^h(p)$ and $e_x = \tilde{e}_x^h(p)$ for calculating values of λ^h. Agreement between the isentropes constructed from Eqs. (26) and (31) and the calculated MG isentropes shows that these equations provide an excellent model for a MG solid.

Calculation of λ

It is clear from Eq. (25) that values of $q(\alpha_1, \ldots, \alpha_c)$ must be known along a particle path before values of λ^h can be calculated from values of p, e^h, v^h, $\tilde{e}_x^h(p)$, and $\tilde{v}_x^h(p)$. It is also clear that $q(\alpha_1, \ldots, \alpha_c)$ will vary along a particle path if the relative composition also varies. This complication is bypassed, however, by assuming that the relative composition is fixed along a particle path and setting $q(\alpha_1, \ldots, \alpha_c) = q(h)$ accordingly.

In this case, the following equation for calculating the reaction coordinate along a particle path,

$$\lambda^h \left[q(h) + \tilde{e}_x^h(p) - \frac{p v_x^h(p)}{k-1} \right] = \tilde{e}_x^h(p) + \frac{p[v^h - \tilde{v}_x^h(p)]}{k-1}$$

$$+ \int_{v_x^H}^{v^h} p \, dv^h - \frac{P_H}{2} (v_x^0 - v_x^H) \quad (32)$$

is obtained by combining the equation for the energy along a particle path

$$e^h - e_x^0 = -\int_{v_x^H}^{v^h} p \, dv^h + \frac{1}{2} P_H (v_x^0 - v_x^H) \quad (33)$$

with Eq. (25).

At this stage, $q(h)$ must be known before Eq. (32) can be used to calculate λ^h from the flow variables and an isentrope of the condensed explosive. Values of $q(h)$ were chosen to make $\lambda^h = 1$ when $\partial\lambda/\partial t = 0$, and were calculated by setting $\lambda^h = 1$ in Eq. (32) at the pressure where the flow satisfied the equation $v\partial p/\partial t = -kp\partial v/\partial t$, expressing the condition that $\partial\lambda/\partial t = 0$.

A Particular Model Solution for Shock Initiation

Values of the group parameters, the flow parameters m, $\bar{\eta}$ and $\hat{\eta}$, and the constitutive parameter r must be

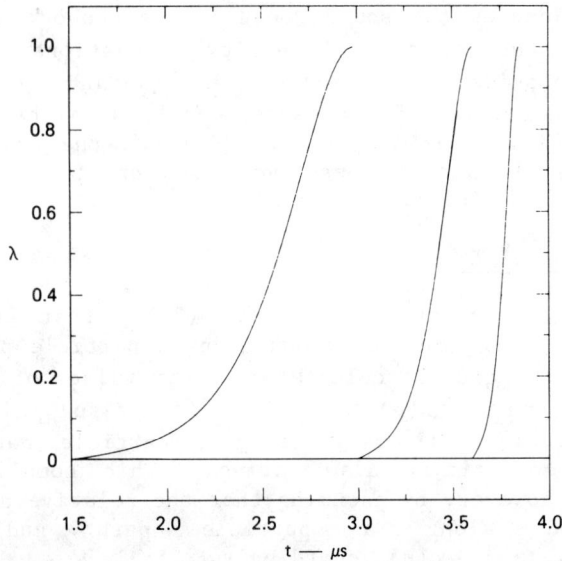

Fig. 5 Reaction coordinate vs time profiles for a reactive shock at 4.7, 10.12, and 12.75 mm. (p_i = 20 kbar, D_i = 3 mm/μs, ε/δ = 2.7, ε/b = 6/7, α = 4 μs, $\bar{\eta}$ = 0.85, $\hat{\eta}$ = 0.6, m = 28, r = 2.75).

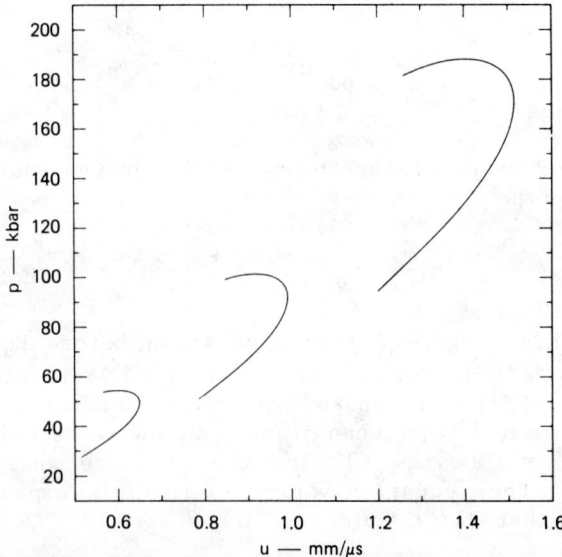

Fig. 6 Lagrange pressure vs particle velocity curves for a reactive shock at 4.7, 10.12, and 12.75 mm. ($\underline{p}_i$ = 20 kbar, D_i = 3 mm/μs, ε/δ = 2.7, ε/b = 6/7, α = 4 μs, $\bar{n}$ = 0.85, $\hat{\eta}$ = 0.6, m = 28).

assigned before the flowfields and global decomposition rate of the explosive can be calculated for a particular solution. Values of p_i, D_i, and ε/δ were assigned first to ensure that shocked states described by the solution are representative of a real explosive. These values were based on the linear relationship $D = 2.3 + 2.15\ u_H$ for cast TNT (Kanel 1977). The values $p_i = 20$ kbar, $D_i = 3$ mm/μs, and $\varepsilon/\delta = 2.7$ were chosen to ensure that shocked states match those given by the linear relationship in the 20-110 kbar region. In this case, $u_i = 0.414$ mm/μs, $v_i/v^o = 0.8620$, and $\bar{B} = 4201$ kbar. The values of $\varepsilon/b = 6/7$ and $\alpha = 4$ μs were then chosen to define the dependence of the shock state on distance. The corresponding values of the other group parameters are

$\delta/b = 2/6.3$, $a/b = 4/6.3$, $c/b = 11.1/18.9$, $\beta = 15.24$ mm.

With the shock defined, values of $\bar{\eta}$, $\hat{\eta}$, and m must be assigned to define the flow behind the shock. The values $\bar{\eta} = 0.85$ and $\hat{\eta} = 0.6$ were chosen to satisfy the condition for a maximum $2\bar{\eta} > 1 + \hat{\eta}$, and the value m = 28 was chosen to

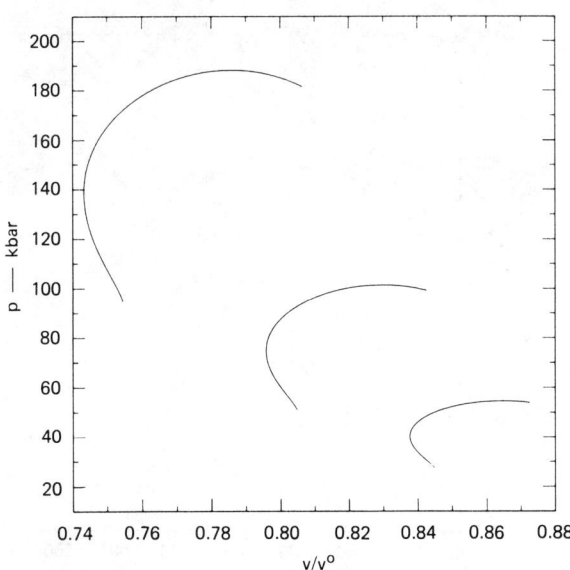

Fig. 7 Lagrange pressure vs relative specific volume curves for a reactive shock at 4.7, 10.12, and 12.75 mm. ($p_i = 20$ kbar, $D_i = 3$ mm/μs, $\varepsilon/\delta = 2.7$, $\varepsilon/b = 6/7$, $\alpha = 4$ μs, $\bar{\eta} = 0.85$, $\hat{\eta} = 0.6$, m = 28.)

ensure first, that a particle is compressed as it leaves the shock, and second that its path in the (p-v) plane lies to the right of the Hugoniot curve.

A value of r compatible with the values of the flow parameters must now be chosen before values of λ^h can be calculated from the flow. To satisfy thermodynamic constraints, a value of r must be chosen to make the Hugoniot curve of the explosive steeper than its isentropes in the (p - v) plane. An equation was thus derived for evaluating the ratio of the slope of the Hugoniot to the slope of an isentrope at a point on the Hugoniot. This equation was derived as

$$\left[\frac{dp/dv_x}{(\partial p/\partial v_x)_s}\right]_H = \frac{\varepsilon/\delta}{r}\left[\frac{1 + (v_x^f - v_x^o)}{(v_x^o - v_x^H)}\right] \qquad (34)$$

with $(v_x^f - v_x^o)/(v_x^o - v_x^H)$ given by Eq. (31) by differentiating Eq. (16) and Eq. (26).

For a given value of ε/δ the value of r can be chosen to make $[(dp/dv_x)/(\partial p/\partial v_x)_s]_H > 1$ at the initial shock pressure because $(v_x^f - v_x^o)/(v_x^o - v_x^H)$ increases along the Hugoniot curve as p_H increases. A limit is set on r at the

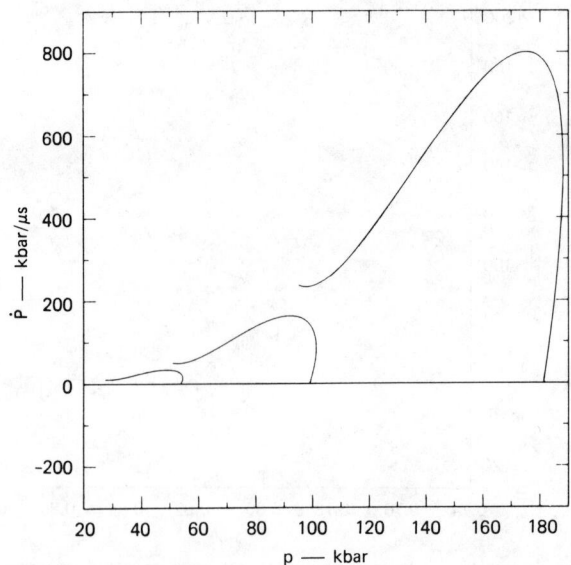

Fig. 8 Lagrange volumetric chemical energy release rate vs pressure curves for a reactive shock at 4.7, 10.12, and 12.75 mm. (p_i = 20 kbar, D_i = 3 mm/µs, ε/δ = 2.7, ε/b = 6/7, α = 4 µs, $\bar{\eta}$ = 0.85, $\hat{\eta}$ = 0.6, m = 28, r = 2.75).

initial shock pressure by setting $[(dp/dv_x)/(\partial p/\partial v_x)_s]_H = 1$ in Eq. (34) and solving Eqs. (34) and (31) for r. Following this procedure with a value of $\varepsilon/\delta = 2.7$ led to the choice of a value of $r = 2.75$.

Plots at the Lagrange distances 4.71, 10.12, and 12.75 mm made from the similarity solution with this set of parameters ($\varepsilon/\delta = 2.7$, $\varepsilon/b = 6/17$, $\alpha = 4$ µs, $m = 28$, $\bar{\eta} = 0.85$, $\hat{\eta} = 0.6$, $r = 2.75$) are shown in Fig. 1 through 8, where $\dot{P}$ denotes the volumetric chemical energy release rate.

Conclusions

Methods of differential geometry and Lie group theory have been used to construct similarity solutions for ZND waves. The governing equations were written as conservation laws so that a constitutive relationship for shocked reacting explosive could be incorporated into the solution and used to calculate the global decomposition rate of the explosive from the flow. The condition that the similarity solution is invariant under the group admitted by the conservation equations determines the equation for the Hugoniot curve of the explosive, but group parameters in this equation can be chosen to model shocked states of condensed explosives over a pressure region covering the interesting stage of the initiation process. A realistic constitutive relationship for shocked reacting explosive was constructed and incorporated into a model solution for the shock initiation process based on Lagrange particle velocity histories recorded in initiating PBX9404 and cast TNT. Values of group, flow, and constitutive parameters were then assigned to obtain a particular solution for this type of flow. Group parameters were chosen to model the Hugoniot curve of cast TNT. The flow parameter m was chosen to make particle paths lie to the right of the Hugoniot curve in the (p-v) plane, and the constitutive parameter r was chosen to make the Hugoniot curve of the explosive steeper than its isentropes in the (p-v) plane.

Acknowledgments

The author thanks B. Y. Lew for programming routines to make plots of the model similarity solution.

References

Cowperthwaite, M. and Adams, G. K., (1967) Explicit solution for steady- and unsteady-state propagation of reactive shocks at constant velocity. Eleventh Symposium (International) on Combustion, p. 703, The Combustion Institute, Pittsburgh, Pa.

Cowperthwaite, M. (1978) Model solutions for shock initiation of condensed explosives. Actes Du Symposium International sur le Compartment des Millieux Denses sous Hautes Pressions Dynamiques, p. 201. Editions du Commissariat a' l'Energie Atomique Centre d'Etudes Nucleaires de Saclay, Paris.

Cowperthwaite, M. (1981) Lagrange gage analysis for calculating chemical energy release rates in shocked explosive. Colloque International Berthelot-Vieille-Mallard-Le Chatelier, First Specialist Meeting (International) of the Combustion Institute, p. 533. Universite de Bordeaux I, France.

Cowperthwaite, M. (1982) Characterization of Initiation and Detonation by Lagrange Gage Techniques. Final Report by SRI International to Lawrence Livermore National Laboratory, Subcontract 9371209, January.

Cowperthwaite, M. (1983) A constitutive model for calculating chemical energy release rates from the flow fields in shocked explosives. Seventh Symposium (International) on Detonation, NSWC MP 82-334, p. 498. Naval Surface Weapons Center, White Oak Laboratory, Silver Spring, Md.

Fickett, W. and Davis, W. C. (1979) Detonation, p. 42. University of California Press, Los Angeles, Calif.

Kanel, G. I. and Dremin, A. N. (1977) Decomposition of cast trotyl in shock waves. Combustion, Explosion, and Shock Waves, 13(1), 71.

Logan, J. D. and Bdzil, J. B. (1982) Self-similar solutions of the spherical detonation problem. Combust. Flame, 46, 253.

Sternberg, H. M., (1970) Similarity solutions for reactive shock waves. Quart. J. Mech. and Appl. Math, 23, 77.

Zel'dovich, I. B. and Kompaneets, A. S. (1960) Theory of Detonation, p. 68. Academic Press, New York.

The Simulation of Shock-Induced Energy Flux in Molecular Solids

Arnold M. Karo,* Franklin E. Walker,† and Thomas M. DeBoni‡
Lawrence Livermore National Laboratory, University of California
Livermore, California
and
John R. Hardy§
Behlen Laboratory of Physics, University of Nebraska, Lincoln, Nebraska

Abstract

Computer molecular dynamics has been used to study the time evolution of the energy of diatomic molecules embedded in a monatomic host lattice when the system is shock loaded. Center-of-mass, rotational, and internal energies were each monitored. For H_2 and CH groups in an iron host, the results demonstrate rapid and violent internal excitation of a totally athermal nature. The origins of this are discussed as are the reasons for the absence of a similar effect for a CH group in a carbon lattice. From these results for diatomic systems, it is argued that large molecules, similarly treated, may easily be excited to the point of rupture. If they are so situated (e.g., at or near a surface) that during, or shortly after, excitation they escape from the lattice, they will rupture rather than de-excite and thus generate molecular fragments (e.g., free radicals) that could, in the case of an explosive system, serve to initiate detonation.

Introduction

This paper presents the results of a large number of molecular dynamics calculations designed to study the behavior and effect of shock wave interaction in condensed systems. These computer simulations show that events occur

---

Paper presented at the 9th ICODERS, Poitiers, France, July 3-8, 1983. Copyright © American Institute of Aeronautics and Astronautics, Inc., 1984. All rights reserved.
*Senior Scientist, Department of Chemistry and Materials Science.
†Deputy Program Director, Non-nuclear Ordnance Program.
‡Consultant to Lawrence Livermore National Laboratory.
§Professor of Physics, Department of Physics.

as the consequence of a sequence of atomic or molecular processes occurring in sub-nanosecond to sub-picosecond times over dimensions of angstroms. In the first of a series of papers reported at the Sixth International Colloquium on Gasdynamics of Explosions and Reactive Systems (Karo et al. 1978), results of computer simulations of the dynamics of shocked crystals were described that indicated that, for nonuniform materials, shocks by themselves could cause structural rearrangements leading to the production of free atoms and atomic or molecular clusters. In this work, one of the most important results was the clear demonstration of shock integrity, i.e., the rather remarkable degree of nonergodic behavior in shock-loaded perfect lattices. It was found that the energy imparted to a perfect lattice structure by shock loading does not degrade to any great extent into random thermal motion, but stays in the shock front until it reaches a surface. Also considered important was the related fact that our calculations showed shocks to be localized on a scale of atomic dimensions. A number of subsequent studies on shock-loaded perfect lattices by other workers (MacDonald and Tsai 1978; Batteh and Powell 1979; Dremin and Klimenko 1979; Klimenko and Dremin 1980) have reached substantially the same conclusions. The two properties of stability and spatial localization consequently enable shocks to deposit their energy selectively to the edge when reaching a lattice boundary. The energy is then expended by the production of small, energetic, spalled fragments of the lattice. In our simulations of perfect crystalline materials, it is by this fracture that the shock energy is most readily degraded.

For lattices containing imperfections and irregularities, a more complex situation develops; and, in subsequent papers (Hardy et al. 1981; Karo and Hardy 1981), work was reported that emphasized two-dimensional systems composed of several species of atoms or containing defects, shock coherence in such systems, and the rather minimal lateral transfer of shock energy that appears to be a general characteristic of lattices without imperfections, regardless of the initial temperature. These papers described the manner in which the energy localized in the shock front can dissipate by interaction with point defects, voids, and crystalline irregularities. It was seen that the role played by imperfections in providing channels for coupling the normal modes of the lattice with the energy localized in the shock wave could be significant, particularly when such imperfections are located near a surface.

In order to quantify and study in detail these general observations, much more general and flexible computer codes were developed that are capable of handling arbitrary atomic arrangements and rearrangements by a sophisticated neighborhood look-up procedure and that incorporate the ability to monitor selectively and in detail the energy flux in any specific "region" of the lattice. This region can be as small as a specific atomic pair or as large as the whole lattice.

The present paper will present the first results obtained from these codes: Specifically, the manner in which energy is taken up from shocks by single diatomic molecules embedded in monatomic host lattices is examined. The basic concern of these studies is to answer the question as to what is the "rise time" of a shock as it transits a molecular group; i.e., what time is required for a major component of the shock energy to be converted into internal energy of the molecule. There is also the obviously related question as to whether or not there is ever a significant conversion of shock energy into internal energy. These questions will be addressed in the following sections.

Discussion of the Current Work

The basic method of molecular dynamics has been described before (Karo et al. 1978; Karo and Hardy 1981) and will not be discussed in detail in the present account. It enables monitoring, by computer solution of the Newton's Second Law equations of motion for the individual atoms, of the time evolution of the system under study. Shock loading is naturally only one of the many phenomena that can be studied by this approach.

In order to address the specific problem of energy flux into and out of the type of region specified in the Introduction, codes have been developed that continuously monitor the center-of-mass energy for the region, its rotational energy (about the center-of-mass), its vibrational energy, and its total energy. With regard to the potential energy of the region, it is possible to include only bonds between atoms within the region or all interactions of atoms within the region, including those with atoms outside the region. Both are valid quantities; however, the former is more relevant since this study is more interested in the storage and flux of energy strictly within the region. This partitioning is illustrated in Fig. 1.

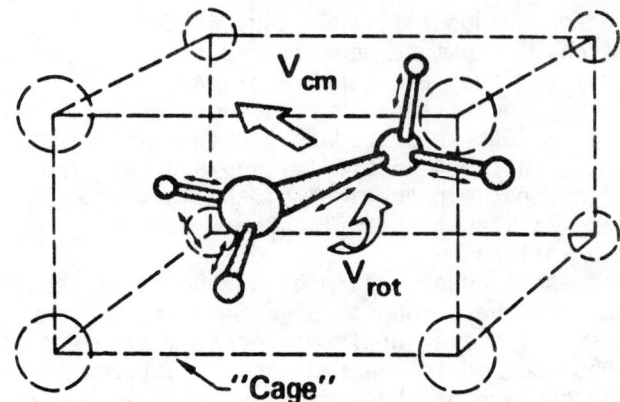

The molecular unit can be treated as isolated or as interacting with the lattice.
Partitioning of the total molecular energy:
- Center of mass energy
- Total vibrational energy: isolated or interacting
- Total rotational energy
- Total internal energy: isolated or interacting

Fig. 1 Schematic illustration of the molecular energy partitioning employed in the present work.

Given this capability, the number of possible studies is practically unlimited. However, for the purposes of this preliminary report, this discussion is restricted to the simplest possible regions that can represent molecules, namely, nearest neighbor pairs or diatoms. For reasons of computational economy, the initial studies have been limited to two-dimensional systems; past experience indicates that the qualitative nature of the conclusions should be unaffected by this restriction in dimensionality. The codes themselves are fully three dimensional; however, to study a system large enough to be free of possible spurious surface effects would be computationally very expensive. Thus, in these first survey studies, a range of two-dimensional systems has been selected for examination.

In order to present a representative sample of the results, three different situations will be discussed: 1) a light symmetric diatom, equivalent to an H_2 molecule, except for bond length, embedded in a heavy (iron) host lattice; 2) a light but highly asymmetric diatom, equivalent to a CH group, except for bond length, in a heavy (iron) host lattice; and 3) an asymmetric diatom, again equivalent to a CH group, in a light (carbon-like) host lattice.

The last situation is a little artificial in that, except for the one strong CH bond, this could almost be regarded as a study of an isolated hydrogen atom in a carbon host. However, it is useful for present purposes as an illustration of general trends to be expected as the acoustic "match" between impurity and host is improved.

In all three cases, shock loading is generated by impact of a plate of host material, from the left, on the host lattice. To illustrate this, the initial configuration is shown in Fig. 2, in which the location of the diatom impurity is also indicated. This location is similarly identical for all three studies. In order that these initial calculations should highlight the effects of differences within the molecules, only the masses of the impurities and the potential between them were changed. Thus, for example, while the two hydrogen atoms of the H_2 molecule interacted with the true H-H potential, the H-Fe interactions were assumed to be the same as the Fe-Fe interactions. Again, this is not a restriction in the code, but was deliberately set initially. Finally, to ensure that all effects were clearly visible, the systems were deliberately shocked hard (shock pressures $\sim$0.5-1.0 Mbar), a procedure that also "overdrives" the somewhat soft Fe-Fe bonds to the point where their stiffness is closer to that of typical organic bonds.

The following subsections describe each history in turn.

H_2 in an Iron Lattice

Figure 3 shows the vibrational, rotational, and center-of-mass energies for this system as functions of time; these quantities are calculated including only the potential energy of the H-H bond. Also shown is the final configuration of the system.

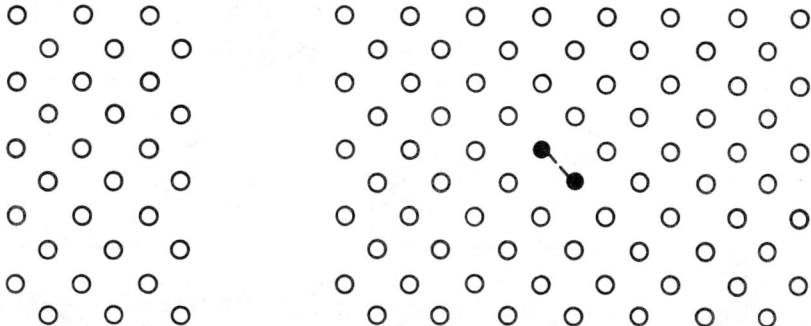

Fig. 2 Illustration of the initial configuration of the plates and lattices, both at 0 K, showing the position of the diatomic impurity.

The pattern that emerges is distinctly interesting. What is apparently happening is that, as the leading edge of the shock reaches the H_2 molecule, there are instantaneous (∿1 femtosecond) pulses of rotational and center-of-mass energies. Subsequently, the vibrational energy rises sharply during a period of the order of 2 fs. It then appears that the center-of-mass and vibrational energies tend to track one another rather closely. However, the initial pulse of rotational energy is rapidly quenched, probably by conversion to vibrational energy. Later, a complex pattern of pulses develops in the rotational energy. Ultimately, the H_2 appears almost to escape from its lattice "cage"; in the process, most of the rotational energy is converted to vibrational energy, probably with additional input of some translational energy from the center-of-mass motion. However, the molecule remains bound and one can see the localized center-of-mass motion at a frequency half that of the internal oscillations. The most interesting feature is that while the molecule is still caged, it is for a period of the order of 6×10^{-14} s (60 fs), vibrationally very "hot" (an equivalent temperature ∿15,000 K), or more than 15 times

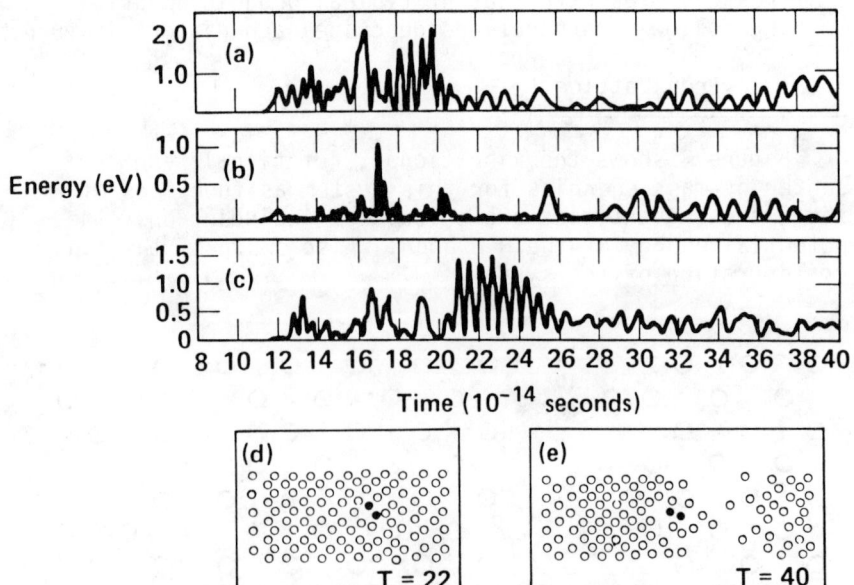

Fig. 3 Energy histories for the H_2 molecule in an iron lattice. a) Center-of-mass energy vs time. b) Rotational energy vs time. c) Vibrational energy vs time (includes only H_2 potential energy). d) Final state configuration. Units of time are in 10^{-14} s.

the melting temperature of the host lattice, which, as can be seen from the final figure, has certainly not melted. It thus follows that over a time ~100 fs in duration, the H_2 molecule has acquired ~20-30 times the energy it would have had if it were in thermal equilibrium with the lattice at the end of the run. This rapid energy transfer and large final state energy are consequences of the highly nonergodic regular evolution of the system displayed by the energy histories. What is startling is the high efficiency of the shock coupling to internal molecular motion. If a similar situation exists in larger molecules, then shock loading should very readily (and rapidly) couple into them sufficient total energy to rupture a single bond.

CH in an Iron Lattice

The energy histories and final configuration for this case are presented in Fig. 4. Again, the results show the rapid and even more efficient coupling of shock energy to the internal motion proceeding by the same mechanisms. There is also considerable evidence of stronger coupling between the center-of-mass and internal vibrational motions. Thus, for the same shock loading (i.e., initial

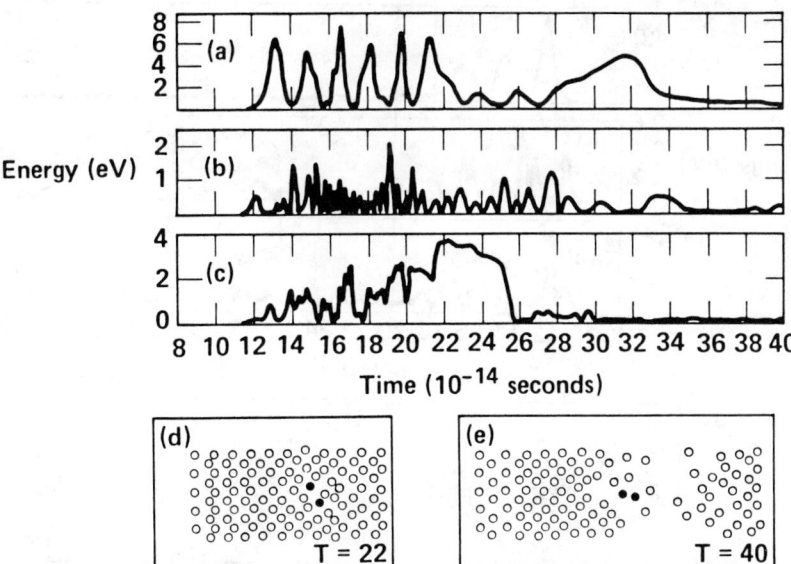

Fig. 4 Energy histories for the CH molecule in an iron lattice. a) Center-of-mass energy vs time. b) Rotational energy vs time. c) Vibrational energy vs time (includes only CH potential energy). d) Final state configuration. Units of time are in 10^{-14} s.

plate velocity), the CH group becomes even "hotter" vibrationally than the H_2 group. Subsequently, this internal energy appears to be converted to center-of-mass energy as the CH group escapes from its cage. This last development is peculiar to this run, as it obviously depends on the initial conditions and CH group location.

The overall result of this study is to reinforce the general conclusions reached from the H_2 study.

CH in a Carbon Lattice

In this case, one further restriction was imposed. While the masses of the lattice atoms were reduced to that of carbon, the bond potentials, both within the lattice and between the lattice atoms and the CH group, were retained as those appropriate to Fe-Fe bonds. Again, this was done in order to study the effect of changing a single variable alone.

The resultant energy histories and final state are shown in Fig. 5. Once again, the initial history is very similar to that of the other two systems, with initial rapid excitation of rotational and internal vibrational motions. However, once the shock has passed the history is

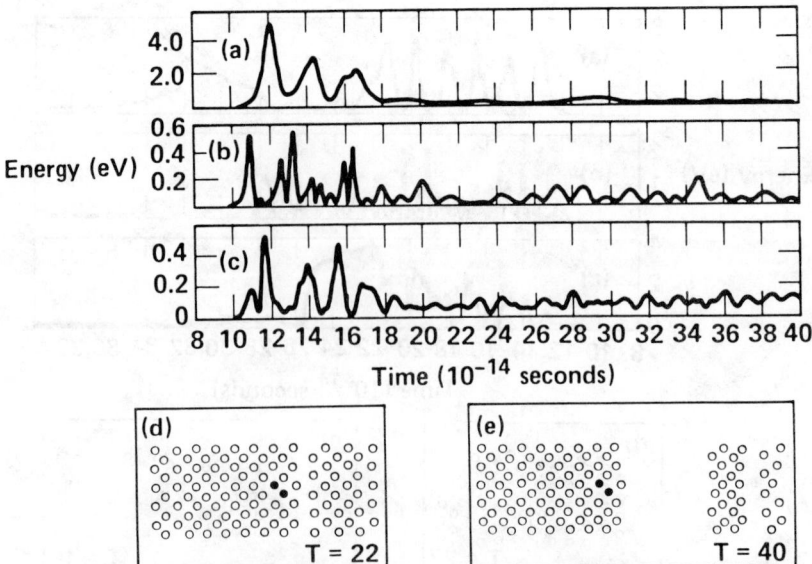

Fig. 5 Energy histories for the CH molecule in a carbon lattice. a) Center-of-mass energy vs time. b) Rotational energy vs time. c) Vibrational energy vs time (includes only CH potential energy). d) Final state configuration. Units of time are in 10^{-14} s.

very different: The rotational energy no longer stays localized in the CH group but rapidly flows out into the host lattice, presumably because of the better acoustic match. As a consequence, there is much less internal vibrational excitation. It is perhaps easier to visualize the situation by realizing that, in this case, the "molecule" is artificial, since the carbon in CH is actually part of the lattice. Thus, the only truly localized energy is that fed into the localized vibrational modes of an H atom in a carbon host. Evidently, this is much less efficient than the rotation-vibration coupling present for cases 1 and 2 where one had genuine diatoms whose confinement in a heavy host ensured that both rotational and vibrational energy remained localized.

Conclusions

The principal result of the present theoretical work is a demonstration that shock fronts can, principally because of their sharpness on an atomic scale, transmit to the internal motion of molecular groups dramatically large amounts of translational energy ($\sim$eV's per bond). This coupling is extremely rapid ($\sim 10^{-12}$-10^{-13} s). The resultant internal energy is of the order of one to two orders of magnitude larger than the thermal energy of the host lattice after the shock has passed. The fact that the rise times obtained from the molecular dynamics are two to three decades shorter than those apparently observed experimentally appears to reflect the fact that experimental techniques with simultaneous sub-picosecond time resolution and angstrom spatial resolution have yet to be developed (Karo et al. 1981). However, recent work on shock-induced electrical polarization in water (Harris and Presles 1982) provides striking evidence that shock front widths and rise times can indeed be much narrower and shorter (e.g., for water, the upper bounds are 50 Å and 1 ps, respectively) than previously reported using standard experimental methods.

The present studies have been restricted to diatomic systems. If the energy uptake per bond is not seriously reduced for polyatomic systems (which seems unlikely), then it would appear that large molecules may readily and rapidly be excited above their dissociation energy by shock loading. However, if they are to undergo actual dissociation, it would appear necessary for them to escape from their host environment; otherwise, energy rapidly gained can be lost equally rapidly. The presence of defects such as surfaces, interfaces, or microscopic voids provides such an escape mechanism. In this way, free radicals could be produced that, in the case of a shock-loaded explosive, could serve to initiate detonation.

Acknowledgments

This work was performed under the auspices of the U. S. Department of Energy by the Lawrence Livermore National Laboratory under Contract W-7405-ENG-48 and the Office of Naval Research under Contract E00014-82-F-0094.

References

Batteh, J. H. and Powell, J. D. (1979) Solitary wave propagation in the three-dimensional lattice. Phys. Rev. Sect. B 20(4), 1398-1409.

Dremin, A. N. and Klimenko, V. Yu. (1979) The effect of the shock-wave front on the origin of reaction. Gasdynamics of Detonations and Explosions: AIAA Progress in Astronautics and Aeronautics (edited by Bowen, Manson, Oppenheim, and Soloukhin), Vol. 75, pp. 253-268. AIAA, New York.

Hardy, J. R., Karo, A. M., and Walker, F. E. (1981) The molecular dynamics of shock and detonation phenomena in condensed matter. Gasdynamics of Detonations and Explosions: AIAA Progress in Astronautics and Aeronautics (edited by Bowen, Manson, Oppenheim, and Soloukhin), Vol. 75, pp. 209-225. AIAA, New York.

Harris, P. and Presles, H.-N. (1982) The shock-induced electrical polarization of water. J. Chem. Phys. 77(10), 5157-5164.

Karo, A. M., Hardy, J. R., and Walker, F. E. (1978) Theoretical studies of shock-initiated detonations. Acta Astronaut. 5(11-12), 1041-1050.

Karo, A. M. and Hardy, J. R. (1981) The study of fast shock-induced dissociation by computer molecular dynamics. Proceedings of the NATO Advanced Study Institute on Fast Reactions in Energetic Systems (edited by C. Capellos and R. F. Walker), pp. 611-643. D. Reidel Publishing Co., Dordrecht, Holland.

Karo, A. M., Walker, F. E., Cunningham, W. G., and Hardy, J. R. (1981) The study of shock-induced signals and coherent effects in solids by molecular dynamics. Gasdynamics of Detonations and Explosions: AIAA Progress in Astronautics and Aeronautics (edited by Bowen, Manson, Oppenheim, and Soloukhin), Vol. 87, pp. 9-21. AIAA, New York.

Klimenko, V. Yu. and Dremin, A. N. (1980) Structure of a shock wave front in a solid. Sov. Phys. Dokl. 25(4), 288-289.

MacDonald, R. A. and Tsai, D. H. (1978) Molecular dynamical calculations of energy transport in crystalline solids. Phys. Rep. 46(1), 1-41.

Detonation Temperatures of Nitromethane Aluminum Gels

Yukio Kato*
Fukui Institute of Technology, Fukui, Japan
and
Christian Brochet†
Université de Poitiers, Poitiers, France

Brightness temperatures of detonation products of nitromethane (NM)-aluminum mixtures were measured with optical pyrometers. Aluminum particles (mean diameter: 10 μm) were suspended in NM with 3 wt.% of gelling agent. The mass fraction of aluminum in the mixtures ranged from 0 to 0.45. Experiments with detonation tubes of different diameters showed the influence of side rarefaction waves on measured brightness temperatures of detonation products. Measurements were also performed with NM-lithium fluoride mixtures to determine the effect of inert particles on NM detonations. The results of numerical calculations indicate that there is negligible heat transfer to aluminum and lithium fluoride particles within the NM reaction zone. Since the particle residence time is of the order of 10 ns, the temperature of detonation products increases because of the aluminum reaction, and a considerable part of aluminum reacts within the first 1 μs. This fast reaction rate of aluminum is in good agreement with other experimental results. Calculation of temperature increase due to the presence of aluminum showed that aluminum particles react only when H_2O is produced by the NM detonation.

Introduction

As aluminum has been extensively used to improve the performances of industrial explosives and solid propellants, several studies have been made to understand the effects

---

Presented at the 9th ICODERS, Poitiers, France, July 3-8, 1983. Copyright © American Institute of Aeronautics and Astronautics, Inc., 1984. All rights reserved.
*Assistant Professor, Department of Environment and Safety Engineering.
†Maitre de Recherche, Laboratoire d'Energétique et de Détonique, E.N.S.M.A.

of aluminum addition on detonation characteristics of heterogeneous explosives. Recent experimental works performed with the mixtures of liquid explosive, particularly nitromethane (NM), and aluminum particles indicate that with such mixtures the complications induced by the existence of voids in solid explosives can be avoided and the effect of aluminum particles can be more easily determined. These mixtures are also easy to handle.

Some interesting results were obtained:

1) The addition of aluminum particles decreases detonation velocity. Aluminum is chemically inert, and heat transfer to aluminum particles in the reaction zone of the explosive component is insignificant (Kato and Brochet 1977; Moulard et al. 1979).

2) Brightness temperature measurements showed that aluminum particles react, after a delay, with the gaseous products produced by the detonation of the liquid explosive. The ignition delay is less than 100 ns, and the duration of aluminum reactions exceeds 2 µs for aluminum particles of mean diameter 10 µm. The time evolution of measured brightness temperatures of detonation products containing aluminized compounds depends strongly on aluminum concentration in the mixtures (Kato et al. 1979; Veyssière et al. 1983).

3) The kinetics of aluminum reactions in detonation products were approached by the aid of a cylinder test. It was shown that aluminum reactions increase the expansion work, and that the duration of aluminum reactions is about 9 µs in case of aluminum particles with a mean diameter of 5 µm (Moulard et al. 1979).

To obtain additional information on the kinetics of aluminum reactions, the brightness temperature measurements by means of optical pyrometers were performed with the mixtures of NM and aluminum particles. Some measurements were made with mixtures of NM and lithium fluoride (LiF) particles. Lithium fluoride was used, as it is chemically inert and its physical properties and shock hugoniot are very similar to those of aluminum. The effect of lithium fluoride on the detonation properties should be very similar to aluminum if it were to behave as an inert.

Experimental

The optical pyrometers were described in detail in a previous work (Bouriannes et al. 1977). One is monochromatic ($\lambda=0.657$ µm) and the other records radiation on four different wavelengths ($\lambda=0.650$, 0.783, 0.915, and 1.008 µm).

Their response time is about 50 ns. The accuracy of measurements is about 100-150 K at 3000-4000 K.

Aluminum particles (mean diameter: 10 μm, purity: 99%) or lithium fluoride particles (mean diameter: 44 μm) are suspended in NM by the addition of 3 wt.% of gelling agent [polymethylmetacrylate (PMMA)]. The composition of NM-aluminum particles mixtures (NM-PMMA/Al) and NM-lithium fluoride particles mixtures (NM-PMMA/LiF) is defined by the mass fraction x of Al and LiF particles. For NM-PMMA/Al mixtures, x ranges from 0 to 0.45. NM-PMMA/Al mixtures are initially opaque, and NM-PMMA/LiF mixtures are semitransparent.

The mixtures were contained in 18-mm-i.d. detonation tubes (Fig. 1). Experiments were also performed with 30-mm-i.d. detonation tubes to verify the influence of side rarefaction waves. The detonation tubes have good optical quality glass plate 10-mm thick on one end. The radiation emitted by detonation products was observed through this glass plate. Because NM-PMMA/Al mixtures are opaque, the observation starts when detonation fronts arrive at the interface with the glass plate (see point A on Fig. 2), and it continues while the glass plate remains transparent (see A and B on Fig. 2). Materials, such as glass, certain plastics, and sapphire retain their transparency under shock loading and do not become opaque until shattered by tension waves (Mallory and McEwan 1961; Urtiew 1974). In the case of a 10-mm-thick glass plate, the observation time is about 2 μs (Fig. 2). Some experiments were performed using two or three glass plates superposed to increase the observation time.

The detonation tubes were placed in a vacuum chamber, so that charges of up to 150 g were possible. All the experimental devices were previously described (Kato et al. 1979).

Results and Discussion

Figure 3 shows the time evolution of brightness temperatures T_b of detonation products of NM-PMMA/Al (x=0.15), deduced from the experiments conducted in various conditions. The time origin t=0 corresponds to the instant when the detonation front arrives at the interface with the glass plate (see point A on Fig. 2). Experiments with three glass plates indicate a slight temperature increase at t≃ 1.7 and 3.4 μs that is produced by shock compression of the air layer between the glass plates. These results suggest that the velocity of compressive waves in glass plates is about 5.9 mm/μs, which is very similar to that of the ramp-

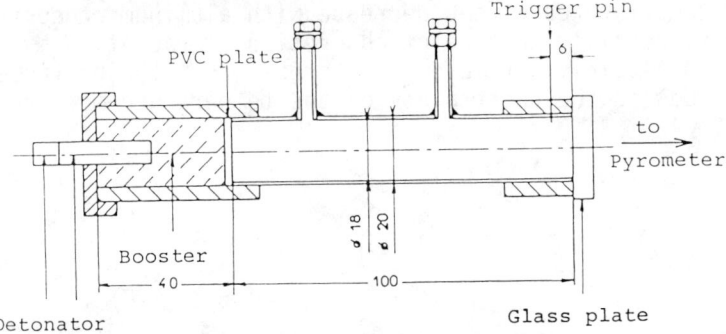

Fig. 1 Detonation tube.

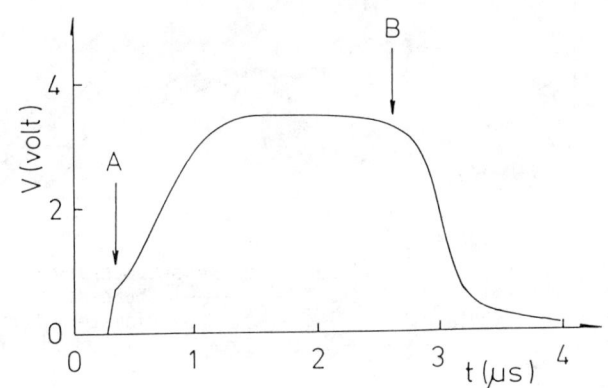

Fig. 2 Typical oscilloscope record: NM-PMMA/Al (x=0.30), λ=0.657 μm.

wave fronts (5.93 mm/μs) observed in fused silica (Barker and Hollenbach 1970). Pyrex glass has an essentially constant shock velocity of 4.8-4.9 mm/μs in the pressure range 0-20 GPa (Kinslow 1970).

Brightness temperatures begin to decrease gradually at t≈1.6 μs for a tube 18 mm in diameter, and at t≈3.0 μs for a tube 30 mm in diameter because of side rarefaction waves. An abrupt decrease of brightness temperature due to the loss of transparency at t≈2.3 μs occurred in experiments with one glass plate. These results indicate that the brightness temperatures of detonation products containing aluminized compounds are essentially constant after t≈1.5 μs in the absence of side rarefaction waves. Figure 4 shows the time evolution of brightness temperatures T_b of detonation products for various NM-PMMA/Al mixtures and NM in 18-mm-i.d. detonation tubes with one glass plate. The brightness temperatures at t=0 decrease with aluminum concentration x, possibly because of absorption of radiation by unreacted aluminum particles. The presence of the particles also adds to the uncertainty of the temperature measurement.

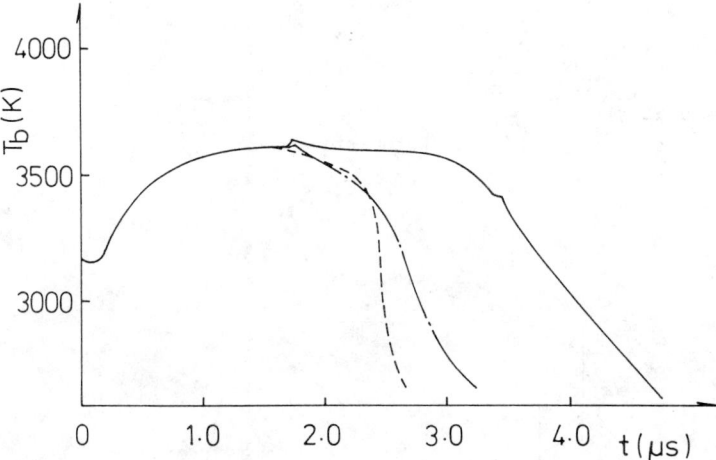

Fig. 3 Time evolution of measured brightness temperatures T_b for NM-PMMA/Al (x=0.15), λ=0.657 μm. (———): tube diameter 30 mm, three glass plates. (———·———): tube diameter 18 mm, three glass plates. (— — — —): tube diameter 18 mm, one glass plate.

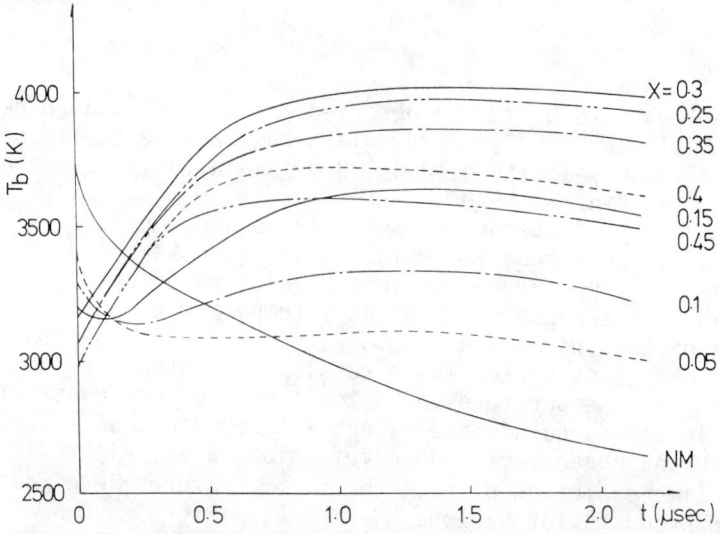

Fig. 4 Time evolution of measured brightness temperatures for various NM-PMMA/Al mixtures and NM.

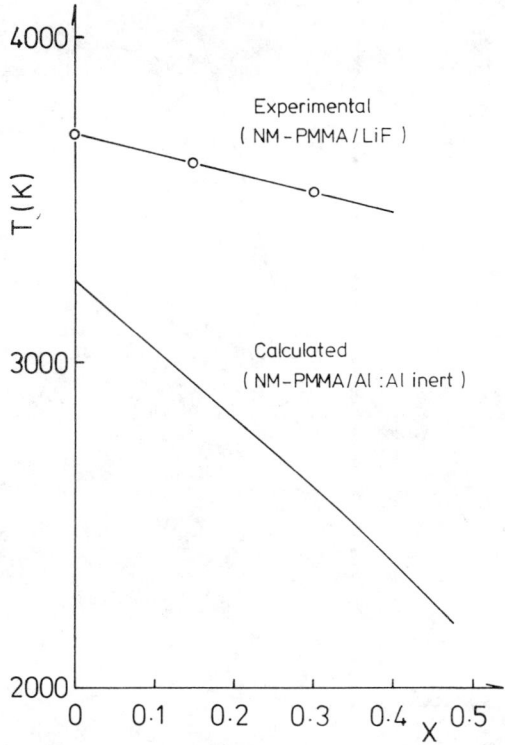

Fig. 5 Detonation front temperatures of NM-PMMA/LiF mixtures and CJ temperatures of NM-PMMA/Al mixtures.

Some experiments were performed with NM-PMMA/LiF mixtures to obtain estimates of detonation front temperatures of NM-PMMA/Al mixtures and to simulate the behavior of aluminum as inert. NM-PMMA/LiF mixtures are semitransparent, which permits precise determination of the brightness temperatures of the detonation front in these mixtures. The measured brightness temperatures of the detonation front in NM-PMMA/LiF (x=0.15 and 0.30) are, respectively, 140 K and 230 K lower than those of NM (Fig. 5). As the time evolution of brightness temperatures is similar to that of NM, one can infer that there are no reactions between LiF and detonation products of NM. Numerical calculations with Kihara-Hikita-Tanaka equation of state (Tanaka 1982) indicate that Chapman-Jouget (CJ) temperatures of NM-PMMA/Al mixtures for "inert" aluminum in thermal equilibrium with the gaseous detonation products decrease more rapidly with x (Tanaka 1983). Calculated CJ temperatures of NM-PMMA/Al (x=0.15 and 0.30) are, respectively, 369 K and 678 K lower than that of NM (Fig. 5). The differences between the observed and predicted temperatures may be due in part to the

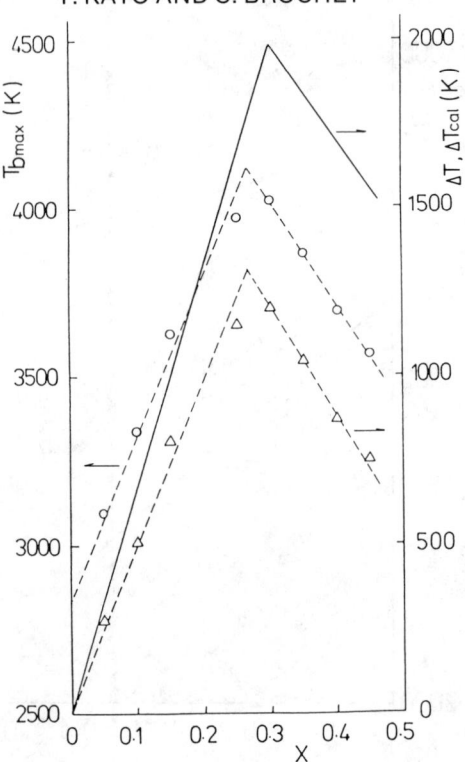

Fig. 6 Variation of $T_{b_{max}}$, ΔT, and ΔT_{cal} in function of aluminum concentration x.

absence of thermal equilibrium between the particles and the gaseous detonation products.

The following salient features are apparent from Figs. 3 and 4: 1) The time evolution of brightness temperatures depend strongly on aluminum concentration x. 2) For all NM-PMMA/Al mixtures studied, the maximum brightness temperature occurs at t=1.0 through 1.5 μs. After which time the nearly constant temperature suggests that side rarefaction waves do not play a significant role.

These results indicate that exothermic reactions between aluminum particles and detonation products of NM increase the temperature of detonation products and that most of the aluminum must have reacted within the first 1 μs. The reaction time of aluminum particles (mean diameter: 10 μm) with product of gaseous explosives is the order of 100 μs (Veyssière et al. 1983). This difference of reaction time of aluminum particles can be attributed to the difference in density of detonation products, as the reaction rate is controlled by diffusion between aluminum and detonation products (McGuire et al. 1979). In mixtures of tet-

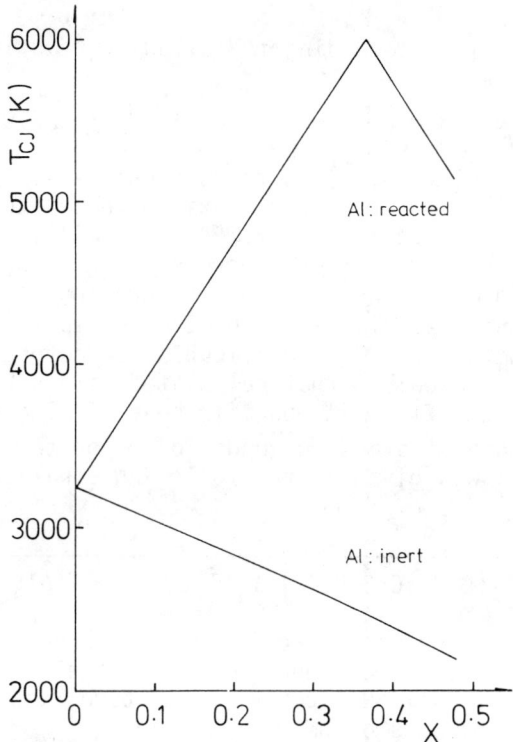

Fig. 7 CJ temperatures of NM-PMMA/Al mixtures, calculated with KHT equation of state (Tanaka 1983).

ranitromethane (TNM) and aluminum particles (mean diameter: 5 μm), reactions of aluminum particles can be completed by the time the CJ state is reached, as TNM has a rather long reaction zone (McGuire et al. 1983).

To estimate the temperature increase by aluminum reaction, differences of brightness temperatures ΔT between NM-PMMA/Al mixtures and NM were calculated at t=1.5 μs, when brightness temperatures attain their maximums for all mixtures studied (Fig. 6). The variation of ΔT and the variation of maximum values of brightness temperatures $T_{b_{max}}$ as functions of aluminum concentration x are presented in Fig. 6. Both ΔT and $T_{b_{max}}$ have their maximum values at x = 0.25 ∿0.30.

Detonation products of NM are mainly composed of H_2O, CO_2, CO, N_2, and solid carbon. Numerical calculations with Becker-Kistiakowsky-Wilson, Lennard-Jones-Devonshire and Kihara-Hikita-Tanaka equations of state indicate that about 1.5 mole of H_2O are produced from 1 mole of NM (Mader 1979; Tanaka 1983). If aluminum reacts only with that 1.5 mole of

H_2O to form Al_2O_3 according to following reaction scheme:

$$2Al + 3H_2O \longrightarrow Al_2O_3 + 3H_2 + 227.0 \text{ Kcal}$$

then NM-PMMA/Al (x=0.30) is the stoichiometric mixture. For NM-PMMA/Al (x<0.30) mixtures, maximum temperatures increase with aluminum concentration x. For NM-PMMA/Al (x>0.30) mixtures, maximum temperatures decrease with x, because unreacted aluminum remains. The temperature increase by aluminum reactions was calculated for the assumptions: 1) Al reacts only with H_2O. 2) Specific heat of detonation products of NM is equal to that calculated with BKW equation of state (Chéret 1971). 3) Specific heat of Al and Al_2O_3 are equal to those at standard conditions. The temperature increase ΔT_{cal} was given by the following equation:

$$\Delta T_{cal} = \frac{\Delta Q}{Y(D.P.)C(D.P.) + Y(Al)C(Al) + Y(Al_2O_3)C(Al_2O_3)}$$

where $Y(i)$ is the mass fraction of i component; $C(i)$ the specific heat of i component; ΔQ the reaction heat ($2Al + 3H_2O \longrightarrow Al_2O_3 + 3H_2$) per kg of mixture; and D.P. the detonation products of NM.

The variation of ΔT_{cal} is shown in Fig. 6. Though ΔT_{cal} is a very rough approximation, good correlation between ΔT and ΔT_{cal} was obtained. Numerical calculations for NM-PMMA/Al mixtures (Fig. 7) indicate that the maximum CJ temperature occurs at x∿0.37, because at chemical equilibrium all available oxygen is associated with aluminum (Tanaka 1983). The experimental results suggest that aluminum reacts only with H_2O produced by NM detonation.

Conclusions

The mechanism of reactions of aluminum particles in detonation products was investigated by mean of optical pyrometers. Results of experiments with NM-PMMA/LiF mixtures, when compared with those of numerical calculations, showed that there is negligible heat transfer to aluminum and lithium fluoride particles in the reaction zone of NM. The temperatures of detonation products of NM-PMMA/Al mixtures increase because of the reactions of aluminum particles, and a considerable part of aluminum must have reacted within the first 1 μs. This fast reaction rate of aluminum particles is consistent with other experimental results obtained with the mixtures of gaseous explosives and aluminum particles and the mixtures of TNM and aluminum particles.

The estimation of temperature increase by aluminum reactions suggested that aluminum particles react only with H_2O produced by NM detonation.

Acknowledgments

The authors are grateful to Y. Sarrazin (Laboratoire d'Energétique et de Détonique, E.N.S.M.A.) for his technical assistance. The authors are also indebted to K. Tanaka (National Chemical Laboratory for Industry, Japan), who offered the results of numerical calculations with KHT equation of state.

References

Barker, L. M. and Hollenbach, R. E. (1970) Shock- wave studies of PMMA, fused silica and sapphire. J. Appl. Phys. 41(10), 4208-4226.

Bouriannes, R., Moreau, M., and Martinet, J. (1977) Un pyromètre rapide à plusieures couleurs. Rev. Phys. Appl. 12(5), 893-899.

Chéret, R. (1971) Contribution à l'étude numérique des produits de détonation d'une substance explosive. Rapport de Commissariat à l'Energie Atomique, R-4122, Paris, France.

Kato, Y. and Brochet, C. (1977) Cellular structure of detonation in nitromethane containing aluminium particles. Proceedings of 6th Symposium on Detonation, pp. 124-131, Office of Naval Research, Arlington, Va.

Kato, Y., Bouriannes, R., and Brochet, C. (1979) Mesure de température de luminance des détonations d'explosifs transparents et opaques. Proceedings of HDP Symposium, pp. 439-449, Commissariat à l'Energie Atomique, Paris, France.

Kinslow, R. (1970) High-Velocity Impact Phenomena. Academic Press, New York.

Mader, C. L. (1979) Numerical Modeling of Detonations. University of California Press, Berkeley, Calif.

Mallory, H. D. and McEwan, W. S. (1961) Transparency of glass and certain plastics under shock attack. J. Appl. Phys. 32(11), 2421-2424.

McGuire, R. R., Ornellas, D. L., and Akst, I. B. (1979) Detonation chemistry: diffusion control in non-ideal explosives. Propel. Explos. 4, 23-26.

McGuire, R. R., Ornellas, D. L., Helm, F. H., Coon, C. L., and Finger, M. (1983) Detonation chemistry: an investigation of fluorine as an oxydizing moiety in explosives. Proceedings of

7th Symposium on Detonation, pp. 940-951, Naval Surface Weapons Center, White Oak, Md.

Moulard, H., Fauquignon, C., Lichtenberger, M., and Lombard, J. M. (1979) Détonation de mélange nitrométhane-Al-NO$_2$. Proceedings of HDP Symposium, pp. 293-307, Commissariat à l'Energie Atomique, Paris, France.

Tanaka, K. (1982) Study of detonation properties of high explosives using the intermolecular potential model (I). Kogyo Kayaku 43(5), 239-248.

Tanaka, K. (1983) Detonation properties of condensed explosives computed using the Kihara-Hikita-Tanaka equation of state. National Chemical Laboratory for Industry, Tsukuba, Japan.

Urtiew, P. A. (1974) Effects of shock loading on transparency of sapphire crystals. J. Appl. Phys. 45(8), 3490-3493.

Veyssière, B., Kato, Y., Brochet, C., Bouriannes, R., and Manson, N. (1983) Pyrometric studies of Al combustion in the wake of two-phase detonations. Archivum Combustionis 3(3), 151-160.

Chapter IV. Explosions

Theory of Vorticity Generation by Shock Wave and Flame Interactions

J. M. Picone,* E. S. Oran,* J. P. Boris,† and T. R. Young Jr.*
Naval Research Laboratory, Washington, D.C.

Abstract

We present detailed numerical and theoretical studies of vorticity generation by the interaction of a weak planar shock with an azimuthally symmetric flame. The calculations are two-dimensional and correspond to a cylindrical flame. To analyze the fluid-dynamic aspects of the problem, we exclude chemical reactions and model the flame as a region of reduced density and elevated temperature. We find that the rotational flows associated with the vorticity distribution are long-lived and can produce a significant distortion of the heated region. The results of our numerical simulations and the estimates of the nonlinear theory are quite consistent. We also compare the nonreactive numerical simulations with the experimental data of Markstein on shock wave and flame interactions in a stoichiometric mixture of n-butane and air. The numerical simulations reproduce most of the major experimental observations. We show that conventional Rayleigh-Taylor instability theory, which assumes a very small initial perturbation, does not provide a viable description of vorticity generation in Markstein's experiment. Instead, a nonlinear treatment based on the finite misalignment and finite interaction time of the pressure gradient associated with the shock and the density gradient of the flame is necessary. The effects of chemical reactions are also clarified through the comparison of numerical simulation and experiment.

Presented at the 9th ICODERS, Poitiers, France, July 3-8, 1983. This paper is declared a work of the U.S. Government and therefore is in the public domain.
*Laboratory for Computational Physics.
†Chief Scientist, Laboratory for Computational Physics.

Introduction

The interaction of pressure waves with density gradients is a fundamental source of long-lived vorticity in fluids (Chu and Kovásznay 1957; Picone and Boris 1983). This mechanism is particularly important in combustion, since the release of chemical energy produces both pressure and density disturbances in the fluid. These disturbances then interact, producing significant vorticity. The rotational motion associated with the vorticity field enhances mixing and introduces additional fluctuations in the flow variables. The scale lengths which characterize these secondary fluctuations are roughly one-half those of the local density gradients which exist prior to interacting with given pressure waves (Picone and Boris 1983). Because the structures and sizes of the fluid interfaces change under the influence of the newly generated vorticity, local reaction rates change, amplifying the effects of the fluctuations and producing more local pressure waves. The process thus continues, progressively reducing the length scales of inhomogeneities in the flowfield. Through this mechanism, a flow which is originally laminar can quickly become turbulent. Such phenomena occur in almost every nonidealized high Reynolds number flow.

In this paper, we focus on the interaction of a weak, planar shock wave with an azimuthally symmetric region characterized by density and temperature distributions similar to those found in an expanding flame. Our calculations are two-dimensional and correspond to a cylindrical flame. This problem contains the essential features of vorticity generation by pressure wave and flame interactions. Ignoring the flame dynamics is acceptable for calculating the large-scale vorticity distribution, since the transit time of the shock across the heated region is quite short compared to the time scale of flame propagation. Markstein studied a similar case experimentally for a reactive medium, using a long, vertical shock tube with a 30-cm combustion chamber at the bottom and a diaphragm 90 cm from the top. Figure 1 shows Schlieren photographs from one of Markstein's experiments. A weak shock (pressure ratio ~ 1.3) passes through a roughly spherical flame approximately 15 cm from the bottom of the chamber, which contains a stoichiometric mixture of n-butane and air. In the first frame, we see the incident shock less than 1 cm from the flame boundary.

SHOCK WAVE AND FLAME INTERACTIONS

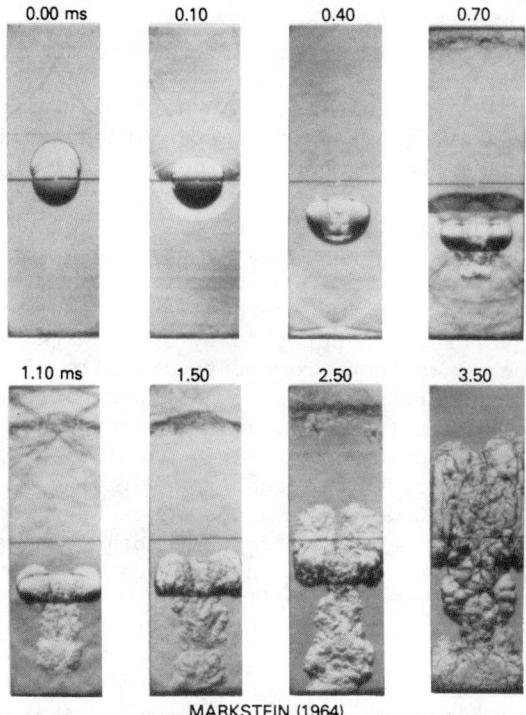

MARKSTEIN (1964)

Fig. 1 "Interaction between a shock wave and a flame of initially roughly spherical shape. Pressure ratio of incident shock wave 1.3; stoichiometric n-butane-air mixture ignited at center of combustion chamber 8.70 ms before origin of time scale." (Rudinger 1958; Markstein 1964) (Reprinted by permission of Pergamon Press, Inc.).

The flame actually appears to be more oblong than spherical. In the second photograph, for which time is defined to be 0.10 ms, we see a curved rarefaction wave moving upward from the flame while the upper flame boundary has been compressed by the flow behind the incident shock. By 0.40 ms, a vortex ring has formed due to the interaction of the shock wave and the flame. The enhanced flow at the center of the ring pulls unreacted gas through the flame, causing the gas to ignite. As the newly ignited gas emerges, a fine-grained turbulent burning zone develops (apparent at 0.70 ms).

To analyze the fluid dynamic aspects of vorticity generation, we shall study the closely related situation in which a planar shock interacts with a hot, low density region in a nonreactive fluid. Past treatments of this problem have relied on the linearization of all or part of the relevant equations or have used a perturbation expansion (Rudinger 1958; Markstein 1964). Those approaches are valid for cases with vanishingly small perturbation amplitudes and time scales but cannot provide a quantitative picture of the large-scale flows which result from the initial asymmetries, for which the characteristic length scale is the flame radius and the interaction time scale is finite.

In the next section, we outline a simple nonlinear theory of vorticity generation by shock wave and flame interactions. We then present the results of a two-dimensional numerical simulation of the nonreactive case. The simulation provides a clear picture of the formation of a vortex ring. The calculation also provides a calibration of constants arising in the theory. Finally, we discuss the shock wave and flame interaction as a mechanism for developing and enhancing turbulence and for determining the set of turbulent scales observed experimentally.

Nonlinear Theory

Equation (1) gives the rigorous, inviscid equation for the time evolution of vorticity,

$$\frac{d\xi}{dt} + \xi \nabla \cdot v = \xi \cdot \nabla v + (\nabla \rho \times \nabla P)/\rho^2 \qquad (1)$$

where v is the fluid velocity, $\xi = \nabla \times v$ the vorticity, ρ the density, and P the pressure. All variables are functions of the position r and time t. Equation (1) provides a direct mechanism for the generation of vorticity by shock wave and flame interactions, since the righthand side contains a source term which is proportional to the cross product of the density and pressure gradients. Whenever the local pressure and density gradients are misaligned, the source term will be nonzero and the production of vorticity will occur. In Fig. 1, the experimental case of interest, the pressure gradient associated with the planar shock is parallel to the axis of the shock tube while the density gradient of the spherical flame is directed approximately radially outward from the center. Large-scale vorticity in the form of a vortex ring results from the interaction.

An estimate of vorticity generation in the nonreactive case by this mechanism requires the integration of Eq. (1) in both space and time. We may, however, simplify our task by working in two dimensions (Cartesian), since the induced rotational flows, which are cylindrically symmetric, differ from those of a vortex ring by geometrical factors of order one [see sentence following Eq. (5)]. We then exploit the fact that after the shock has passed through a circular region of density variation in the x-y plane, the residual flowfield consists of a pair of vortices of strength $\pm \kappa_z \equiv \pm \kappa$ (Picone and Boris 1983). Figure 2 defines the notation for this section and the Appendix. For the purpose of integrating Eq. (1) analytically, we align the x axis with the direction of propagation of the shock and place the origin at the center of the density depression. The quantity S_0 denotes the radius of the heated region, and the vortices will be centered at $(x,y) \approx (0, \pm \bar{y})$, where $\bar{y} \lesssim S_0$. That is, in the case of a planar shock propagating downward through a circular heated region in a two-dimensional vertical chamber, there will be one vortex on either side of the vertical axis of the chamber. The pair of vorticies will be roughly centered on the heated region, and the associated rotational flows will be oppositely directed. The strength of one vortex at time t is equal to the integral of the vorticity over the half plane containing the vortex. For the case of

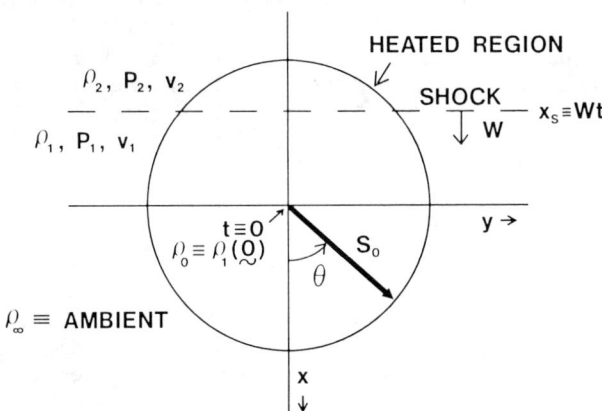

Fig. 2 Diagram defining terms for calculation of vortex strength in the section on Nonlinear Theory and in the Appendix.

interest, the vortex strength is

$$\kappa_z(t) = \int_0^\infty dy \int_{-\infty}^\infty dx \; \xi_z(x,y,t) \qquad (2)$$

which satisfies the equation (see the Appendix)

$$\frac{d\kappa_z}{dt} \approx \int_0^\infty dy \int_{-\infty}^\infty dx \; [(\nabla\rho \times \nabla P)_z / \rho^2] \qquad (3)$$

In a formal sense, the residual vortex strength is then

$$\kappa \equiv \lim_{t \to \infty} \kappa_z(t) \qquad (4)$$

In practice, the generation of vorticity occurs over a finite-time interval $(-\tau/2, \tau/2)$. For an azimuthally symmetric density depresssion with a characteristic radius S_0, the time interval τ is approximately equal to $2S_0/W$, where W is the average velocity with which the shock propagates across the region. Because this time is much shorter than the time required for the resulting rotational flows to affect the density gradient appreciably, we may ignore the feedback effects of the vorticity on the generation process.

Given the value of κ, we may compute the fluid velocity along the line bisecting the vortex pair by summing the flows induced by each vortex separately. Because the azimuthal velocity of an isolated vortex filament has the magnitude $|v_\theta| = |\kappa|/2\pi r$, where r is the radial distance from the center of the filament, we obtain

$$|v_x(x,0)| \approx |\kappa| \; \bar{y} / \pi(x^2 + \bar{y}^2) \qquad (5)$$

Notice that the maxiumum value is $|\kappa|/\pi\bar{y}$, which agrees closely with $|\kappa|/2\bar{y}$, the corresponding velocity for a vortex ring. We may define the mixing time scale τ_{mix} as the interval over which a fluid element travels from $(-S_0, 0)$ to $(S_0, 0)$ under the influence of the residual flows. Integrating Eq. (5) gives

$$\tau_{mix} = 2\pi S_0 (S_0^2/3 + \bar{y}^2) / |\kappa|\bar{y} \qquad (6)$$

In Fig. 1, we notice that the compression of the hot region by the flows behind the shock has reduced the distance that the fluid elements must travel to reach the lower edge of the heated region. Thus our τ_{mix} will be larger than the actual transit time.

In the Appendix, we compute the integral in Eq. (4) for the case of a planar shock passing through a cylindrically symmetric density depression like that in Fig. 1. We find that, prior to the return of the reflected shock from the end of the shock tube,

$$\kappa \approx 2 v_2 (1 - v_2/2W) S_0 \ln(\rho_\infty / \rho_0) f$$

$$\approx 2 v_2 (W - v_2/2) (\tau/2) \ln(\rho_\infty / \rho_0) f \qquad (7)$$

where $f \leq 1$ is a form factor which varies with the initial density profile. In Eq. (7), v_2 is the flow velocity behind the shock in the laboratory frame; W is the shock velocity; ρ_∞ is the ambient density of the unburned medium; and ρ_0 is the density at the center of the spherical flame. For a shock of pressure ratio 1.3 and $\gamma = 1.4$, the Rankine-Hugoniot relations give us $v_2 \approx$ W/6 and W $\approx 1.12 c_s$, where c_s is the ambient sound speed prior to passage of the shock. In our simulation $S_0 \approx 2.5$ cm, $\rho_\infty/\rho_0 \approx 9.5$, and $c_s \approx 3.4 \times 10^4$ cm/s. Equation (7) gives us a vorticity strength of $\kappa \leq 6 \times 10^4 \text{cm}^2/\text{s}$. For $\bar{y} \leq S_0$, the time scale τ_{mix} for a fluid element to be pulled from one side of the heated region to the other ($\Delta x = 2S_0$) under the influence of the residual vorticity is ≤ 900 µs. This is consistent with the simulation results (Fig. 5) and with Markstein's experiment (Fig. 1), which shows that fluid is pulled through the center of the burning zone prior to the arrival of the reflected shock approximately 600 µs after the vorticity is generated. We note again that in the spherical case the fluid velocity through the center of the vortex ring is 50% higher than that for the two-dimensional model which we have used here. Thus the shorter time scale (τ_{mix}) of Markstein's experiment is consistent with the estimate given above. Because the region of density variation is no longer azimuthally symmetric, the reflected shock should increase the number of vortex centers.

Numerical Simulation

We have used the two-dimensional reactive shock model (Oran et al. 1983), which is based on the code FAST2D (Book et al. 1981), to simulate the interaction between a weak planar shock and a cylindrically symmetric density depression in air. The algorithm employs time-step splitting in conjunction with flux corrected transport (FCT) (Boris and Book 1976) to solve the equations for conservation of mass, momentum, and energy. The reactive shock model also contains algorithms for chemical kinetics, energy release, and thermal and molecular diffusion. We have calibrated the code extensively through studies of shocks on wedges in nonreactive fluids (Book et al. 1981) and studies of detonation cell structure, which required models of energy release (Oran et al. 1983). In the present simulations, we used Cartesian geometry and did not include chemical reactions or diffusive transport processes.

The Cartesian grid consisted of 150 x 50 cells of dimension 0.2 cm on each side. The grid remained fixed throughout the calculation. Thus the simulated chamber was 30 x 10 cm, as in Markstein's experiment. We have aligned the x axis with the axis of the shock tube. The simulation had inflow boundary conditions with $\underline{v} = (v_2, 0)$ at the side from which the shock propagated, where v_2 is the velocity behind the shock in the laboratory frame. The other boundaries were reflecting. A typical time-step was $\Delta t_s \approx 1$ μs, where the subscript $\underline{s}$ indicates simulated time. Figure 3 is a density contour diagram approximately 90 μs (simulated time) after the simulation began. The density contours run from 3.0×10^{-4} to 1.26×10^{-3} g/cm^3 in equal increments. Figure 4 shows the initial density and temperature profiles more clearly. The density profile had the convenient functional form

$$\rho(r) = \rho_\infty + (\rho_0 - \rho_\infty)/[1+(r/S_0)^2]^2 \qquad (8)$$

The shock pressure ratio was 1.3, again similar to Markstein's experiment (Rudinger 1958; Markstein 1964), and the shock moved from top to bottom. The ambient pressure was 1.0 atm, and the ratio of specific heats $\gamma (\equiv c_p/c_v)$ was constant at the value 1.4. The ambient density (ρ_∞) and temperature (T_∞) were 1.17×10^{-3} g/cm^3 and 300 K, respectively, while the minimum density (ρ_0) and peak temperature (T_0) were 1.24×10^{-4} g/cm^3 and 2840 K. To compare simulation times (t_s) with those in Markstein's experiment, one should subtract approximately 150 μs from the simulated time.

Figures 5-7 show the evolution of the density, pressure, and vorticity, respectively. The x axis is

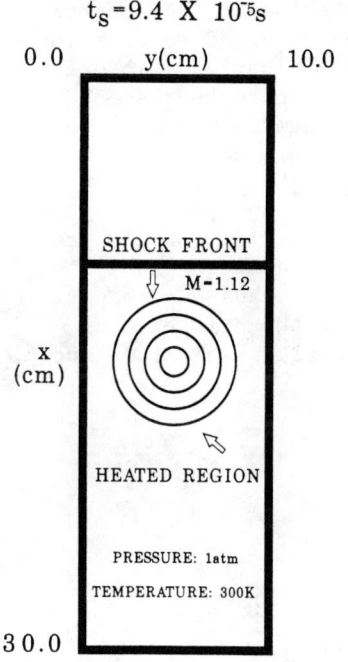

Fig. 3 Density contour diagram shortly after the simulation began. The dimensions of the chamber are similar to those in Fig. 1. Gas flows into the chamber from the top boundary.

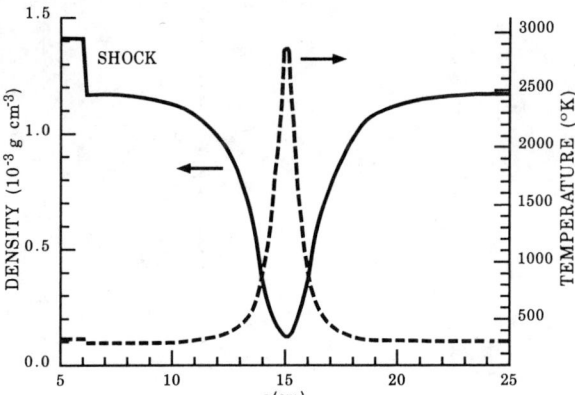

Fig. 4 The density and temperature distributions at simulation time $t_s = 0$ show the shock at the left and the heated region in the center of the grid.

vertical to facilitate comparisons with Fig. 1; thus the shock initially propagates downward in the figures. There are striking similarities to the experimental photographs (Fig. 1). Figure 4 shows that the density distribution is compressed by the shock and the entire

heated region is pulled toward the bottom of the chamber
by the flows behind the shock. The reflected rarefaction
wave and curved transmitted shock are clearly visible in
Fig. 6. The curvature of the latter occurs because of
the increased sound speed inside the heated region. An
interesting feature is the Mach structure caused by
reflection of the curved shock by the chamber. The
sequence of density diagrams clearly demonstrates the
effects of the vortex pair, which has pulled ambient gas
from the top of the heated region to the bottom in a time
$\tau_{mix} \lesssim 1$ ms, consistent with both Fig. 1 and our theory.
The vorticity contours in Fig. 7 show the existence of
two long-lived vortices which are equal in magnitude and

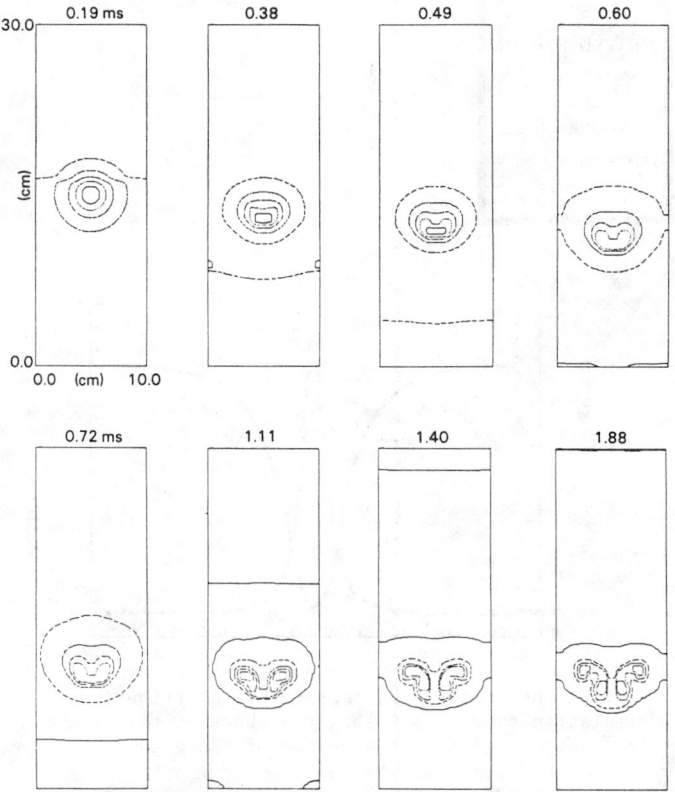

Fig. 5 Density contour diagrams at given times t_s. In
the first three frames the six contours range from 3×10^{-4} to 1.5×10^{-3} g/cm^3 in equal increments while in the
last five, the contours range from 3×10^{-4} to 1.6×10^{-3} g/cm^3. Notice the Mach structure at the channel
boundary at $t_s = 384$ μs.

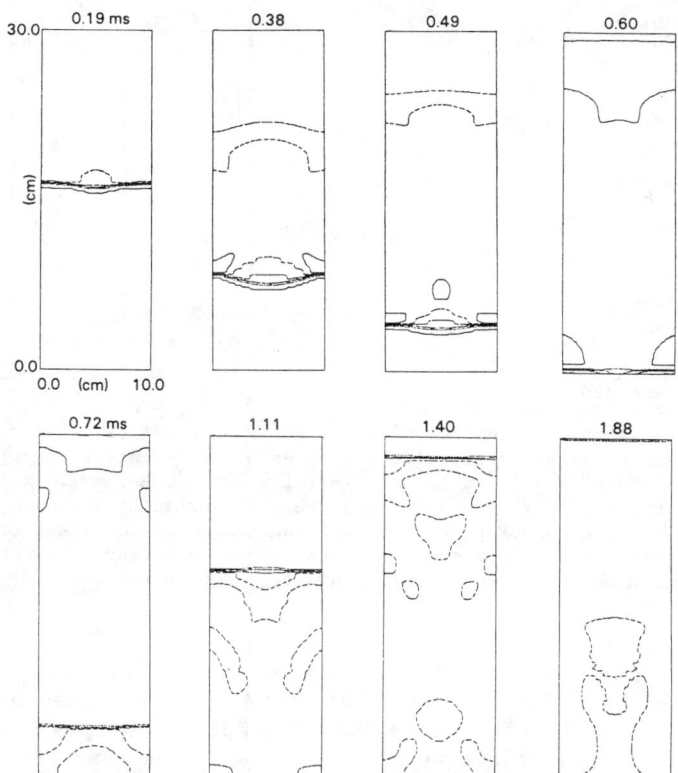

Fig. 6 Pressure contour diagrams at given times t_s. In the first three frames the contours range from 1.02×10^6 to 1.35×10^6 dyne/cm^2 in equal increments. In the last five the range is 1.3×10^6 to 1.8×10^6 dyne/cm^2. Again notice the Mach structure prior to reflection at the bottom boundary.

oppositely directed. Sometime after the shock has passed through the heated region ($t_s \sim 500$ μs), the simulation gives a vortex strength $\kappa \sim 3 \times 10^4$ cm^2/s, which is consistent with the theoretical estimate.

Another point of considerable interest is the enhancement of the vorticity by subsequent shocks. The vorticity has a maximum magnitude of approximately 1.1×10^4 s^{-1} just before the reflected shock strikes the heated region ($t_s = 722$ μs), and 1.8×10^4 s^{-1} afterward ($t_s = 1.4$ ms). Note the reduction in size and the increased number of the vortex centers between those times in Fig. 7, indicating the role of successive pressure waves in the evolution of a structure more like that found in turbulent flows.

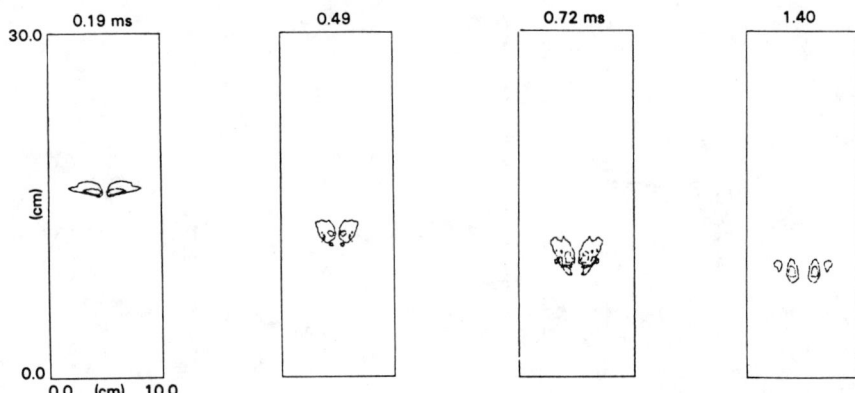

Fig. 7 Vorticity contour diagrams at given times t_s. The range of vorticity values (respectively from earliest to latest times) are $\pm 3 \times 10^3$ s^{-1}, $\pm 1.5 \times 10^4$ s^{-1}, $\pm 1.1 \times 10^4$ s^{-1}, and $\pm 1.8 \times 10^4$ s^{-1}. The number of vortex centers and the peak vorticity increase after the reflected shock passes through the heated region (last frame), as anticipated. Corresponding vortices in each frame are oppositely directed and oriented to pull fluid from top to bottom.

Because the release of chemical energy was not included in the calculation and because of the resolution limitations of the calculation, several noteworthy differences exist between the experiment and the simulation. First we notice in the density contour diagrams that the heated region spreads and that the innermost contour disappears by t_s = 600 µs, indicating that cooling occurs as the ambient gas is pulled through the center by the vortices. In a reactive medium, the ambient gas would ignite, at least maintaining the temperature and producing pressure waves, which would hasten the transition to turbulence. The reactions would also alter the reflected shock. Figure 1 (t = 0.70 ms) shows that the reflected shock wave actually emerges as a dark band above the vortex ring. Finally, the limited resolution of the calculation and the omission of three-dimensional effects have prevented much of the cascade to small-scale turbulence, which appears in Fig. 1. Despite these limitations, the simulation has revealed major fluid dynamic effects in shock wave and flame interactions.

Comparison with Rayleigh-Taylor Instability Theory

Previous analyses of shock wave and flame interactions have used or have suggested using a modified version of conventional Rayleigh-Taylor instability theory [see, for example, Rudinger (1958) and Markstein (1964)]. Indeed, Markstein (private communication) has pointed out that such an analysis, with proper choices of parameters, gives a result similar to Eq. (7). In the usual Rayleigh-Taylor picture, the length scale of the initial perturbation is infinitesimal or "small", while the source term in Eq. (1),

$$\underset{\sim}{S} = \underset{\sim}{\nabla}\rho \times \underset{\sim}{\nabla}p/\rho^2 \qquad (9)$$

is nonzero for a "long" period of time compared to the growth rate of the perturbation. Markstein (1964) modified this to account for the short time scale of the shock passage through a perturbed (rippled) flame boundary. He demonstrated the phenomenon experimentally by passing a smooth flame through a wire grid to perturb the flame boundary prior to the arrival of a planar shock. As expected, the small ripples in the flame surface were amplified by the perturbed flowfield caused by the shock. These features were, however, soon overwhelmed by nonlinear effects. Because the unperturbed flame was hemispherical, the passage of the planar shock produced large-scale vorticity, which obliterated the smaller features.

In this paper, we have been concerned with the more general, nonlinear description of vorticity generation. Thus we are addressing situations in which the length scale of the perturbation is finite (flame radius) and the source term is also nonzero over the time period during which the shock passes through the entire flame. Because both scales are finite, we require a nonlinear integration of Eq. (1), which covers the entire spectrum of inviscid vorticity-generating phenomena. We note that, even for perturbations of small spatial and temporal extent, the analysis of the flow field over long time periods must include nonlinear effects.

Summary

We have presented both an analytic theory and a numerical simulation of the vorticity generated by a planar shock wave passing through a hot region similar to that of a flame. In order to extract the basic fluid dynamic mechanism in this interaction, we have restricted our work to two-dimensional Cartesian coordinates and have excluded chemical reactions. This is a good approximation for the calculation of the large-scale vorticity distribution, since the laminar flame moves much slower than the shock. Our analysis has shown that, after the shock passes through the flame, a residual rotational flowfield remains. This flowfield is equivalent to that of a vortex pair in the two-dimensional Cartesian case (or a vortex ring in the two-dimensional axisymmetric case) characterized by a circulation κ [Eq. (7)]. The separation $\bar{y}$ of the filaments is $\lesssim S_0$, the radius of the heated region. We have also defined and computed a mixing time scale for the vorticity distribution which is consistent with Markstein's experiment (Fig. 1) and the simulation.

Our numerical simulation of a planar shock interacting with a cylindrically symmetric region of reduced density used our reactive shock model, which is based on the FAST2D computer code. The computer model was identical to that used previously to study detonation structure (Oran et al. 1983); only the boundary and initial conditions were different. From the simulations, we have determined the time development of the flowfield as well as the flows which remain after the shock has passed through the flame. The simulations agree well with experiment and theory, and show that conventional Rayleigh-Taylor instability theory does not provide an adequate description of vorticity generation. The source of large-scale vorticity for the case studied here is the initial large scale misalignment of pressure and density gradients present in the fluid.

The general framework presented in this paper applies to vorticity generation by any pressure perturbation interacting with a density gradient. Examples of this phenomenon include sound waves or shocks passing through flames, hot spots, or boundary layers; the key ingredient is the misalignment of the pressure and density gradients as the interaction occurs. We conclude that pressure waves produced in one portion of a reacting medium will generate or enhance turbulence as the waves encounter density fluctuations within the medium. The scale lengths of the additional turbulence will be less than the spatial size of the original density fluctuations. We have thus described and demonstrated a mechanism for producing turbulence in reactive systems and have developed a framework for calculating the strength and the scales on which the turbulence is generated.

Appendix: Planar Shock Interaction

We shall analyze the case shown in Fig. 2 by considering (in two dimensions) a planar shock interacting with a circular heated region. The symbol W denotes the velocity of the shock through the burning zone, and ρ_2, v_2, and P_2 are the density, fluid velocity, and pressure behind the shock. The corresponding quantities ahead of the shock are ρ_1, v_1 (≈ 0), and P_1. The hot gas has a radius S_0, and the time $t \equiv 0$ occurs when the shock reaches the center of the burning zone. Here we shall assume that the shock is weak (pressure ratio $P_2/P_1 \approx 1.3$) and that the pressure and density ratios do not change significantly when the shock enters the burning zone. The circular shape of the hot flame and the elevated speed of sound in the interior of the flame cause the shock to become curved upon passing through the heated region. Our analysis will not account for this curvature. We represent the flowfield of the shock as follows:

$$\begin{bmatrix} \rho \\ P \\ v \end{bmatrix} \equiv \begin{bmatrix} \rho_1 \\ P_1 \\ v_1 \end{bmatrix} + \begin{bmatrix} \rho_2 - \rho_1 \\ P_2 - P_1 \\ v_2 - v_1 \end{bmatrix} f_\alpha(s, \theta, W, t) \tag{A1}$$

where s and θ define the position of a point relative to the center of the heated region, α is a parameter defining shock thickness, f_α satisfies the relation

$$\lim_{\alpha \to 0} f_\alpha \equiv \begin{cases} 1 - h(\cos\theta), & t = 0 \\ h(s\cos\theta/Wt - 1), & t < 0 \\ 1 - h(s\cos\theta/Wt - 1), & t > 0 \end{cases} \tag{A2}$$

and h(u) is the unit step function,

$$h(u) \equiv \begin{cases} 0, & u < 0 \\ 1, & u > 0 \end{cases} \tag{A3}$$

To insure the consistency of f_α in the limit $\alpha \to 0$, we choose $h(0) = 1/2$.

Having set the initial conditions, we must now derive an equation for the circulation or vortex strength $\kappa(t)$ from Eq. (1) in the main text. We note that the vorticity is generated in the region containing the heated gas, denoted here by F, which corresponds to the flame. The region F and the associated residual vorticity then move with the fluid behind the incident shock after the shock has passed through the flame. Given a volume of fluid $\Omega(t)$ containing the region F, we have (Meyer, 1982)

$$\int_{\Omega(t)} \left(\frac{d\vec{\xi}}{dt} + \vec{\xi}\, \vec{\nabla} \cdot \vec{v} \right) dV = \frac{d}{dt} \int_{\Omega(t)} \vec{\xi}\, dV \qquad (A4)$$

In this paper, we are considering only variations in the x-y plane, so that $\xi_z \neq 0$, $\xi_x = \xi_y \equiv 0$ and $\vec{\xi} \cdot \vec{\nabla}\vec{v} \equiv 0$. Since the integrands do not vary with z, Eqs. (1) and (A4) give us

$$\frac{d\kappa_z}{dt}(t) \equiv \frac{d}{dt} \int_{A(t)} \xi_z\, dA(t)$$

$$= \int_{A(t)} [(\vec{\nabla}\rho \times \vec{\nabla}P)_z / \rho^2]\, dA(t) \qquad (A5)$$

To determine the vorticity generated by a shock passing through F, we need to integrate eq. (A5) over the interval τ during which the shock interacts with F. In the case which we are considering, F has a radius $S_0 \sim 2.5$ cm. The interaction (shock transit) time will then be quite short compared to the time scales of flame propagation and vortex-induced rotational motion. The area A(t) containing F and nonzero vorticity will change due to the motion of the fluid behind the shock and the compression of fluid elements by the shock. For weak shocks these will also be small effects, and we shall assume that $A(t) \approx A_0$, where A_0 is the area containing F prior to the arrival of the shock. In other words, we assume that the Jacobian matrix of the transformation from dA(t) to dA_0 is approximately the identity matrix. Thus we obtain

$$\kappa_z(t) = \int_{-\tau/2}^{t} dt' \int_{A_0} dx\, dy\, [(\vec{\nabla}\rho \times \vec{\nabla}P)_z / \rho^2] \qquad (A6)$$

To compute $\kappa_z(t)$ for a single vortex in the upper half plane, we use

$$\kappa_z(t) = \int_{-\infty}^{t} dt \int_{-\infty}^{\infty} dx \int_{0}^{\infty} dy \, [(\nabla\rho \times \nabla P)_z / \rho^2] \quad (A7)$$

In Eq.(A7), we have extended the limits of the integrals to the entire half plane and $t \to -\infty$ for convenience in performing the theoretical calculations. This is reasonable as long the intensity of the interaction drops off rapidly as time decreases from $t = -\tau/2$.

We shall compute the vortex strength in Eq.(A7) in the limit that the shock has infinitesimal thickness ($\alpha \to 0$). We assume for convenience that the density profile of the heated region is given by

$$\rho(\underset{\sim}{s}, t) = \rho_\infty \exp[-g(s/S_0, t) \ln(\rho_\infty/\rho_0)] \quad (A8)$$

where $\underset{\sim}{s}$ is the displacement from the center of the region and $g(s/S_0, t)$ is a function defined so that $g(0, t) = 1$ and the limit of $g(s/S_0, t)$ as s approaches infinity is zero. For brevity we will use $g(s) \equiv g(s/S_0, t)$ in subsequent equations. In the limit $\alpha \to 0$, the source term in Eq. (1) for $t < 0$ becomes

$$\frac{1}{\rho^2} \nabla\rho \times \nabla P = -\underset{\sim}{e}_z \sin\theta \frac{\partial g}{\partial s} \ln(\frac{\rho_\infty}{\rho_0}) \delta(\frac{s \cos\theta}{Wt} - 1)$$

$$\times [\frac{-v_2 s \cos\theta}{Wt^2} + \frac{v_2^2 h(\frac{s \cos\theta}{Wt} - 1)}{Wt}] \quad (A9)$$

since the derivative $h'(u) = \delta(u)$. To derive Eq. (A9), we have computed ∇P from the equation of motion

$$\frac{d\underset{\sim}{v}}{dt} = -\frac{1}{\rho} \nabla P \quad (A10)$$

In computing κ from Eqs. (A7) and (4) (main text), we notice that the integral over the interval $(-\infty, 0)$ for $t < 0$ is identical to that over the interval $(0, \infty)$ for $t > 0$, so that we need only double the result for either interval to determine the residual vortex strength. [In practice the time integral is negligible outside the interval $(-\tau/2, \tau/2)$, where τ is the time scale of the shock and flame interaction.]

Equation (A7) now gives us

$$\kappa \approx 2 \ln\left(\frac{\rho_\infty}{\rho_0}\right) \int_{-\infty}^{0} dt \int_{\pi/2}^{\pi} d\theta(-\sin\theta) \int_{0}^{\infty} ds\, s$$

$$\times \left\{ \frac{\partial g}{\partial s} \delta\left(\frac{s \cos\theta}{Wt} - 1\right) \left[\frac{v_2^2 h\left(\frac{s \cos\theta}{Wt} - 1\right)}{Wt} - \frac{v_2 s \cos\theta}{Wt^2} \right] \right\} \quad (A11)$$

With a change of variables to $\beta = s \cos\theta/Wt - 1$, Eq. (A11) becomes

$$\kappa \approx 2 \ln\left(\frac{\rho_\infty}{\rho_0}\right) \int_{-\infty}^{0} dt \int_{\pi/2}^{\pi} d\theta(-\sin\theta) \frac{\partial g}{\partial \beta}\bigg|_{\beta=0}$$

$$\times \left[\frac{v_2^2 h(0)}{Wt} - \frac{v_2}{t} \right] \frac{Wt}{\cos\theta} \quad (A12)$$

Application of the chain rule of differentiation results in

$$\frac{\partial g}{\partial \beta}\bigg|_{\beta=0} = \frac{Wt}{\cos\theta} \frac{\partial g}{\partial s}\bigg|_{s=Wt/\cos\theta} \quad (A13)$$

which permits a change of variables to $\zeta = Wt/\cos\theta$. Equation (A12) is then

$$\kappa \approx 2 \ln\left(\frac{\rho_\infty}{\rho_0}\right) \int_{-\infty}^{0} dt\, Wt \left[\frac{v_2^2 h(0)}{Wt} - \frac{v_2}{t}\right] \int_{-Wt}^{\infty} d\zeta\, \frac{\partial g}{\partial \zeta}$$

$$= 2 \ln\left(\frac{\rho_\infty}{\rho_0}\right) v_2 [W - v_2/2] \int_{0}^{\infty} dt\, g(Wt) \quad (A14)$$

Because $W > v_2$ and $\rho_\infty > \rho_0$, we see that the direction of the vorticity in Eq. (A14) agrees with that in Figs. 1

and 5. We may define a form factor f by

$$f = \frac{2}{\tau} \int_0^\infty dt\, g(Wt) \qquad (A15)$$

where $\tau \equiv 2 S_0/W$, so that

$$\kappa \approx 2 v_2 (W - v_2/2) \left(\frac{\tau}{2}\right) \ln\left(\frac{\rho_\infty}{\rho_0}\right) f$$

$$\approx 2 v_2 (1 - v_2/2W) S_0 \ln\left(\frac{\rho_\infty}{\rho_0}\right) f \qquad (A16)$$

Note that we have previously defined $g(s) \equiv g(s/S_0)$. For a Gaussian form,

$$g(s/S_0) = \exp(-s^2/S_0^2) \qquad (A17)$$

we have $f = \sqrt{\pi}/2$, while for a square wave

$$g(s/S_0) = \begin{cases} 1, & s/S_0 \leq 1 \\ 0, & s/S_0 > 1 \end{cases} \qquad (A18)$$

$f = 1$. For the Bennett profile used in the simulation, Eq. (8), f will be less than $\sqrt{\pi}/2$ (Picone and Boris 1983).

Acknowledgments

The authors gratefully acknowledge support for this work by the Office of Naval Research. We also acknowledge support by the Defense Advanced Research Projects Agency (DARPA Order No. 4395) which funded the initial theoretical and numerical investigations leading to the work reported here. The comments of Mr. John Gardner, Dr. Raafat Guirguis, and Dr. K. Kailasanath have been most helpful. In addition, we thank Dr. George Markstein of Factory Mutual Research Corporation for his comments on his experiment and his kind assistance in obtaining a print of Fig. 1.

References

Book, D.L.,Boris, J.P., Kuhl, A.L., Oran, E.S., Picone, J.M., and Zalesak, S.T. (1981) Simulation of complex shock reflections from wedges in inert and reactive gas mixtures. Proceedings of the Seventh International Conference on Numerical Methods in Fluid Dynamics, Springer-Verlag, New York, pp. 84-90.

Boris, J.P. and Book, D.L. (1976) Solution of continuity equations by the method of flux-corrected transport. Methods in Computational Physics, Vol. 16, Academic Press, New York, pp. 85-129.

Chu, B. - T. and Kovásznay, L.S.G. (1957) Non-linear interactions in a viscous heat-conducting compressible gas. J. Fluid Mech.3, 494-514.

Markstein, G.H. (1964) Experimental studies of flame-front instability. Nonsteady Flame Propagation, AGARDograph No.75, Pergamon Press, Oxford, pp. 75-100.

Meyer, R.E. (1982) Introduction to Mathematical Fluid Dynamics Dover Publications, Inc., New York, Chapt. 1.

Oran, E.S., Young, T.R. , Boris, J.P. , Picone, J.M., and Edwards, D.H. (1983) A Study of detonation structure: the formation of unreacted gas pockets. Proceedings of the Nineteenth Symposium (International) on Combustion, The Combustion Institute, Pittsburgh, PA, pp. 573-582.

Picone, J. M. and Boris, J.P. (1983) Vorticity generation by asymmetric energy deposition in a gaseous medium. Phys. Fluids 26, 365-382.

Rudinger, G. (1958) Shock wave and flame interactions. Combustion and Propulsion, Third AGARD Colloquim, Pergamon Press, New York, pp. 153-182.

Interaction of Explosively Produced Shock Waves with Internal Discontinuities and External Objects

M. A. Fry*
Science Applications, Inc., McLean, Virginia
and
D. L. Book†
Naval Research Laboratory, Washington, D.C.

Abstract

Shock waves and blast wave phenomena include reactive and two-phase flows resulting from the motion of chemical explosion products. The structure of the blast wave includes contact discontinuities, rarefaction waves and backward-facing shocks (in nonplanar geometries). When the blast wave interacts with structural surfaces (external discontinuities), multiple reflections and refractions occur from both external and internal discontinuities. The ability to accurately model the wave interaction is dependent upon the choice of numerical algorithm and the representation of the two phases (air and chemical explosive products). The most recent version of the flux corrected transport (FCT) convective-equation solver has been used both in one and two dimensions to simulate chemical explosive blast waves reflecting from planar structures for yields ranging from 8 lb to 600 tons. One can relate the strength of the second reflected peak to the sharpness of the contact discontinuity, and thus measure the capability to predict all the salient features of the blast wave. Comparison with static pressure-time data taken from PBX 9404 experiments shows excellent agreement and yields new insight into the physical structure of the contact discontinuity. The flow patterns obtained in the

---

Presented at the 9th ICODERS, Poitiers, France, July 3-8, 1983. Copyright © American Institute of Aeronautics and Astronautics, Inc., 1984. All rights reserved.
*Senior Research Scientist.
†Senior Research Physicist, Laboratory for Computational Physics.

600-ton calculations reveal four different vortices, two forward and two reversed.

I. Introduction

In this paper we describe a series of calculations carried out as part of an ongoing effort aimed at studying explosion effects. The phenomena of chief interest to us include the following: peak overpressures and pressure histories on the ground as functions of yield, range, and height of burst (HOB), both at early times (prior to and during transition to Mach reflection) and at late times (after shock breakaway, with peak pressure in the range of tens of psi); velocity fields, particularly those associated with the toruses (both forward and reverse) in the neighborhood of the rising fireball; and the distribution of dust lifted off the ground by the winds and the structure of the cloud at the time of stabilization. We are interested in studying the nature of the gas dynamic discontinuities which appear, the vortices (both forward and reverse), and the pressure histories at stations along the ground.

The technique we have employed for this purpose is numerical modeling. One- and two-fluid hydrocodes based on the flux-corrected transport (FCT) shock-capturing techniques (Boris and Book 1976) have been used to simulate air-blast phenomena in one and two dimensions. FCT refers to a class of state-of-the-art fluid computational algorithms developed at NRL in the course of the past ten years with supersonic gas dynamic applications expressly in mind. We model two cases: an 8-lb charge of PBX at a HOB of 51.6 cm, and a 600-ton ammonium nitrate + fuel oil (ANFO) charge at a HOB of 166 ft, and analyze the results. In order to validate, initialize, and interpret these two-dimensional simulations, a number of ancillary calculations (mostly one-dimensional) were undertaken. The results are most conveniently exhibited in terms of plots of peak overpressure vs range and time, station histories, contour plots of combustion product and total density, velocity vector plots, and tracer particle trajectories. Examples of these are presented to illustrate our results and conclusions.

The plan of the paper is as follows: In the next section we discuss our numerical techniques and validation procedures. In Section III we discuss the free-field (one-dimensional) solution and indicate the salient features. Sections IV and V describe the 8-lb and 600-ton two-dimensional HOB calculations, respectively. In Section VI we summarize our conclusions and discuss their domain of validity.

II. Numerical Treatment

As described by Boris and Book (1976), FCT is a finite-difference technique for solving the fluid equations in problems where sharp discontinuities arise (e.g., shocks, slip surfaces, and contact surfaces). It modifies the linear properties of a second- (or higher-) order algorithm by adding a diffusion term during convective transport, and then subtracting it out "almost everywhere" in the antidiffusion phase of each timestep. The residual diffusion is just large enough to prevent dispersive ripples from arising at the discontinuity, thus ensuring that all conserved quantities remain positive. FCT captures shocks accurately over a wide range of parameters. No information about the number or nature of the surfaces of discontinuity need be provided prior to initiating the calculation.

The FCT routine used in the present calculations, called JPBFCT (an advanced version of ETBFCT), consists of a flexible, general transport module which solves one-dimensional fluid equations in Cartesian, cylindrical, or spherical geometry (Boris 1976). It provides a finite-difference approximation to conservation laws in the general form:

$$\frac{\partial}{\partial t} \int_{\delta V(t)} \phi dV = - \int_{\delta A(t)} \phi(\underline{u}-\underline{u}_g) \cdot d\underline{A} + \int_{\delta A(t)} \tau dA \qquad (1)$$

where ϕ represents the mass, momentum, energy, or mass species in cell $\delta V(t)$; $\underline{u}$ and $\underline{u}_g$ represent the fluid and the grid velocities, respectively; and τ represents the pressure/work terms. This formulation allows the grid to slide with respect to the fluid without introducing any additional numerical diffusion. Thus, knowing where the features of greatest interest are located, one can concentrate fine zones where they will resolve these features most effectively as the system evolves.

Let us describe the basic FCT algorithm we have used. We want to solve the one-dimensional continuity equation

$$\frac{\partial \rho}{\partial t} = - \frac{\partial}{\partial x} (\rho v) \qquad (2)$$

To advance one timestep on a uniform mesh with a given flow velocity v, we suppose the old cell-centered values ρ_j^0 and v_j are known.

A general three-point conservative finite-difference operator acting on ρ_j^0 can be written in the form

$$\rho_j^T = \rho_j^0 - \varepsilon_{j+\frac{1}{2}} \rho_{j+\frac{1}{2}} + \varepsilon_{j-\frac{1}{2}} \rho_{j-\frac{1}{2}}$$
$$+ \gamma_{j+\frac{1}{2}}(\rho_{j+1}^0 - \rho_j^0) - \gamma_{j-\frac{1}{2}}(\rho_j^0 - \rho_{j-1}^0) \quad (3)$$

where $\rho_{j+\frac{1}{2}} = \frac{1}{2}(\rho_j + \rho_{j+1})$. To obtain a first-order approximation in Eq. (1), we must define $\varepsilon_{j+\frac{1}{2}}$ to be the Courant number $v_{j+\frac{1}{2}} \delta t/\delta x$, where $v_{j+\frac{1}{2}} = \frac{1}{2}(v_j + v_{j+1})$; likewise, for a conventional scheme with second-order (spatial) accuracy, we must let $\gamma_{j+\frac{1}{2}} = \varepsilon_{j+\frac{1}{2}}^2$. In FCT, however, a numerical dif-fusion is applied to ρ_j^T according to

$$\rho_j^{TD} = \rho_j^T + \nu_{j+\frac{1}{2}}(\rho_{j+1}^0 - \rho_j^0) - \nu_{j-\frac{1}{2}}(\rho_j^0 - \rho_{j-1}^0) \quad (4)$$

Evidently the γ's and the ν's combine in the same way in ρ_j^{TD}, so we leave both unspecified for the moment.

If we now applied another diffusion operation, this time with coefficients $\mu_{j+\frac{1}{2}}$ but having the opposite sign, it would just tend to cancel the diffusion terms already present in ρ_j^{TD}. Instead we write for the new densities ρ_j^n produced by this "antidiffusion" operation

$$\rho_j^n = \rho_j^{TD} - \phi_{j+\frac{1}{2}}^C + \phi_{j-\frac{1}{2}}^C \quad (5)$$

where the "corrected" fluxes $\phi_{j+\frac{1}{2}}^C$ differ from the "raw" fluxes $\phi_{j+\frac{1}{2}} = \mu_{j+\frac{1}{2}}(\rho_{j+1}^0 - \rho_j^0)$ in two ways: we use $\{\rho_j^T\}$ instead of $\{\rho_j^0\}$ in the definition of the raw fluxes (this allows us to define γ so as to optimize some property in the difference scheme as will be seen shortly); and we <u>correct</u> the fluxes, so that the antidiffusion process (which evidently tends to make all gradients steeper) can enhance no extrema already present in $\{\rho_j^{TD}\}$ nor introduce any new ones. The simplest formula for achieving this in all possible situations is that used in "strong flux limiting":

$$\phi^c_{j+\frac{1}{2}} = S \cdot \{\max[0, \min(|\phi_{j+\frac{1}{2}}|, \Delta_{j-\frac{1}{2}}, \Delta_{j+3/2})]\} \quad (6)$$

where

$$S = \text{sign } \phi_{j+\frac{1}{2}} \text{ and } \Delta_{j+\frac{1}{2}} = \rho^{TD}_{j+1} - \rho^{TD}_j$$

At most points in a gently varying profile no correction is required and $\phi^c_{j+\frac{1}{2}} = \phi_{j+\frac{1}{2}}$. In that case we can perform a Von Neumann analysis, calculating the complex propagator or amplification factor $A = A_r + iA_i = \rho^n_j/\rho^o_j$ for a sinusoidal density profile $\rho^o_j = \exp(ijk\,\delta x)$, where k is the wave number. Writing $\beta = k\delta x$ and $\varepsilon = v\delta t/\delta x$, we expand in powers of β the amplification

$$|A| = 1 + A_2\beta^2 + A_4\beta^4 + \cdots \quad (7)$$

and the relative phase error

$$R = (1/\beta\varepsilon) \tan^{-1}(A_i/A_r) - 1 = R_2\beta^2 + R_4\beta^4 + \cdots \quad (8)$$

It is easy to show that the condition that A_2 vanish is

$$\gamma + \nu - \mu = \tfrac{1}{2}\varepsilon^2 \quad (9)$$

The condition that R_2 vanish is

$$\mu = \frac{1}{6} - \frac{1}{6}\varepsilon^2 \quad (10)$$

There remains one free parameter which can be chosen so as to build some desirable property into the algorithm (this is the reason for introducing γ). The choice $\gamma = \tfrac{1}{2}\varepsilon^2$ confers no particular benefit. Letting $\gamma = 0$ as in ETBFCT (Boris 1976) reduces the operation count and yields a useful general-purpose algorithm. There are two other obvious candidates: $\gamma = \tfrac{1}{4}\varepsilon^2$, which implies $A_4 = 0$, and

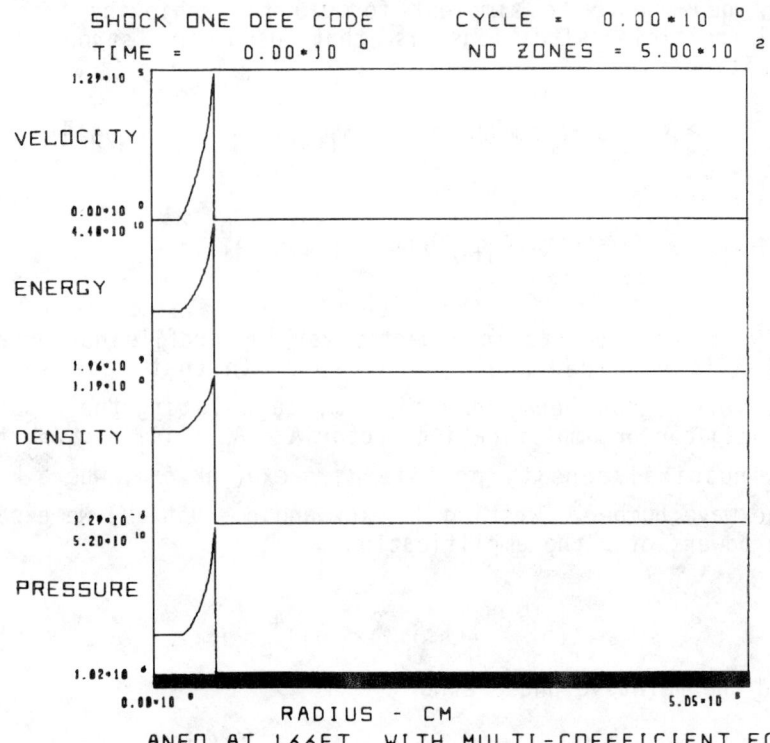

Fig. 1 Profiles of velocity, energy density, mass density, and pressure vs radius out to 5.05×10^5 cm as calculated with the Chapman Jouguet routine used to initialize the 600-ton problem. Variable ranges are $0 \leq v \leq 1.29 \times 10^5$ cm/s; $1.96 \times 10^9 \leq E \leq 4.48 \times 10^{10}$ erg/cm^3; $1.29 \times 10^{-3} \leq \rho \leq 1.19$ g/cm^3, $1.02 \times 10^6 \leq p \leq 5.20 \times 10^{10}$ dyne/cm^2.

$\gamma = 1/5 + (1/5)\varepsilon^2$, which implies $R_4 = 0$. We have elected to use the latter in the initialization (one-dimensional) stages of our airblast calculation because its superior dispersive properties and slightly higher residual diffusion at $\varepsilon = 0$ greatly reduce the size of the oscillations in the slope of the profiles of density, pressure, etc., in the rarefaction region of the free-field blast wave solution.

The same transport routine with $\gamma = 0$ was employed in the two-dimensional r-z code (called FAST2D) via coordinate splitting. A Jones-Wilkins-Lee (JWL) equation of state (EOS) was used for the detonation products and a real-air EOS was used outside the HE-air interface (Kuhl et al. 1981). The EOS routine was written in the form of a table lookup, using interpolation with logarithms to the base 16 computed by means of logical shifts (Young 1983). By thus taking account of the architecture of the machine (in these calculations, a 32-bit-word two-pipe Texas Instruments ASC) it was possible to generate a very efficient vector code, decreasing the time required for EOS calculations to a small fraction of that required for the hydro. The EOS specifies pressure as a function of density and internal energy. In mixed cells the combined pressure was calculated according to Dalton's law. The pressure terms in the momentum and energy equations are obtained by straightforward differencing and are added in immediately before the diffusion step is applied.

Also, at very late times (typically tens of seconds), gravity must be included in the vertical momentum equation to describe the effect of buoyancy. The equation we advance has the form

$$\frac{\partial(\rho v)}{\partial t} + \frac{\partial}{r\partial r}(r\rho uv) + \frac{\partial}{\partial z}(\rho v^2) + \frac{\partial}{\partial z}(p - p_a)$$
$$+ (p - p_a)g = 0 \qquad (11)$$

The ambient quantities p_a and ρ_a, which are taken from standard tables (Young 1983), are the same as those used in the boundary conditions. This guarantees that the ambient quantities represent a hydrostatic balance, which truncation errors will not destabilize.

The initial conditions were taken to be the self-similiar flowfield used by Kuhl et al. (1981), corresponding to a spherical Chapman Jouguet detonation at the time the detonation wave reaches the charge radius (Fig. 1). This was propagated with the one-dimensional spherical code until the detonation front attained a radius just smaller than the HOB, at which time the solution was laid down on the two-dimensional mesh (Fig. 2).

The boundary conditions were chosen to enforce perfect reflection on the ground and on the axis of symmetry $[(df/dn)_{bc} = 0, f = \rho, p, v^t,$ and $v^n_{bc} = 0]$, where t and n denote tangential and normal components, respectively, and outflow on the outer and the top boundaries $[(df/dn)_{bc} = 0, f = \rho, p, v^t, v^n]$.

For the two-dimensional calculations the mesh was typically 200 x 100. Fixed gridding was used to minimize numerical errors. The zone sizes for the two cases were 0.1 x 0.1 cm and 2.1 x 2.1 m, respectively. For the late-time 600-ton calculations, a fixed mesh with 100 zones in

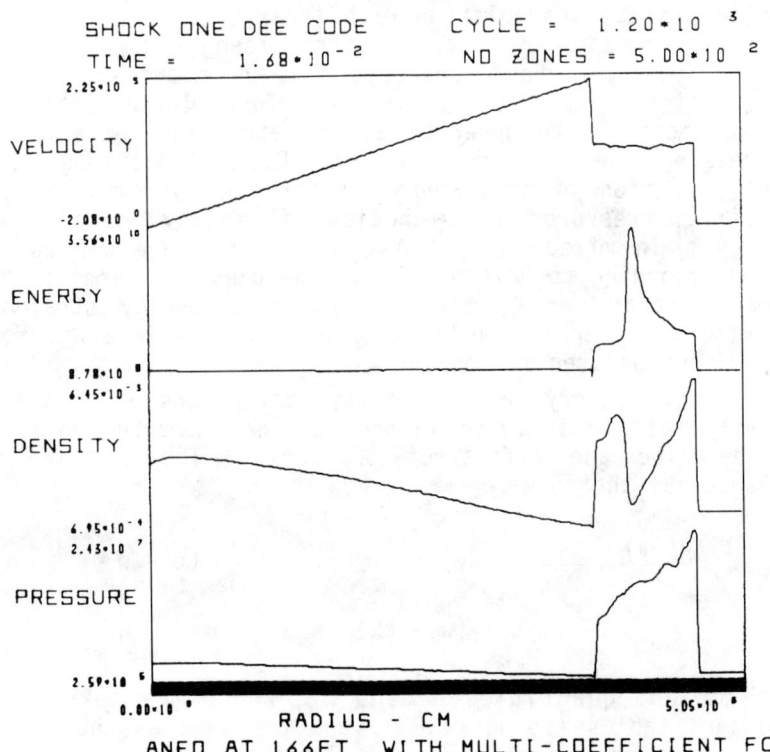

Fig. 2 As in Fig. 1, after being advanced 1200 timesteps using the algorithm described in 2 with $\gamma = 1/5 - (1/5)\varepsilon^2$. Ranges are $-2.08 \leq v \leq 2.25 \times 10^5$ cm/s, $8.78 \times 10^8 \leq E \leq 3.56 \times 10^{10}$ erg/cm^3, $6.95 \times 10^{-4} \leq \rho \leq 6.45 \times 10^{-3}$ g/cm^3, $2.59 \times 10^5 \leq p \leq 2.43 \times 10^7$ dyne/cm^2.

the radial and 200 zones in the vertical direction was used, with all cells of dimension 4.2 x 4.2 m. Time-centered pressures and velocities were obtained by calling the transport routine with half of the total timestep. Running times were approximately 6 µs/zone for each call to the one-dimensional convective equation solver (i.e., per equation per coordinate) plus an additional 20% for the equation-of-state routine.

III. Free-Field Solution

The well-known Sedov (1959) similarity solution for a point blast consists of a strong shock (postshock pressure much larger than ambient pressure) followed by a rarefaction wave (Fig. 3). The density distribution is extremely concave and approaches zero at the origin. The pressure approaches a constant as $r \to 0$, so that the temperature diverges strongly.

The solution used to initialize the HE problem, however, contains a number of features which are absent in the Sedov solution. These are most conspicuous in the density profile (Fig. 2), which exhibits a contact discontinuity between the HE products and shocked air, a secondary shock facing inward within the detonation products, and a gentle maximum near the origin. Because of this last feature, temperatures in the fireball are nowhere divergent.

It follows that the speed of sound in the fireball is much smaller in the Sedov case. This has two immediate consequences, one physical and one numerical. The first is that shocks propagate much more slowly through the fireball. The second is that the upper limit on the computational time-step, set by the Courant criterion

$$\max \left[(|v|+c/\delta x)\delta t \right] \leq 1 \qquad (12)$$

which usually is determined by conditions in the fireball, is much larger relative to the shock time scale τ_s = HOB/v_{shock} than it would be for a Sedov blast. Nevertheless, many tens of timesteps are required for the reflected shock to build up to its peak values. This, coupled with the property of FCT known as clipping, makes the rise time of the reflected shock substantially longer than it should be.

Clipping (Boris and Book 1976) is observed whenever an isolated sharp peak is propagated using FCT. The top of the peak is chopped off because of the action of the flux limiter, and the values on either side increase until a flat table-like structure three zones across appears, which is then propagated stably. When a shock reflects from a solid boundary, the values of the fluid quantities rise in the zone closest to the boundary toward their reflected values. The rise is delayed, however, by their tendency to "hold back" until a plateau three zones across can form. When it finally emerges, all three points rise in unison, eventually attaining the correct value.

When the reflected shock begins propagating back toward the origin it encounters the contact discontinuity, where it is partly transmitted and partly reflected. The reflected shock then proceeds outward until it reaches the ground (the end of the grid in the one-dimensional calculation), producing a second peak in the station history. The

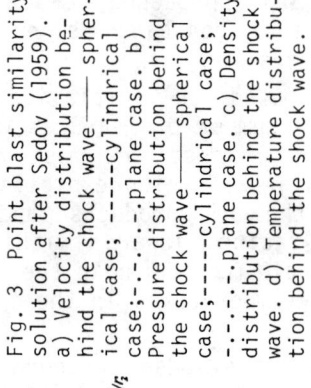

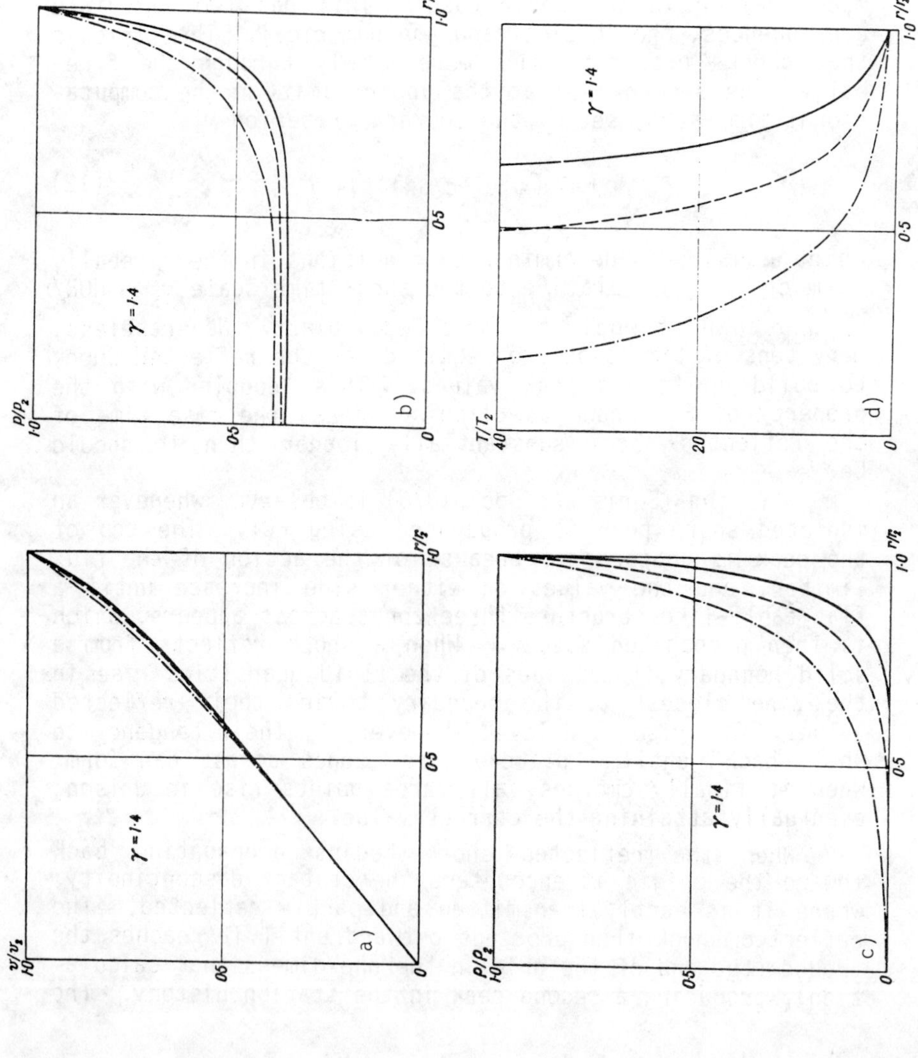

Fig. 3 Point blast similarity solution after Sedov (1959). a) Velocity distribution behind the shock wave —— spherical case; -----cylindrical case;-.-.-.-.plane case. b) Pressure distribution behind the shock wave —— spherical case;-----cylindrical case; -.-.-.-.plane case. c) Density distribution behind the shock wave. d) Temperature distribution behind the shock wave.

shock wave transmitted through the contact surface passes unhindered through the fireball at high speed until it reaches the upper boundary of the fireball, whereupon it reflects back.

IV. 8-lb PBX Explosion at 51.6 cm

The first two-dimensional calculation we carried out was a simulation of one of Carpenter's (1974) HOB experiments which used spherical 8-lb charges of PBX 9404 at 51.6 cm. We employed the Chapman Jouguet routine described in Kuhl et al. (1981) with parameters appropriate to PBX 9404. The results obtained by propagating this initial state with FCT on a mesh of 200 zones are shown in Fig. 4a. The fine-zone one-dimensional calculation was mapped onto the two-dimensional grid just prior to the onset of reflection. The solution was then advanced in time, with pressure being calculated from a real-time equation of state and a JWL equation of state for the combustion products.

A pressure sensor located beneath the burst point records one peak at the instant of arrival of the incident shock and subsequent peaks when the secondary (reflected) shock arrives at that point. Figure 5a shows the pressure history at a sensor located on the ground directly beneath the burst. Note that the second peak, produced when the reflected shock reflects off the contact surface back toward the ground, is considerably weaker than the primary peak.

To confirm this interpretation and test how the pressure signals depend on the numerical accuracy of our solution, we repeated the calculation using doubled resolution (400 zones) in the one-dimensional stage of the calculation. The resulting profiles, shown in Fig. 4b, evidently resolve the contact surface significantly better than do those produced by the 200-zone calculation. It would be surprising if reflection off this surface were not more efficient than with the coarsely resolved profiles.

When we lay the 400-zone solution down on the two-dimensional mesh and propagate it with FAST2D, we find that this expectation is borne out (Fig. 5b). Both of the calculated peaks are higher, but the second is now almost the same as the first (note that the time scales in Figs. 5a and 5b are different). Improving the resolution still further produces essentially no change in this result but it introduces a new phenomenon. The secondary shock can, after reflecting from the ground, reflect again off the contact surface if the latter is sufficiently well resolved. In fact, we have observed a third peak in some

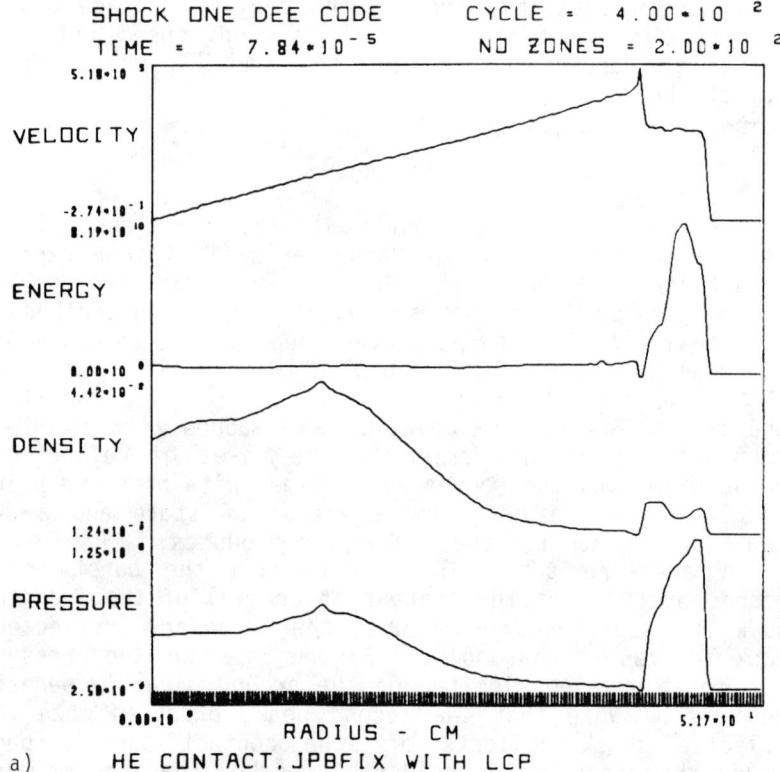

Fig. 4 Profiles of velocity, energy density, mass density, and pressure profiles for $0 \leq v \leq 51.2$ cm calculated as in Fig. 2 for the 8-16 case. a) Using 200 zones (ranges are $-0.274 \leq v \leq 5.18 \times 10^5$ cm/s, $0 \leq E \leq 8.19 \times 10^{10}$ erg/cm^3, $1.24 \times 10^{-3} \leq \rho \leq 4.42 \times 10^{-2}$ g/cm^3; $2.50 \times 10^6 \leq p \leq 1.25 \times 10^9$ dyne/cm^2); and b) using 400 zones (ranges are $-0.894 \leq v \leq 4.72 \times 10^5$, $6.28 \times 10^8 \leq E \leq 1.14 \times 10^{11}$ erg/cm^3, $1.29 \times 10^{-3} \leq \rho \leq 5.51 \times 10^{-2}$ g/cm^3, $3.63 \times 10^6 \leq p \leq 1.31 \times 10^9$ dyne/cm^2).

of our calculations, including that described in the next section. This process can repeat indefinitely until the rise of the fireball causes the contact surface to move away from the ground. To our knowledge, tertiary and higher-order peaks have not yet been observed experimentally, which may imply that turbulent effects smear out the contact surface.

In the present case, we imployed the sliding-zone (continuous regridding) capability of FAST2D to concentrate zones near the shock front until the transition from regular to Mach reflection occurs. The front of the blast

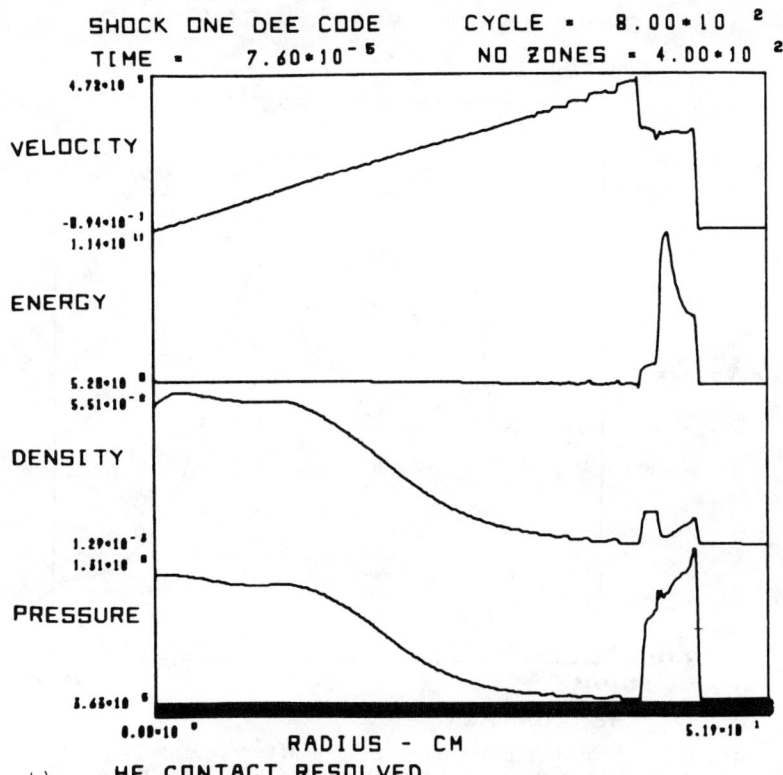

Fig. 4 Profiles of velocity, energy density, mass density, and pressure profiles for $0 \leq v \leq 51.2$ cm calculated as in Fig. 2 for the 8-16 case. a) Using 200 zones (ranges are $-0.274 \leq v \leq 5.18 \times 10^5$ cm/s, $0 \leq E \leq 8.19 \times 10^{10}$ erg/cm^3, $1.24 \times 10^{-3} \leq \rho \leq 4.42 \times 10^{-2}$ g/cm^3; $2.50 \times 10^6 \leq p \leq 1.25 \times 10^9$ dyne/cm^2); and b) using 400 zones (ranges are $-0.894 \leq v \leq 4.72 \times 10^5$, $6.28 \times 10^8 \leq E \leq 1.14 \times 10^{11}$ erg/cm^3, $1.29 \times 10^{-3} \leq \rho \leq 5.51 \times 10^{-2}$ g/cm^3, $3.63 \times 10^6 \leq p \leq 1.31 \times 10^9$ dyne/cm^2).

wave was captured in a finely gridded region which moved outward horizontally. Special care was taken to ensure that the grid moved smoothly. The resulting solution, particularly the curve of peak overpressure vs range, was within 15% of Carpenter's (1974) experimental data.

V. 600-Ton ANFO Explosion at 166 ft

The yield and HOB (600 tons and 166 ft, respectively) in the second two-dimensional calculation were chosen to equal the values used in the Defense Nuclear Agency Direct

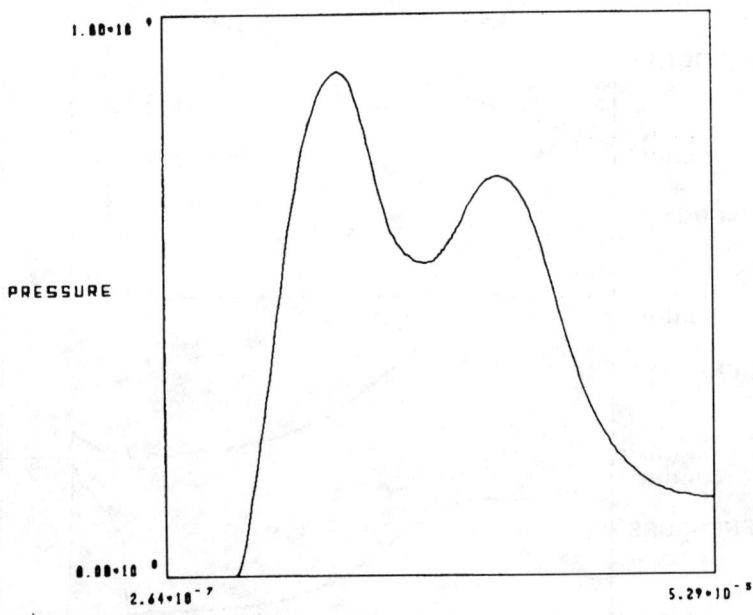

a) HE CONTACT

Fig. 5 Station histories of pressure after the profiles of Fig. 4 reflect from a solid wall at the HOB. a) Using 200 zones, and b) using 400 zones. Times are $2.64 \times 10^{-7} < t < 5.29 \times 10^{-5}$ s and $9.19 \times 10^{-5} < t < 1.21 \times 10^{-4}$ s, respectively; pressure scale is $0 < p < 10^9$ dyne/cm^2.

Course experiment, which we were simulating. The Chapman Jouguet parameters used to initialize the spherical free-field calculation were taken to be those for the NH_4NO_3-fuel oil (ANFO) mixture used as the explosive. This case scales to an 8-lb charge at a HOB of 95.2 cm, i.e., almost twice that in the previous calculation if we ignore differences in the nature of the two explosives.

Figures 6a-c show the contours of HE density and internal energy per unit mass and the velocity arrow plot at t=0, just before the reflection at ground zero occurs. Figures 6d-f show the corresponding plots 54 ms later, while Figs. 6g-i show them after 345 ms. Note the reflected shock proceeding upward, as shown by the arrows in Fig. 6f. It reflects again off the fireball and propagates back in a downward and outward direction. This is most clearly seen in the outer (roughly spherical) contours of Fig. 6h.

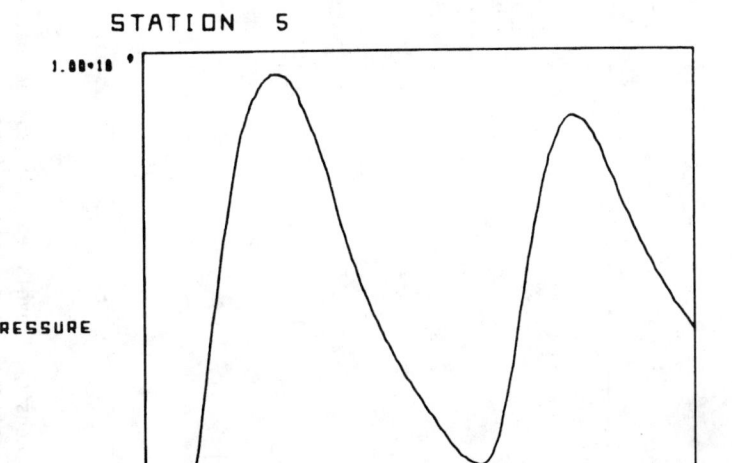

Fig. 5 Station histories of pressure after the profiles of Fig. 4 reflect from a solid wall at the HOB. a) Using 200 zones, and b) using 400 zones. Times are $2.64 \times 10^{-7} \leq t \leq 5.29 \times 10^{-5}$ s and $9.19 \times 10^{-5} \leq t \leq 1.21 \times 10^{-4}$ s, respectively; pressure scale is $0 \leq p \leq 10^9$ dyne/cm^2.

(The scale in Fig. 6h differs from that in the rest of Fig. 6 by a factor of 2.) The interaction of this shock with the radially inward flow near the ground generates the reverse vortex, which is clearly seen in Fig. 6i. Note also the positive vortex forming near the top of the grid in the same plot. The latter results when the upward-propagating reflected shock interacts with the radially outward flow near the top of the fireball; it is not produced by the buoyant rise of the fireball, which at these early times has scarcely begun.

To look at the evolution of the fireball at late times, we reinitialized on a larger, coarser grid representing a cylinder 400 m in radius and 800 m high. The first 300 cycles approximately reproduce the early-time results. The spherical shock breaks away and leaves the mesh. The flows remaining on the grid are now subsonic

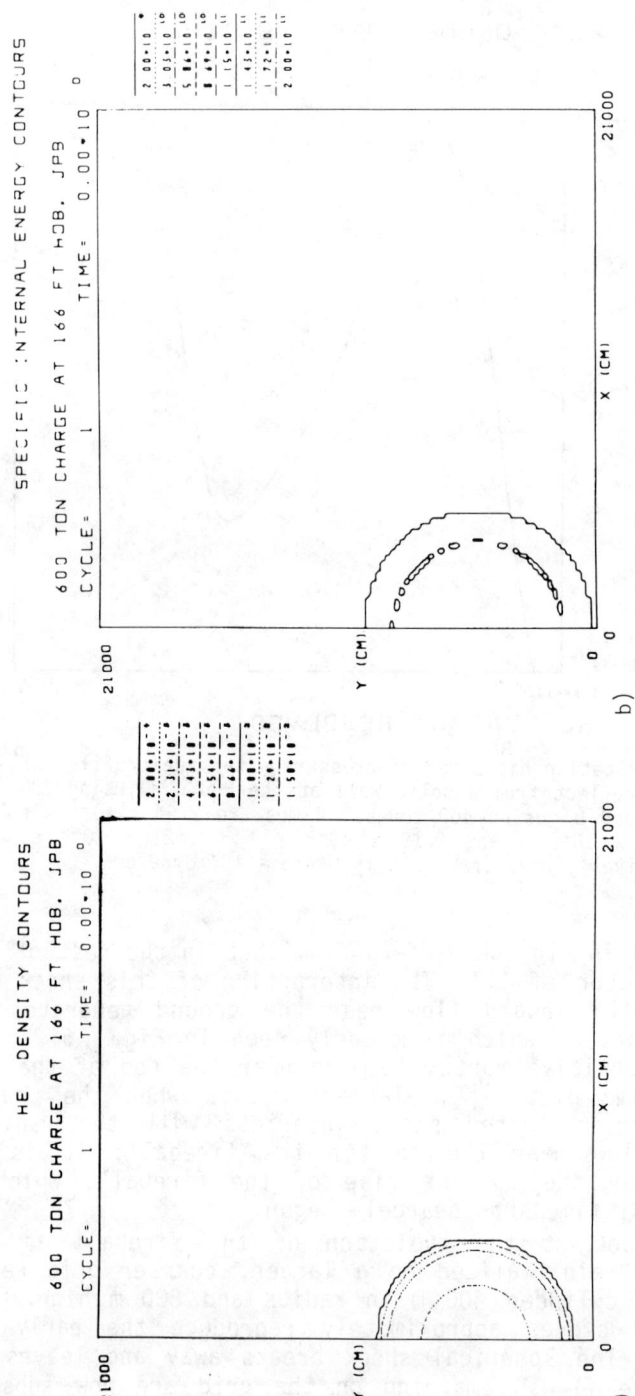

Fig. 6 a), d), and g) contours of constant detonation product mass density (levels equally spaced between 2.00×10^{-4} and 1.50×10^{-2} g/cm^3); b), e), and h) contours of constant internal energy per unit mass (2.00×10^9 to 2.00×10^{11} erg/g); c), f), and i) velocity vector plots. Region shown in plots is $2.1 \times 10^4 \times 2.1 \times 10^4$ cm except in h), which is $10^4 \times 10^4$ cm.

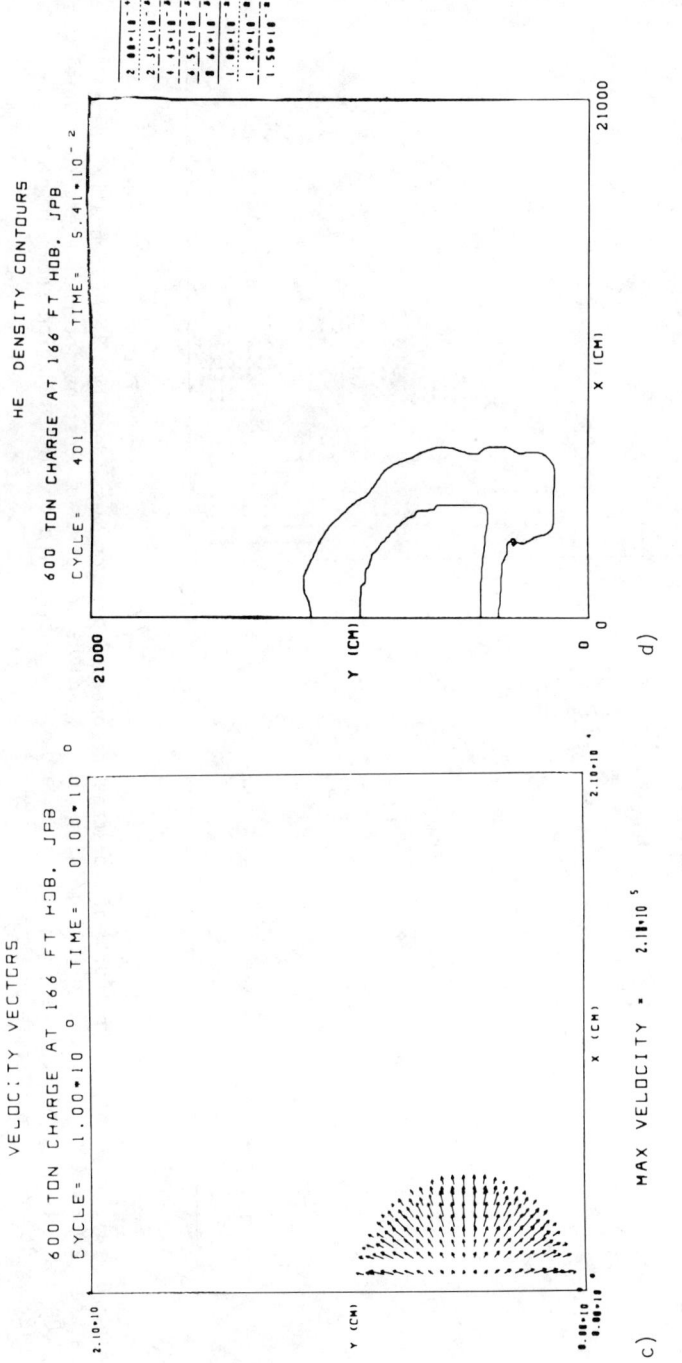

Fig. 6 a), d), and g) contours of constant detonation product mass density (levels equally spaced between 2.00 x 10⁻⁴ and 1.50 x 10⁻² g/cm³); b), e), and h) contours of constant internal energy per unit mass (2.00 x 10⁹ to 2.00 x 10¹¹ erg/g); c), f), and i) velocity vector plots. Region shown in plots is 2.1 x 10⁴ x 2.1 x 10⁴ cm except in h), which is 10⁴ x 10⁴ cm.

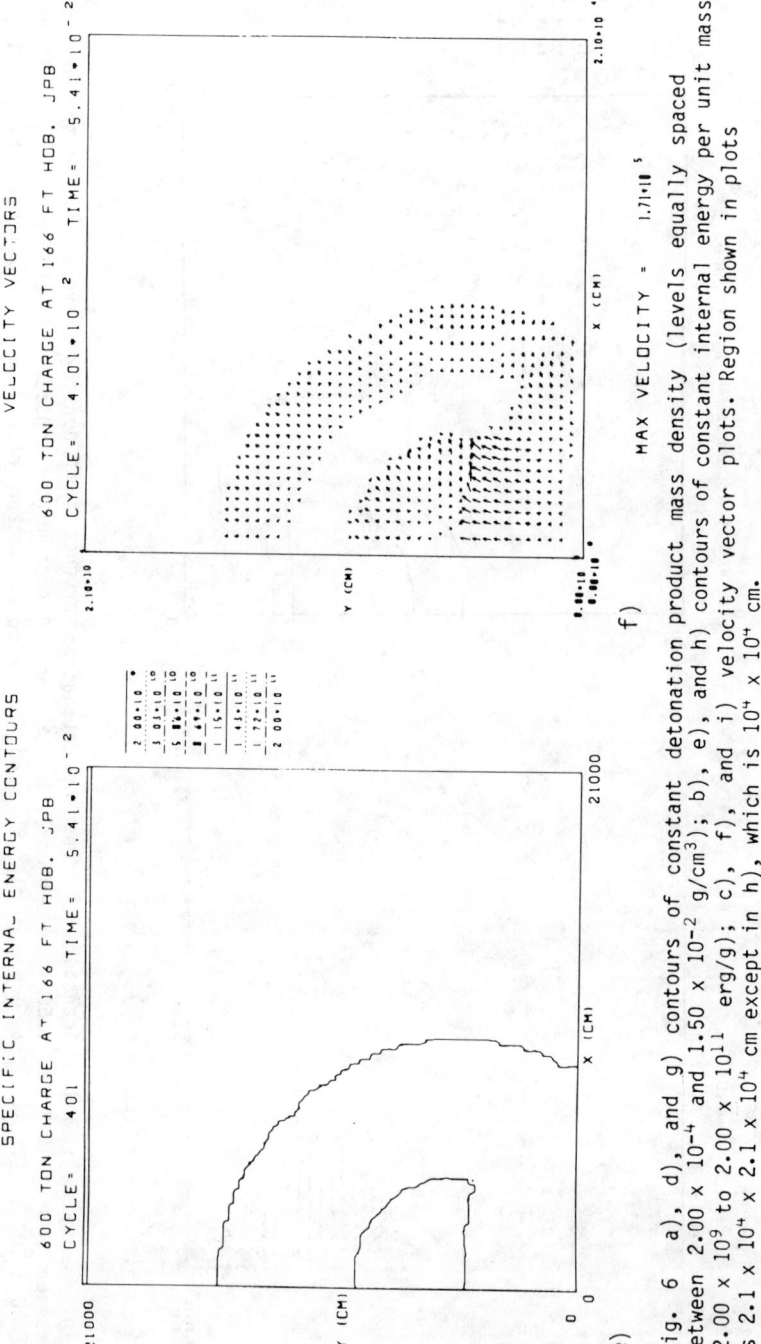

Fig. 6 a), d), and g) contours of constant detonation product mass density (levels equally spaced between 2.00×10^{-4} and 1.50×10^{-2} g/cm^3); b), e), and h) contours of constant internal energy per unit mass (2.00×10^9 to 2.00×10^{11} erg/g); c), f), and i) velocity vector plots. Region shown in plots is $2.1 \times 10^4 \times 2.1 \times 10^4$ cm except in h), which is $10^4 \times 10^4$ cm.

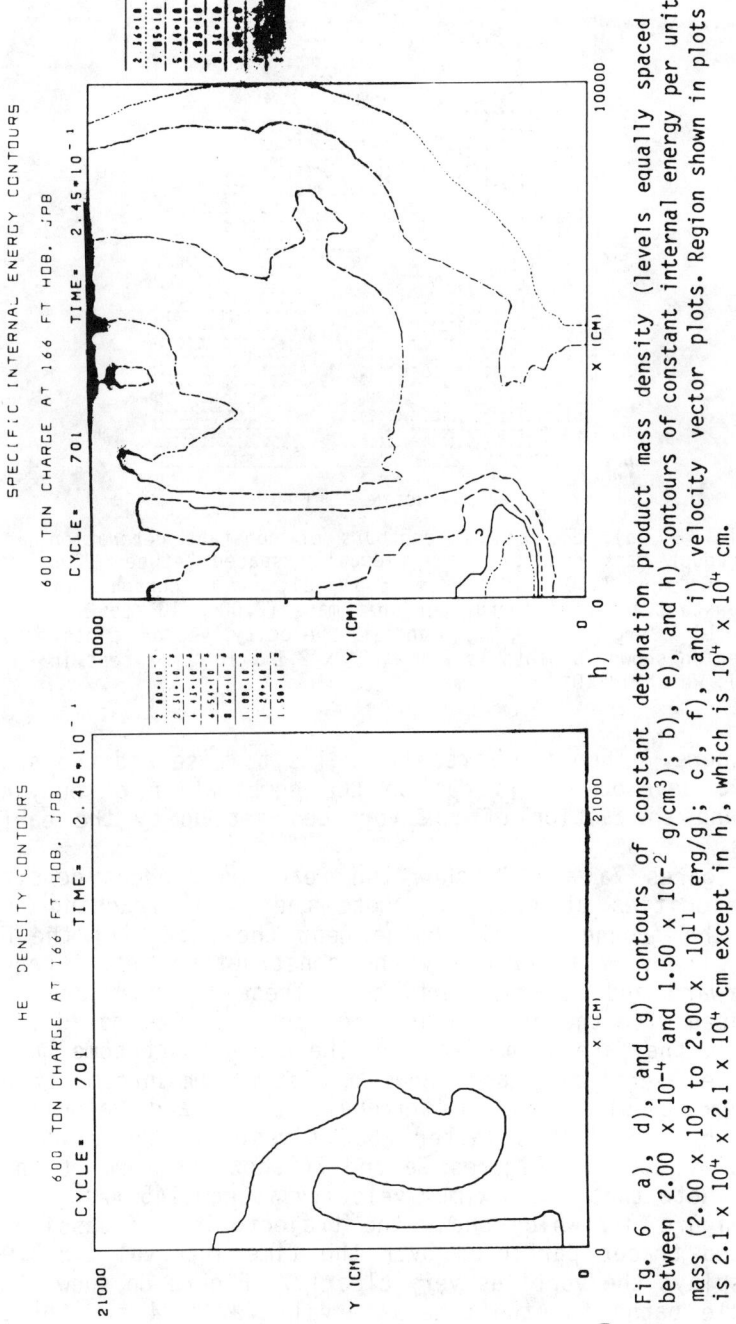

Fig. 6 a), d), and g) contours of constant detonation product mass density (levels equally spaced between 2.00×10^{-4} and 1.50×10^{-2} g/cm^3); b), e), and h) contours of constant internal energy per unit mass (2.00×10^9 to 2.00×10^{11} erg/g); c), f), and i) velocity vector plots. Region shown in plots is $2.1 \times 10^4 \times 2.1 \times 10^4$ cm except in h), which is $10^4 \times 10^4$ cm.

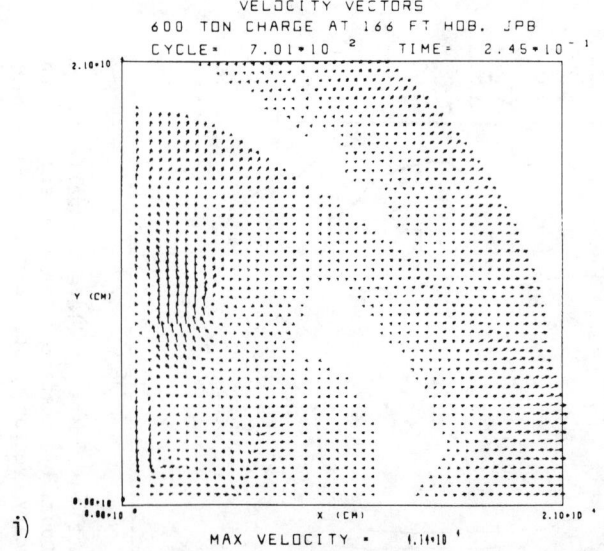

i)

Fig. 6 a), d), and g) contours of constant detonation product mass density (levels equally spaced between 2.00×10^{-4} and 1.50×10^{-2} g/cm^3); b), e), and h) contours of constant internal energy per unit mass (2.00×10^9 to 2.00×10^{11} erg/g); c), f), and i) velocity vector plots. Region shown in plots is $2.1 \times 10^4 \times 2.1 \times 10^4$ cm except in h), which is $10^4 \times 10^4$ cm.

everywhere. Then the fireball begins to rise and the subsequent development is due to the combination of buoyant rise and the action of the vortices set up by the early shocks.

Figures 7a and 7b show the reaction product density and velocities at 0.93 s. Note the "toe" reaching out along the ground and the bulge near the bottom of the HE product density produced by the constructive interference of forward and reverse vortices. These features are accentuated with the passage of time; in Figs. 7c and 7d (t = 2.70 s), they are even clearer. The cloud has become quite elongated vertically and shows a distinct mushroom shape. Development slows as the fireball cools and velocities diminish. By 7.34 s (after 2600 timesteps) the cloud is almost at 600 m. Figures 7e and 7f show its form at this time. Note that the maximum velocity is now 145 m/s.

Figure 8a, which shows the trajectories of passively advected tracer particles over the time interval 1.8-3.97 s, displays the vortices very clearly. Figure 8b shows the particle paths for the time interval 3.97-7.34 s. Notice that there are <u>four</u> vortices visible in the plot: two

EXPLOSIVELY PRODUCED SHOCK WAVES 469

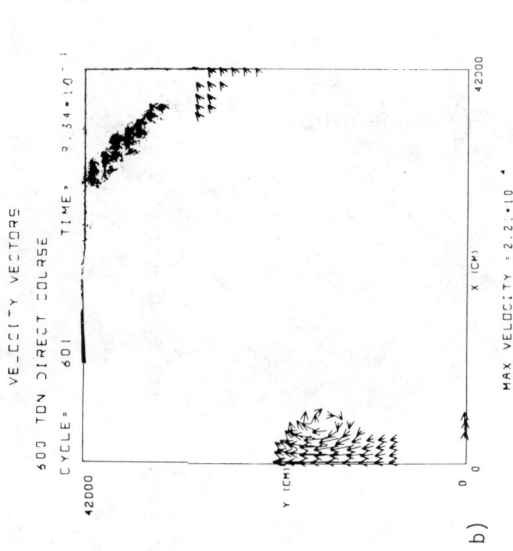

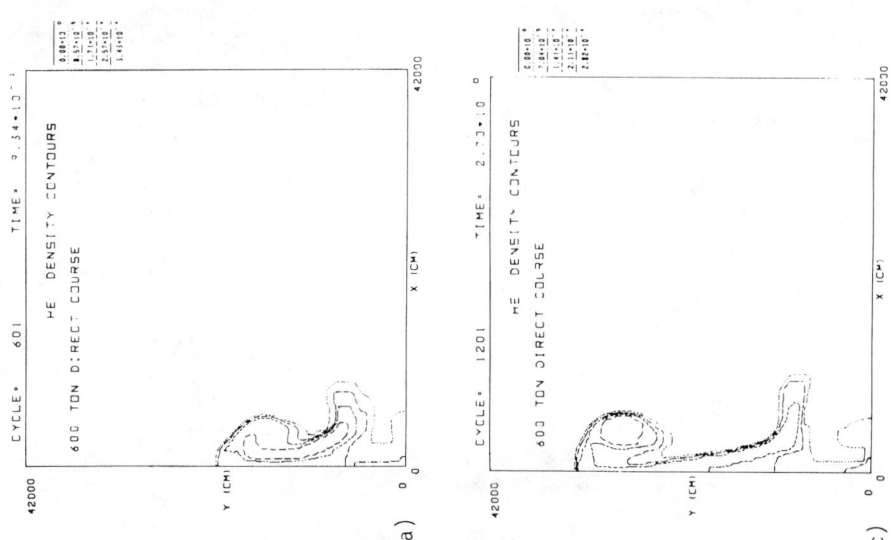

Fig. 7 Detonation product density contours a), c), and e) and velocity vector plots b), d), and f) at three times for the 600-ton calculation. Level lines in contour plots are equally spaced: a) 0-3.43 × 10^{-4} g/cm^3; c) 0-2.82 × 10^{-4} g/cm^3; and e) 0-2.60 × 10^{-4} g/cm^3.

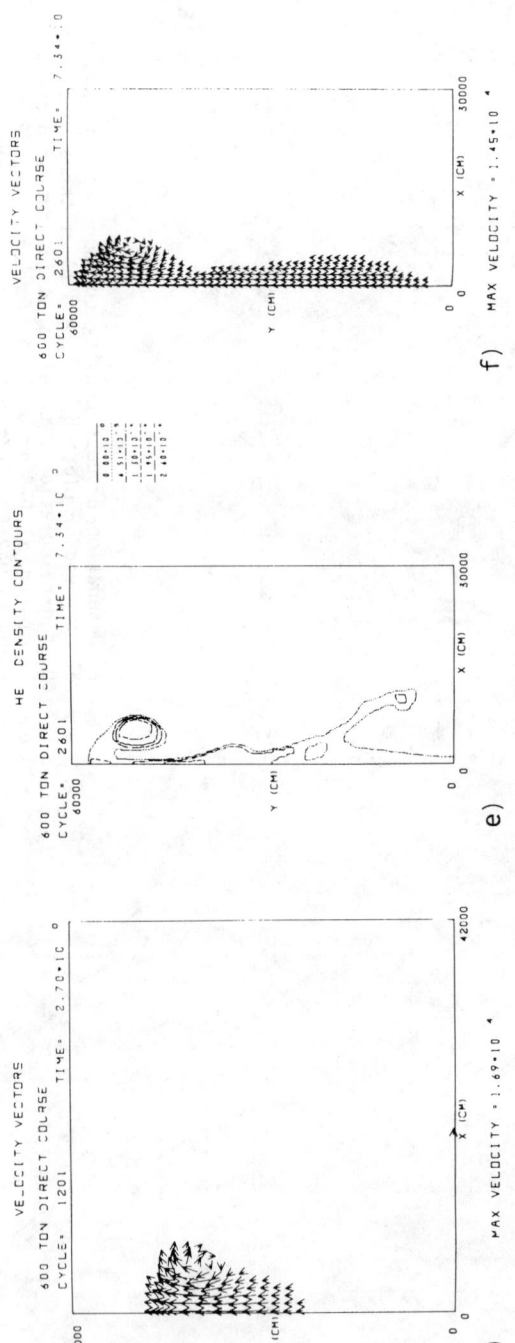

Fig. 7 Detonation product density contours a), c), and e) and velocity vector plots b), d), and f) at three times for the 600-ton calculation. Level lines in contour plots are equally spaced: a) 0-3.43 × 10^{-4} g/cm³; c) 0-2.82 × 10^{-4} g/cm³; and e) 0-2.60 × 10^{-4} g/cm³.

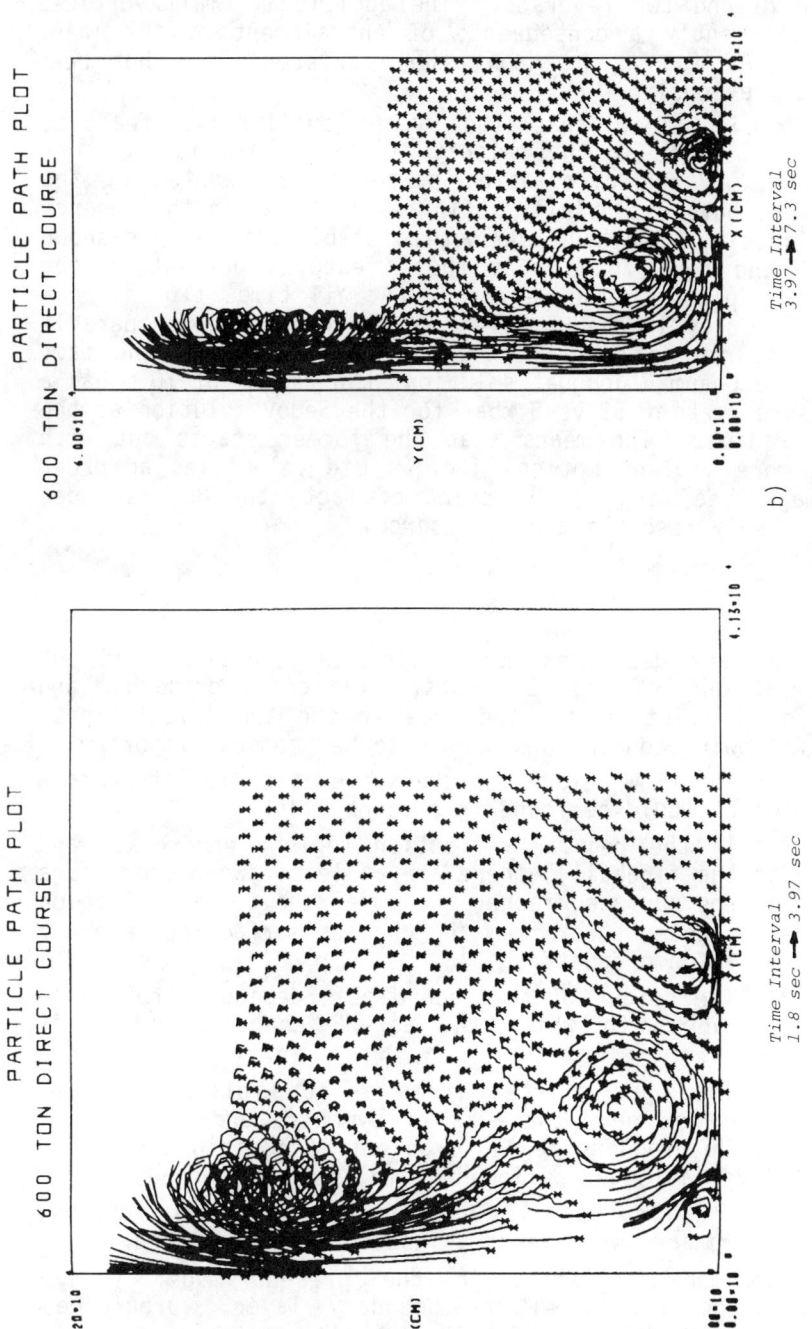

Fig. 8 Trajectories of marker particles, advected with calculated wind fields produced by 600-ton explosion over a) intermediate and b) later time intervals.

positive and two reversed. The additional small vortices are apparently a consequence of entrainment by the major ones. As far as we know, their existence has not been noted previously.

It is clear that the major qualitative features persist longest in the velocity plots. This is not surprising, as the circulation patterns represented by the vortices have essentially infinite lifetimes in the absence of viscosity. We have run out to stabilization (not shown here) and have found that these features persist in the velocity plots to the very end. At all times t>0 the peak flow velocity in the HE case exceeds that in the comparable point source solution. This is a reflection of the fact that the Chapman Jouguet solution at a radius of 10 m has a pressure peak of 52 vs 3 kbar for the Sedov solution at the same radius. The means that the former starts out with much more violent motion, i.e., fluid velocities an order of magnitude larger. In point of fact, the HE case does not closely resemble a point source.

VI. Conclusions

We have described numerical simulations carried out for 8-lb and 600-ton HE bursts. The code, gridding, and method of solution are the same in the two calculations. The following conclusions appear to be the most important.

1) The flow velocities are systematically larger than those which would be found in a point blast.

2) In the regular reflection region (underneath and close to the fireball), there are at least two overpressure peaks at the surface, rather than one, due to re-reflection of the reflected shock from the contact surface between air and detonation products.

3) The height of the second peak depends strongly on the accuracy with which the air-HE contact surface is resolved.

4) The upper vortex forms first, followed by the reverse vortex near the axis of symmetry and the ground. Adjacent HE products begin to be entrained into a positive vortex over a longer period of time, several seconds.

5) The flow establishes a pattern of four vortices, two forward and two reversed, instead of one of each.

Phenomena neglected in the present model (e.g., terrain, conditions in the boundary layer, turbulence, humidity, etc.) are unlikely to alter the above conclusions, which mainly depend on the characteristics of the solutions in the interior of the mesh and over long periods of time. Further study of the tracer particle motions we have calculated is likely to be illuminating, particularly when we take into account drag and inertial effects, rather than (as here) merely advecting the tracers passively. In closing, it is appropriate to emphasize the far-reaching significance of the role played by the HE-air interface in the dynamics of both airblast and cloud rise phenomena and the importance for numerical simulation of correctly treating this interface.

Acknowledgment

This work was supported by the Defense Nuclear Agency under Subtask N99QAXAH/MISSILE SYSTEMS/X-Ray Hardness, Work Unit # 00061, Work Unit Title "Cloud Calculations," and under Subtask Y99QAXSG/AIRBLAST AND THERMAL PREDICTIONS Work Unit #00057, Work Unit Title "Airblast Calculations."

References

Boris, J. P. and Book, D. L. (1976) Methods in Computational Physics (edited by J. Killeen), Vol. 10, p. 85. Academic Press, New York.

Boris, J. P. (1976) Flux-corrected transport modules for solving generalized continuity equations. NRL Memo Report 3237.

Carpenter, H. J. (1974) Height-of-burst blast at high overpressure. 4th International Symposium on Military Applications of Blast Simulation. Southend-on-Sea,UK.

Kuhl, A. L., Fry, M. A., Picone, J. M., Book, D. L., and Boris, J. P. (1981) FCT simulation of HOB airblast phenomena. NRL Memo Report 4613.

Sedov, L. I. (1959) Similarity and Dimensional Methods in Mechanics. Academic Press, New York.

Young, T. R. (1983) Table look-up: An effective tool on vector computers. Tenth Conference on Numerical Simulation of Plasmas, San Diego, Calif., Jan. 4-6.

Flame Propagation and Pressure Buildup in a Free Gas-Air Mixture Due to Jet Ignition

Manfred Schildknecht,* Werner Geiger,† and Manfred Stock*
Battelle-Institut e.V., Frankfurt am Main, Federal Republic of Germany

Abstract

From recent experiments it is known that jet ignition of partially confined gas-air mixtures can lead to explosion pressures which are up to two orders of magnitude larger than for point ignition. Since during accidental gas release in an industrial environment jet ignition is conceivable, this problem is of practical interest. Jet ignition of unconfined gas clouds has not been investigated before. Therefore experiments in model scale have been performed in order to investigate pressure buildup and flame propagation during jet ignition of an unconfined gas cloud.

For the experiments ethylene-air mixtures have been used. The scale of the experiments compared to typical dimensions of industrial buildings was about 1:20. The free jet was produced by an explosion in a vented rectangular chamber (size 0.5 x 0.5 x 1 m). To change the characteristics of the free jet (velocity, duration, and diameter of the jet) orifices of different shape (square-shaped and slot-shaped) and size have been used. For large orifices a turbulence generator has been used in some experiments to increase the burning rate inside the chamber.

The pressure buildup inside the chamber as well as the pressure buildup and flame propagation in the free gas cloud (size: 4 x 1 x 1 m) has been investigated for free jets with different properties (underexpanded free jet, free jet at approximately sonic conditions, subsonic free jet). For the chamber without turbulence generator the largest value for the overpressure in the free gas cloud (P_e = 0.87 bar) was produced by an approximately sonic free jet (orifice ration F_v/F_t = 0.1). For orifice ratios F_v/F_t lar-

---

Presented at the 9th ICODERS, Poitiers, France, July 3-8, 1983. Copyright © American Institute of Aeronautics and Astronautics, Inc., 1984. All rights reserved.
*Senior Scientist, Safety Technologies Department.
+Deputy Manager, Safety Technologies Department.

ger than 0.2 the overpressure in the free gas cloud is significantly increased by a turbulence generator in the chamber (P_e = 1.3 bar). In the most adverse case the overpressure is raised by a factor of 5 (P_e = 1.3 bar compared to the same chamber without turbulence generator where P_e = 0.25 bar).

Nomenclature

a,b	= length and width of square-shaped orifice
c_*	= critical speed
D	= equivalent orifice diameter
F_t	= total area of the vent
F_v	= area of the vent
F_v/F_t	= orifice ratio
I	= maximum momentum per unit area in the external cloud
I(x)	= momentum per unit area at position x
P_e	= maximum value of peak overpressure in the external cloud
$P_{e,max}(x)$	= peak overpressure in the external cloud at position x
$P_e(T, s)$	= time-dependent overpressure in the external cloud at position x
$P_{i,max}$	= peak overpressure within the vented chamber
$P_i(t)$	= time-dependent overpressure within the vented chamber
$P_1, P_2, \ldots P_5$	= position of pressure transducer no. 1, 2, ...5
P_*	= critical pressure
T_I	= duration of "cold" release phase
T_{II}	= duration of "hot" release phase
V	= flow velocity in the orifice (jet velocity near the orifice)
V_F	= flame speed
V_{max}	= maximum flow velocity (jet velocity) in the "hot" release phase
V_p	= pressure wave speed
x	= distance from orifice

Subscripts

I	= "cold" release phase
II	= "hot" release phase
*	= critical condition (sonic flow)

Introduction

Laboratory experiments by Lee et al. (1977) with an experimental setup consisting of two interconnected pipes of different diameter separated by an orifice (Fig. 1, upper part) have shown that jet ignition of partially confined gas-air mixtures can lead to explosion pressures which are two or three orders of magnitude larger than with point ignition. A dramatic increase of overpressure due to jet ignition has also been found in large-scale experiments by Eckhoff et al. (1980) and Moen et al. (1982) where an ignition tube (0.5 m in diameter) had been connected to a large tube of 2.5-m diam, and in very small scale (soap bubble experiments) by Charuel et al. (1981).

In an industrial environment scenarios are conceivable where jet ignition may occur. Because of the high pressure which may be generated this is of practical concern. Jet ignition may happen for example, if after an accidental release gas penetrates into a building such that a flammable gas-air mixture is formed inside and outside of the building. If the mixture is ignited inside the explosion pressure will be relieved through openings, doors, and windows which act as vents. During the venting process a free jet of unburned gas-air mixture is formed first. It is followed by a free jet of hot reaction products which ignites the gas cloud outside. Within the gas cloud there may be additional buildings or installations providing some partial confinement. A possible scenario is sketched in the lower part of Fig. 1.

In spite of its practical interest jet ignition of unconfined gas clouds seems to have not been investigated before, apart from the experiments in very small scale by Charuel et al. (1981). The objective of the experiments reported here was to investigate pressure buildup and flame propagation in an unconfined gas-air mixture after jet ignition. Jets of different strength (characterized by the velocity, duration, and diameter of the jet) are produced by explosion venting of a rectangular chamber. The scale chosen was about 1:20 compared to typical dimensions of industrial buildings.

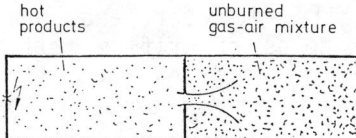

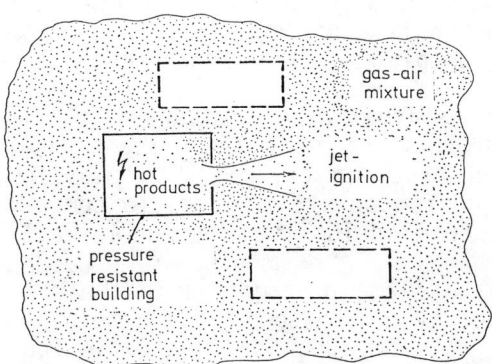

Fig. 1 Above: Setup used by Lee et al. (1977). Below: Scenario of jet ignition in an industrial environment.

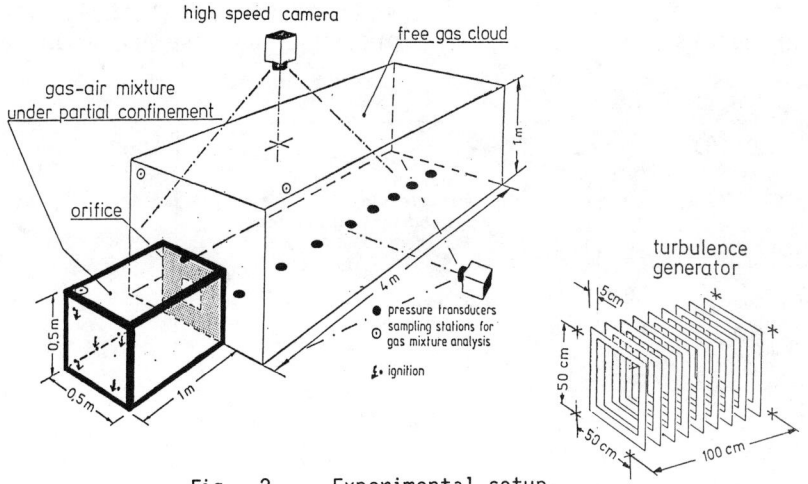

Fig. 2　Experimental setup.

Experimental Setup and Measuring Technique

The experiments were carried out in the setup shown in Fig. 2. The jet is produced by means of a pressure-resistant rectangular chamber with a size of 1 x 0.5 x 0.5 m. The chamber has an interchangeable front plate to permit insertion of orifices of different size and shape and thus to generate jets with different velocity, diameter, and duration. Orifice plates with square holes of orifice ratios $0.01 \leq F_v/F_t \leq 0.61$ and with slot-shaped holes of orifice ratios $0.1 \leq \overline{F_v}/F_t \leq 0.3$ have been used (F_v area of the vent, F_t total area of the front plate). To increase the flame speed for large vent areas a turbulence generator could be used. The generator is also sketched in Fig. 2. The mixture inside the chamber is ignited by 5 spark plugs which are mounted on the back plate of the chamber opposite to the orifice.

The dimensions of the external gas cloud are 1 x 1 x 4 m. To hold the gas-air mixture which simulates the external cloud a steel rail construction covered with plastic foil was used. Ethylene-air of about stoichiometric composition was used for the mixture inside the chamber as well as in the external cloud.

During an experiment the flame propagation in the external gas cloud is ovserved by a high-speed film camera of type Hycam with a framing rate of 4000 fps. The explosion pressure is measured side-on by 11 piezoelectric pressure transducers Type Kistler (1 pressure transducer inside the vented chamber and 10 pressure transducers outside). The transducers for measuring the pressure outside are mounted

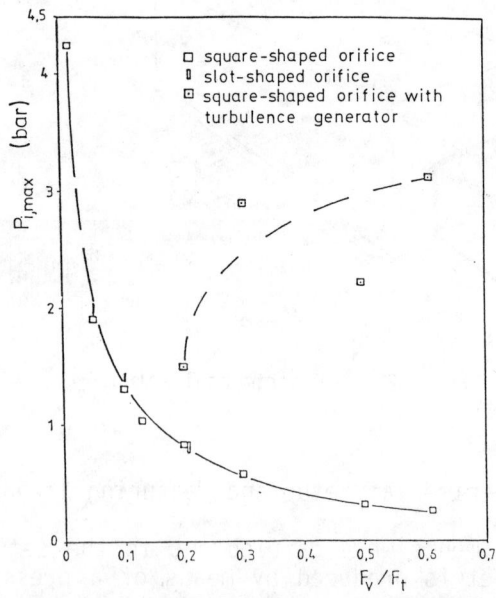

Fig. 3 Peak overpressure in the vented chamber vs orifice ratio for setups with turbulence generator (dashed curve) and without (solid curve).

in a rail. The rail is positioned in the external gas cloud on the ground parallel to the axis of the free jet. The pressure signals are recorded on magnetic type. To correlate the film frames with the pressure signals a foto flash is fired and its trigger signal is recorded on magnetic type. The composition of the ethylene-air mixture is checked by gas chromatographic analysis of gas samples which could be taken at three different locations inside the setup.

Results

Pressure Buildup Inside the Chamber

In Fig. 3 peak overpressure $P_{i,\,max}$ in the vented chamber is plotted vs F_v/F_t for square-shaped and slot-shaped vents (solid curve). If no turbulence generator is used, peak overpressure decreases with increasing orifice ratio, as expected. Within the range of orifice ratios investigated peak overpressure is independent of the shape of the vent. When a turbulence generator is used, the dependence of peak overpressure on orifice ratio is different. For orifice ratios larger than about 0.1, peak overpressure increases with increasing orifice ratio. For example, for an orifice ratio $F_v/F_t = 0.3$, peak overpressure exceeds that in a chamber without turbulence generator by a factor

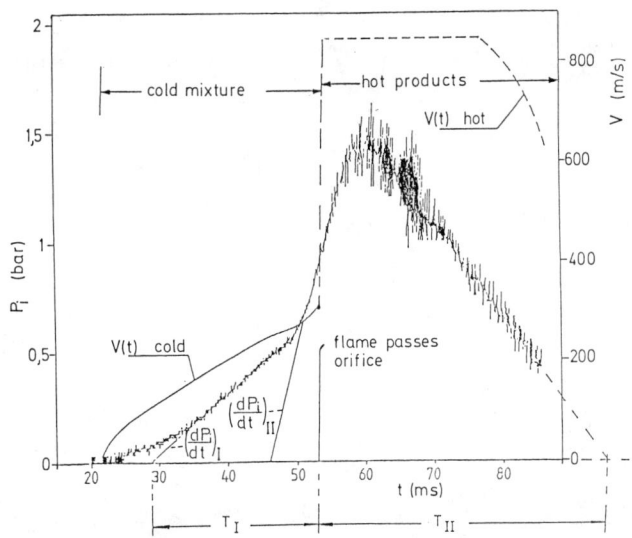

Fig. 4 Overpressure signal measured in the chamber and corresponding flow velocity (calculated) in the orifice for $F_v/F_t = 0.1$, without turbulence generator.

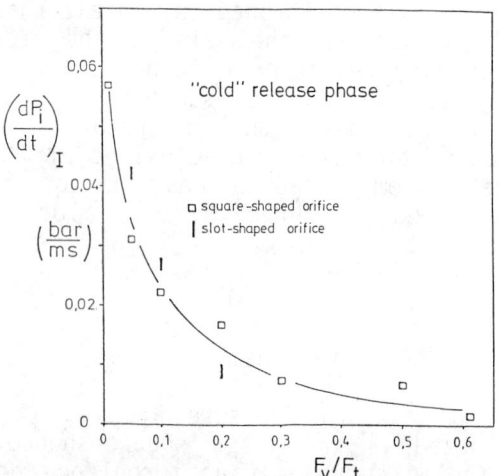

Fig. 5 Pressure rise $(dP_i/dt)_I$ during the "cold" release phase vs orifice ratio.

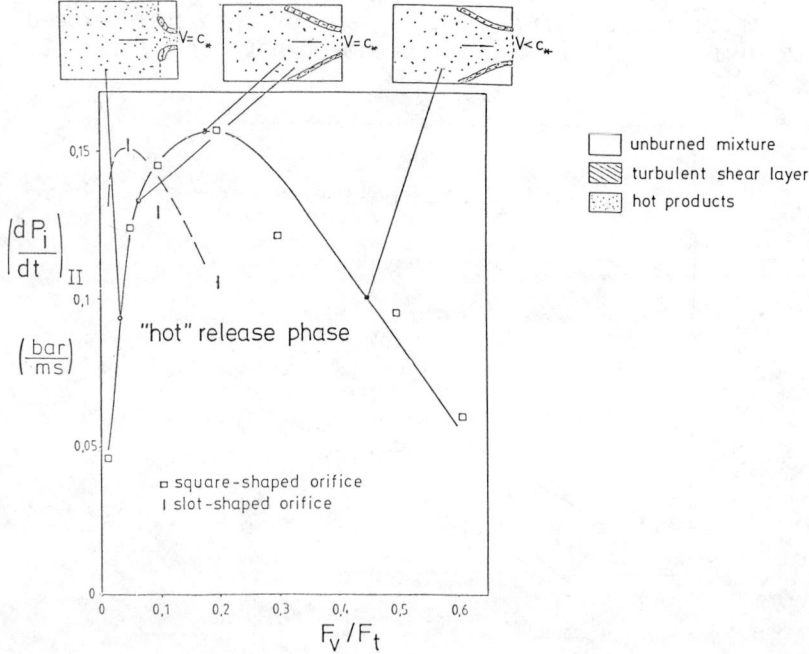

Fig. 6 Pressure rise $(dP_i/dt)_{II}$ during the "hot" release phase vs orifice ratio.

of 4-5. For orifice ratios larger than 0.3 peak overpressure is grossly independent from orifice ratio (see dashed curve in Fig. 3).

In Fig. 4 a typical overpressure signal, measured inside the chamber without turbulence generator for $F_v/F_t = 0.1$, is depicted. During the phase of pressure buildup two distinct periods with different slopes in pressure rise $(dP_i/dt)_I$ and $(dP_i/dt)_{II}$ appear. In Fig. 4 also the corresponding flow velocity in the orifice, calculated from the overpressure values, is plotted. It is obvious that during the period with slope (dP_i/dt) cold mixture is pushed out of the vent. Shortly before the hot products leave the orifice an increase in burning rate occurs which increases the slope of the pressure-time signal to the value (dP_i/dt). This increase is believed to be caused by the formation of a turbulent shear layer between the unburned gas and the hot products in a region near the orifice inside the chamber. (The shear layer formation is sketched in Fig. 8, below left.)

When the hot products pass the orifice the slope $(dP_i/dt)_{II}$ is approximately maintained. Since the sound speed of the hot products is about twice that of the unburned fas due to their higher temperature, the volume flow is approximately doubled after the hot products reach the orifice. As a consequence of the faster venting a decrease in pressure rise would be expected. Hence, the slope $(dP_i/dt)_{II}$ can only be maintained if a second increase in burning rate appears. The second increase in burning rate is thought to be caused by an increased turbulence production in the shear layer inside the chamber due to the increase in flow velocity.

The pressure rise $(dP_i/dt)_I$ during the "cold" release phase and $(dP_i/dt)_{II}$ during the "hot" release phase are plotted vs F_v/F_t in Figs. 5 and 6, respectively. It can be seen that $(dP_i/dt)_I$ drops continuously with increasing orifice ratio while $(dP_i/dt)_{II}$ has a maximum at $F_v/F_t = 0.05$ for slot-shaped orifices and at $F_v/F_t = 0.2$ for square-shaped orifices.

To explain the variation of $(dP_i/dt)_{II}$ three sketches are shown at the top of Fig. 6. For small orifice ratios corresponding to the increasing part of the curve for $(dP_i/dt)_{II}$ the peak pressure reaches supercritical values which lead to sound velocity in the orifice independent from the orifice ratio. An increase in orifice ratio leads to an increase of the volume which is occupied by the turbulent shear layer. The increase in shear layer volume again results in an increase in burning rate.

At the decreasing part of the curve for $(dP_i/dt)_{II}$ the peak pressure inside the chamber is subcritical. This means that flow velocities in the orifice decrease with increasing orifice ratio. For this part of the curve it can be inferred, that the velocity gradient in the shear layer between unburned gas and hot products which is responsible for turbulence production is reduced with increasing orifice ratio. This results in a decrease in burning rate with increasing orifice ratio.

To characterize the properties of the free jet the maximum velocity of the jet V_{max}, the equivalent diameter D of the orifice and the duration T_{II} of the "hot" release phase are plotted in Fig. 7 vs F_v/F_t for chambers with and without turbulence generator. The equivalent diameter of orifice is defined as $D = (ab/\pi)^{1/2}$, where a and b are length and width of the orifice, respectively.

Pressure Buildup and Flame Propagation in the Free Cloud

The principal effect of jet ignition may be illustrated by one experiment in the series without turbulence generator where the largest overpressure was observed in the free cloud ($F_v/F_t = 0.1$, square orifice). An overview of the event is given in Fig. 8. After spark ignition in the cham-

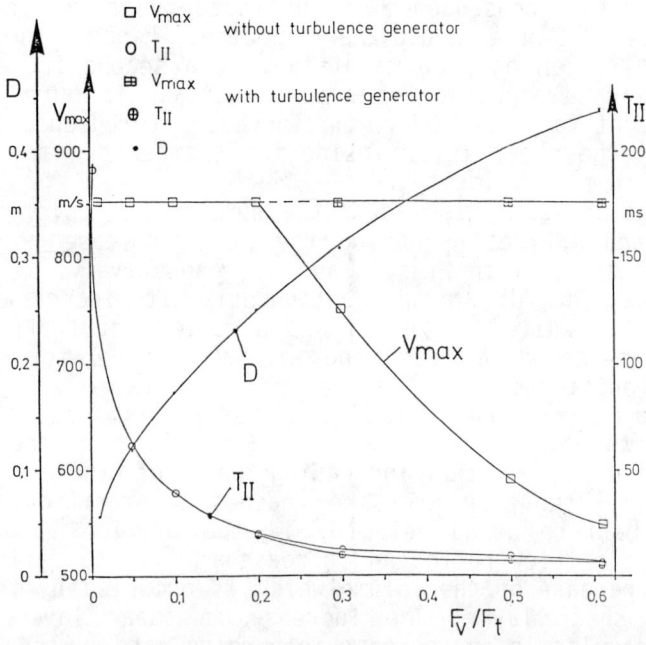

Fig. 7   Maximum velocity V_{max} of the jet in the "hot" release phase, equivalent diameter D of the vent, and duration T_{II} of the "hot" release phase vs orifice ratio.

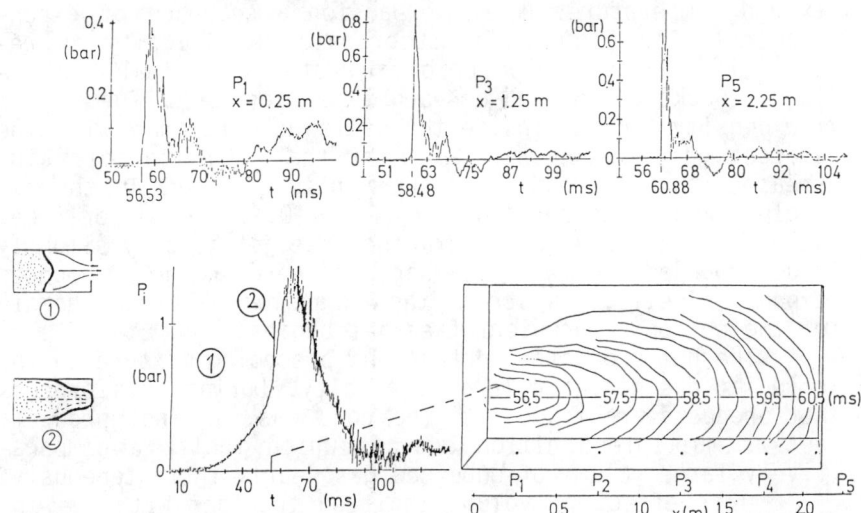

Fig. 8 Overview of the events connected with jet ignition. Above: Pressure signals at locations P_1, P_3, P_5 in the external cloud. Below: Overpressure signal measured within the chamber, flame front contours in the free cloud, and positions of the pressure transducers.

ber cold unburned gas is pushed out of the orifice due to the pressure buildup inside. Since the flow velocity increases with increasing pressure inside the chamber an unsteady free jet is formed in the external cloud. At the end of the "cold" release phase the free jet attains approximately sound speed.

Along the turbulent free jet, the gas vented from the chamber is mixed with the ambient gas. Turbulence is produced by the mixing process in the gas cloud which will increase the burning velocity of the following flame. After the end of the "cold" release phase hot products are issued from the vent. The critical sound velocity and the volume flow of the hot products are about two times as high as that of the cold mixture. The hot products again form an unsteady free jet. At the beginning of the "hot" release phase the pressure inside exceeds the critical pressure P_* in this example. This means that the flow velocity in the orifice cannot increase any more (the maximum possible velocity is the critical sound velocity c_*). However, with increasing pressure inside the jet becomes more and more underexpanded. This leads to an expansion of the jet in the free gas cloud to ambient pressure and correspondingly to an increase of cross-sectional area and of flow velocity (achieving supersonic flow) and to a decrease of gas density and temperature. During expansion a sequence of expansion cells is formed. The supersonic jet velocity is reduced to subsonic velocity by a system of normal and oblique shocks at the end of each expansion cell. Due to its low density the gas inside the cells does not mix with the ambient gas. Number and length of the expansion cells increase with the pressure in the vented chamber. In the example under consideration (F_v/F_t = 0.1, square orifice, without turbulence generator) the free jet is only slightly underexpanded ($P_{i,max} \approx 1.3$ bar) so there are no pronounced expansion cells. Downstream the expansion cells the domain of the subsonic turbulent free jet begins.

The hot products overtake the preceding mixture in the cold jet due to their larger velocity. During this process hot products are mixed into the cold gas. The hot products act as multiple ignition sources which ignite a comparatively large volume of unburned gas nearly instantaneously.

This effect of volume ignition together with the increase of burning velocity due to the turbulence induced by the jet causes very large energy release rates (flame speeds) and correspondingly large overpressures in the free cloud. The flame contours at successive times are shown in Fig. 8, below right, and the pressure signals at several locations in the free cloud in Fig. 8, above. The corresponding trajectory of the flame front, the trajectory of the pressure wave and the distribution of peak overpressure in the free gas cloud are depicted in Fig. 9.

In Fig. 10 flame speed V_F and pressure wave speed V_p are plotted as a function of distance x from the orifice.

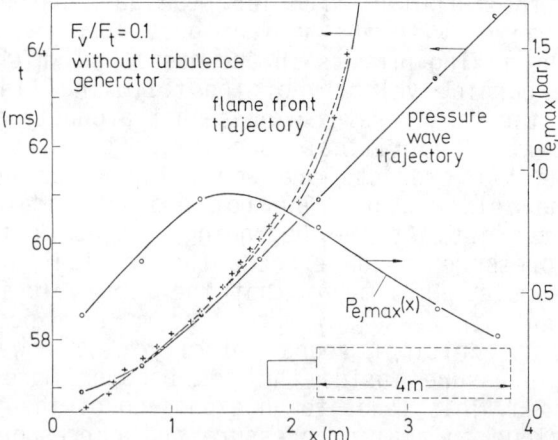

Fig. 9   Flame front trajectory, pressure wave trajectory, and peak overpressure vs distance from orifice (ignition at t = 0).

Over a considerable distance the flame speed is very large, with maximum values of about 550 m/s near the orifice (for comparison: the velocity of a steady-state "cold" free jet under the conditions of Fig. 10 is about 310 m/s, the velocity of a steady-state "hot" free jet about 830 m/s). With increasing distance to the orifice the flame speed decreases in a similar way as the velocity of a free jet. It can be inferred that, in spite of the initially high-energy release rates enforced by the free jet, the flame cannot maintain the originally very large velocity by own production of turbulence. The pressure wave front propagates with large supersonic velocity near the orifice. This is explained by the fact that there the pressure wave is convected by the flowfield of the free jet. With increasing distance to the orifice the velocity of the pressure wave front decreases to sound speed.

Due to the high-energy release rate enforced by the jet, a pressure wave with a correspondingly large peak value is generated. The variation of peak overpressure $P_{e,max}$ and of the corresponding momentum per unit area I in the external cloud with distance is shown in Fig. 11. There is a maximum in pressure and momentum at x = 1.7 m. Obviously at this location the largest energy release rate does occur.

A second example concerns a free jet which is strongly underexpanded (F_v/F_t = 0.01, without turbulence generator). The variation of flame speed and of pressure wave speed

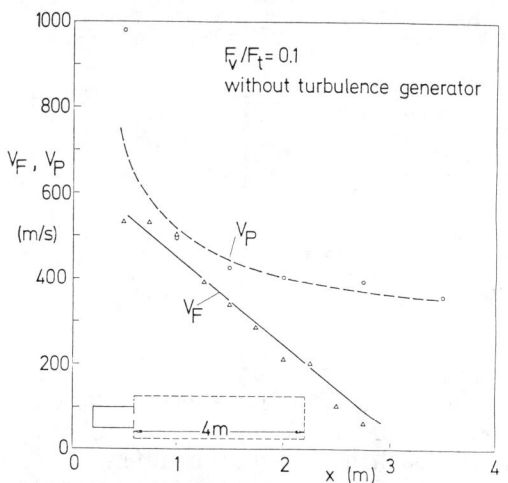

Fig. 10   Flame speed V_F and pressure wave speed V_p vs distance from orifice for orifice ratio F_v/F_t = 0.1.

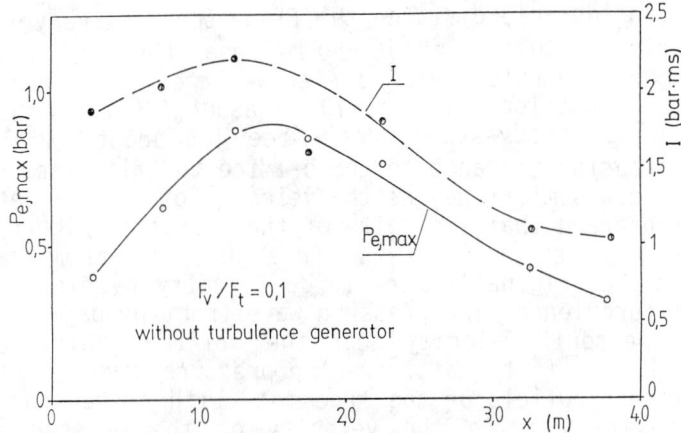

Fig. 11   Peak overpressure and momentum per unit area vs distance from orifice for orifice ratio $F_v/F_t = 0.1$.

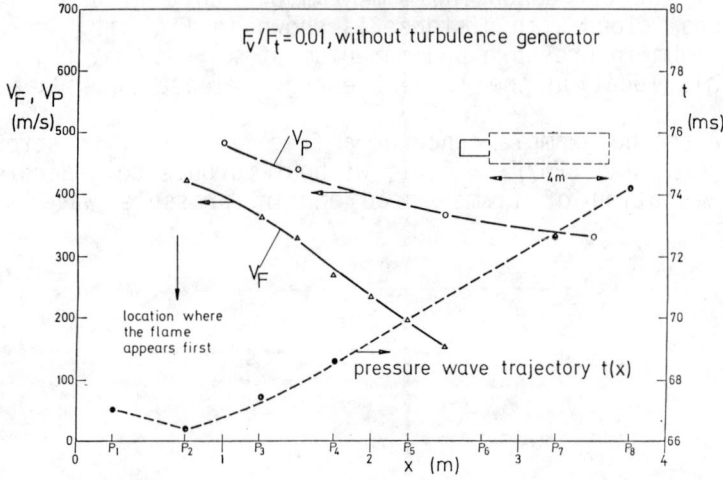

Fig. 12   Flame speed V_F and pressure wave speed V_p vs distance from orifice as well as trajectory $t(x)$ of the pressure wave for orifice ratio $F_v/F_t = 0.01$.

with distance as well as the trajectory of the pressure wave is shown in Fig. 12. The flame front appears first at the large distance of about 14 orifice diameters downstream because of the far-reaching expansion cells where no mixing occurs. (In contrast, in the experiment discussed before the flame front appeared already at a distance of about 3 orifice diameters downstream.) The pressure wave trajectory

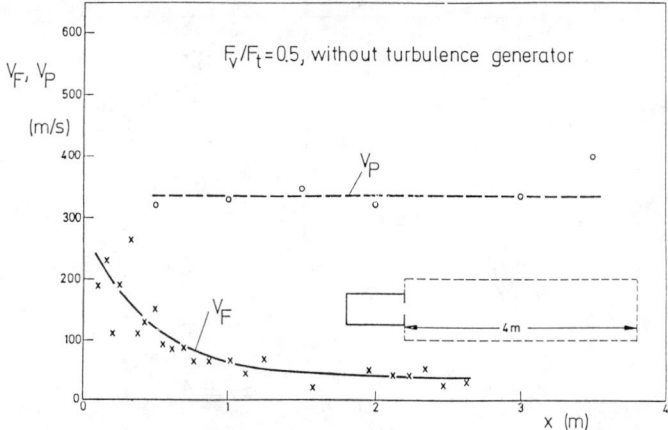

Fig. 13　Flame speed V_F and pressure wave speed V_p vs distance from orifice for orifice ratio $F_v/F_t = 0.5$.

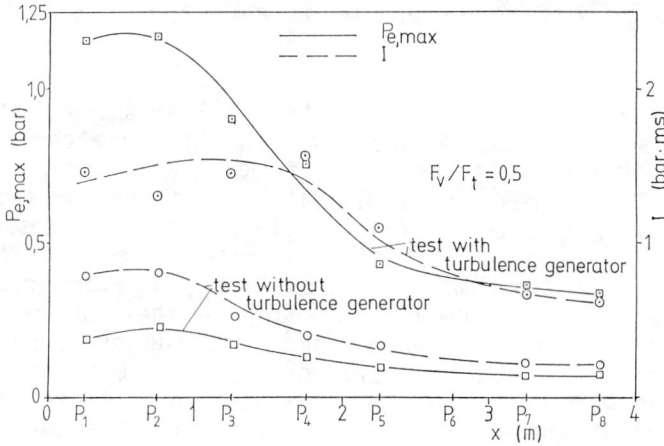

Fig. 14　Peak overpressure and momentum per unit area vs distance from orifice for area ratio $F_v/F_t = 0.5$, with and without turbulence generator.

indicates that the pressure wave reaches position P_2 earlier than position P_1. This supports the conclusion drawn from the time of first appearance of the flame front, namely that in the expansion cells no mixing occurs.

A third example concerns a free jet from a large orifice ($F_v/F_t = 0.5$) with subsonic velocity ($V_{max} = 550$ m/s). The corresponding variation of flame speed and of pressure wave speed is shown in Fig. 13. Because of the large orifice the "hot" release phase is only of short duration,

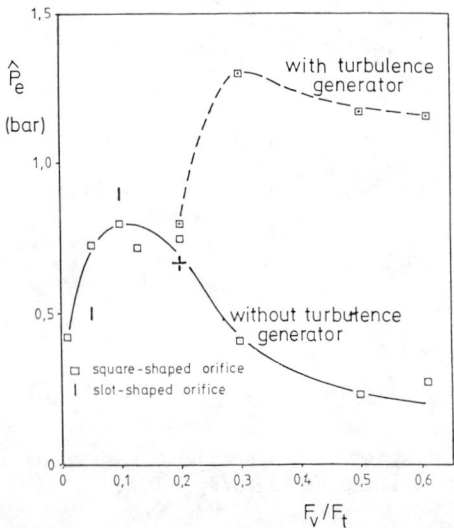

Fig. 15 Maximum value of peak overpressure in the free gas cloud vs orifice ratio, with and without turbulence generator.

hence the jet is strongly unsteady. From the variation of flame speed it can be seen that in this case the increase in burning rate is restricted to the region near the orifice. The corresponding distribution of overpressure and momentum is plotted in Fig. 14. In Fig. 14 overpressure and momentum are shown also for the experiment with the same orifice ratio but with a turbulence generator inside the chamber. Because the burning rate in the chamber is increased by the turbulence generator, the pressure in the free cloud exceeds the value observed in the case of the chamber without turbulence generator by a factor of 4-5. In Fig. 15 the maximum value of peak overpressure in the free cloud is plotted vs the orifice ratio for experiments with and without turbulence generator (the distance from the orifice at which the maximum of overpressure appears decreases with increasing orifice ratio and is smaller for tests with turbulence generator than without). When no turbulence generator is used the largest overpressure in the free cloud is observed for orifice ratios at which the peak pressure in the chamber slightly exceeds the critical pressure. For orifice ratios larger than about 0.2, the maximum overpressure in the case of no turbulence generator decreases continuously. In contrast, when a turbulence generator is used, the maximum overpressure for orifice ratios larger than 0.2 first increases strongly with increasing orifice ratio and decreases only slowly for $F_v/F_t > 0.3$.

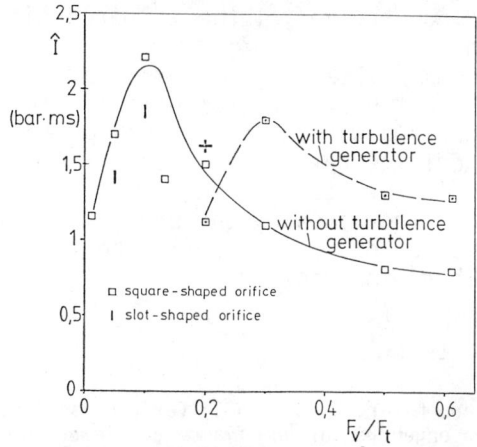

Fig. 16 Maximum momentum per unit area vs orifice ratio, with and without turbulence generator.

Figure 16 shows the corresponding variation of the maximum specific momentum with orifice ratio. The curves for the maximum momentum are similar in shape to those for the maximum overpressure (Fig. 15). A remarkable difference consists, however, in the effect of the turbulence generator. While the maximum overpressure may be increased by the turbulence generator by a factor of 4-5 (for $F_v/F_t > 0.3$), the maximum momentum is increased only by about 50%.

Conclusions

The pressure buildup and flame propagation in the external cloud due to jet ignition has been shown to depend strongly on the properties of the free jet (underexpanded free jet leading to supersonic velocities after complete expansion, free jet at approximately sonic conditions, subsonic free jet). When a turbulence generator is not used in the vented chamber supersonic velocity conditions for the jet are obtained only for orifice ratios smaller than about 0.2. When a turbulence generator is used, pressure buildup in the chamber is so fast that supersonic velocity conditions for the jet persist over the whole range of orifice ratios investigated (up to $F_v/F_t = 0.6$). Accordingly, with a turbulence generator the maximum overpressure in the free cloud after jet ignition from a vented chamber of a given orifice ratio is strongly increased for orifice ratios $F_v/F_t > 0.2$ (the increase of maximum momentum per unit area is less pronounced). Nevertheless, also without a turbulence generator in the chamber the overpressures generated in the free cloud are very large compared with the case of weak ignition.

The maximum value of peak overpressure generated in the cloud is about 0.87 bar (for orifice ratio 0.1, approximately sonic flow) without turbulence generator, and about 1.3 bar (for orifice ratio 0.3) with a turbulence generator, see Fig. 15. For orifice ratios larger than 0.3, maximum overpressure is increased by the presence of a turbulence generator by a factor of 4-5, while the increase of maximum momentum is less pronounced.

References

Charuel, P. and Leyer, J. C. (1981) Caractéristiques du champ de pression engendré par une flamme acceleree en espace libre: Effet d'un confinement partiel placé au voisinage su point d'inflammation. Université de Poitiers, Laboratoire d'Energetique et de Détonique. Report final ECI 699-80-F.

Eckhoff, R. K., Fuhre, K., Krest, O., Guirao, C. M., and Lee, J. H. S. (1980) Some recent large scale gas explosion experiments in Norway. The Chr. Michelsen Institute, Report CMI Nr. 790 750-1.

Lee, J. H., Pangritz, D., and Wagner, H. Gg. (1977) Beschleunigung instationärer Flammen in Rohren durch Blenden. Mas Planck Institut für Strömungsforschung, Göttingen, Berict 18/1977.

Moen, I. O., Lee, J. H. S., Hjertager, B. H., Fuhre, K., and Eckhoff, R. K. (1982) Pressure development due to turbulent flame propagation in large-scale methane-air explosions. Comb. Flame 47, 31-52.

Flame Acceleration by a Postflame Local Explosion

Manfred Stock,* Manfred Schildknecht,* and Werner Geiger†
Battelle-Institut e.V., Frankfurt, Federal Republic of Germany

Abstract

Delayed local explosion of segregated parts of an explosive gas mixture behind the flame front can lead to flame front acceleration and hence to increased overpressures. The effect of a local explosion on flame speed and overpressure during deflagration of a stoichiometric ethylene-air mixture was investigated in two different experimental configurations. In the first experimental setup, the interaction of a spherical shock front with a planar flame front was studied in a 1-m-diam tube. The effect of flame front acceleration was indicated by an increase of overpressure from 0.1 to 0.5 bar. In a second experimental setup, the interaction of a spherical shock front with a spherical flame front in an unconfined configuration was studied. The flame front, which moved with an initial velocity of about 10 m/s, became strongly folded and torn up during the interaction with the shock front. As a result, a large increase in the overall flame speed was observed. Different parts of the reaction zone attained velocities in the range between 75 and 125 m/s. The folding of the flame front due to the occurrence of the Markstein-Taylor instability is thought to be the main cause for this large increase of combustion rate and overall flame speed.

Introduction

This paper presents results of an experimental study within the subprogram Gas Explosions that has been initiated within the German light water reactor safety program as part of the project External Events. A central part of

---

Presented at the 9th ICODERS, Poitiers, France, July 3-8, 1983. Copyright © American Institute of Aeronautics and Astronautics, Inc., 1984. All rights reserved.
*Senior Scientist, Safety Technologies Department.
†Deputy Manager, Safety Technologies Department.

the subprogram Gas Explosions is the question which mechanism could possibly initiate detonationlike explosion modes (quasidetonations, fast deflagrations) in unconfined vapor clouds.

One possible mechanism that can lead to flame front acceleration and hence to increased overpressures is the delayed local explosion of segregated parts of the explosive gas mixture behind the flame front (Lee 1977). Such segregated parts within an unconfined vapor cloud might be the result of inhomogeneous concentration or large vortices formed by the wind field during the cloud formation process. In addition, segregated parts of the cloud can be produced by the explosion-induced flowfield at the exit of lanes between buildings or--as shown in Fig. 1--behind obstacles in the flame path. The basic conditions for highly increased combustion rates in the explosion of segregated parts ("pockets") of the explosive gas mixture were investigated by Lee and Moen (1980) and Chan et al. (1983) at McGill University. The work presented here deals with the question of which level of overpressure generated by the delayed explosion of a vapor cloud pocket will lead to a significant increase of the overall flame speed and hence to a fast deflagration of the vapor cloud.

Experimental Configurations

The effect of a local explosion on flame speed and overpressure during deflagration of a stoichiometric ethylene-air mixture was investigated in two different geometrical configurations: 1) planar flame in a tube and 2) hemispherical flame in an unconfined configuration. In

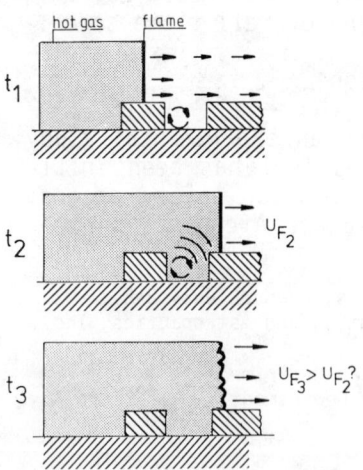

Fig. 1 Possible scenario for flame front acceleration after the delayed local explosion of a segregated part of the explosive gas mixture behind the flame.

both configurations, the local explosion behind the flame front was simulated in the same way. A spherical glass balloon with a volume of about 3 liters (shock wave generator) was filled with an explosive gas-oxygen mixture. (Though, in fact, a blast wave with immediate decay of overpressure behind the front is generated, the term shock wave will be used throughout.) The ~0.5-mm-thick glass protected the mixture from the propagating flame front. After a delay, a detonation of this segregated gas mixture was initiated by a strong ignition source (detonator or spark) to produce a shock wave behind the flame front.

In the first experimental setup, which is shown in Fig. 2 the interaction of a spherical shock with a planar flame was studied in a tube 2 m long and 1 m in diameter. The flame was ignited by seven sparks at the closed end of the tube. The blast wave generator was situated in the central part of the tube. Pressure transducers and a window for optical observation of the flame were situated near the open end of the tube. In a second experimental setup, the interaction of a spherical shock with a spherical flame was investigated in an unconfined configuration, as shown in Fig. 3. A stoichiometric mixture of ethylene-air was prepared in a volume of 4 m^3 enclosed in plastic foil. The mixture was ignited in the center using a heated wire around the glass balloon. After the flame had traveled some distance, the local explosion (detonation) was ignited in its back. Different pressure transducers were used to measure the shock pressure and the overpressure generated by the flame. The flame propagation was observed

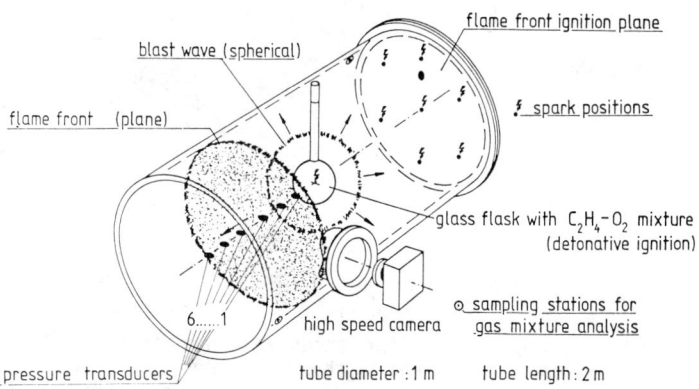

Fig. 2 Experimental setup for study of flame front acceleration due to local explosions behind the front. Planar flame in a tube 2 m long and 1 m in diameter ignited at the closed end of the tube.

with a high-speed camera and, in addition, by ionization probes and miniature thermistors.

Experimental Results

Planar Flame in a Tube

The propagation of the planar flame in the tube is shown in the (x,t) diagram of Fig. 4. Without generation of a shock wave, the flame attained in all experiments a nearly constant speed in the range of 80-100 m/s toward the open end of the tube.

Figure 5 shows typical pressure records with a broad plateau of 20-30 mbars followed by a nearly simultaneous rise to a pressure peak of about 0.1 bar. The plateau corresponds to the maximum overpressure generated by the flame when it passes the transducer in the tube. This low value (compared with the value for the flame speed) shows that pressure relief from the open end of the tube is effective during flame propagation in the tube. When the flame propagates into the open space, a maximum overpressure of about 0.1 bar is generated, which is the cause for the second,

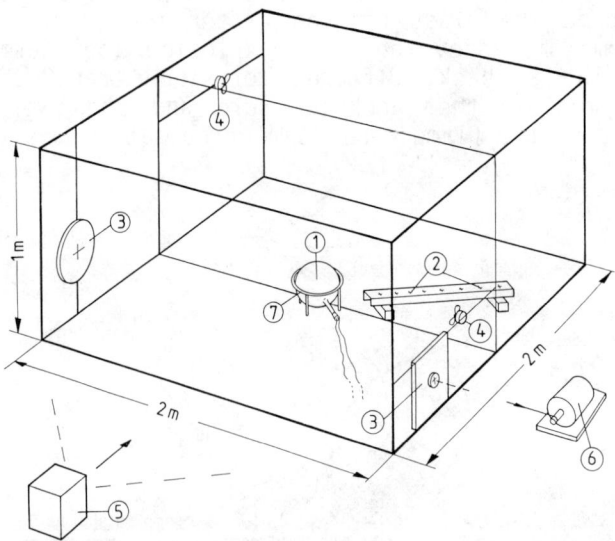

Fig. 3 Experimental setup for study of flame front acceleration due to local explosions behind the front hemispherical flame in an unconfined volume of 4 m³. (1) Blast wave generator (glass flask with C_2H_2-O_2 mixture); (2) pressure transducers; (3) (6) setup with motor to strip of a plastic tube filled with C_2H_4; (4) fans for preparation of gas mixture; (5) HYCAM high-speed camera; (7) ignition by a circular wire.

more pronounced peak in the pressure records. The pressure wave propagates from the open end back to the pressure transducers inside the tube.

In Fig. 6, pressure records are shown with and without interaction of a shock wave with the flame front. In the experiments without shock wave interaction, the signals at different transducers were quite similar and did not exceed a maximum overpressure of 90 mbars. In the two experiments with shock wave interaction that were fully recorded, the maximum overpressure increased significantly up to 0.5 bar after a shock wave of about 1-bar peak overpressure had passed the flame front. The pressure signal of the shock wave is shown in an expanded time scale in the left upper part of Fig. 6. An optical measurement of the flame speed after the interaction failed, because the flame radiation was masked by the radiation of the detonation.

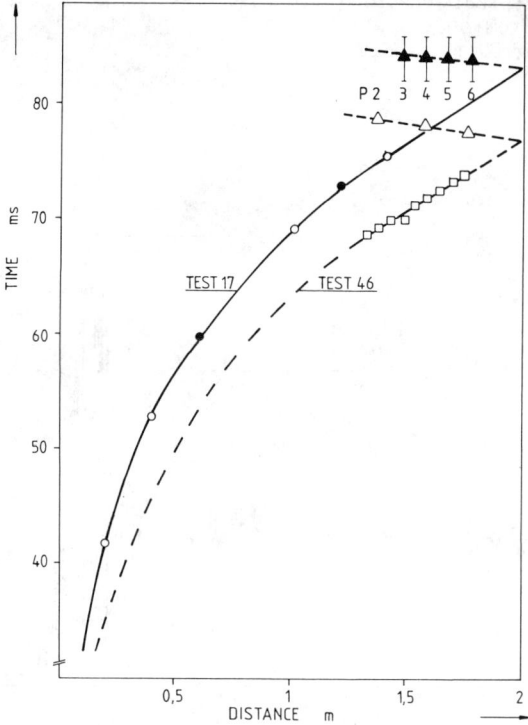

Fig. 4 (x,t) diagram of the flame front in the tube without local explosion. In test 17, ionization probes (o) and miniature thermistors (●) are used, while in test 46 the flame was recorded by high-speed photography (□). In all tests, the curred after the maximum overpressure at the pressure transducers (△,▲) occurred after the flame had left the tube.

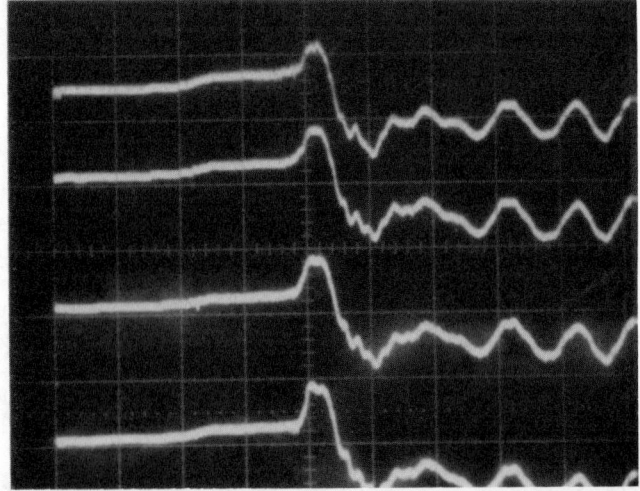

Fig. 5 Typical pressure records (test 17) at the transducers No. 3-6 (from above), separated in steps of 0.1 m in the tube. x-division: 20 ms; y-division: 0.1 bar.

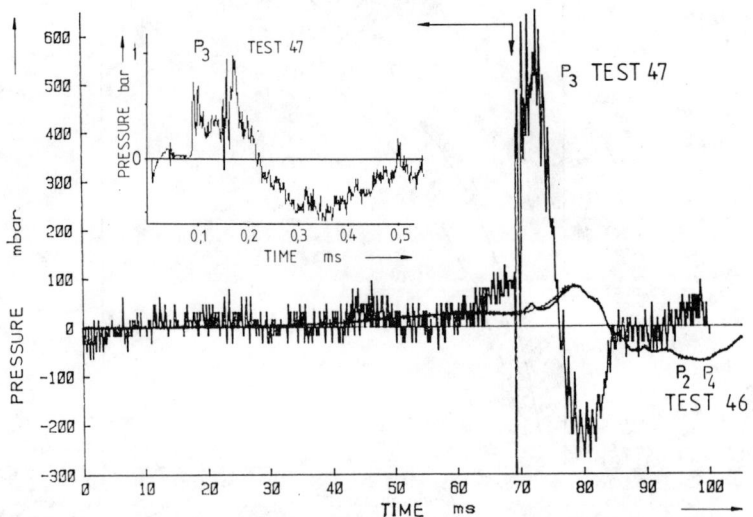

Fig. 6 Overpressure vs time without (transducers 2 and 4 of test 46) and with blast wave generated behind the flame (transducer 3 of test 47, filtered curve). The blast wave overpressure starting at $t_o = 69$ ms is shown separately with extended time scale (unfiltered curve).

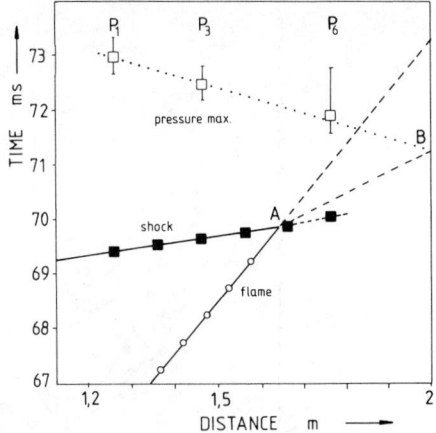

Fig. 7 (x,t) diagram of an experiment (test 47) with a shock wave (■) generated behind the flame (o), shock flame interaction at A, maximum overpressure after shock wave (□). A-B is the extrapolated trajectory of the flame after the interaction.

Figure 7 shows an (x,t) diagram with the shock wave, the flame front recorded until the time of interaction, and the time of arrival of the pressure maximum at the location of the pressure transducers. The extrapolation of the trajectory of the pressure maximum (dotted line) to the open end of the tube leads to point B. This yields a flame speed after the interaction between 200 and 250 m/s derived from the (x,t) slope A-B. This increase in flame front velocity corresponds to the large increase in the maximum overpressure. A decay of the shock wave overpressure could not be determined in the tube because of multiple reflected shocks.

Hemispherical Flame in an Unconfined Configuration

The decay of the shock wave overpressure with distance has been measured only in the unconfined experimental setup.

Figure 8 shows experimental and numerical results for the peak overpressure vs normalized distance for three cases. Curve A: shock wave coupled into hot burnt gas and traveling in the hot gas. Curve B: shock wave coupled into hot burnt gas, traveling in cold unburnt gas after having passed the interface between hot gas and unburnt mixture. The larger acoustic impedance of the cold gas let the shock amplitude increase by a factor of about 1.5. Curve C: shock wave coupled into cold unburnt gas and traveling in the cold gas. The shock amplitude is about twice as high as in curve A.

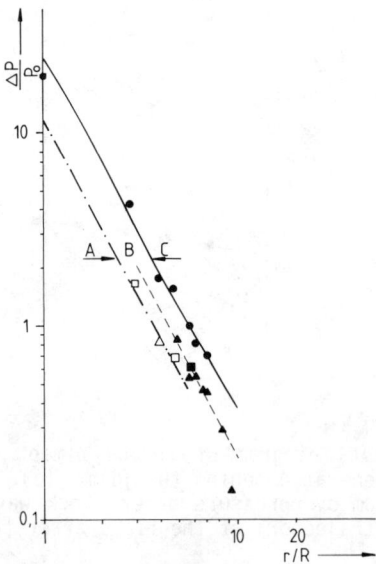

Fig. 8 Peak overpressure (p_o = 1 bar) vs reduced distance for the detonation of the glass sphere (R = 0.09 m) filled with stoichiometric C_2H_2-O_2 mixture. Curve A: Blast wave coupled into hot gas (burnt reaction products), experimental results of tests 72 (□) and 75 (△) and numerical results (BKWAVE — · —). Curve B: Blast wave after having passed the interface between hot and cold gas, experimental results of tests 72 (■) and 75 (▲). Curve C: Blast wave coupled into cold gas, experimental results of tests 61 and 73 (●) and numerical results (BKWAVE unbroken line).

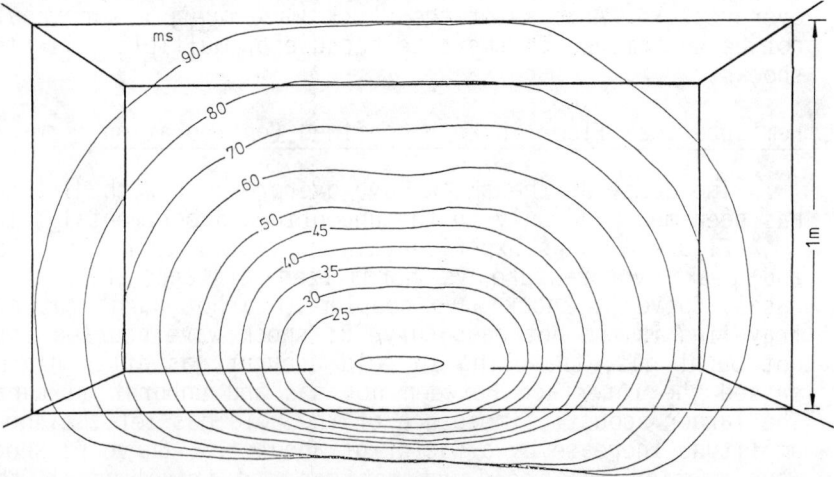

Fig. 9 Flame front contours (in ms) without shock wave interaction (test 74).

Curves A and C are calculated according to the Brinkley and Kirkwood (1947) method using the BKWAVE-code of Briscoe and Fogg (1978).

A slightly distorted hemispherical flame front was observed when no shock wave was generated (Fig. 9) with a flame speed of the order of 10 m/s. Two experiments with shock wave-flame interaction were fully recorded in unconfined configuration.

In Fig. 10, flame front contours before and after the generation of a shock wave are shown. The nearly hemispherical flame front becomes strongly folded and torn up during the interaction with the shock wave. The (x,t) diagram (Fig. 11) shows the resulting order of magnitude increase of the overall flame speed from u_0 = 10 m/s before the local explosion to about 100 m/s average afterwards. Different parts of the flame front move with velocities in the range between 75 and 125 m/s some milliseconds after the interaction process. The shock wave that caused such a large acceleration of the flame front had an overpressure of about 1.6 bars in the hot reaction products. In the second fully recorded experiment with interaction in unconfined configuration, a shock overpressure of nearly 1 bar caused a flame front acceleration of the same order of magnitude. In this case, the flame propagation after the interaction has only been observed at the edge of the experimental setup, which is not adequate to study the effects of weaker shocks. The distance over which the enhanced combustion rate and flame speed and,

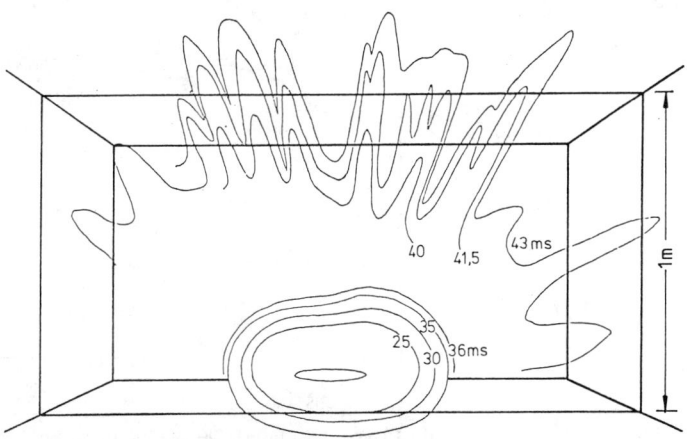

Fig. 10 Flame front contours (in ms) before and after a local explosion (t = 36.7 ms; test 75).

hence, an increased explosion pressure is maintained could not be determined within the limits of the project.

Discussion

In the tube experiments, the maximum overpressure is generated when the flame front leaves the tube. The flame enters a highly turbulent region produced by the gas flow ahead of the flame forming large vortices outside the tube. The time of arrival of this pressure peak at the transducers, as well as the fivefold increase in overpressure (compared to the case of no shock), indicate that a shock wave of 1 bar increases the flame speed from 100 m/s to more than 200 m/s. (The relationship between overpressure and flame speed outside the tube is assumed to be qualitatively similar to that valid for spherical geometry.) In the unconfined configuration, the increase of the flame speed from 10 to 125 m/s caused by a shock wave of similar magnitude has been directly observed.

The observed flame front acceleration by a local explosion behind the flame front can be explained by the following interaction mechanisms: 1) the increase of the particle velocity at the time of shock passage, 2) the increase of the reaction rate and burning velocity due to

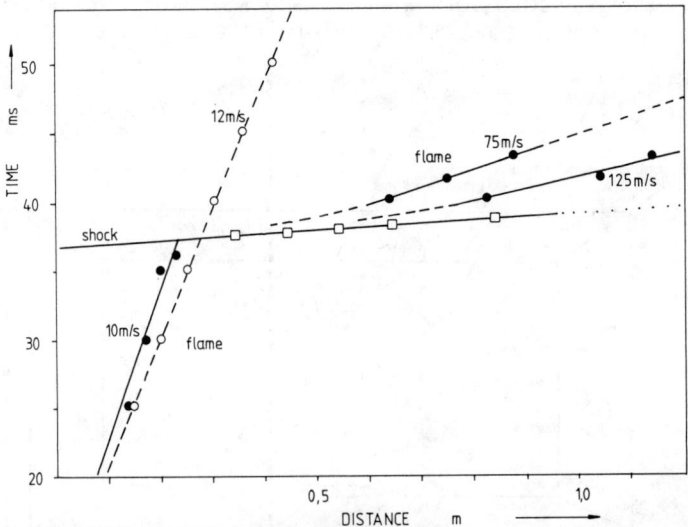

Fig. 11 (x,t) diagram of flame propagation without local explosion (o, test 74) and before and after a local explosion (●, test 75). After the shock front (□) has passed a velocity distribution is observed in the flame front.

temperature increase in the shock front, and 3) the increase of the reaction rate and flame front area due to flame front instabilities.

The first mechanism may explain the large distance the flame front is displaced a short period after the local explosion in the unconfined configuration (see Fig. 11). A shock front of 2-bar overpressure in the unburnt mixture corresponds to a peak particle velocity of about 300 m/s. The high particle velocities that lead to a displacement of the flame front are effective only for a short period of some microseconds. The same holds for the increase of the burning velocity during the interaction between flame and shock front due to the sudden change of temperature and pressure, according to Kurylo et al. (1979). The increase in flame speed by a factor of 2 in the tube experiments is of the order expected by this effect, whereas, in the unconfined configuration, the observed factor of increase is more than 10. The further propagation of the flame front at still high velocity must be explained by other mechanisms.

Flame front instabilities are thought to be the main cause for the large increase in combustion rate and in overall flame speed. Because of the large difference in the densities of cold gas (ρ_1) and hot gas (ρ_2), the interface formed by the flame front is instable. Taylor (1950) has shown that any slight distortion of the shape of this interface defined by a wavenumber K and an amplitude A, described by

$$\xi_0(x) = A \cos(Kx) \tag{1}$$

will be enhanced when the interface is accelerated in the direction toward the denser medium (the coordinate x lies in the plane of the interface, while the distortion ξ is perpendicular to it).

This means that accelerations of the interface toward the cold gas will lead to an increase of the perturbation with time. The effect of an acceleration a on shape and velocity will be

$$\xi(x,t) = A \cos(Kx) \cosh(\beta t) \tag{2a}$$

$$u(x,t) = A \beta \cos(Kx) \sinh(\beta t) \tag{2b}$$

with

$$\beta = \left(Ka \frac{\rho_1 - \rho_2}{\rho_1 + \rho_2}\right)^{1/2} \tag{2c}$$

Markstein (1957) modified Taylor's approach for accelerations, with short duration τ leading to an increase U = aτ of the flame speed. In the system of the interface, the velocity distribution can then be discribed as

$$u(x) = AkU(\rho_1 - \rho_2)/(\rho_1 + \rho_2) \qquad (3)$$

This mechanism has been mentioned by Lee and Moen (1980) as a powerful means to generate flame folds. Such a flame fold has, in fact, been observed in our experiments (see Fig. 10). To explain the observed excessive folding of the flame front and the resulting large increase of flame speed, small initial distortions (A ≤ 4 mm) are sufficient. (The shock wave generator also produces a veil of flying fragments that may produce a secondary effect that can not be distinguished from the shock wave effects.)

In both experimental configurations, the flame speed is increased by the same increment after interaction with a shock wave of the order of 1-bar overpressure:

1) Tube: initial velocity = 100 m/s; final velocity = 200 m/s.
2) Unconfined: initial velocity = 10 m/s; average final velocity = 100 m/s.

This observation that the increase of flame speed for given shock strength is independent of the initial velocity is consistent with the Markstein-Taylor instability mechanism.

Summary and Conclusion

The experimental results have shown that local explosions behind the flame front can lead to a significant flame front acceleration and hence to increased overpressures. The increase of flame speed and overpressure is observed to be accompanied by a strong folding of the flame front, which can be explained with the occurrence of the Markstein-Taylor instability after shock-flame interaction.

The experiments were performed in two different configurations with initial flame speeds of 10 and 100 m/s, and shock overpressures of the order of 1 bar. The difference between initial and final flame speed is of the same order of magnitude in both configurations (100 m/s). It was not possible in the experiments to vary the shock overpressure at the flame front significantly. The effect of weaker shocks is a matter for speculation at the present time. Shock overpressures that cause a high and lasting flame speed should be of the order of 1 bar (as used in the experiments) or more. In accidental explosions, shock waves of this magnitude in the rear of the flame front would be possible only in very special configurations (e.g., when a pressure vessel bursts behind a flame front).

Acknowledgment

This research was supported by Gesellschaft für Reaktorsicherheit, Cologne, FRG.

References

Brinkley, S. R. and Kirkwood, J. G. (1947) Theory of the propagation of shock waves. Phys. Rev. 71, 9, 606.

Briscoe, F. and Fogg, G. F. (1978) A guide to the use of BKWAVE. SRD-Report 104. Safety and Reliability Directorate, Culcheth Warrington, UK.

Chan, C., Moen, I. O., and Lee, J. H. S. (1983) Influence of confinement on flame acceleration due to repeated obstacles. Combust. Flame 49, 27-39.

Lee, J. H. S. (1977) Initiation of gaseous detonation. Annual Review of Physical Chemistry, 75-104. 28 Annual Reviews Inc., Palo Alto Calif.

Lee, J. H. S. and Moen, I. O. (1980) The mechanism of transition from deflagration to detonation in vapor cloud explosions. Prog. Energy Combust. Sci., 6, 4, 359-389.

Kurylo, J., Dwyer, H. A., and Oppenheim, A. K. (1979) Numerical analysis of flow fields generated by accelerating flames. AIAA Paper 79-0290, 17th Aerospace Sciences Meeting, New Orleans, La.

Markstein, G. H. (1957) Flow disturbances induced near a slightly wavy contact surface or flame front, traversed by a shock wave. J. Aeronaut. Sci. 24, 238.

Taylor, G. (1950) The instability of liquid surfaces when accelerated in a direction perpendicular to their planes. Proc. R. Soc. London, Ser. A 201 (1046), 192-196.

Flame Acceleration of Propane-Air in a Large-Scale Obstructed Tube

B. H. Hjertager,* K. Fuhre,†
S. J. Parker,† and J. R. Bakke†
Chr. Michelsen Institute, Bergen, Norway

Abstract

Propane-air explosion experiments have been performed in a large-scale 50-m^3 obstructed tube of 2.5-m diam. and 10-m length and with one end fully open to the atmosphere (vent area 4.7-2.4 m^2). The weak match head ignition source was located at the opposite, closed end. The obstructions inside the tube consisted of regularly spaced orifice rings which blocked off either 16, 30, or 50% of the free tube cross-sectional area, i.e., the blockage ratios were either 0.16, 0.3, or 0.5. It was found that a blockage ratio of 0.16 generated peak overpressures as high as 3 bar. The highest peak overpressure obtained in the experiments was 13.9 bar, using five equally-spaced rings of blockage ratio 0.5. In general the peak overpressures were approxiamtely three times higher than those obtained in the methane-air explosions of Moen et al. (1982) in identical geometries. The arithmetic means of three pressure measurements at 0.8, 4.8, and 9.6 m from the ignition tube were in general about twice as high as those found by Moen et al. (1982) for methane-air. The difference between methane-air and propane-air are explicable in terms of the different ignition delay times of the two gases. The highest terminal flame speed of approximately 650 m/s was obtained with 5 equally-spaced orifice rings of blockage ratio 0.3. The analysis of the flame-arrival-time data from the ionization gap probes located at various positions inside the tube revealed that the

---

Presented at the 9th ICODERS, Poitiers, France, July 3-8, 1983. Copyright © by the American Institute of Aeronautics and Astronautics, Inc., 1984. All rights reserved.
    *Senior Scientist, Department of Science and Technology.
    +Scientist, Department of Science and Technology.

highest flame front speeds occurred in the shear layers behind the orifice rings. Peak blast overpressures outside the tube are found to reach maximum at a flame speed of approximately 550 m/s and compare well with predictions based on the theory of Brinkley and Kirkwood (1947).

Introduction

The present work is a continuation of previous work (see Eckhoff et al. 1980; Moen et al. 1982; and Hjertager 1982 a and b) concerned with the effects of turbulence produced by obstacles on the flame and pressure development in accidental gas explosions. In the evaluation of the explosive properties of natural gas, it has in the past been common practice to assume that its behavior would be close to that of pure methane, the main constituent of natural gas. With regard to sensitivity to detonation, methane is classified among the least hazardous fuels (Matsui and Lee 1978). However, fairly recently it has been shown that addition of even small quantities of higher alkanes to the methane, e.g., propane, has markedly increased the sensitivity to detonation (Bull et al. 1979).

The possibility of a similar effect on combustion acceleration by turbulence has been, thus far, more or less ignored, since the thermodynamic properties of various hydrocarbon-air mixtures are roughly the same and the laminar flame speed is almost identical for a number of alkane-air mixtures. Studies of turbulent deflagration waves in hydrocarbon-air mixtures in the past have also indicated that the flame propagation is related only to the laminar flame speed and turbulence parameters, and not to the spontaneous ignition delay which is relevant for detonation sensitivity rating.

Although it has been known for many years that wall-friction-induced turbulence can cause dramatic acceleration of gas explosions in long tubes, the general implication of that fact has not been appreciated until fairly recently. The work by Eckhoff et al. (1980) and subsequently by Moen et al. (1982) has shown that turbulence has a strong influence on the pressure buildup in partly-confined large-scale experiments. Pressures as high as approximately 9 bar can readily by achieved with combustion of methane-air in a well vented large-scale system. Furthermore, Hjertager (1982 a and b) demonstrated by comprehensive computer simulations that turbulent shear-layer combustion is the likely, governing mechanism. The question then arises whether the flame propagation in such situations is influenced significantly by the chemical combustion kinetics operating at high pressures, or by the laminar flame speed and turbulence parameters encountered at ambient pressure. The answers to these questions will no doubt influence the explosion hazard forecast in practical situations.

The present work will report an experimental investigation of flame acceleration properties of stoichiometric propane-air mixture in a large-scale (50-m^3) obstructed tube. The experiments were performed at Raufoss, Norway during a three-week period in late August and early September of 1981.

Experimental Arrangement

The combustion tube used in these experiments had a 2.5-m diam and was 10-m long. One end of the tube was fully open and the other end was closed by a shell equipped with a flanged 0.46-m diam opening. A 4.5-m long and 0.48-m internal diam. ignition tube was fitted to the flange at the closed end of the combustion tube. The schematic tube facility is shown in Fig. 1. In the present experiments the ignition tube was not used; instead 19 electrically-fired match heads (ICI CeMg) were uniformly distributed over the cross-sectional area at the closed end of the large tube to simulate a planar ignition source.

The tube was filled with a premixed stoichiometric mixture of propane and air. A thin plastic film covered the tube exit during filling and was removed just prior to ignition. The mixture composition was monitored at the inlet and outlet of the combustion tube by using an infrared (i.r.) gas analyzer (URAS 7N, Hartman and Brown). Filling continued until the propane concentration at the inlet and outlet was within 0.1% of the desired stoichiometric concentration (4.25%).

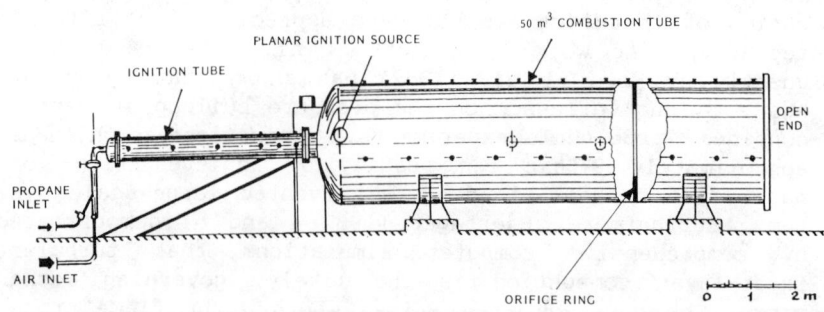

Fig. 1 Schematic diagram of explosion tube facility.

Synchronization of the ignition, resetting and calibration of pressure transducers, and startup of camera were controlled by a 10 channel programmed timer (UP timer, Xanadu Controls). The signals were recorded on a 14 channel analog tape recorder (PR2230, Ampex Corporation). After each test the records were displayed with a multichannel uv-recorder (Autograph 8, Bryans). Three types of diagnostic probes were used to monitor the explosion propagation. Flame position in the tube was recorded by using eight regularly-spaced ionization gap probes mounted on two rods. The pressure inside the tube was measured by four pressure transducers (603B, Kistler) mounted at various positions along the tube wall. The outside blast was monitored by a blast wave transducer (LC33, Celesco) placed 10 m from the tube exit at an angle of approximately 10 deg from the tube axis.

Twenty explosion tests were performed using various obstacle configurations inside the tube. The obstacles consisted of steel rings with outside diameter (o.d.) equal to the tube diameter (D=2.5 m), and the inside diameter (i.d.) was varied according to Table 1.

These rings were placed at different locations along the tube to subject the flame to obstacles which provided a fixed blockage ratio, $BR = 1-(d/D)^2$ to the flow. The number of rings regularly spaced along the tube varied from one to six. Fig. 2 gives the details of the various geometries used, and locations of the diagnostic probes.

Results and Discussion

Pressure and Flame Development Inside the Tube

Generally it was found that for the same geometrical conditions the violence of propane-air explosions was much stronger than for the corresponding methane-air explosions reported by Moen et al. (1982). The most violent explosion was observed with five orifice plates and a blockage ratio of BR = 0.5. In this particular test the average maximum overpressure was 7 bars; the outside blast at 10 m

Table 1 Obstacle configurations

d (m)	Blockage ratio (BR)	Height of ring H (m)
2.26	0.16	0.1
2.06	0.3	0.2
1.74	0.5	0.37

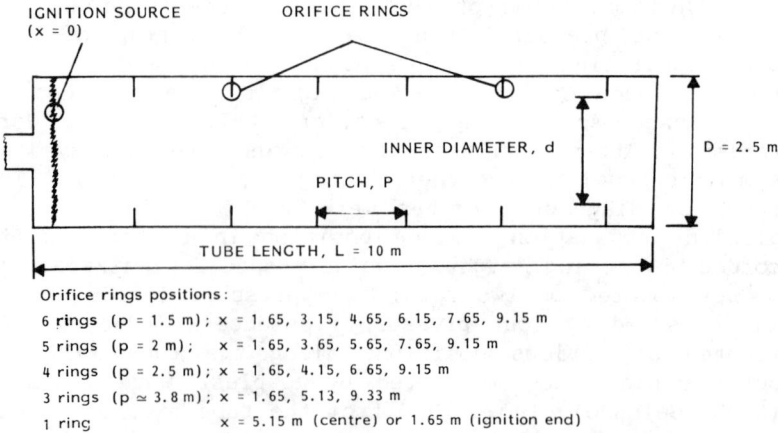

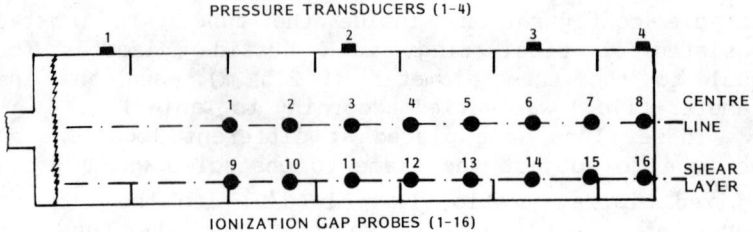

Transducer positions:

a. **Pressure:**
$x_1 = 0.8$ m, $x_2 = 4.8$ m, $x_3 = 7.8$ m, $x_4 = 9.6$ m

b. **Centre line ionization gaps:**
$x_1 = 2.61$ m, $x_2 = 3.61$ m, $x_3 = 4.61$ m, $x_4 = 5.61$ m, $x_5 = 6.11$ m, $x_6 = 7.61$ m, $x_7 = 8.61$ m, $x_8 = 9.61$ m

c. **Shear layer ionization gaps:**
$x_9 = 2.61$ m, $x_{10} = 3.61$ m, $x_{11} = 4.61$ m, $x_{12} = 5.61$ m, $x_{13} = 6.61$ m, $x_{14} = 7.61$ m, $x_{15} = 8.61$ m, $x_{16} = 9.61$ m

Fig. 2 Schematic diagram showing the obstacle configurations and diagnostic probe positions.

was 0.61 bars; and the maximum peak overpressure at the tube exit was 13.9 bars. The corresponding pressures for methane-air were: 4.0 bars, 0.39 bars, and 4.3 bars, respectively.

Fig. 3 shows the variation of average maximum overpressures with blockage ratio with number of rings as a parameter. The trend is the same as that reported for methane-air by Moen et al. (1982). A strong variation with blockage ratio can be observed. Also increasing the number of rings increases the average maximum pressure in the tube. Fig. 4 shows a comparison of the maximum aver-

FLAME ACCELERATION OF PROPANE-AIR

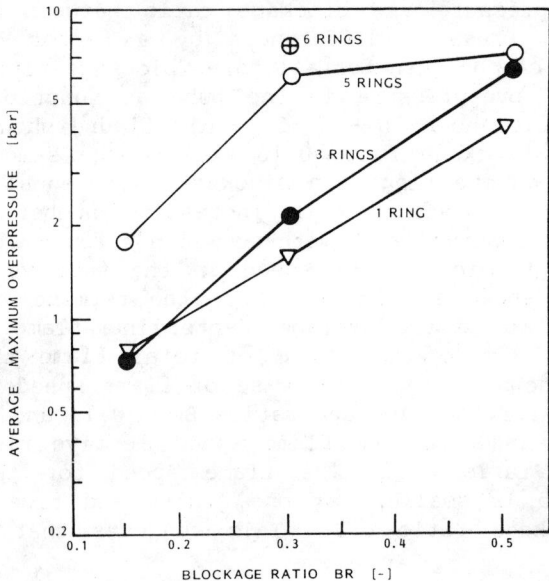

Fig. 3 Average maximum overpressure near the ignition end ($x = 0.8$ m), in the middle ($x = 4.8$ m), and near the open end ($x = 9.6$ m) as function of blockage ratio $BR = 1-(d/D)^2$.

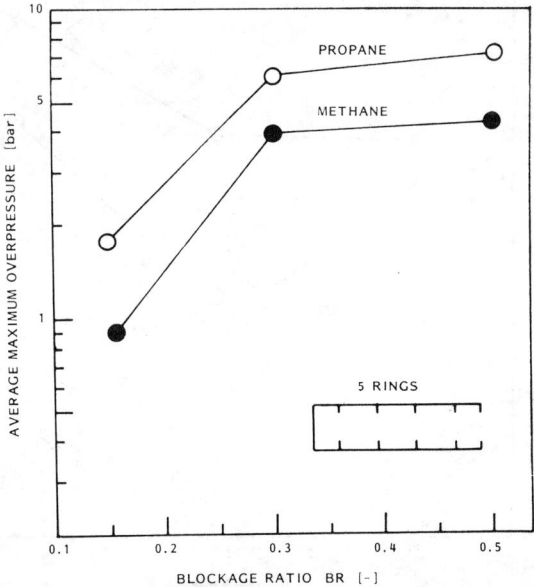

Fig. 4 Comparison of average maximum overpressure for propane and methane (Moen et al. 1982) for different blockage ratios.

age overpressure vs blockage ratio between methane and propane. Pressure differences by a factor of approximately 2 can be seen for all three blockage ratios tested.

Peak overpressure in the tube as function of number of rings is shown in Fig. 5. This figure shows that five rings of blockage ratio 0.16 exhibit the same peak overpressure as one ring with blockage ratio equal to 0.3. A decrease in pressure with increasing number of orifice plates as observed with methane was not found with propane.

The terminal flame speed in the tube vs number of rings is shown in Fig. 6. Here the terminal flame speed is taken to be the average centerline flame speed over the last 3 m of the 10 m of total flame travel. The figure shows a steady increase of flame speed with number of rings for the blockage ratios BR = 0.16 and 0.5, whereas there is a maximum flame speed at five rings for the blockage ratio 0.3. The flame speed for the blockage ratio 0.5 is smaller for one, three and five rings than for blockage ratio 0.3. This suggests that an optimum

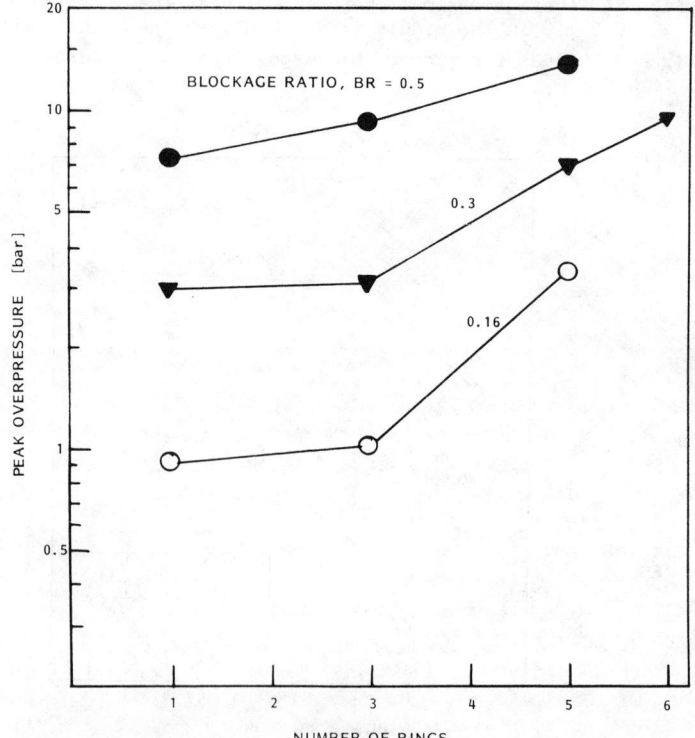

Fig. 5   Peak overpressure inside the tube as function of number of rings for blockage ratios BR = 0.16, 0.3, and 0.5.

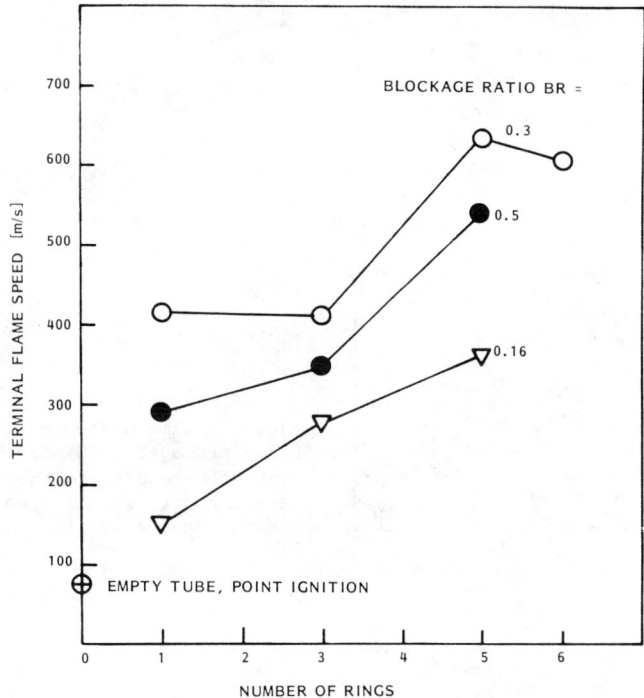

Fig. 6   Terminal flame speed (between x_5 = 6.61 m and x_8 = 9.61 m) vs number of rings for various blockage ratios.

flame speed exists at a certain number of rings and a certain blockage ratio.

Fig. 7 shows the arrival time of the flame vs. distance from ignition along the centerline and along the shear layers of the orifice rings. The figure shows three different tests with the same geometry and mixture conditions and gives the degree of repeatability of the flame propagation process. An interesting finding is depicted in Figs. 7b and 7c. These figures show that the flame at the shear layer behind a ring accelerates so fast that the flame at the centerline leaves the tube exit later than at the shear layer. This suggests shown also by the prediction method of Hjertager (1982 a), that the largest rates of combustion occur in the shear layers. Fig. 7 also shows the same trend as that which was found for methane by Moen et al. (1982), namely, that the maximum centerline terminal flame speed is reached before the flame has travelled halfway down the tube length. This was the general result of most of the geometrical arrangements tested.

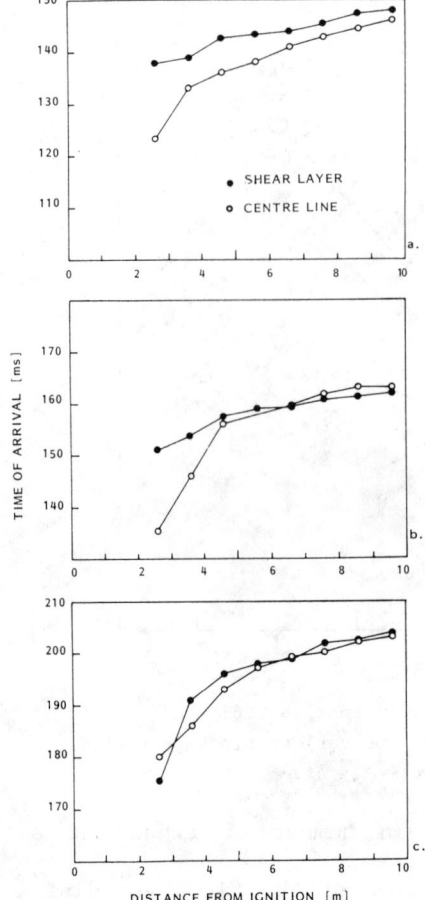

Fig. 7 Time distance plots of the flame propagation along the centerline and along the shear layers for three different tests in the same geometry; BR = 0.3 and 5 rings.

Comparisons with Predictions. A prediction method has recently been proposed and tested (Hjertager 1982 a) against the methane-air data of Moen et al. (1982). Hjertager (1982 b) has subsequently found that this original combustion rate model had to be slightly modified to correctly predict the differences between the previous methane data and the present propane data. It was obvious that differences in the thermodynamics and laminar flame speed data could not explain the dramatic difference between the two gases. Hence, the role played by the chemical kinetics of the two systems had to be taken into account. The most probable phenomenon which comes into play is the differences in ignition delay times known to govern both the sensitivity to ignition by a hot airstream

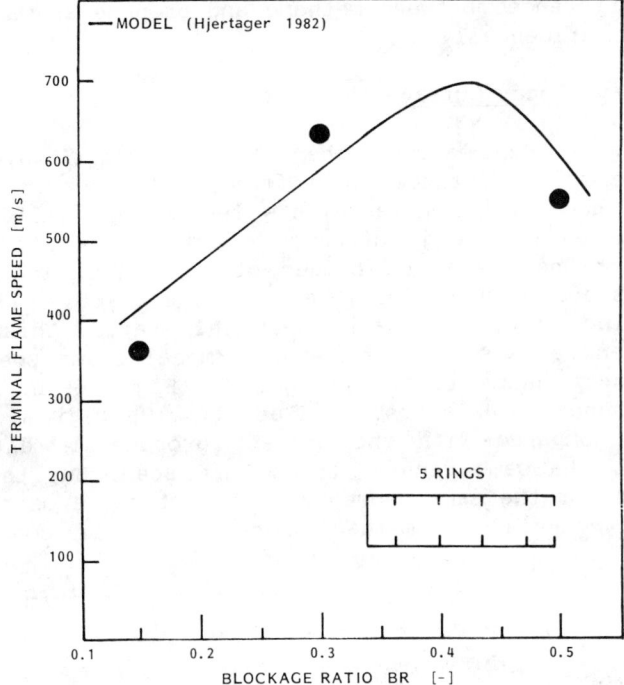

Fig. 8 Comparison between measured and predicted (Hjertager 1982) variations of terminal flame speed with blockage ratio.

(Mullins 1953) and detonation initiation sensitivity of these two fuels (Bull 1979). An ignition/ quenching criteria was proposed which made the rate of combustion only finite whenever the turbulent eddy dissipation (or mixing) time, τ_e, exceeded a constant multiplied by some chemical time, $D*\tau_{ch}$. When the opposite occurs, the combustion rate is set equal to zero, thus indicating either failure to ignite or quenching of an already established combustion wave.

The chemical times for the methane-air and propane-air are taken from Burcat et al. (1971 and 1972). Fig. 8 shows a comparison between the computation model and the experiments of the terminal flame speed as function of blockage ratio. It is seen that the agreement is satisfactory and that the model predicts the optimum flame speed at a blockage ratio approximately equal to 0.4. The observed differences in peak overpressures between methane and propane are compared with the corresponding predictions in Fig. 9. The computation model predicts almost the

514   B.H. HJERTAGER ET AL.

same differences between methane and propane as was determined experimentally.

Blast Wave Decay Outside the Tube

Fig. 10 summarizes the measured blast wave overpressure at a distance 10 m from the tube exit for the various geometrical arrangements tested. The results show that there is strong influence on the shock strength of both blockage ratio and number of obstacles. The largest overpressure occurs for the blockage ratio 0.3, thus suggesting an optimum value near this ratio. This is the same trend as was found for the terminal flame speed as a function of geometrical arrangement (Figs. 6 and 8). If the methane-air data from Moen et al. (1982) given in Fig. 11 are compared with the present propane-air data from Fig. 10, the same general trends are seen. But the shock strength for the same geometry is almost twice as high for propane compared to methane for almost all conditions.

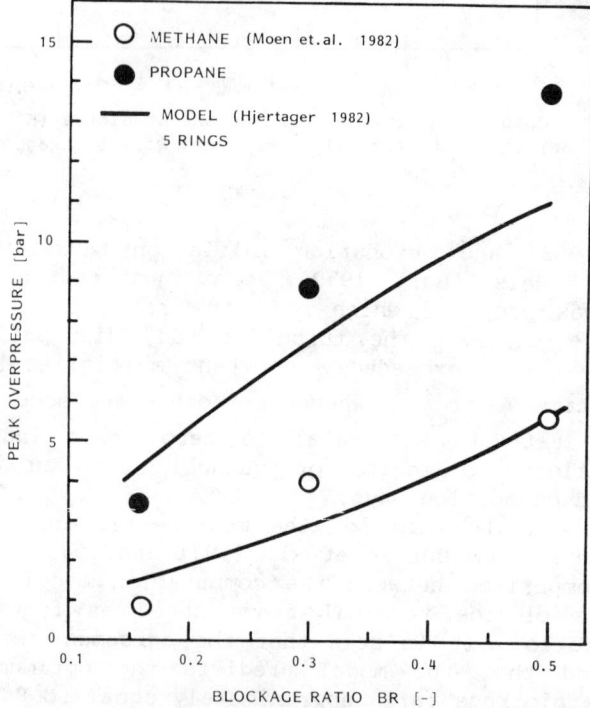

Fig. 9   Comparison of predicted (Hjertager 1982) and measured variation of peak overpressure with blockage ratio for methane (Moen et al. 1982) and propane.

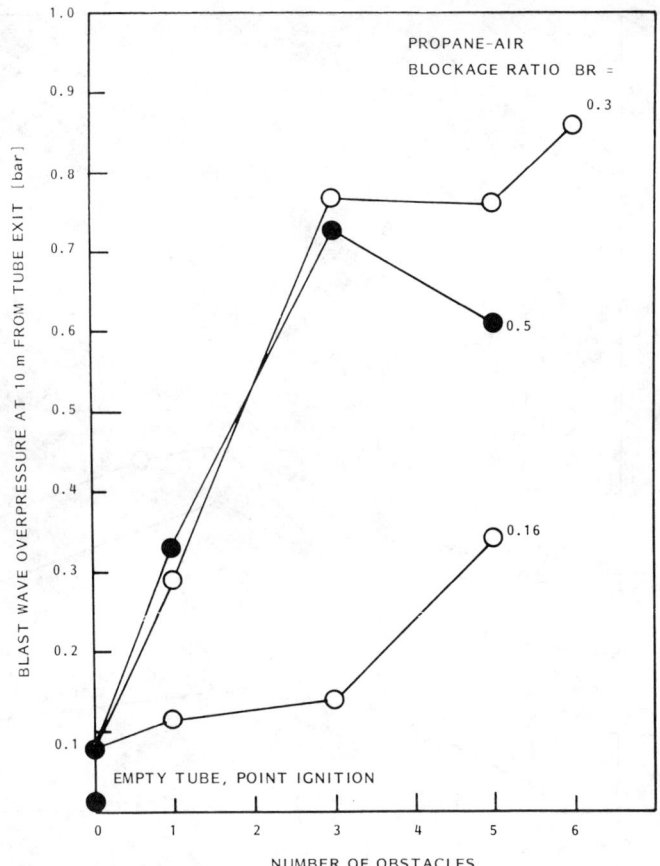

Fig. 10 Blast overpressure at 10 m from tube exit as function of number of obstacles for various blockage ratios.

This is to be expected, given the differences in average maximum pressure previously shown inside the tube in Fig. 4.

Comparisons with the Blast Wave Theory. The theory of Brinkley and Kirkwood (1947) determines the air blast overpressure and blast wave energy as a function of radial distance from the solution of two ordinary differential equations with two initial conditions. One initial condition for this theory is determined by the method of Guirao et al. (1976), which gives the shock wave overpressure and its radial location as function of a constant flame speed. In addition, the initial blast wave energy is taken as a constant fraction of the total combustion energy in the exploding gas mixture. In accordance with

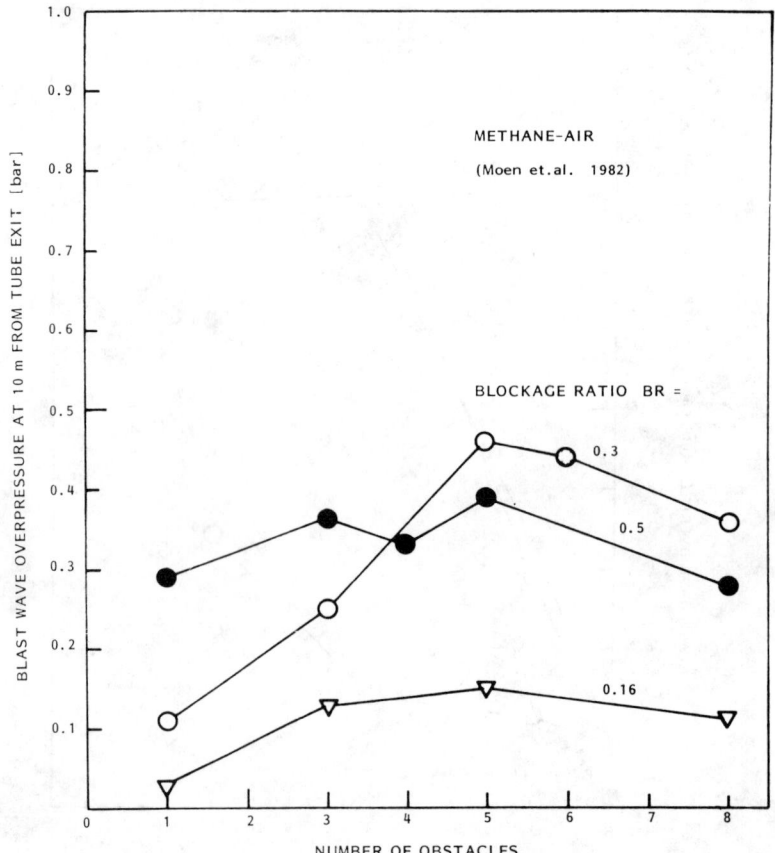

Fig. 11 Blast overpressure at 10 m from tube exit as function of number of obstacles for various blockage ratios (Moen et al. 1982).

the work of Wiekema (1979) this fraction was put equal to 0.18. A more detailed description of the method used is given by Bakke and Hjertager (1981) and Lee et al. (1977).

To apply the above theory, the 50-m^3 fuel-air mixture in the tube is transformed to a hemispherical gas cloud of the same volume. The experimentally-determined terminal flame speed (Fig. 6 and Moen et al. 1982) is the input parameter, and the resulting calculated air blast at 10 m from the tube exit is compared with the experimentally-observed overpressures in Fig. 12 for propane-air and in Fig. 13 for the methane-air of Moen et al. (1982).

Both the theory and the experiments show that for both cases there is a maximum peak pressure for a given flame speed. This can be explained by considering two

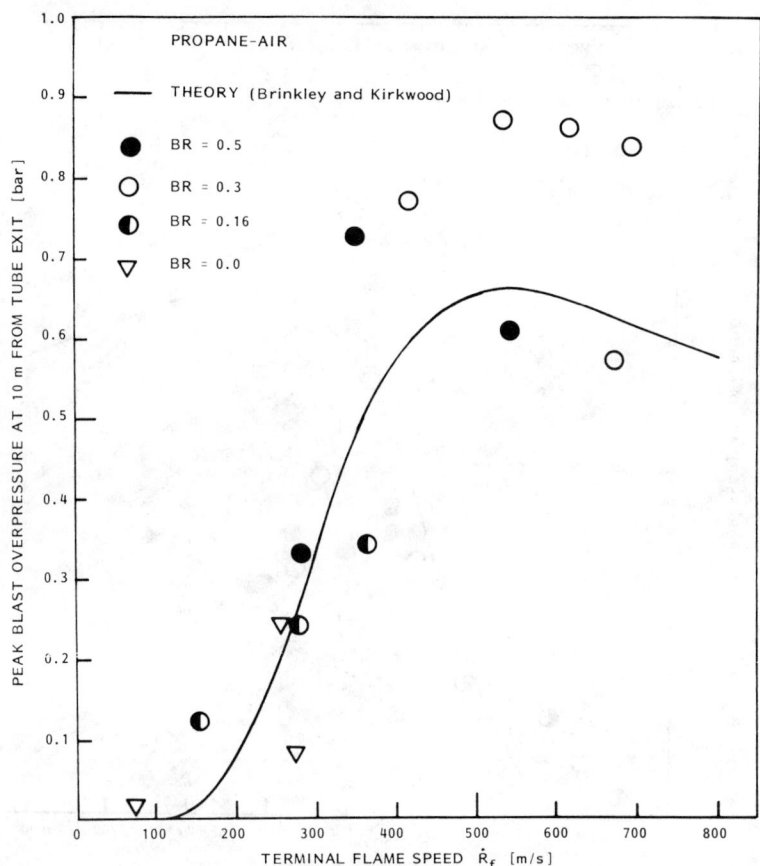

Fig. 12   Comparison between measured and predicted blast overpressures at 10 m from tube exit vs terminal flame speed.

competing mechanisms. First, increased flame speed will increase the initial shock strength (Guirao et al. 1976). Second, increased flame speed will decrease the initial radial position (due to compression between the shock and the flame) where this initial shock will begin to decay. Fig. 12 shows that there is good agreement between theory and experiments for the propane-air explosions, whereas Fig. 13 shows that the theory overpredicts the blast pressure at the highest flame speeds.

Further Considerations

It has become evident that the largest rates of combustion occur in localized volumes behind the ring ob-

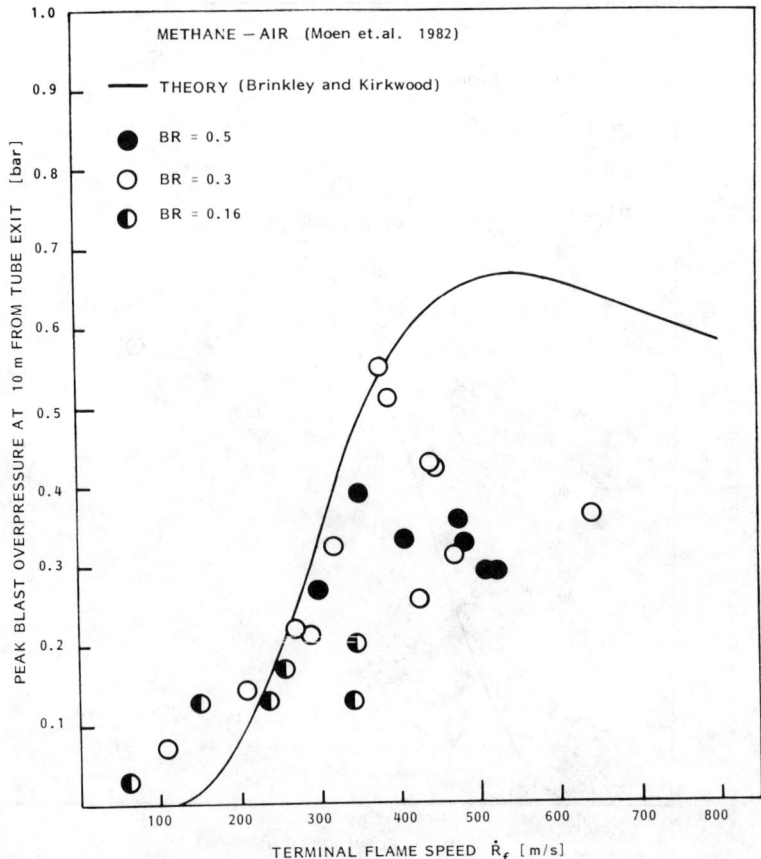

Fig. 13 Comparison between measured and predicted blast overpressure at 10 m from tube exit vs terminal flame speed.

stacles. To explain some of the observed properties of flame propagation, it is important to establish some characteristics of these volumes or, rather, mixing layers. From the theory of simple mixing layers (Bradshaw 1978) it is known that the thickness varies linearly with distance from the origin. Also, it is known that the largest gradients are at the start, and then gradually decrease along the flow path. This again implies that turbulence is large close to the origin and then decays, because turbulence velocity is approximately proportional to gradients in the velocity field. It is therefore possible to estimate the turbulence-influenced regions of the flow as a ring torus with triangular cross-section stretching from one obstacle to the next. Thus the total volume may be estimated as

$$V_t \sim p^2 \, d \, (U_c/u_t)^2 \, n$$

where p is the pitch, d the inner diameter of the ring orifices, U_c the flow velocity in the center of the ring passage, u_t the turbulence velocity, and n the number of obstacles.

From the above expression it is seen that when the inner diameter is decreased, the volume also decreases. This means that, for choked flow through the ring orifices the integrated rates of reaction should decrease and as a consequence the decrease should reduce the pressure generated by combustion. But on the other hand, as the blockage ratio increases, the pressure drop due to flow through the orifice will also increase. This suggests, as shown in Figs. 3 and 4, that the pressure produced by increasing the blockage ratio seems to level off.

For a given blockage ratio, an increase in number of plates also increases the pressure to some level. If the turbulence-influenced volume is expressed as

$$V_t \sim L^2/nd \, (U_c/u_t)^2$$

increasing the number of plates, n, along a given length of flame travel, L, <u>diminishes</u> the volume. This indicates that an increase in the number of plates may decrease the combustion-produced pressure. This is found in the methane-air experiments by Moen et al. (1982). For the present experiments this decrease in produced pressure was not observed. This can be explained by the difference in ignition delay times between the two fuels. Methane-air will ignite at a longer distance from the obstacles than the propane-air. Thus, for the same total shear layer volume in the tube, more of this volume will burn for propane as compared to methane. The result of this effect for a given number of orifice rings is shown in Fig. 9.

Since the integrated rate of combustion is smaller for methane compared to propane, the observed differences between peak blast overpressures outside the tube is easier to understand. Figs. 12 and 13 show that for the same flame speeds the measured methane explosion blast pressure is smaller than the propane explosion blast. This indicates that the flame speed may not be the essential parameter, since it contains contributions both from flow and burning velocities. It is evident that since the rate of burning is largest in the shear layers, the flame speed is mainly a measure of centerline flow velocity, not burning velocity. Hence, the distribution of turbulence domains, together with the influence of chemical kinetics, is also the key to the prediction of blast waves emitted from explosions.

Conclusions

Propane-air explosion experiments have been performed in a large-scale 50-m^3 obstructed tube. Pressures as high as approximately 14 bar were produced with five orifice rings which blocked off 50% of the free tube area. The observed pressures in propane-air were two to three times larger than the corresponding methane-air explosions of Moen et al. These differences were explained with an ignition/quenching model which used the ignition delay times for the two gases as typical chemical times. The differences in laminar flame speeds and thermodynamic properties could not explain the large differences observed.

Maximum flame speeds were found for blockage ratio 0.3 and five equally-spaced orifice rings. Measurement of flame positions along the tube indicates that the largest rates of combustion are found in the shear layers. Additional measurements of the turbulence velocities in the layers must be performed to further verify these observations. Peak blast overpressures outside the tube are found to have an optimum at a flame speed of approximately 550 m/s. This observation compares well with the predictions of the Brinkley and Kirkwood theory.

The observed differences between methane-air and propane-air explosions indicate that the gas type must be considered when gas explosion hazard analyses are performed. This suggests that natural gas-air mixtures containing higher alkanes, when ignited, will produce higher pressures than pure methane-air, but lower pressures than pure propane-air. The experimental data have made a major contribution to the development of the computer simulation model, because the chemical aspects have been introduced and their validity verified.

Acknowledgments

This work has been financially supported by BP Petroleum Development Ltd., Elf Aquitaine Norge A/S, Esso Exploration and Production Norway Inc., Mobil Exploration Norway Inc., Norsk Hydro, and Statoil. The authors acknowledge the technical support during the experiments of Mr. H. G. Thorsen and Mr. T. Håland. The experiments were performed on a test site belonging to Raufoss Ammunisjonsfabrikker A/S, Raufoss, Norway.

References

Bakke, J. R. and Hjertager, B. H. (1981) The blast wave decay from an exploding hemispherical gas cloud. Chr. Michelsen Institute, CMI No. 813403-4.

Bradshaw, P. (1978) Turbulence (second corrected and updated edition), p. 144. Springer Verlag, Berlin.

Brinkley, S. K. and Kirkwood, J. G. (1947) Theory of the propagation of shock waves. Phys. Rev. 71, 606.

Bull, D. C. (1979) Concentration limits to the initiation of unconfined detonation in fuel-air mixtures. Trans. Inst. Chem. Eng. 57, 219-227.

Bull, D. C., Elsworth, J. E. and Hooper, G. (1979) Susceptibility of methane-ethane mixtures to gaseous detonation. Combust. and Flame 34, 327-330.

Burcat, A., Crossley, R.W. and Scheller, K. (1972) Shock tube investigation of ignition in ethane-oxygene-argon mixtures. Combust. and Flame 18, 115-123.

Burcat, A., Scheller, K. and Lifshitz, A. (1971) Shock tube investigation of comparative ignition delay times for C_1-C_5 alkanes. Combust. and Flame 16, 29-33.

Eckhoff, R. K., Fuhre, K., Krest, O., Guirao, C. M. and Lee, J. H. S. (1980) Some recent large-scale experiments in Norway. NTNF Project 18306500, gas explosions on offshore platforms, flame propagation and pressure development. Chr. Michelsen Institute Rept., CMI No 790750-1.

Guirao, C. M., Bach, G. G. and Lee, J. H. S. (1976) Pressure waves generated by spherical flames. Combust. and Flame 27, 341.

Hjertager, B. H. (1982a) Simulation of transient compressible turbulent reactive flows. Combust. Sci. Technol. 27, 159-170.

Hjertager, B. H. (1982b) Numerical simulation of turbulent flame and pressure development in gas explosions, SM Study No. 16, pp. 407-426, University of Waterloo Press, Ontario, Canada.

Lee, J. H. S., Guirao, C. M., Chiu, K. W. and Bach, G. G. (1977) Blast effects from vapour cloud explosion. Loss Prev. 11, 59.

Matsui, H. and Lee, J. H. S. (1978) On the measure of the relative detonation hazards of gaseous fuel-oxygen and air mixtures. 17th Symposium (International) on Combustion, pp. 1269-1289, Combustion Institute, Pittsburg, Pa.

Moen, I. O., Lee, J. H. S., Hjertager, B. H., Fuhre, K. and
    Eckhoff, R. K. (1982) Pressure development due to turbulent
    flame propagation in large-scale methane-air explosions.
    Combust. and Flame 47, 31-52.

Mullins, B. P. (1953) Studies on the spontaneous ignition of fuels
    injected into a hot air stream. Fuel, 32, 211.

Wiekema, B. J. (1979) Methods for the calculation of the physical
    effects of the escape of dangerous material (liquid and
    gases), vapour cloud explosion. Ch.8, Report of the
    Committee for the Prevention of Disaster, Directorate General
    of Labour, Ministry of Social Affairs, the Netherlands.

Initiation of Unconfined Gaseous Detonation by Diffraction of a Detonation Front Emerging from a Pipe

Aziz Ungut,* Philip J. Shuff,† and John A. Eyre‡
Shell Research Ltd., Chester, Great Britain

Abstract

Critical pipe diameters have been estimated for stoichiometric ethane-air and propane-air mixtures when the diffraction of a steady Chapman-Jouguet (CJ) detonation established in a circular pipe initiates detonation in an unconfined mixture of the same composition and pressure. For each fuel/O_2/N_2 mixture critical nitrogen dilution was measured for pipe diameters of 51.3, 101.4, and 139.7 mm. Critical pipe diameters for stoichiometric fuel-air mixtures were then obtained by extrapolation of the results. Most of the experiments were carried out at one atmosphere but low-pressure studies of stoichiometric ethane/O_2/N_2 mixtures are also reported. Detonation velocity was measured in the pipe and near the wall of the semi-spherical detonation chamber by several pressure transducers and light detectors. This information has been used to verify the establishment of steady CJ detonation in the pipe and the consequent transition to detonation or failure to deflagration. A multibeam laser Schlieren time-of-flight anemometer has been developed to measure shock velocity at the pipe exit with good spatial resolution. High-speed frame photographs were also taken to study the transition phenomena. Soot plate records of the cellular

---

Presented at the 9th ICODERS, Poitiers, France, July 3-8, 1983. Copyright © Shell Research Ltd. 1984. Published by the American Institute of Aeronautics and Astronautics with permission.
*Senior Scientist, Thornton Research Centre.
†Senior Technician, Thornton Research Centre.
‡Group Leader, Thornton Research Centre.

structure at critical transition to detonation provide the cell size data and display the pulsating nature of the critical reinitiation process. The correlation between the critical pipe diameter and the steady cell size has been investigated. Results generally support the previous findings that the critical pipe diameter is approximately 13 times the standard cell width.

Introduction

Direct initiation of detonation in unconfined, homogeneous stoichiometric mixtures of most of the common hydrocarbon fuels and air requires that the power density of the high-energy initiation is several orders-of-magnitude higher (Bull 1979) than that which can be envisaged in most credible accident scenarios. However, detonation of a partially confined fuel-air mixture could conceivably result from acceleration of the deflagration wave. Emergence of such a detonation wave from confinement (vessel, pipe, duct, etc.) into an unconfined fuel-air mixture presents a possible, albeit unlikely, source of initiation of unconfined detonation.

For a planar detonation wave emerging from a pipe, the critical parameter that determines whether the transition to spherical detonation occurs, is the pipe diameter (Zel'dovich et al. 1956). Since the effects of shape and characteristics of the high-energy initiator are eliminated, the study of planar-to-spherical transition is also of considerable importance in the fundamental investigation of the direct initiation of detonation.

Mitrofanov et al. (1964) have established that for acetylene-oxygen mixtures at low pressures the critical dimension of the exit is approximately thirteen times the cell width for tubes and ten times the cell width for planar channels. Recent studies (see Lee et al. 1977; Matsui et al. 1978; Edwards et al. 1981; Edwards et al. 1979) suggest that this correlation between the critical tube diameter and standard cell size may be of a more fundamental nature and holds for all stoichiometric fuel-oxygen-nitrogen mixtures. However, it has also been suggested (see Urtiew et al. 1981; Vasil'ev et al. 1980) that for each mixture the number of cells required along the critical pipe diameter for a successful transition to detonation varies with the initial pressure and the nitrogen dilution.

Knystautas et al. (1982) have recently reported a detailed study of most of the common fuel-air mixtures. Their results show that approximately 13 cells are required along the critical tube diameter for a transition to detonation in all the mixtures investigated and this correlation appears to be valid for low initial pressures as well.

The main objective of the present study is to investigate the validity of the correlation between the critical pipe diameter and the cell width for stoichiometric fuel-oxygen-nitrogen mixtures in view of the range of opinions described above. It has also been felt that further studies of the mechanism of direct initiation of detonation, and particularly investigation of the possible "pulsating" nature (see Bull et al. 1978; Ul'yanitskii 1980) of reinitiation, demanded more detailed measurements of shock velocity at the pipe exit with good spatial resolution.

Experimental

Fig. 1 shows the schematic diagram of the Thornton Research Centre detonation chamber (Atkinson et al. 1980). Total volume of the chamber is approximately 1.5 m^3. A stainless steel pipe of 139.7-mm i.d. and 1870-mm length is accomodated at one of the 152-mm portholes of the vessel. The largest diameter of the semi-spherical vessel is 1460 mm. Compared with previous experiments (see Edwards et al. 1979; Knystautas et al. 1982) the dimensions of the detonation chamber are much larger than the pipe diameter, which eliminates completely the effect of any possible interaction with the walls during the transition process. The optical access is through two 305-mm portholes with toughened glass windows.

Mixtures of fuel (99% purity), oxygen (99.7% purity), and nitrogen (99.9% purity) were prepared by monitoring partial pressures of gases with a Texas Instruments model 145 precision pressure gage. The accuracy of the final composition of the mixture was estimated to be 0.1% vol. A high throughput (150 liter/min) oil-free diaphragm pump was used to recirculate the mixture via the pipe and an external loop up to approximately 4-vol. displacements in order to achieve good mixing.

Detonation was initiated at the closed end of the pipe by means of a Tetryl charge of approximately 1-g weight which was detonated by a plastic bridge wire detonator containing 70 mg of explosive. The propagation of detonation was measured at three locations along the pipe by monitoring the arrival of the shock front with Kistler Type 601A quartz pressure transducers mounted on the wall and coupled with Kistler Type 5007 charge amplifiers. An independent light detection method was used to measure

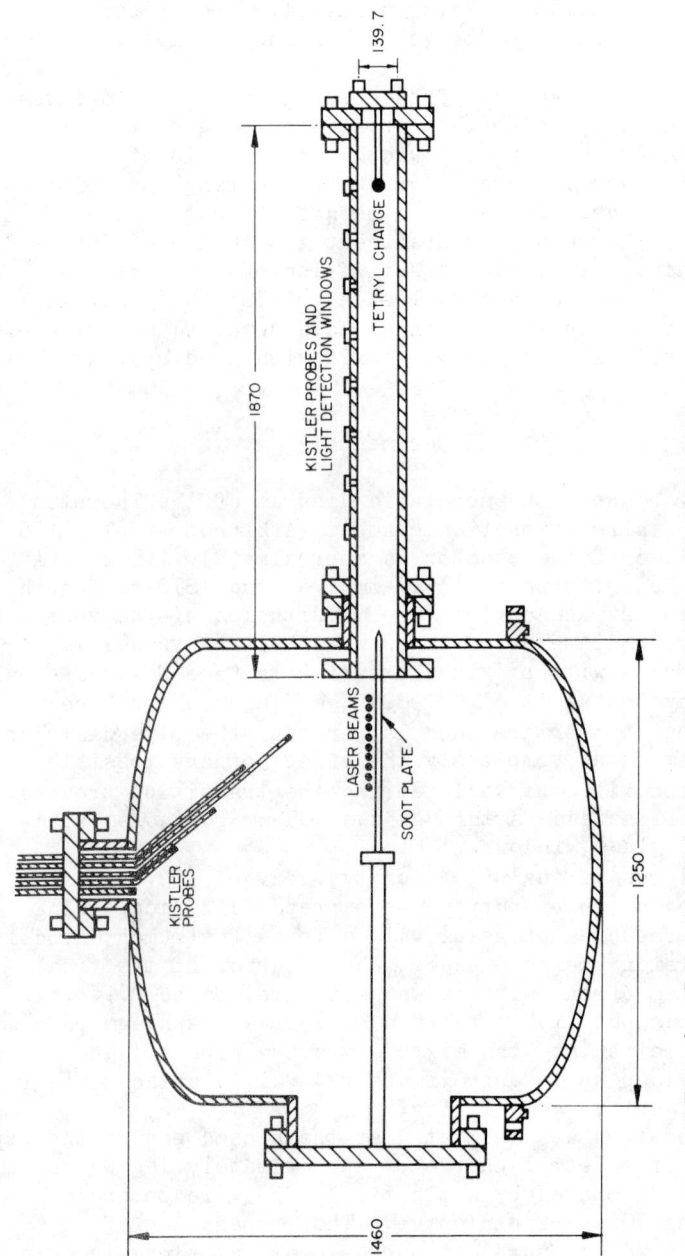

Fig. 1   Schematic diagram of Thornton Research Centre detonation chamber.

detonation velocity by monitoring the arrival of the flame front at three locations along the pipe. All analog signals were digitized and stored for further analysis in a fast transient recorder (Datalab DL922).

In all experiments detonation velocity along the pipe remained constant at CJ values after approximately five pipe diameters from the Tetryl charge. A Kistler pressure transducer mounted on the opposite wall of the vessel along the pipe axis monitored the overall average velocity of detonation between the time of initiation and arrival at the wall of the vessel. Five Kistler pressure transducers mounted radially on steel tubes were introduced from the top porthole and provided shock velocity measurements at several locations between 295 mm from the pipe exit and the vessel wall. Time-of-flight measurements in the vessel provided the confirmation of a successful transition to detonation.

A multibeam laser Schlieren time-of-flight anemometer developed in Thornton Research Centre was used for measurement of the shock front velocity along a distance of approximately two pipe diameters from the pipe exit with a spatial resolution of 28 mm.

Aluminium soot plates with a length of 610 mm were introduced from the porthole on the opposite side of the vessel to the pipe. An Imacon 790 high-speed camera was used to take frame photographs at a rate of 10^5 frames/s.

The exit diameter of the pipe was reduced by inserting a specially designed cylindrical wedge (Edwards et al. 1979) (see Fig. 2) at the end of the pipe. For pipe diameters of 101.4 and 139.7 mm, critical nitrogen dilutions for a successful transition to detonation of stoichiometric ethane-oxygen and propane-oxygen mixtures were determined at atmospheric pressure. For the pipe diameter of 51.3 mm, experiments were carried out at lower pressures and results were subsequently extrapolated to atmospheric pressure. For the pipe diameter of 139.7 mm, critical transition to detonation for ethane-oxygen-nitrogen mixtures of lower pressures has also been investigated in detail. Table 1 summarizes the mixtures investigated. Critical dilution for a successful transition to detonation was established with a resolution of 0.1 moles of nitrogen. At critical conditions experiments were repeated with soot plates to examine cell structures.

Results and Discussion

Figs. 3-5 show the variation of the shock velocity as a function of distance from the pipe exit for a mixture of

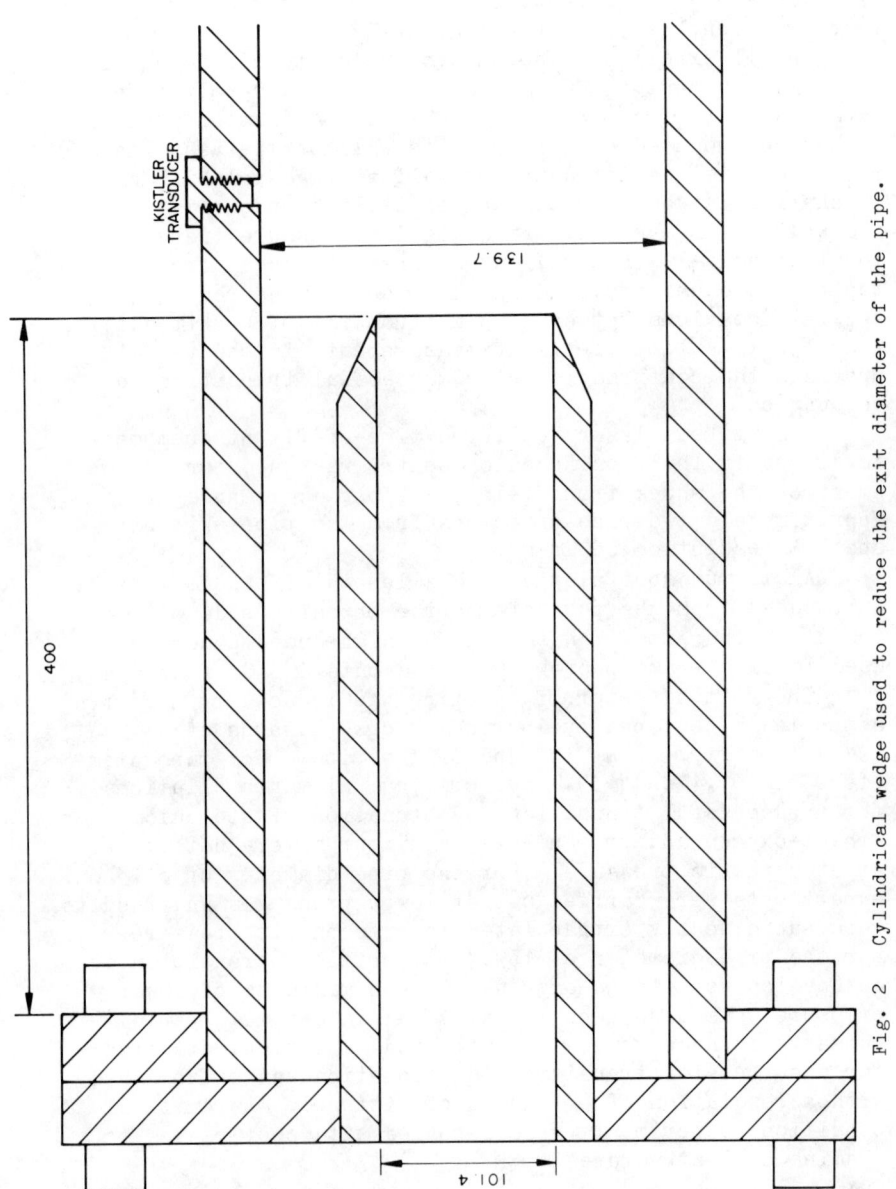

Fig. 2 Cylindrical wedge used to reduce the exit diameter of the pipe.

Table 1 Critical pipe diameters for stoichiometric mixtures of ethane and propane

Stoichiometric mixture	Dilution (Z), moles N_2	Initial pressure, mbar	Cell width (b), mm	Critical pipe diameter (d_c), mm	$d = 13 \cdot b$, mm	$\dfrac{d_c - d}{d_c} \times 100$, %	$N_c = \dfrac{d_c}{b}$
$C_2H_6 + 3.5 O_2 + ZN_2$	2	260	8.8 ± 30%	139.7	114.4	18	15.9 ± 30%
	3.5	400	9.8 ± 18%	139.7	127.4	9	14.3 ± 18%
	3.16[a]	1013	...	51.3	...	...	...
	5.2	1013	6.8 ± 24%	101.4	88.4	13	14.9 ± 24%
	6.9	1013	9.9 ± 11%	139.7	128.7	8	14.1 ± 11%
$C_3H_8 + 5 O_2 + ZN_2$	4.54[a]	1013	...	51.3	...	...	...
	7.9	1013	7 ± 26%	101.4	91	10	14.5 ± 26%
	8.95	1013	9.9 ± 19%	139.7	128.7	8	14.1 ± 19%

[a] Extrapolated from low-pressure experiments.

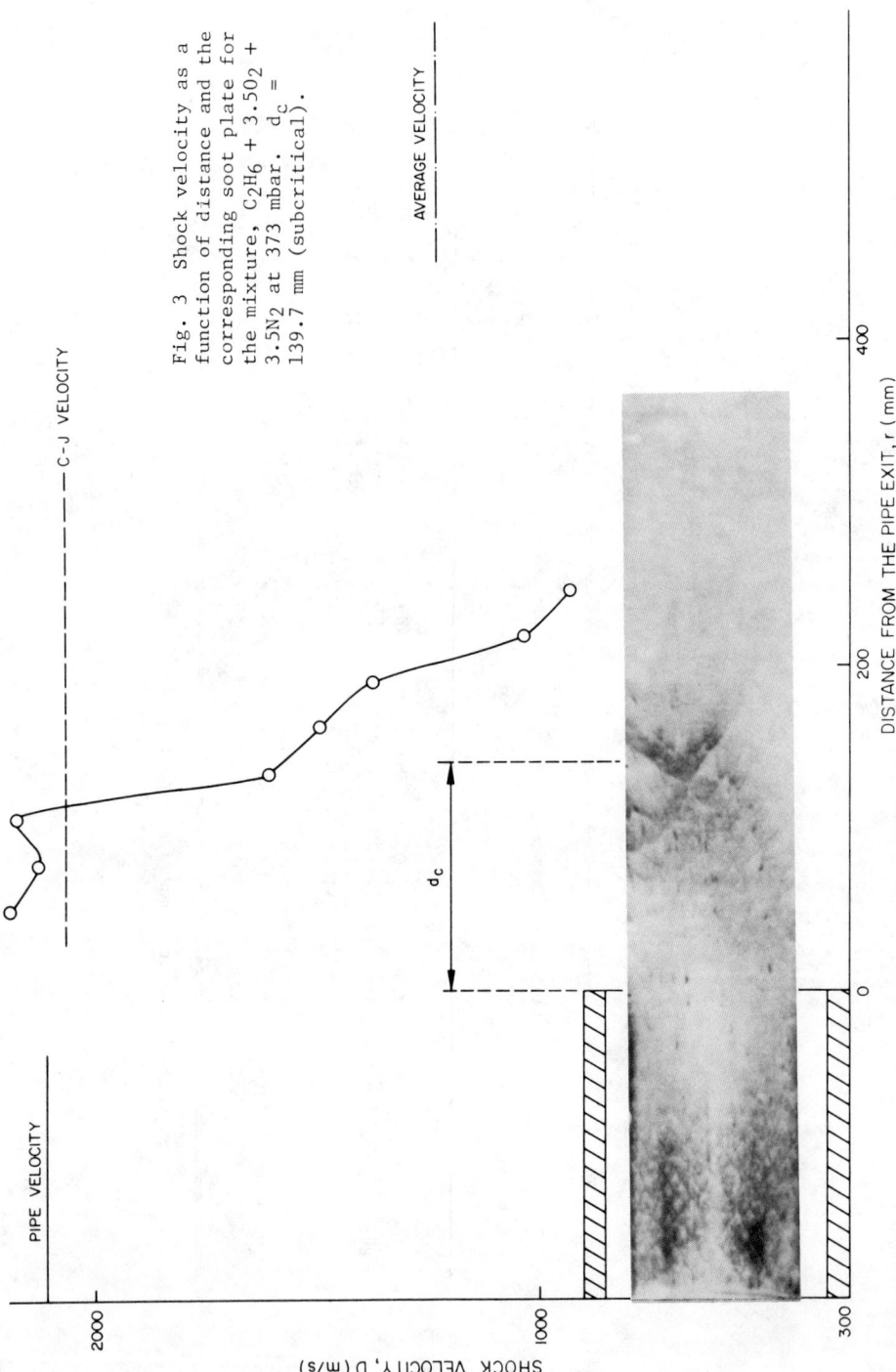

Fig. 3 Shock velocity as a function of distance and the corresponding soot plate for the mixture, $C_2H_6 + 3.5O_2 + 3.5N_2$ at 373 mbar. $d_c = 139.7$ mm (subcritical).

INITIATION OF UNCONFINED GASEOUS DETONATION 531

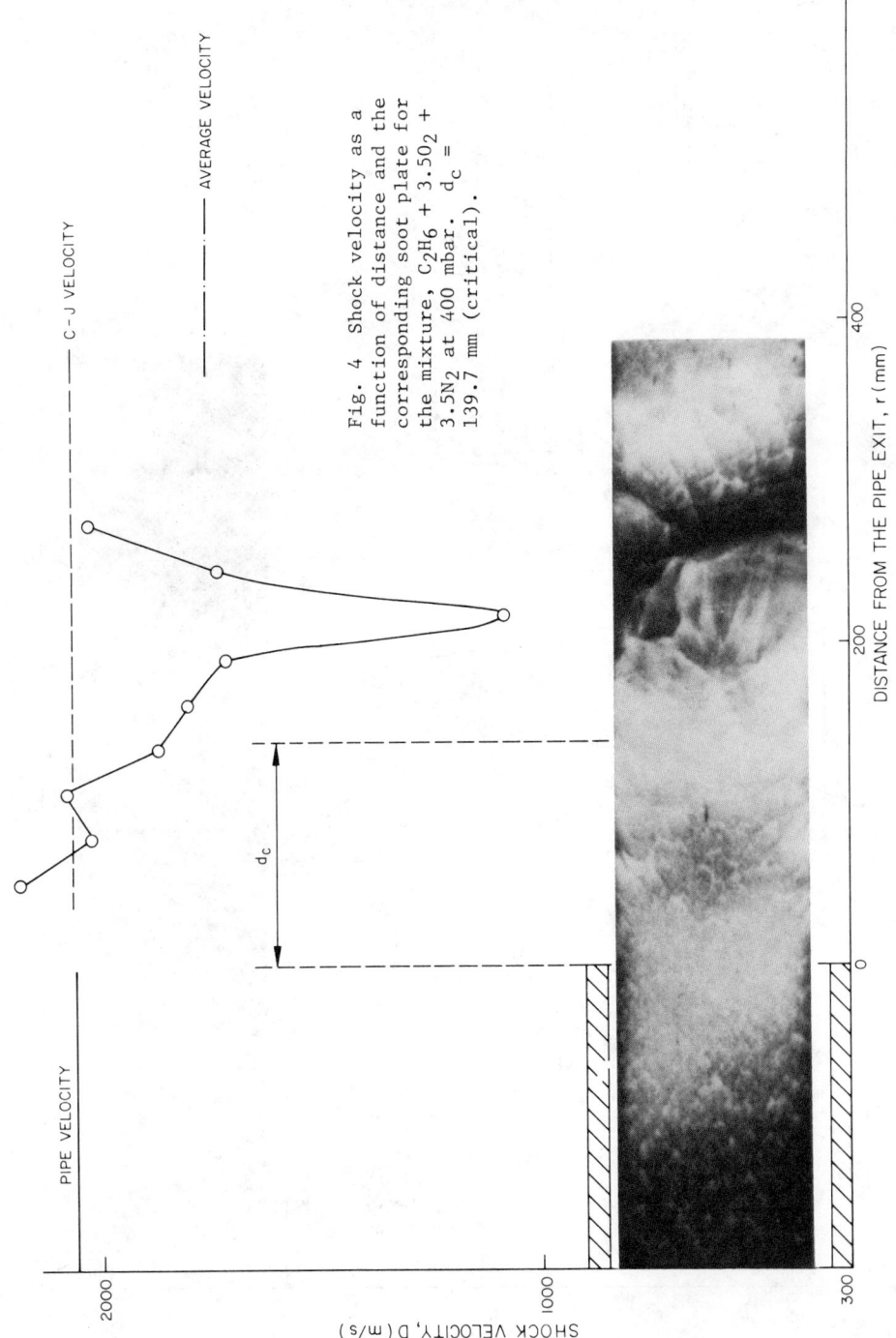

Fig. 4 Shock velocity as a function of distance and the corresponding soot plate for the mixture, $C_2H_6 + 3.5O_2 + 3.5N_2$ at 400 mbar. $d_c = 139.7$ mm (critical).

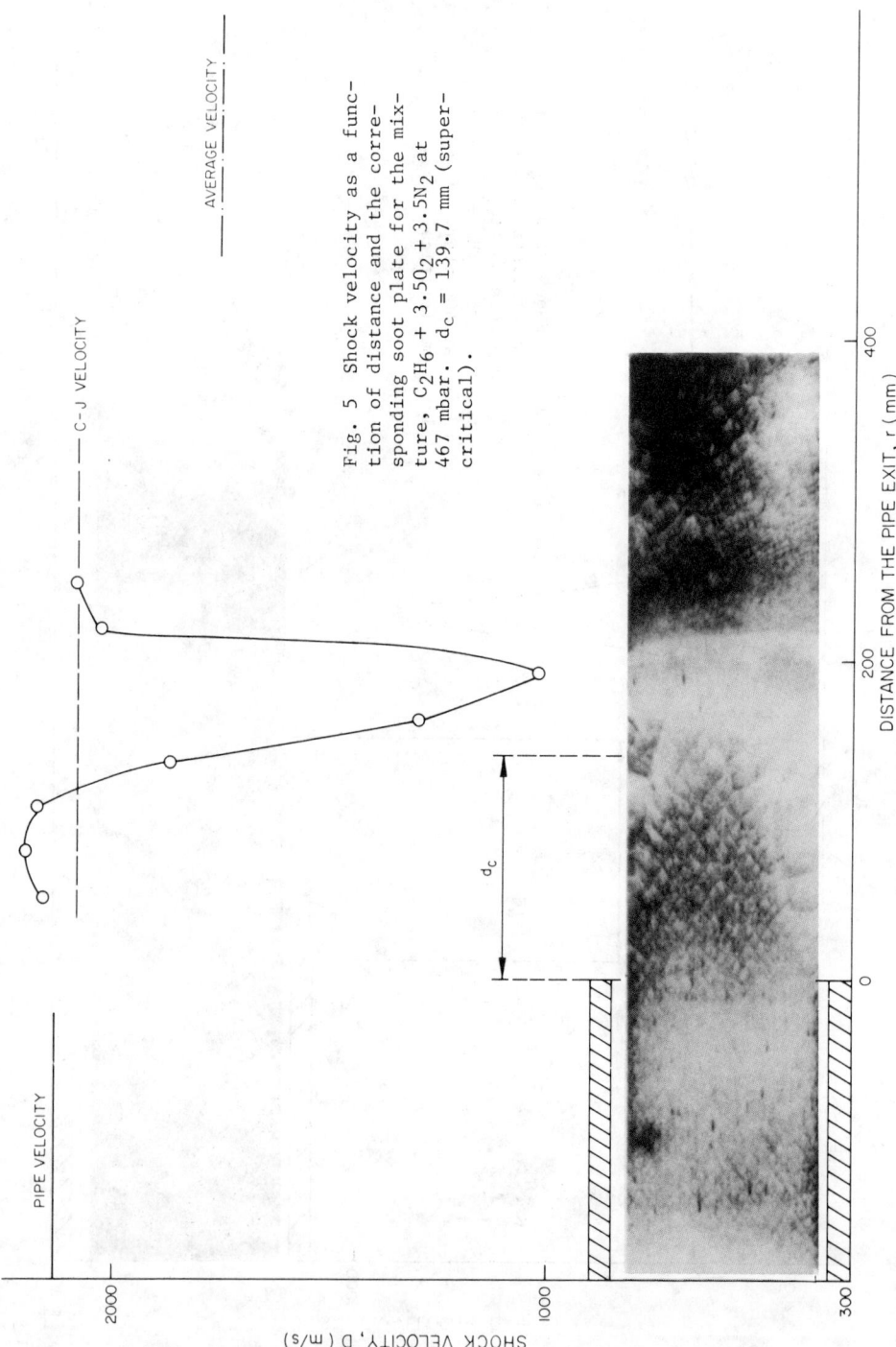

Fig. 5 Shock velocity as a function of distance and the corresponding soot plate for the mixture, $C_2H_6 + 3.5O_2 + 3.5N_2$ at 467 mbar. $d_c = 139.7$ mm (supercritical).

stoichiometric ethane-oxygen diluted with 3.5 moles of nitrogen at subcritical, critical, and supercritical pressures (373, 400, and 467 mbar). Soot plates obtained simultaneously are also shown in order to correlate the velocity measurements with the cell structure. Measurements of detonation velocity in the pipe and the average velocity along the pipe and the vessel are also shown. There is a very good agreement between the detonation velocity in the pipe and the theoretical CJ velocity.

For a subcritical initial pressure of 373 mbar (Fig. 3) cellular structure observed on the soot plate disappears gradually as the expansion fan originating at the pipe wall propagates towards the pipe axis. This is seen on the soot plate as a well-defined cone of cellular structure, the apex of which corresponds to the complete failure (Edwards et al. 1979) of the planar detonation front. Figure 3 also shows that the shock velocity along the pipe axis remains at CJ levels for about one pipe diameter after the pipe exit and following the complete decoupling of shock and flame fronts near the cone apex, decays very rapidly.

Investigation of Figs. 3-5 shows that the emerging planar detonation for critical (400 mbar) and supercritical (467 mbar) transitions fails in a similar way before the reinitiation of spherical detonation. For critical transition the reinitiation of spherical detonation is marked by the reappearance of cellular structure at about two pipe diameters from the pipe exit and the corresponding jump of the shock velocity to CJ levels. The soot plate record of the critical transition also shows that the process of failure-reinitiation is repeated at least once more. This pulsating (Ul'yanitskii 1980) or "galloping" reinitiation of detonation at critical conditions has also been observed (Elsworth et al. 1983) for direct initiation of spherical detonation by Tetryl charges in marginal ethane-oxygen mixtures diluted with nitrogen. Bull et al. (1978) have also suggested the possibility of several attempts to recouple before the establishment of a steady CJ detonation. Figure 5 shows that for a supercritical detonation there is only one gallop and the cellular structure reappears along a well defined front at about 1.5 pipe diameter from the pipe exit. Edwards et al. (1978) have observed similar features on smoked foil records for supercritical and critical initiation of spherical detonation in oxyacetylene mixtures using high-energy exploding-wire sources.

Figs. 6 and 7 show the shock velocity as a function of distance from the pipe exit for the same pipe diameter (139.7 mm) at atmospheric pressure under subcritical and

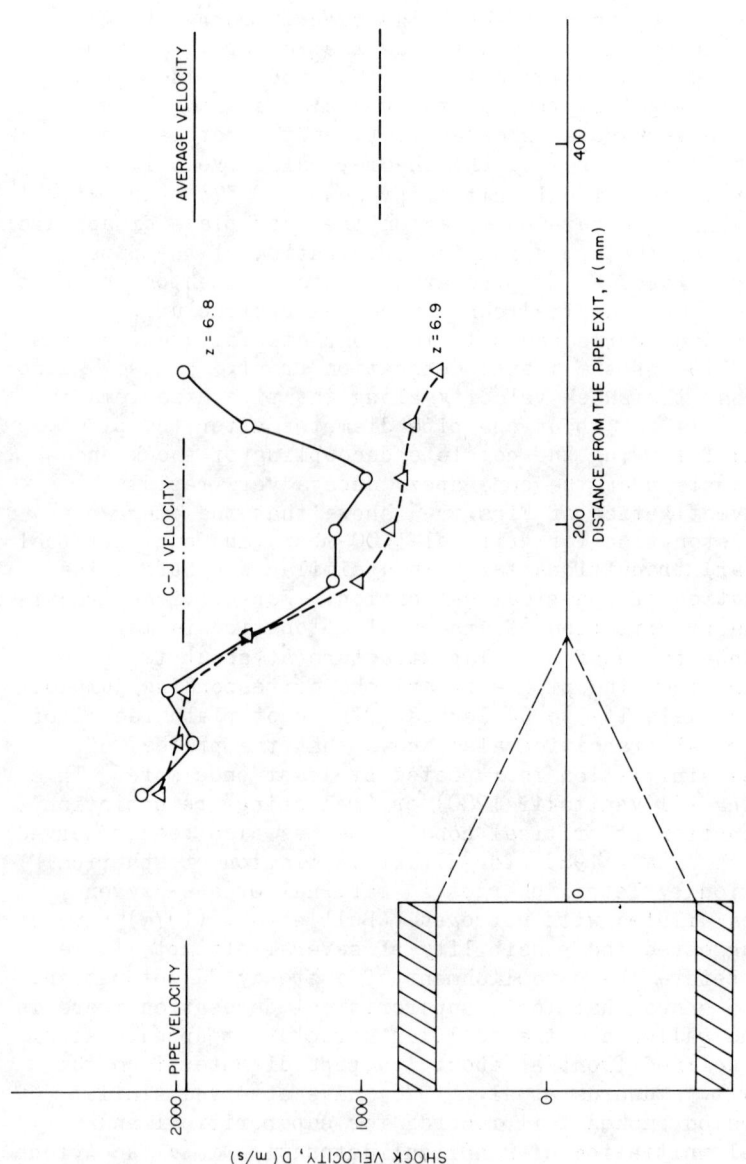

Fig. 6 Shock velocity as a function of distance for the mixture, $C_2H_6 + 3.5O_2 + Z N_2$ at atmospheric pressure for subcritical ($Z = 6.9$) and critical ($Z = 6.8$) dilutions. $d_c = 139.7$ mm.

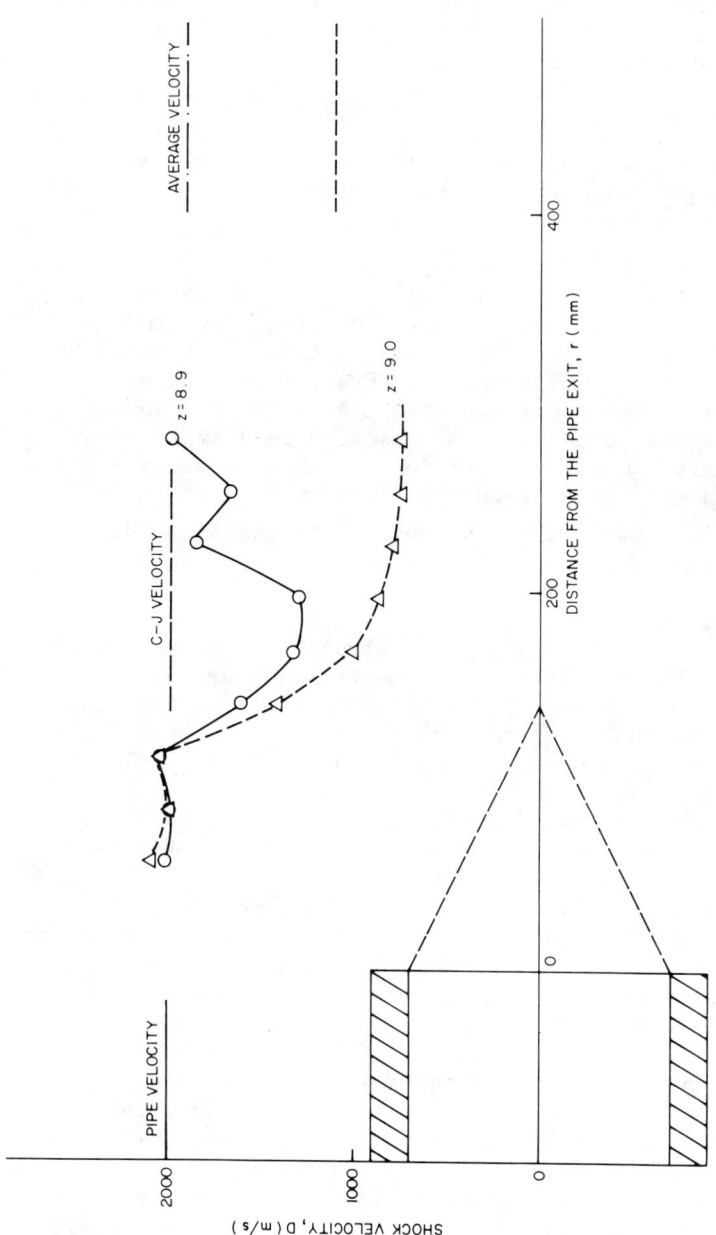

Fig. 7 Shock velocity as a function of distance for the mixture $C_3H_8 + 5O_2 + ZN_2$ at atmospheric pressure for subcritical (Z = 9.0) and critical ($\bar{Z}$ = 8.9) dilutions. d_c = 133.7 mm.

critical dilutions for stoichiometric ethane-oxygen and propane-oxygen mixtures. Comparisons with the low-pressure results indicate that the transition mechanisms for both cases are very similar.

Fig. 8 shows the variation of shock velocity as a function of distance from the pipe exit for a pipe diameter of 101.4 mm at atmospheric pressure under subcritical, critical, and supercritical dilutions for a stoichiometric ethane-oxygen mixture.

Fig. 9 shows the high-speed (10^5 frames/s) frame photographs taken simultaneously under the critical transition ($Z = 5.2$) conditions. The first five frames show the failing planar detonation as the bright core of detonation gradually disappears. Frame 7 shows the first signs of the local explosion which takes place at an offaxis position of approximately one pipe diameter from the pipe exit. Subsequent frames display the propagation of the resulting flame front.

Fig. 10 shows the comparison of the positions along the pipe axis of the leading shock front measured by the laser technique and the flame front measured from the frame photographs as a function of time elapsed since the initiation. Observations of cellular structure recorded on a soot plate obtained in a similar experiment are also summarized in Fig. 10. For a distance of about one pipe diameter from the pipe exit where the shock velocity remains at CJ values corresponding to the failing planar detonation, measurements of flame front positions agree very well with the position of the shock front. At approximately one pipe diameter from the exit, shock and flame fronts decouple, the leading shock front decays, and the cellular structure disappears. The flame front decelerates at first but, following the local explosion, accelerates very rapidly through the induction region of hot and compressed combustible mixture and catches up with the decaying shock front. The reappearance of cellular structure near this point indicates that the shock and flame fronts recouple and propagate as a multifront detonation wave. Cell sizes are initially considerably smaller than that of the steady detonation but they increase in size rapidly. This may suggest that the reinitiated multifront detonation wave is initially overdriven (Edwards et al. 1979) for a very short time. This point also corresponds to the sudden increase in the shock velocity measured by the laser technique. Investigation of the shock and flame front velocities in comparison with the soot plate recording indicates that the steady CJ detonation is established after a few secondary "gallops."

INITIATION OF UNCONFINED GASEOUS DETONATION 537

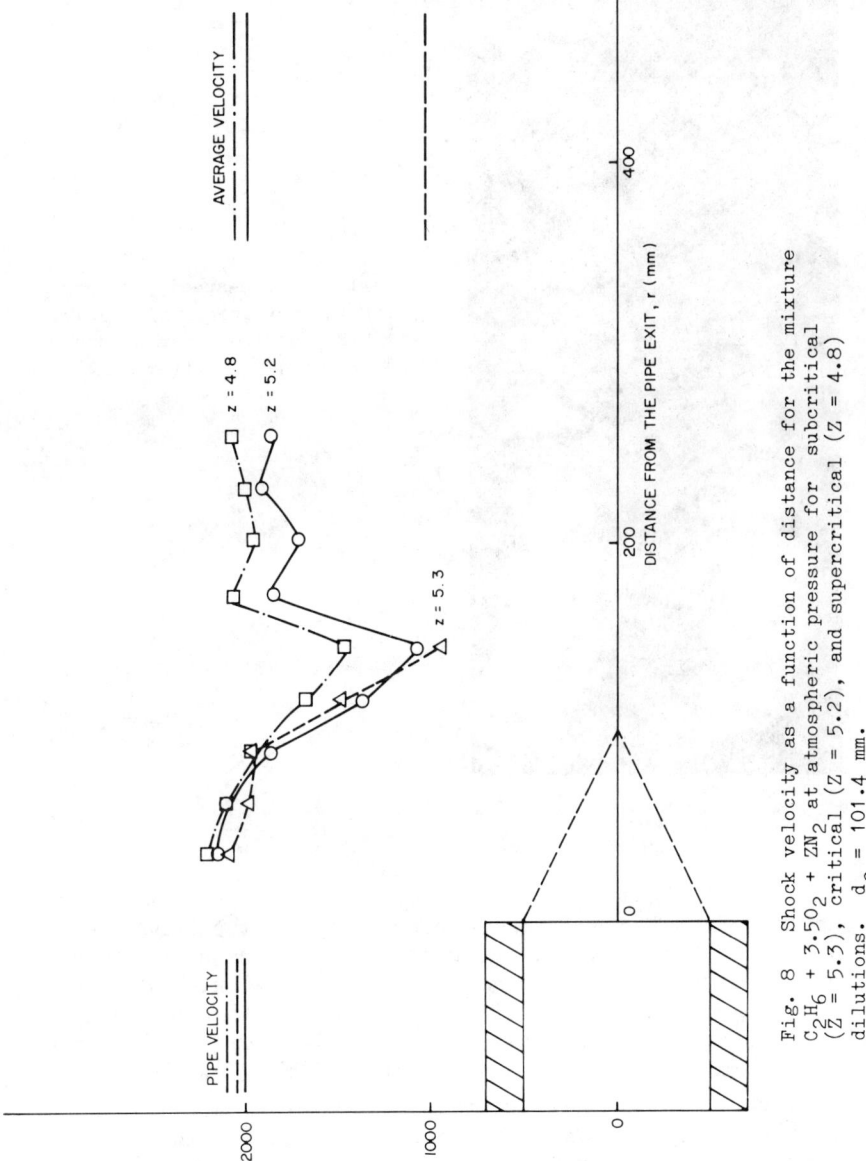

Fig. 8 Shock velocity as a function of distance for the mixture $C_2H_6 + 3.5O_2 + ZN_2$ at atmospheric pressure for subcritical (Z = 5.3), critical (Z = 5.2), and supercritical (Z = 4.8) dilutions. d_c = 101.4 mm.

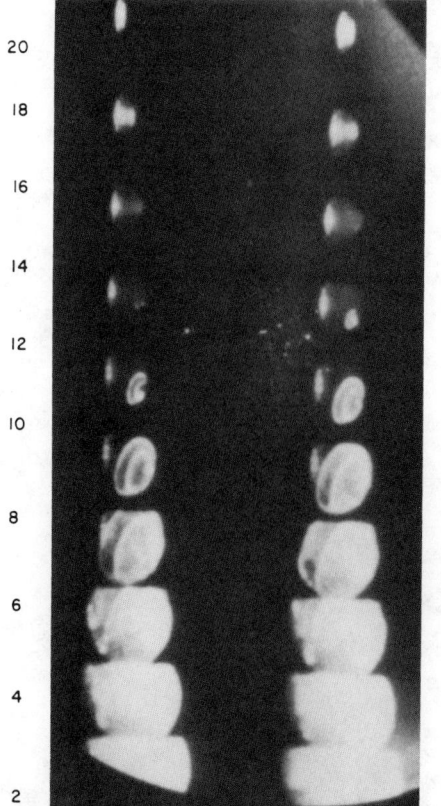

Fig. 9 High-speed photographs for the mixture. $C_2H_6 + 3.5O_2 + 5.2N_2$ at 1.013-bar pressure. $d_c = 101.4$. 10^5 frames/s.

Fig. 11 shows the behavior of the shock velocity as a function of distance from the pipe exit for the same pipe diameter of 101.4 mm and under atmospheric pressure for a stoichiometric propane-oxygen mixture diluted with nitrogen under subcritical and critical conditions. It can be observed that the mechanism of reinitiation discussed above is very similar for ethane and propane mixtures.

Fig. 12 shows the variation of the critical pipe diameter as a function of nitrogen dilution (Z) for stoichiometric fuel-oxygen-nitrogen mixtures of ethane and propane. Results of Knystautas et al. (1982) are also shown for comparison. For the mixtures investigated, there is good agreement between the two sets of measurements. Table 1 summarizes the present measurements for all the mixtures and initial pressures investigated. Critical

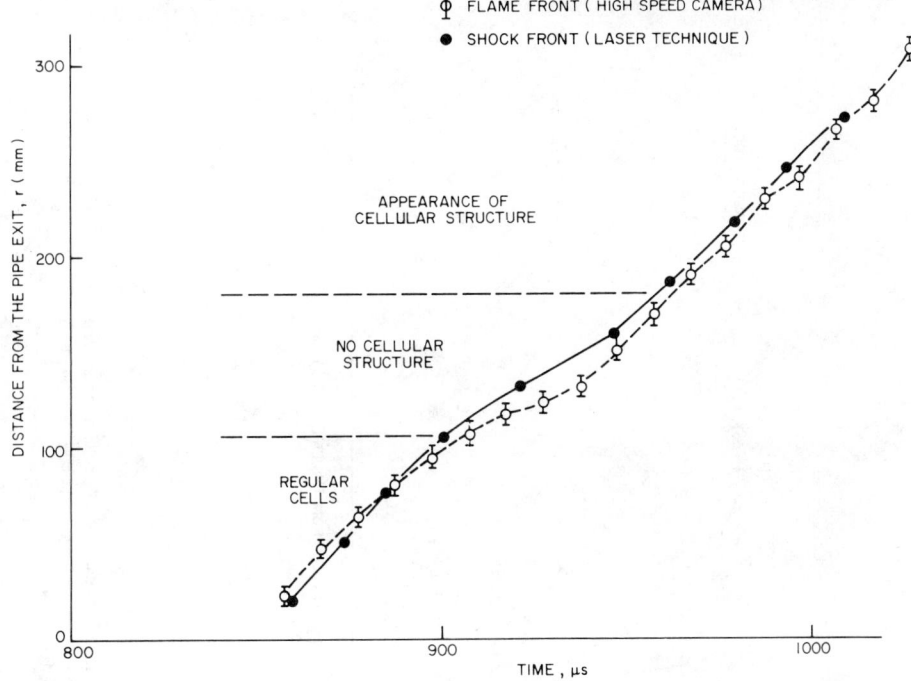

Fig. 10 Positions of the flame and the shock fronts measured as a function of time for the mixture, $C_2H_6 + 3.5O_2 + 5.2N_2$ at 1.013-bar pressure. $d_c = 01.4$.

number of cells (N_c), calculated by using the measured values of the critical pipe diameter (d_c) and the standard cell width (b), vary between 14-16 with an average value of 14.6. The percent standard variation in the measurement of the cell width, mainly due to the irregularity of cells and the ambiguities caused by the fine cellular structure (Bull et al. 1982) varies between ±11 and ±30% with an average value of ±21%. On the other hand in all the mixtures and at all of the pressures investigated the difference between the experimentally determined critical pipe diameter (d_c) and the diameter predicted by the correlation, $d = 13 \cdot b$, is less than 13% except for the low initial pressure of 260 mbar where it is 18%. Therefore, it may be concluded that for the mixtures investigated so far, the agreement between the critical number of cells measured and the previously reported values of 13 is good and well within the experimental accuracy of our measurements.

A curve fitting method based on the two-dimensional model (Vasil'ev et al. 1976) of a detonation cell has been used in order to extrapolate the available data to sto-

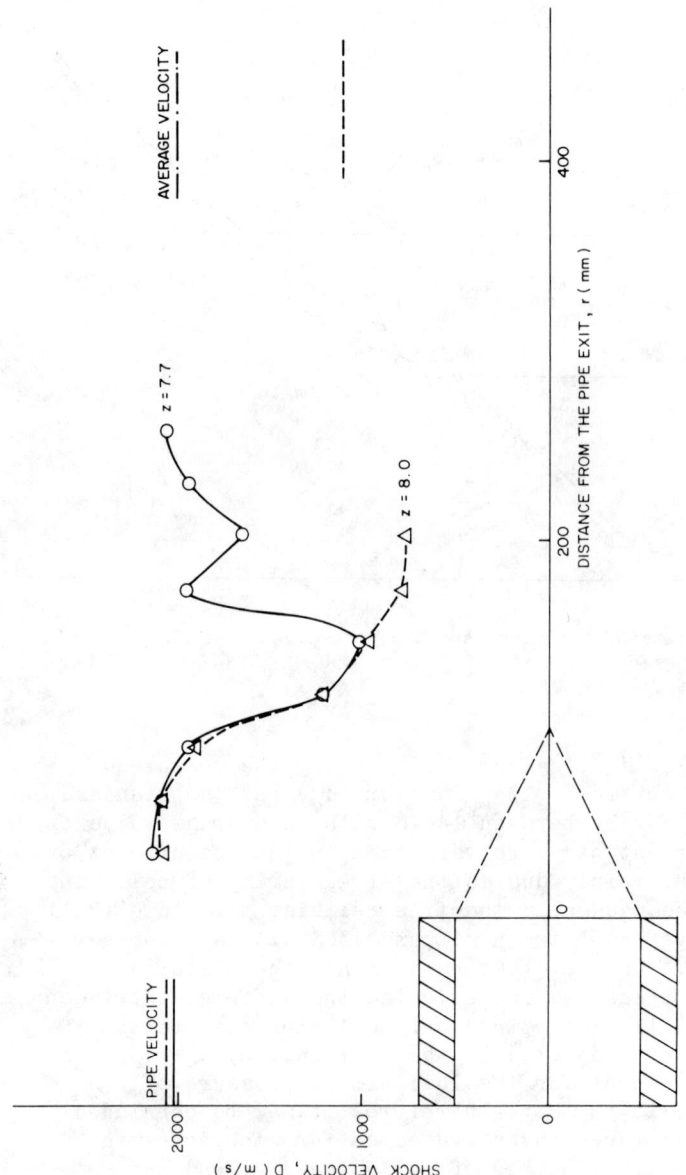

Fig. 11 Shock velocity as a function of distance for the mixture $C_3H_8 + 5O_2 + ZN_2$ at 1.013-bar pressure for subcritical ($Z = 8$) and critical ($Z = 7.7$) dilutions. $d_c = 101.4$ mm.

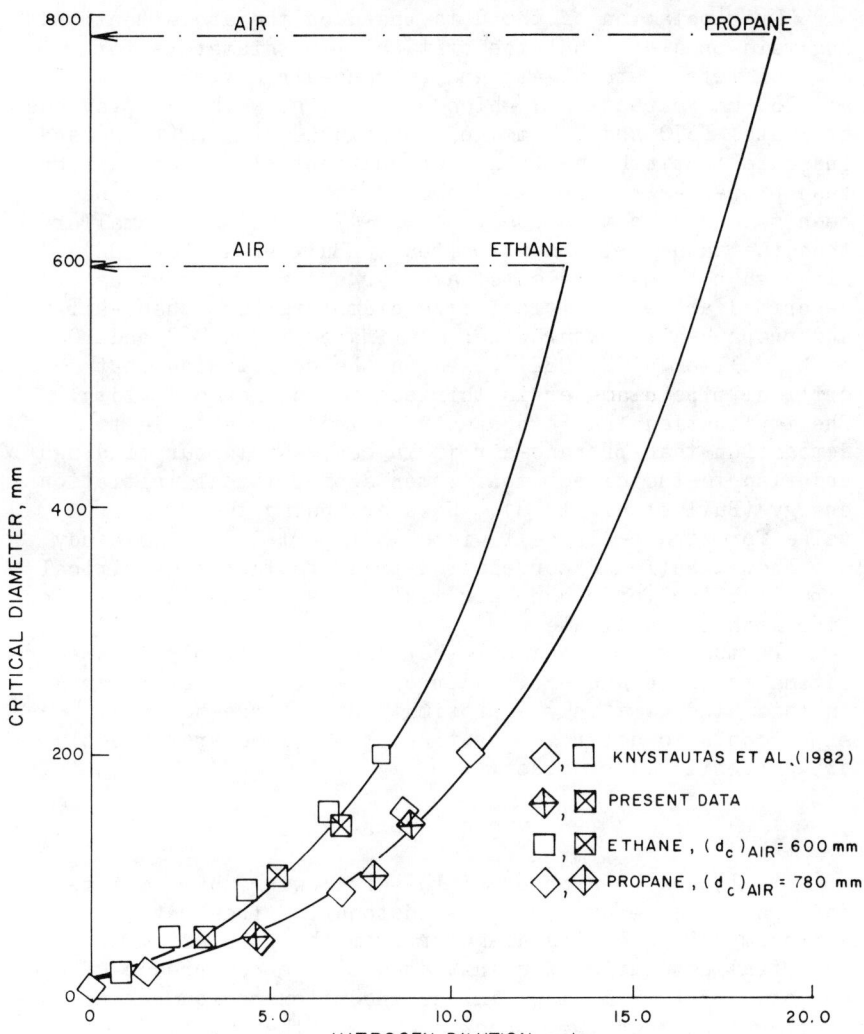

Fig. 12 Variation of critical pipe diameter as a function of dilution.

ichiometric ethane-air and propane-air mixtures (see Appendix). Existing computer codes in Thornton Research Centre for CJ and steady shock front calculations were modified in order to carry out the above mentioned calculations. Kinetic constants for ethane and propane mixtures were taken from Burcat et al. (1972). Figure 12 shows the theoretical calculations for stoichiometric ethane and propane mixtures which provide excellent fits to the available experimental data.

Extrapolation of the data based on the above mentioned analysis predicts that the critical pipe diameters for stoichiometric ethane-air and propane-air mixtures are 600 and 780 mm respectively which are in line with the previous estimates (670 and 700 mm) of Knystautas et al. (1982) and indicate that ethane-air is more susceptible to detonation than propane-air. Cell width for ethane-air mixture has been measured in this study to be ~47 mm which is smaller than the value (~58 mm) reported by Bull et al. (1982) although both measurements have large error margins as described above. Critical pipe diameters for ethane-air and propane-air mixtures were estimated to be 760 and 600 mm by Bull et al. (1982) based on the correlation that the critical pipe diameter is thirteen times the cell width. The implication that ethane-air is less susceptible to detonation than propane-air is in contrast to our findings and also to the direct measurements of critical initiation energy (Bull et al. 1978). However, using the smaller value for ethane-air cell width as reported in this study the above mentioned correlation predicts that the critical pipe diameter for ethane-air mixture is 611 mm which is in line with our estimates.

It must be pointed out that the critical pipe diameter estimates for ethane-air and propane-air mixtures reported in this study are only approximations. Large-scale experiments to determine critical pipe diameters directly are currently in progress.

Conclusions

1) It has been observed that following the complete failure of the emerging planar detonation front at approximately one pipe diameter from the pipe exit, spherical detonation was initiated by the occurrence of one or more local explosions in the induction region behind the leading (but decaying) shock front and the subsequent acceleration of the flame front. It has also been observed under critical conditions that the reinitiated spherical detonation might fail again with increasing cell sizes and may subsequently be followed by another initiation. This failure-reinitiation process, which is referred to as "galloping" detonation, may repeat itself several times before the establishment of steady detonation.

2) Present measurements of critical pipe diameters for stoichiometric ethane and propane mixtures diluted with nitrogen agree well with the measurements of Knystautas et al. (1982).

3) Critical pipe diameters for stoichiometric ethane-

air and propane-air at atmospheric pressure have been estimated by extrapolation of the present results in connection with the results of Knystautas et al. (1982) as 600 and 780 mm, respectively.

4) Results generally support the previous findings that the critical pipe diameter is approximately 13 times the standard cell width within an error margin of 21%.

Appendix

Assuming that the critical number of cells (N_c) and the cell length-width ratio (a/b) are constant for the whole range of nitrogen dilution, the ratio of the critical pipe diameter for a stoichiometric fuel-oxygen-nitrogen mixture to that of the same fuel-air mixture may be expressed as

$$\frac{d_c}{(d_c)_{Air}} = \frac{b}{b_{Air}} = \frac{a}{a_{Air}} \qquad (1)$$

By using the expression,

$$a \simeq 4 \cdot D \cdot \tau \cdot (E_a/R_g T)$$

for the cell length proposed by Vasil'ev et al. (1976), $d_c/(d_c)_{Air}$ can be expressed as

$$\frac{d_c}{(d_c)_{Air}} \simeq \frac{D}{D_{Air}} \frac{T_{Air}}{T} \frac{\tau}{\tau_{Air}} \qquad (2)$$

where D is the theoretical CJ velocity, T the particle temperature behind a shock front moving with CJ velocity, τ the induction period of the mixture behind the shock front, E_a the activation energy, and R_g the gas constant.

Assuming that the induction period of a stoichiometric fuel-oxygen-nitrogen mixture can be approximated by a single "global" kinetic expression over the whole range of nitrogen dilution, τ/τ_{Air} may be expressed as

$$\frac{\tau}{\tau_{Air}} = \left(\frac{C_f}{(C_f)_{Air}}\right)^m \left(\frac{C_{O_2}}{C_{O_2 Air}}\right)^n \frac{\exp(E_a/R_g T)}{\exp(E_a/R_g T_{Air})} \qquad (3)$$

where C_f is the fuel concentration, C_{O_2} the oxygen concentration, and m and n are kinetic constants.

Equations 1-3 may be used to calculate $d_c/(d_c)_{Air}$ for a fuel-oxygen-nitrogen mixture as a function of nitrogen dilution provided that kinetic constants m, n, and E_a are known. By varying the value $(d_c)_{Air}$ theoretical calculations for d_c can then be optimized to give a best fit to the experimental data of d_c vs nitrogen dilution.

References

Atkinson, R., Bull, D. C., and Shuff, P. J. (1980) Initiation of spherical detonation in hydrogen/air. Combustion and Flame 39, 287-300.

Bull, D. C., Elsworth, J. E., and Hooper, G. (1978) Initiation of spherical detonation in hydrocarbon/air mixtures. Acta Astronautica 5, 997-1008.

Bull, D. C. (1979) Concentration limits to the initiation of unconfined detonation in fuel/air mixtures. Trans. I Chem E 57, 219-227.

Bull, D. C., Elsworth, J. E., and Shuff, P. J. (1982) Detonation cell structures in fuel/air mixtures. Combustion on Flame 45, 7-22.

Burcat, A., Crossley, R. W., and Scheller, K. (1972) Shock tube investigation of ignition in ethane-oxygen-argon mixtures. Combustion and Flame 18, 115-123.

Edwards, D. H., Hooper, G., Morgan, J. M., and Thomas, G. O. (1978) The quasi-steady regime in critically initiated detonation waves. J. Phys. D: Appl. Phys. 11, 2130-2117.

Edwards, D. H., Thomas, G. O., and Nettleton, M. A. (1979) The diffraction of a planar detonation wave at an abrupt area change. J. Fluid Mech. 95, 79-96.

Edwards, D. H. and Thomas, G. O. (1981) Diffraction of a planar detonation in various fuel/oxygen mixtures at an area change. Gasdynamics of Detonations and Explosions: Progress in Astronautics and Aeronautics (edited by J. R. Bowen, N. Manson, A. K. Oppenheim, and R. I. Soloukhin), V0l. 75, pp. 341-357. AIAA, New York.

Elsworth, J. E., Shuff, P. J., and Ungut, A. (1983) "Galloping" gas detonations in the spherical mode. Paper submitted to 9th ICOGER, Poitiers, France.

Knystautas, R., Lee, J. E., and Guirao, C. M. (1982) The critical tube diameter for detonation failure in hydrocarbon-air mixtures. Combustion and Flame 48, 63-83.

Lee, J. H. and Matsui, H. (1977) A comparison of the critical energies for direct initiation of spherical detonations in acetylene-oxygen mixtures. Combustion and Flame 28, 61-66.

Matsui, H. and Lee, J. H. (1978) On the measure of the relative detonation hazards of gaseous fuel-oxygen and air mixtures. Seventeenth Symposium (International) on Combustion, pp.1269-1279, The Combustion Institute, Pittsburgh, Pa.

Mitrofanov, V. V. and Soloukhin, R. I. (1964) The diffraction of multifront detonation waves. Sov. Phys. - Dokl. 9, 1055.

Ul'yanitskii, V. Yu. (1980) Closed model of direct initiation of gas detonation taking account of instability. Fizika Goreniya i Vzryva 16, 101-113.

Urtiev, P. A. and Tarver, C. M. (1981) Effects of cellular structure on the behaviour of gaseous detonation waves under transient conditions. Gasdynamics of Detonations and Explosions: Progress in Astronautics and Aeronautics (edited by J. R. Bowen, N. Manson, A. K. Oppenheim, and R. I. Soloukhin), Vol. 75, pp.370-384. AIAA, New York.

Vasil'ev, A. A. and Nikolaev, Yu. A. (1976) Model of the nucleus of a multifront gas detonation. Fizika Goreniya i Vzryva 12, 744-754.

Vasil'ev, A. A. and Grigor'ev, V. V. (1980) Critical conditions for gas detonation in sharply expanding channels. Fizika Goreniya i Vzryva 16, 117-125.

Zel'dovich, Ia. B., Kogarko, S. M., and Simonov, N. N. (1956) An experimental invetigation of spherical detonation of gases. Sov. Phys. Tech. Phys. 1, 1689-1713.

Large-Scale Experiments on the Transmission of Fuel-Air Detonations from Two-Dimensional Channels

W. B. Benedick*
Sandia National Laboratories, Albuquerque, New Mexico
and
R. Knystautas[†] and John H. S. Lee[‡]
McGill University, Montreal, Canada

Abstract

Results of large-scale experiments on the transmission of hydrogen-air and ethylene-air detonations from a two-dimensional channel into a large volume are reported. These data confirm the surprising observations of Liu et al who reported recent laboratory scale results which indicate that successful transmission of a detonation emerging from a rectangular orifice with a large aspect ratio $L/W > 5$, requires that the minimum width of the channel need be only about three detonation cell diameters, i.e., $W_c \approx 3\lambda$. These results are surprising in view of the previous findings of Soloukhin and Edwards, where for square channels ($L/W = 1$), they observed that $W_c \approx 10\lambda$. The apparatus for the large-scale experiments consisted of a two-dimensional confined channel (1.83-m wide x 2.44-m long) with a variable width orifice, to the end of which was attached an inflated polyethylene bag of much larger dimensions to provide an unconfined environment for detonation transmission. Observations were made with high-speed cinematography and with smoked foils to measure the detonation cell size. Our results indicate that a minimum channel width of 3λ is required for successful transmission from channels whose aspect ratios varied from 34 to 8. For channel aspect ratios of less

---

Presented at the 9th ICODERS, Poitiers, France, July 3-8, 1983. This paper is a work of the U.S. Government and therefore is in the public domain.
*Technician Staff, Shock Wave and Explosives Physics Division.
[†]Associate Professor, Mechanical Engineering Department.
[‡]Professor, Mechanical Engineering Department.

than about 8, the minimum channel width for successful detonation transmission increases continuously from 3λ to 10λ as the aspect ratio approaches 1.

Introduction

During the past few years, there have been many investigations to determine the critical tube diameter for the transmission of confined planar detonations to unconfined spherical detonations. Both laboratory experiments (Matsui and Lee 1978; Knystautas et al. 1982) as well as large-scale field tests have been carried out for a variety of hydrocarbon-air and oxygen enriched mixtures over a range of initial pressures (see Guirao, et al 1982; Moen, et al 1982; Rinnan 1982). The importance of the critical tube diameter is that it is directly related to the dynamic detonation parameters, such as critical initiation energy and detonability limits (Lee et al. 1982). Knowledge of the critical tube diameter permits these important dynamic parameters of the explosive to be estimated. The critical tube diameter d_c is also found to be directly related to the cell size λ of the detonation front via the simple empirical formula $d_c \simeq 13\lambda$. This observation was first made by Mitrofanov and Soloukhin (1965) who observed that for low-pressure acetylene-oxygen detonations in circular tube $d_c \simeq 13\lambda$. The significance of this went unnoticed until Edwards et al (1979) repeated the experiments, confirmed the results, and suggested that the law should be valid for other gas mixtures. Subsequently, there have been a number of experimental demonstrations of the validity of this simple empirical law for many different fuel-oxidizer combinations at various initial pressures. Despite the fundamental significance and simplicity of this empirical law, no theoretical justification has appeared. However, extensive calculations by Westbrook and Urtiew show a strong correlation between the critical tube diameter (or equivalently the cell size) and the reaction zone thickness as determined by the detailed kinetics of the oxidation process (see Westbrook 1982; Westbrook and Urtiew 1982a; Westbrook and Urtiew 1982b; Westbrook 1980).

To test the range of validity of this simple empirical law, experiments have been carried out recently by Liu et al (1983) for detonation transmission through orifices of different geometries. They found that the transmission through an orifice is identical to that through a tube of the same cross-sectional area. For the two-dimensional case of a rectangular orifice of large aspect ratio (i.e., $L/W > 5$), Liu obtained the rather surprising result that the critical channel width W_c for transmission is only about three cell diameters. This result is in disagreement with that reported by Mitrofanov and Soloukhin (1965), as well as Edwards et al (1979) who gave $W_c \simeq 10\lambda$ for the two-dimensional case. However, both Soloukhin's and Edwards' experiments were based on square channels, whereas Liu's two-dimensional orifice experiments covered a range of aspect ratios $1 < L/W < 15$. For $L/W = 1$, Liu's results agree with Mitrofanov and Soloukhin (1965) and Edwards et al (1979), $W_c \simeq 10\lambda$.

Due to the fundamental significance of this two-dimensional result reported by Liu, we felt it was desirable to perform large-scale experiments to confirm that $W_c \simeq 3\lambda$. Whereas Liu's experiments were based on rectangular orifice plates of a maximum $L/W \simeq 15$, two-dimensional channels with large aspect ratios (e.g., $L/W \simeq 30$) can readily be used in field experiments to give a truly two-dimensional geometry. The present paper reports the results of a series of such field tests recently carried out at Sandia National Laboratories (New Mexico) to verify the empirical law $W_c \simeq 3\lambda$.

Experimental Method

The experimental detonation channel consisted of two 25-mm-thick steel plates 1.33-m wide and 2.44-m long. The channel width W (the spacing between the plates) was varied by using four spacer blocks placed at each of the four corners of the plates (Fig. 1a). One of the side walls of the channel is covered by a sheet of clear plastic to permit photographic observations of the detonation propagation in the channel, while the opposite side wall of the channel is covered by a plywood sheet. The closed end of the channel was also covered by a sheet of plywood onto which are taped thin strips of sheet explosive for initiating the gas detonation. The quantity of strip explosive ranged from 50-140 g depending upon the width of the channel. The high explosive strips are centrally initiated by an exploding bridgewire detonator. The plastic and plywood side walls and the initiation end of the channel are blown off during the experiment and have to be replaced for each test. The open end of the channel is connected to a large plastic bag fabricated from 100-μm-thick polyethylene sheets which are taped to the steel plates. The plastic bag was about 3.5 m in length. The filling procedure is to first retract the entire plastic bag into the channel by removing almost all of the air in the system with a vacuum pump; then a premixed hydrogen-air mixture is introduced from a large underground storage tank. The plastic bag is extruded

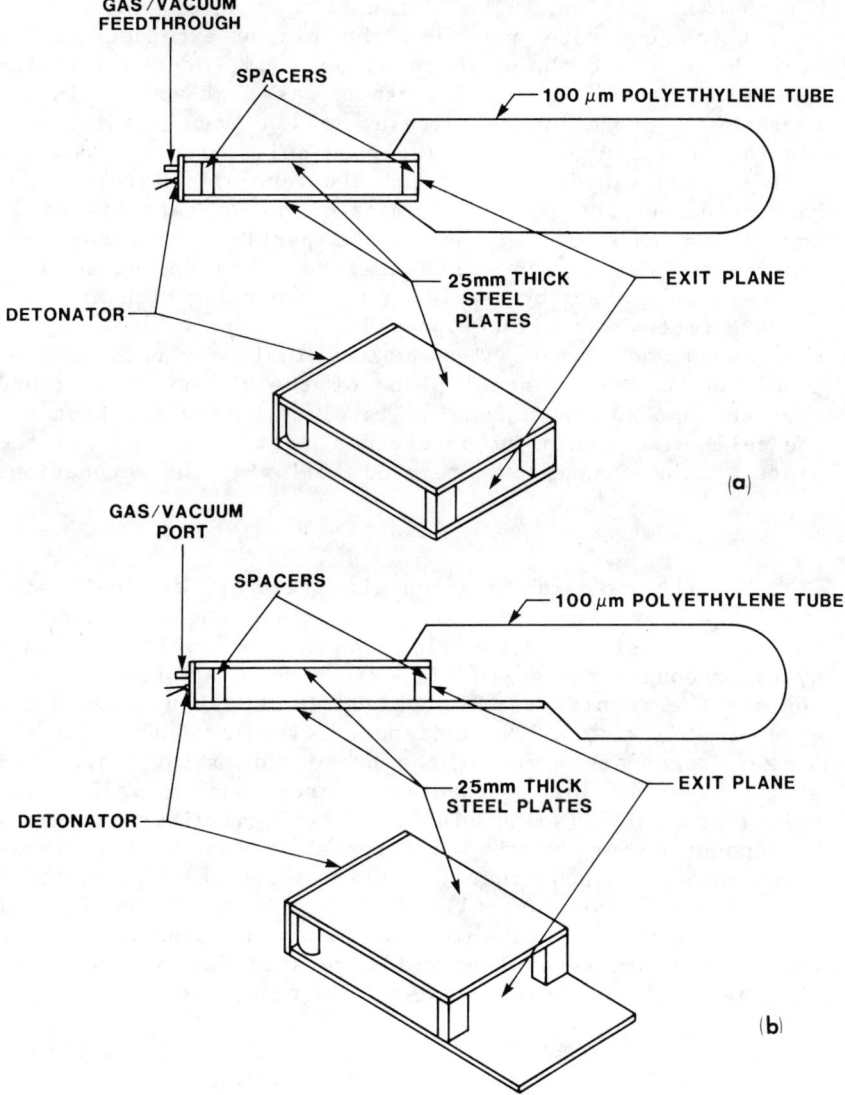

Fig. 1 Two-dimensional channel apparatus with: (a) unconfined exit plane; (b) 1 degree of confinement at the exit plane.

from the channel as the system is being filled with the explosive gas mixture. The final pressure in the system is slightly above ambient so that the plastic bag remains taut. On windy days or when the hydrogen concentration is rich enough such that the buoyancy of the bag is significant, thin nylon ropes are used to anchor the plastic bag in the horizontal position. The detonation channel is located on a thick concrete pad with the plates extended 0.55 m over the edge of the slab to allow room for the inflated plastic bag. The only diagnostic was high-speed cinematography. Detonation velocity as well as successful transmission or failure can be discerned quite readily from the movie record. Hydrogen-air and ethylene-air mixtures have been used and the desired composition was established in the mixing tank by the method of partial pressures. To measure the detonation cell size, soot was deposited on a thin aluminum sheet by a welding torch burning rich mixtures of MAPP (methyl acetylene, propadiene, propane, butane, and propylene) and oxygen. The smoked aluminum sheet was then taped to the bottom steel plate of the channel. We found that the smoked sheet remains in place during the test and the cell size could be determined from it after the top plate of the channel was removed following the detonation.

Results

For the case of hydrogen-air mixtures, the cell size λ as determined from the smoked aluminum sheets placed in the two-dimensional detonation channel is plotted against hydrogen-concentration in Fig. 2. Also shown for comparison are the results for λ obtained at McGill by Knystautas et al (1982) with a 300-mm-diameter circular tube. For the same hydrogen-air mixture, the present data for λ are generally slightly larger than the corresponding values obtained at McGill (solid line). The slight discrepancy may be accounted for by the difference in the ambient pressures between Albuquerque ($\sim$ 1600-m altitude) and Montreal ($\sim$ 100 m). Since the cell size is directly proportional to the induction time which is inversely proportional to the initial pressure, the McGill results can be scaled to the lower ambient pressure of Albuquerque, i.e.,

$$\lambda_{Albuquerque} = (P_{Montreal}/P_{Albuquerque})(\lambda_{Montreal})$$

Using this correction, there is better agreement between measured cell sizes and the rescaled McGill data (shown as the dotted curve).

Rich mixtures of hydrogen-air were not used in the present series of experiments since the objective was only to verify the empirical law $W_c \simeq 3\lambda$ for two-dimensional channels. Extensive experiments have already been carried out for the rich hydrogen-air mixtures for circular tubes and have been reported previously (Guirao et al 1982). It should be noted that the cell size is indicative of the

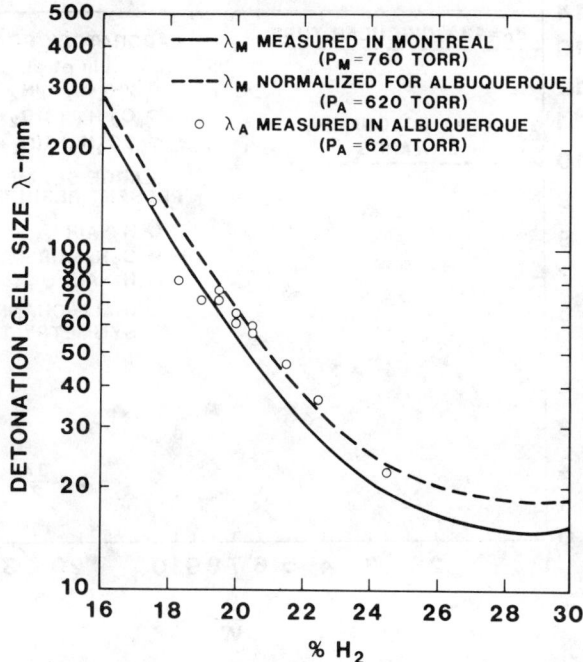

Fig. 2 Detonation cell size measured in Montreal, normalized for lower ambient pressure in Albuquerque, and as measured in channel experiments in Albuquerque, all as a function of H_2:Air mixture.

rate processes in the complex detonation structure and is a very useful parameter. Detonation velocities deduced from the high-speed movies were in agreement with the Chapman-Jouguet values within our estimated experimental error.

The normalized critical channel height W_c/λ measured in the present experiments is plotted against the aspect ratio of the channel L/W in Fig. 3. Note that for large aspect ratios, $L/W > 10$, the present results give a value $W_c/\lambda \simeq 3$ in approximate agreement with the results obtained by Liu using rectangular orifices instead of a full two-dimensional channel. Experiments with aspect ratios up to 35 yield the same value for $W_c \simeq 3\lambda$. Using different hydrogen concentrations and hence different λ's and different Chapman-Jouguet states, also gives the same results, $W_c/\lambda \simeq 3$. This observation indicates that the critical channel height W_c scales according to the cell size λ, and within the present experimental accuracy does not depend explicitly on the Chapman-Jouguet states. Several experiments were also carried out for 6% ethylene-air mixtures with aspect ratios of 14.5-18.5. The result $W_c/\lambda = 3$ was obtain-

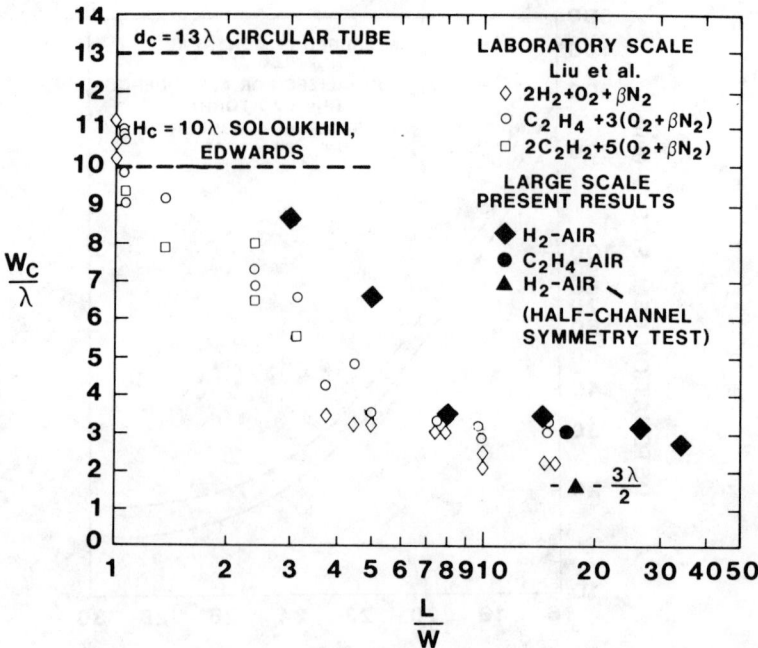

Fig. 3 Critical channel width normalized to detonation cell size, W_c/λ, as a function of the channel aspect ratio, L/W, for detonations emerging from the exit plane of the channel into an unconfined volume. The data point $3\lambda/2$ is for detonation supported by one plane after exiting the channel.

ed for ethylene as well, reinforcing our conjecture that the scaling law for channels is based only on the cell size λ as has been demonstrated in the case of circular tubes (Knystautas et al 1982).

In the present two-dimensional geometry, it was also possible to test the symmetry of the transmission process about the centerline of the channel. By retracting the top plate of the channel relative to the bottom plate so that the emergent detonation is supported by one plane, we constructed a half channel (Fig. 1b). By symmetry, transmission should now be possible for a spacing between the plates corresponding to only 1.5λ. Several experiments were conducted using 20-20.5% hydrogen-air mixtures with channel widths from 10-10.5 cm and a channel length of 1.82 m, yielding $17.5 \leq L/W \leq 18.2$. The go/no-go results bracketed a critical spacing between the plates of close to 1.5λ, confirming that the two-dimensional transmission mechanism is symmetrical about the channel center line. For aspect ratios $L/W < 8$, the present results indicate that the value for W_c/λ rises sharply. A similar trend was also obtained by Liu for the case of transmission through rectangular orifices, but the increase does not occur until L/W < 5. Why three-dimensional effects begin to become important at larger aspect ratios for the channel than for the orifice plates has not been resolved. Other than this discrepancy, the present channel experiments are in qualitative agreement with the rectangular orifice plate experiments of Liu.

To explain the surprising result that the critical condition for the successful transmission of detonation changes from $W_c = 10\lambda$ for a square channel to $W_c = 3\lambda$ for a rectangular channel of large aspect ratio (L/W > 8), one can invoke the wave curvature concept proposed recently by Lee et al (1984). Lee's arguments suggest that when a planar detonation wave propagating in a confining linear tube or channel emerges suddenly into an unconfined geometry, it is subject to a gasdynamic expansion process by virtue of the the rarefaction waves that are generated at the edge which then penetrate into the wave front and impose a curvature on the wave. He then defines the critical condition for transmission in terms of the minimum radius of curvature of the diffracted wave, namely; if the rarefraction waves which give rise to a curvature of the diffracted detonation and exceed a certain critical value, then failure of the wave results. For a three-dimensional diffraction process the minimum radius of curvature would be of the order of the hydrodynamic thickness Δ_H so that $R=\Delta_H=6.5\lambda$. Thus for the diffraction of detonation through a square opening, the expansion process is three-dimensional and the curvature imposed on the diffracting wave would be spherical in nature corresponding to some effective radius of curvature related to the average of the minimun and maximum dimensions of the square opening $R \propto (W+ \sqrt{2}W)/2$. Liu et al (1983) has shown that for this case $W_c \approx 10\lambda$ in agreement with the experimental results of Mitrofanov and Soloukhin (1965) and Edwards et al (1979).

For rectangular channels of high aspect ratio (L/W>8) the planar wave diffracting through the opening transforms into a cylindrical wave. For the same critical curvature, by simple geometric considerations, namely that $\kappa_{cyl}=\kappa_{sph}=(1/A)dA/dR$, one can show that the radius of curvature for the cylindrical wave would be one half that for the spherical case so that $R_{cyl} = R_{sph}/2 = \Delta H/2 = 3.25\lambda$ which is in agreement with the present experimental results.

Conclusion

The present series of large-scale tests confirm the smaller laboratory experiments of Liu et al that transmission from confined planar detonations to unconfined cylindrical detonations requires a critical channel height of only about three cell diameters. Furthermore, the channel results are almost identical to the results from a rectangular orifice. This suggests that the re-initiation mechanism is very local and does not depend strongly on the flow structure of the product gas emerging into confined space behind the leading front. The onset of three-dimensional effects at surprisingly large aspect ratios (i.e., $L/W \simeq 5$) suggests that quenching by the rarefaction waves may not be a dominant mechanism.

Acknowledgments

This work performed at Sandia National Laboratories supported by the U.S. Department of Energy under contract #DE-AC04-76DP00789 for the U.S. Nuclear Regulatory Commission, Office of Research. The authors acknowledge the stimulating discussions with Y. K. Liu and J. Cummings, the experimental assistance of P. Prassinos, C. Daniel, and J. Fisk, and the enthusiastic support and encouragement of M. Berman.

References

Edwards, D. H., Thomas, G. O., and Nettleton, M. A. (1979) The diffraction of a planar detonation wave at an abrupt area change. J. Fluid Mechanics 95, 79-96.

Guirao, C. M., Knystautas, R., Lee, J. H., Benedick, W., and Berman, M. (1982) Hydrogen-air detonations. 19th Symposium (International) on Combustion, pp. 583-590. The Combustion Institute, Pittsburgh, Pa.

Knystautas, R., Lee, J. H., and Guirao, C. M. (1982) The critical tube diameter for detonation failure in hydrocarbon-air mixtures. Combustion and Flame 48, 63-83.

Lee, J. H., Knystautas, R., and Guirao, C. M. (1982) Proceedings of the International Conference on Fuel-Air Explosions (edited by J. H. S. Lee and C. M. Guirao) SM Study No. 16, pp. 157-187. University of Waterloo Press, Waterloo, Ontario, Canada.

Lee, J. H. (1984) Dynamic parameters of gaseous detonations. Annual Review of Fluid Mechanics 16, (in press).

Liu, Y. K., Lee, J. H., and Knystautas, R. (1983) The effect of geometry on the transmission of detonation through an orifice. (submitted to Combustion and Flame).

Matsui, H. and Lee, J. H. (1978) 17th Symposium (International) on Combustion, pp. 1269-1280. The Combustion Institute, Pittsburgh, Pa.

Mitrofanov, V. V. and Soloukhin, R. I. (1965) The diffraction of multifront detonation waves. Soviet Physics Doklady, 9 (12), 1055-1058.

Moen, I. O., Murray, S. B., Bjerketvedt, D., Rinnan, A., Knystautas, R. and Lee, J. H. (1982) Diffraction of detonation from tubes into a large fuel-air explosive cloud. 19th (International) Symposium on Combustion, pp. 635-644, The Combustion Institute, Pittsburgh, Pa.

Rinnan, A. (1982) Proceedings of the International Conference on Fuel-Air Explosions (edited by J. H. S. Lee and C. M. Guirao) SM Study No. 16, pp. 553-564. University of Waterloo Press, Waterloo, Ontario, Canada.

Westbrook, C. K. (1980) Chemical kinetics of hydrocarbon oxidation in gaseous detonations. Combustion and Flame 49, 191-210.

Westbrook, C. K. (1982) Proceedings of the International Conference on Fuel-Air Explosions (edited by J. H. S. Lee and C. M. Guirao) SM Study No. 16, pp. 189-242. University of Waterloo Press, Waterloo, Ontario, Canada.

Westbrook, C. K., and Urtiew, P. A. (1982) Chemical kinetic predictions of critical parameters in gaseous detonations. 19th Symposium (International) on Combustion, pp. 615-623. The Combustion Institute, Pittsburgh, Pa.

Westbrook, C. K., and Urtiew, P. A. (1982) Inhibition of hydrocarbon oxidation in laminar flames and detonation by halogenated compounds. 19th Symposium (International) on Combustion, pp. 127-141. The Combustion Institute, Philadelphia, Pa.

Air Blast from Unconfined Gaseous Detonations

J. Brossard*
University of Orléans, Bourges, France
J. C. Leyer,† D. Desbordes,‡ and J. P. Saint-Cloud‡
University of Poitiers, Poitiers, France
S. Hendrickx§
Electricité de France, Paris, France
J. L. Garnier**
Centre de l'Energie Atomique
Fontenay aux Roses, France
A. Lannoy††
Electricité de France, Saint-Denis, France
and
J. L. Perrot‡‡
Centre de l'Energie Atomique, Le Barp, France

Abstract

Numerous experimental results for blast waves generated by spherical and hemispherical unconfined gaseous detonations are presented. They concern different mixtures of hydrocarbons with oxygen or air at various dilutions. The pressure field is investigated as far as the symmetry was preserved and up to $\bar{R}$ ($=R/\sqrt[3]{E/p_o}$) ∼ 16. The scaling parameter $\bar{R}$ is defined with the distance R (in meters) from the center of explosion and the energy E (in joules) of the spherical charge at initial pressure $p_o = 10^5$ Pa. From the comparison of the

---

Presented at the 9th ICODERS, Poitiers, France, July 3-8, 1983. Copyright © by the American Institute of Aeronautics and Astronautics, Inc., 1984. All rights reserved.
*Professor, Laboratoire de Recherche Universitaire.
†Professor, Laboratoire d'Energétique et de Détonique, E.N.S.M.A.
‡Assistant Professor, Laboratoire d'Energétique et de Détonique, E.N.S.M.A.
§Engineer, Direction de l'Equipement.
**Engineer, Départment de Sureté Nucléaire.
††Engineer, Direction des Etudes et Recherches.
‡‡Engineer, Centre d'Etudes Scientifiques et Techniques d'Aquitaine.

various and numerous data over a large range of volume (5.2×10^{-4} to 1.45×10^4 m^3) a good coherence of the different characteristics of the pressure signal is observed. The peak overpressure, positive duration, positive impulse, maximum negative overpressure, negative duration, and negative impulse are clearly correlated with the scaling parameter $\bar{R}$. A number of curves similar to those well-known TNT curves are defined independently of equivalency considerations.

Introduction

This study is part of an effort to assess the mechanical effects of an accidental explosion of a hydrocarbon-air cloud. Analysis of actual accidents has shown that the most likely explosion regime in a free field is something like deflagration (Strehlow 1973). However, because of the hazards resulting from a potential detonation of a large gas cloud (Desbordes et al. 1978), a French working group (Brossard et al. 1978) was established to collect a large body of experimental data on the pressure wave generated by the ignition of an unconfined gaseous charge with as close to spherical symmetry as possible (that is, minimal ground effects). The objective is to determine whether Sachs-Hopkinson's similarity conditions exist (Baker et al. 1983).

Experimental Results

Data from three different experiments have been incorporated into the analysis, as summarized in Table 1. They correspond to different authors, to very different experimental techniques and explosive charges, and finally, to a very wide volume ranging from 5.2×10^{-4} m^3 up to $1.45 \times 10^{+4}$ m^3.

The ENSMA and DISTANT PLAIN trials are clearly described in Desbordes et al.(1978) & Reisler et al.(1971).

The AMEDE trials developed on the CESTA site, concerned large spherical or hemispherical volumes ranging from 1.6 to 510 m^3, slightly confined in a latex or mylar envelope (See table). The detonation of air-hydrocarbon mixtures was initiated by small charges of solid explosive located at the center of symmetry either at ground level or at some elevation h_o. The homogeneity of the mixture was the consequence of the mechanical turbulence generated by the experimental device which introduced the constitutive gases and was checked by two probes just before each experiment. The propagation of the detonation wave was followed by means of different streak and framing high-speed cameras. The pressure field was investigated by means of specially profiled piezoelectric pressure gages located at the level of the center of explosion, and associated with a wide bandwidth electronic device.

In the case of spherical charges of a radius R_o, the pressure data of the pressure field investigation were considered as far as the symmetry of the phenomena was preserved; that is, without reflection on the ground (radial distance $R \leq h_o$). It is known that the presence of an envelope, as thin as it is, defining the limits of air/explosible mixture interface, perturbs the pressure

Table 1 Conditions for experiments

Experiment	Mixture Symmetry Ignitor	Hydrocarbon %	Volume m^3	Envelope	h_o m	ρ_f kg/m^3	E_f MJ/kg	$\bar{R}_o$ $(J/m^3 \cdot Pa)^{-1/3}$
ENSMA Desbordes et al.(1978), Desbordes (1983)	$C_2H_2-O_2$ $C_2H_4-O_2$ Spherical Exploding wire	28.6 25.0	5.24×10^{-4}	Soap bubble	∞	1.35 1.39	11.83 10.65	0.115 0.117
DISTANT PLAIN Reisler et al. (1971)	$C_3H_8-O_2$ Hemispherical Detonator	22.2	$1.45 \times 10^{+4}$	Mylar	0	1.42	8.44	0.126
AMEDE Brossard et al. (1978) Brossard (1979, 1982)	C_2H_2-air Spherical Detonator	16.2	1.56 4.0 11.4 18.0 26.8 54.0	Latex 150 μm	3.5	1.27	2.76	0.189
	C_2H_2-air Spherical Detonator	16.2	6.8	Triplex 63 μm	2.5 3.2 4.5	1.27	2.76	0.189
	C_3H_8-air Spherical Detonator + 0.5 and 1 kg plastic	5.0	32.0	Triplex 63 μm	3.5	1.41	2.59	0.187
	C_2H_4-air Spherical Detonator + 10 g plastic	8.0	3.0 9.0 26.8 6.8 31.4 215	Latex 150 μm Triplex 63 μm	3.5 3.5	1.28 1.28	2.88 2.88	0.187 0.187
	C_2H_4-air Hemispherical Detonator + 10 g plastic Detonator + 100 g plastic	8.0	2.9 14.9 510.0	Triplex 63 μm Mylar 30 μm	0 0 0	1.28 1.28 1.28	2.88 2.88 2.88	0.187 0.187 0.187
	C_2H_4-air-O_2 Hemispherical Detonator + 10 g plastic	6.9	116.0	Triplex 63 μm	0	1.29	3.13	0.182

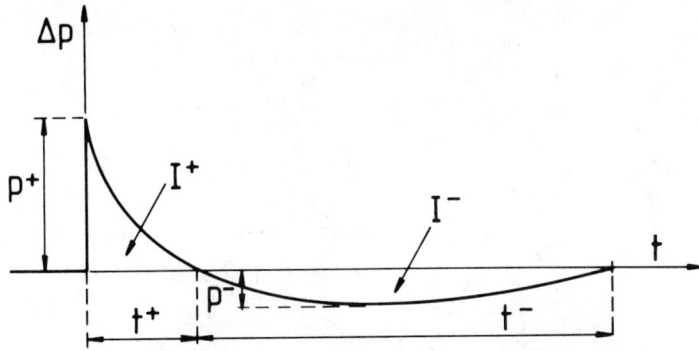

Fig. 1 Scheme of the overpressure profile.

signal and sometimes induces a secondary peak overpressure close to the blast wave front, mainly in the near field. Moreover, the presence of a secondary shock in the flowfield generated by the blast wave front sometimes causes difficulties in the assessment of the pressure signal characteristics, particularly of the positive and negative phases durations (Reisler et al. 1971; Brossard 1979).

Analysis

The main parameters characterizing the pressure profile are as follows (Fig. 1):

1) The positive peak overpressure p^+ of the wave front.
2) The positive overpressure duration t^+.
3) The positive overpressure phase impulse

$$I^+ = \int_0^{t^+} p\, dt.$$

4) The maximum negative pressure signal p_-.
5) The negative pressure phase duration t^-.
6) The negative pressure phase impulse

$$I^- = \int_{t^+}^{t^+ + t^-} p\, dt.$$

Figures 2-7 present the experimental values of p^+/p_o, $t^+/\sqrt[3]{E}$ (ms/MJ$^{1/3}$); $I^+/\sqrt[3]{E}$ (bars·ms/MJ$^{1/3}$); $(p^+ + |p^-|)/p_o$, $(t^+ + t^-)/\sqrt[3]{E}$ (ms/MJ$^{1/3}$); and

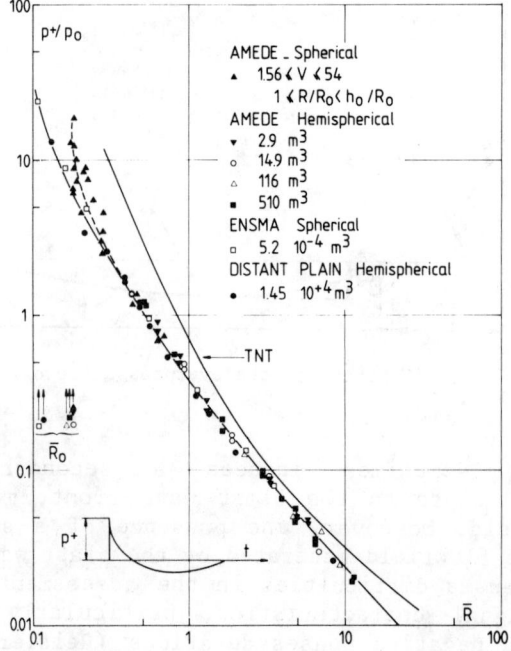

Fig. 2 Reduced peak overpressure vs the reduced distance.

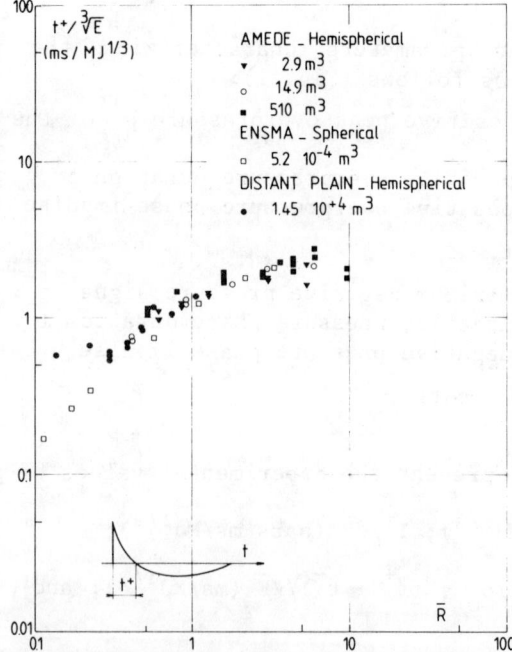

Fig. 3 Reduced positive time vs the reduced distance.

AIR BLAST FROM UNCONFINED GASEOUS DETONATIONS

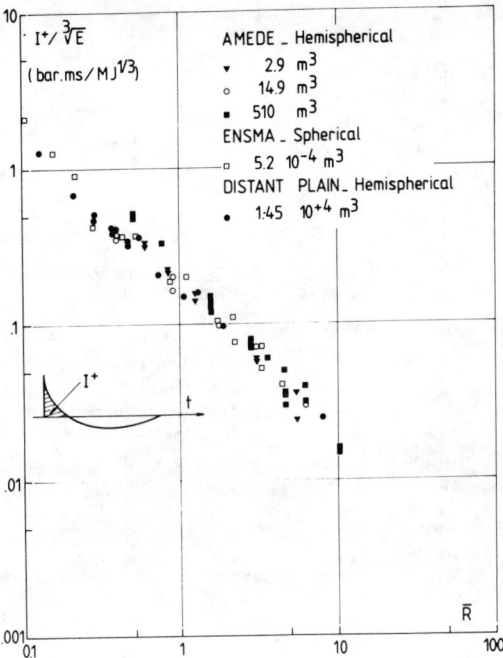

Fig. 4 Reduced positive impulse vs the reduced distance.

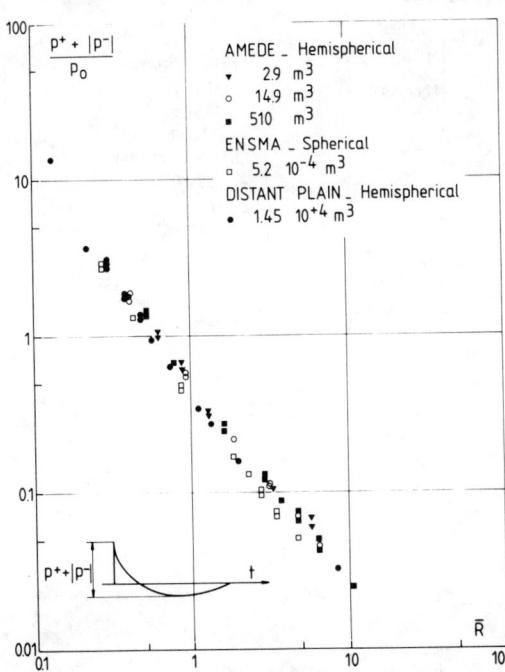

Fig. 5 Reduced total amplitude of the pressure signal vs the reduced distance.

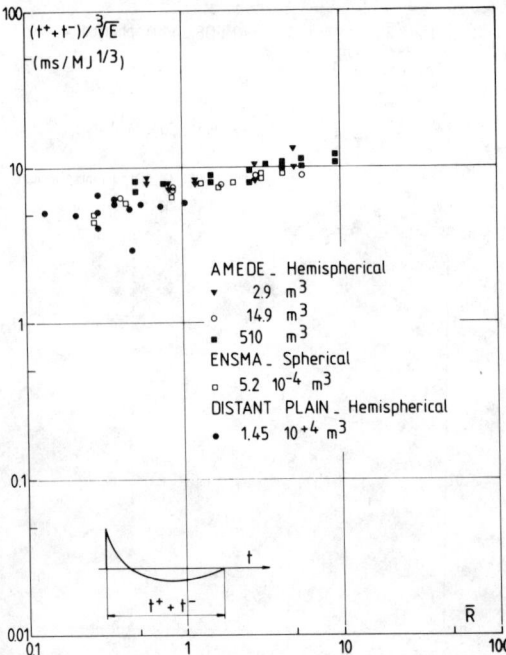

Fig. 6 Reduced total time duration of the pressure signal vs the reduced distance.

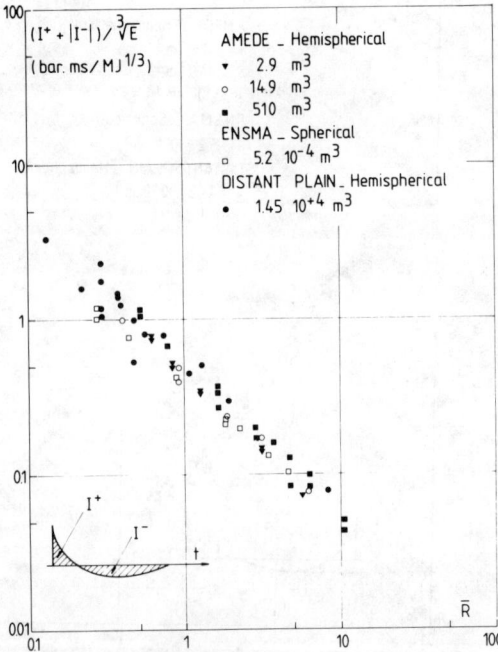

Fig. 7 Reduced total absolute impulse vs the reduced distance.

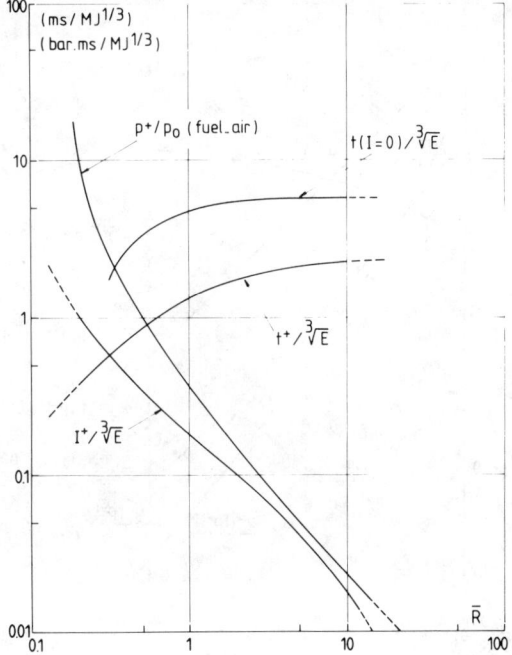

Fig. 8 Characteristics of the positive phase of the pressure profile vs the reduced distance.

$(I^+ + |I^-|)/\sqrt[3]{E}$ (bars·ms/MJ$^{1/3}$) as functions of the unique similarity parameter $\bar{R} = R/\sqrt[3]{E/p_o}$, for which R (m) is the radial distance of the shock front and E (in joules) is the energy included in the R_o radius sphere, whether the charge is spherical or hemispherical. It is assumed that the effects of the explosion of an hemispherical charge (radius R_o) on the ground in a half space are identical to those of a spherical charge (radius R_o) in a whole space; that is, far enough from the ground to preserve symmetry of the phenomena. Consequently, $E = (4/3)\pi R_o^3 \rho_f E_f$, for which ρ_f and E_f are respectively the density and energy (enthalpy of reaction at initial conditions p_o = 1 atm and T_o = 298 K) released by the explosive reaction. $\bar{R}_o$ is the particular value of $\bar{R}$ on the envelope interface (R = R_o); that is, $\bar{R}_o = \left[(4/3)\pi \rho_f E_f/p_o\right]^{-1/3}$, which characterizes the explosive mixture. The $\bar{R}_o$ values are approximately 0.18 for hydrocarbon-air mixtures and 0.12 for the hydrocarbon-oxygen mixtures, and, consequently, the p^+/p_o values as a function of $\bar{R}$ lead to different limits with decreasing $\bar{R}$ to the minimum $\bar{R}_o$. The absolute values $|p^-|/p_o$ are not presented here, but one would

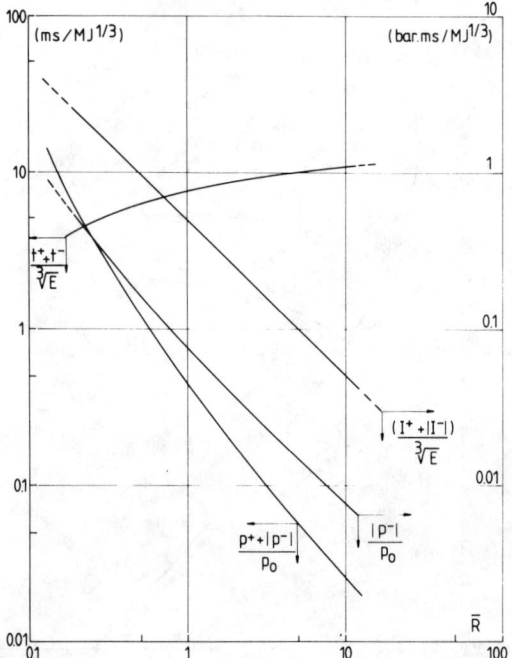

Fig. 9 Characteristics of the pressure profile, taking into account the negative part, vs the reduced distance.

notice a good correlation of the numerous values vs $\bar{R}$. In Fig. 5, excellent agreement of the variation of the total amplitude of the signal $(p^+ + |p^-|)/p_0$ with $\bar{R}$ is observed. A plot of the relative impulse $(I^+ + I^-)/\sqrt[3]{E}$, not included in this work, shows a fairly large scattering of experimental points because of the irregularities of the pressure profile.

Finally, the Figs. 2 — 7 show a good agreement of the results leading to the experimental curves (Figs. 8 and 9) which are merely "eye ball" curve fits of data. But from the numerous experimental data, we have deduced the least square polynomials. For instance, the relations:

$$\ln(p^+/p_0) = -0.9126 - 1.5058(\ln \bar{R}) + 0.1675(\ln \bar{R})^2 - 0.0320(\ln \bar{R})^3$$

$$\ln(t^+/\sqrt[3]{E}) = +0.2500 + 0.5038(\ln \bar{R}) - 0.1118(\ln \bar{R})^2$$

$$\ln(I^+/\sqrt[3]{E}) = -1.5666 - 0.8978(\ln \bar{R}) - 0.0096(\ln \bar{R})^2 - 0.0323(\ln \bar{R})^3$$

are considered as representative in the range of
$0.3 < \bar{R} < 12$ for fuel-air explosions. The closeness of the relationship between p_+/p_o and $t_+/\sqrt[3]{E}$ is also demonstrated. Fig. 8 gives the orders of magnitude of the time t where the impulse I equals zero.

Conclusions

Because of the good agreement of the numerous experimental data that characterize the blast wave generated by the donation of gaseous charges in free field, it can be concluded that similarity of physical phenomena exists, provided that the data are taken under conditions for which symmetry is preserved. This similitude seems clearly established over a large range of volume as on a wide range of scaled distances $\bar{R}$ ($0.40 \leq \bar{R} \leq 12$). Furthermore, for detonations, a set of practical curves are shown that enable one to foresee the effects created by this detonation.

References

Baker, W. E., Cox, P. A., Westine, P. S., Kulesz, J. J., and Strehlow, R. A. (1983) Explosion hazards and evaluation. Fundamental Studies in Engineering,5, Elsevier, Amsterdam, pp. 115-118.

Brossard, J. (1979) Analyse des traitements SIDEX des essais AMEDE 2/4 - 2/5 - 2/6 et 2/7. ENSMA, Poitiers,France. Note technique EA 79.14.

Brossard, J. (1982) Les essais AMEDE 2/8 et premiers résultats de synthèse. ENSMA, Poitiers, France. Note technique EA 79.15.

Brossard, J., Perrot, J., and Manson, N. (1978) Pressure waves generated by detonating spherical gaseous charges. Communication (unpublished) at the 17th Symposium (International) on Combustion, Leeds, England.

Desbordes, D. (1983) Explosions non confinées. Note technique ENSMA, Poitiers, France.

Desbordes, D., Manson, N., and Brossard, J. (1978) Explosion dans l'air de charges sphériques non confinées de mélanges réactifs gazeux. Acta Astronaut. 5, Nov-Dec, pp. 1009-1026.

Kogarko, S. M., Aduschkin, V. V., and Liamin, A. G. (1965) Investigation of spherical detonation on gas mixtures. Combust. Explos. Shock Waves USSR 1, (2), pp. 22-34.

Lannoy, A. (1983) Méthodes probabilistes et déterministes d'estimation du risque industriel appliquées à la prévision des effets des explosions non confinées air-hydrocarbure. Doctor Engineer Thesis, Université de Poitiers, France.

Lee, J. H., Guirao, C. M., Chiu, K. W., and Bach, G. G. (1977) Blast effects from vapour cloud explosions. AIchE Loss Prevention Symposium, Houston, March.

Reisler, R. E., Ethridge, N. H., Lefebvre, D. P., and Giglio-Tos, L. (1971) Air blast measurements from the detonation of an explosive gas contained in a hemispherical balloon. Memorandum Report 2108.BRL.

Strehlow, R. A. (1973) Unconfined vapour cloud explosions: An overview. 14th Symposium (International) on Combustion, pp. 1189-1200. The Combustion Institute, Pittsburgh, Pa.

Chapter V.   Interactions

Collapse of Gas-Filled Cavities in Water

Maurice Holt* and M. J. Djomehri†
University of California, Berkeley, California

Abstract

The calculation, using a similarity solution, of the collapse of an empty spherical cavity in water, is extended to apply to cavities that contain air or other gases. The flow field outside the cavity wall is largely unaffected, but that in the gas within the cavity requires the fitting of a converging shock wave.

Introduction

The collapse of an empty spherical cavity in water is dominated in its final stages by compressibility effects and can be calculated by a similarity technique. This was first carried out by Hunter (1960) and modified by Holt and Schwartz (1963), Holt, Sakurai, and Kawaguti (1967), Holt (1967), and Holt and Ballhaus (1970). Hunter's calculation determined the essential features of the collapse process, including the exact value of the parameter defining the similarity variable. However, since assumption of similarity restricts the number of physical constants that can be considered, Hunter's solution requires that velocity of sound at the cavity wall be zero. In order to take account of the finite value of this velocity, the other papers mentioned all deal with perturbations of Hunter's solution.

In practice, cavities are not empty but are filled with low-density air or gas. The present paper considers the effect of this change on solutions previously obtained. To the first order of approximation, the presence of gas inside the cavity does not influence the motion outside of

---

Presented at the 9th ICODERS, Poitiers, France, July 3-8, 1983. Copyright © American Institute of Aeronautics and Astronautics, Inc., 1984. All rights reserved.
  *Professor, Department of Mechanical Engineering.
  †Assistant Research Engineer, Department of Mechanical Engineering.

it or that of the cavity wall itself. This is still determined by Hunter's similarity solution, and the similarity parameter for motion inside the cavity is therefore taken at Hunter's value. In fact, the collapsing cavity acts as a spherically imploding piston driving the gas inside, compressing it up to a concentric shock wave, and separating the disturbed gas from an undisturbed core. The gas flow is nonisentropic and is treated by a technique similar to that used by Welsh (1967) for spherical impositions. The calculation is more far reaching than that based on a perturbation technique applied to the same problem by Holt (1971).

Conservation Equations for Spherical Flow Inside a Cavity

The equations of motion for spherically symmetric unsteady compressible flow may be written

$$\rho_t + u\rho_r + \rho(u_r + 2\frac{u}{r}) = 0$$

$$u_t + uu_r + \frac{1}{\rho} p_r = 0 \qquad S_t + uS_r = 0 \qquad (1)$$

where t is time, r is radial distance, and u, p, ρ, and S denote fluid velocity, pressure, density, and entropy, respectively. It is then assumed that the gas in the cavity is polytropic with constant specific heats. Thus

$$S = c_v \log \frac{p}{\rho^\gamma} + \text{constant} \qquad c^2 = \left(\frac{\partial p}{\partial \rho}\right)_S = \frac{\gamma p}{\rho} \qquad (2)$$

If Eqs. (1) and (2) are made dimensionless with respect to the standard values of the physical variables in water (e.g., sound speed, density, etc.) and we rearrange the equations, we find

$$(c^2)_t + u(c^2)_r + (\gamma-1)c^2(u_r + 2\frac{u}{r}) = 0$$

$$u_t + uu_r + \frac{1}{\gamma-1}(c^2)_r - \frac{1}{\gamma(\gamma-1)c_v} S_r c^2 = 0$$

$$S_t + uS_r = 0 \qquad (3)$$

where c is the local speed of sound. Letting $\tilde{\phi} \equiv \frac{1}{\gamma-1} \frac{S}{c_v}$, from Eq. (2) we get

$$\tilde{\phi} \equiv \frac{1}{\gamma-1} \frac{S}{c_v} = \frac{1}{\gamma-1} \log \frac{p}{\rho^\gamma} + \text{const} = \log \frac{c^{(\frac{2\gamma}{\gamma-1})}}{p} + \text{const} \qquad (4)$$

Similarity Solutions

Assuming that the gas motion in the cavity is governed by a similarity solution in the sense of Guderley (1942), the variable $\lambda = A(-t)^n$ may be introduced, representing the distance of the shock front from the cavity center and

$$\xi = -\left(\frac{\lambda}{r}\right)^{1/n} = (A^{1/n} t)/(r^{1/n})$$

is a similarity variable. Then the value of the similarity parameter n is the same as that obtained by Hunter (1960) when considering the similarity solution in water just outside the cavity wall.

For convenience, the constant A is chosen so that the similarity variable ξ has value $\xi = -1$ on the converging shock. The similarity equations are then integrated towards the outer flow, starting at the shock wave, and terminate the integration at the cavity wall, determining the value of ξ at that station. If desired, conversion can subsequently be made to the similarity variable of Hunter's solution (which has a value of 1 at the cavity wall) and the value at the shock wave can be determined. This would give the position of the shock wave in terms of Hunter's similarity varible.

We assume that strong shock conditions are valid for the concentric shock wave inside the cavity. Thus

$$\overset{*}{u} = \frac{2}{\gamma+1}\overset{*}{U} \qquad \overset{*}{c}^2 = \frac{2\gamma(\gamma-1)}{(\gamma+1)^2}\overset{*}{U}^2 \qquad \overset{*}{\phi} = k\left(1-\frac{1}{n}\right)\log\lambda + \overset{*}{\phi}_0$$

$$k = \frac{2}{\gamma-1} \qquad (5)$$

where

$$\overset{*}{U} = \dot{\lambda} = \text{shock speed} = -(nA^{\frac{1}{n}})\lambda^{(1-\frac{1}{n})}$$

(see Welsh, 1967). Express u, c^2 and $\tilde{\phi}$ in terms of the similarity functions f, g, and ϕ, respectively, and the similarity variable ξ gives

$$u = \dot{\lambda} f\left(\frac{r}{\lambda}\right) \equiv -(nA^{\frac{1}{n}}) r^{(1-\frac{1}{n})} F(\xi)$$

$$c^2 = \dot{\lambda}^2 g\left(\frac{r}{\lambda}\right) \equiv (n^2 A^{\frac{2}{n}}) r^{(2-\frac{2}{n})} G(\xi)$$

$$\tilde{\phi} = k(1-\frac{1}{n})\log r + \phi(\frac{r}{\lambda}) + \phi_0^*$$

$$\equiv k(1-\frac{1}{n})\log r + \phi(\xi) + \phi_0^* \qquad (6)$$

Substitute Eqs. (6) into Eqs. (3) to obtain the following ordinary differential equations for F, G, and ϕ:

$$(\gamma-1)\xi GF' + (1+\xi F)G' + (1+n+\gamma-3n\gamma)FG = 0$$

$$(\gamma-1)(1+\xi F)F' + \xi G' - \frac{1}{\gamma}\xi G\phi' - (n-1)[(\gamma-1)F^2 + 2G - \frac{k}{\gamma}G] = 0$$

$$(1+\xi F)\phi' - k(n-1)F = 0 \qquad (7)$$

From the third of Eqs. (7),

$$\phi' = \frac{k(n-1)}{(1+\xi F)} F \qquad (8)$$

If Eq. (8) is substituted into the first and second of Eqs. (7) and rearranged, the following ordinary differential equations are obtained for F and G only:

$$[(1+\xi F)^2 - \xi^2 G]F' = \frac{(1+n+\gamma-3n\gamma)}{\gamma-1}\xi FG + \frac{k(n-1)}{\gamma(\gamma-1)} FG$$

$$+ (\frac{n-1}{\gamma-1})(1+\xi F)[(\gamma-1)F^2 + 2G - \frac{k}{\gamma}G]$$

$$[(1+\xi F)^2 - \xi^2 G]G' = -(1+\xi F)(1+n+\gamma-3n\gamma)FG - \frac{k(n-1)}{\gamma(1+\xi F)}\xi^2 G^2 F$$

$$- (n-1)\xi G[(\gamma-1)F^2 + 2G - \frac{k}{\gamma}G] \qquad (9)$$

The above system of ordinary differential equations may be integrated for given initial values. These conditions can be derived from Eqs. (6) and from the strong shock relation, Eq. (5), at $r = \lambda$. Thus, at $\xi = -1$

$$F(-1) = \frac{2}{\gamma+1}$$

$$G(-1) = \frac{2\gamma(\gamma-1)}{(\gamma+1)^2}$$

$$\Phi(-1) = 0 \qquad (10)$$

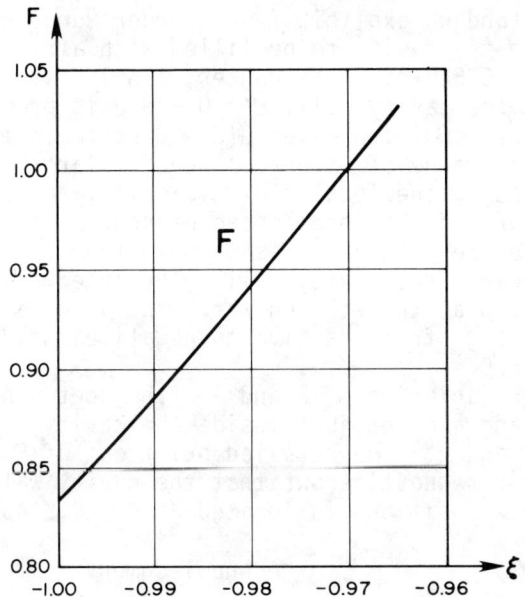

Fig. 1 Variation of dimensionless velocity behind the shock.

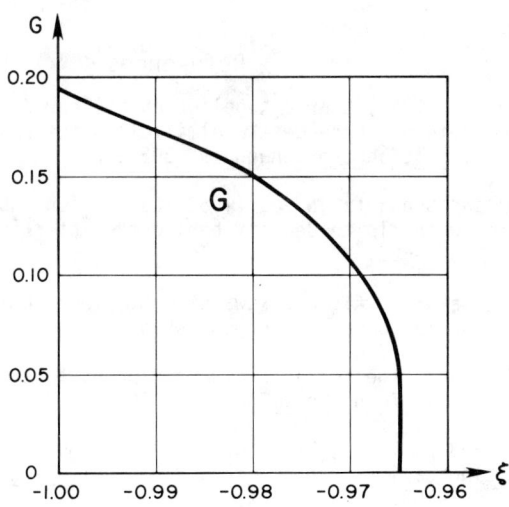

Fig. 2 Variation of dimensionless sound speed behind the shock.

The system of ordinary differential equations [Eqs. (9)] holds for values of ξ lying between those of the shock front and the cavity boundary. Equations (9) are integrated numerically with initial values [Eqs. (10)] employing a standard explicit fourth-order Runge-Kutta method. Assuming the cavity to be filled with air, we take $\gamma = 1.4$ and $n = 0.5552$ (given by Hunter, 1960).

At the cavity wall, $c^2 = 0$ and u is proportional to λ. The first condition gives $G(\xi) = 0$ at the wall and this fixes the value of ξ there. The similarity functions F, representing the local fluid velocity, and G, representing the square of the local speed of sound, will increase and decrease, respectively, as we move from the shock wave outward towards the cavity wall. The integration process is terminated at the station where the value of $G = c^2/\lambda^2$ changes sign from positive to negative inside the interval $-1 \leq \xi < 0$.

The variation of F and G with ξ between -1 (the shock point) and a point just inside the cavity wall is shown in Figs. 1 and 2. The detailed print-out of F and G (available upon request) shows that the cavity wall (in terms of the shock position) is located at $\xi = -0.9645$.

Acknowledgment

This work was supported by the Office of Naval Research, Mechanics Division, Dr. Chuong M. Lee, Scientific Officer.

References

Guderley, G. (1942) Starke kugelige un zylindrische Verdichtungsstösse in der Nähe des Kugelmittelpunktes bzw. der Zylinderachse. Luftfahrtforschung 19, 302-312.

Holt, M. and Schwartz, N. J. (1963) Cavitation bubble collapse in water with finite density behind the interface. Phys. Fluids 6(4), 521-525.

Holt, M., Sakurai, A., and Kawaguti, M. (1967) University of California, Berkeley (unpublished).

Holt, M. (1967) The collapse of an imploding spherical cavity. Rev. Roum. Sci. Tech. Mec. Appl. 12, 407.

Holt, M. and Ballhaus, W. F. Jr. (1970) Singular points in the collapsing cavity problem. Rev. Roum. Sci. Tech. Mec. Appl. 15 (1), 91-104.

Holt, M. (1971) Application of similarity methods to collapsing cavity problems. Proceedings of IUTAM Symposium on Non-steady Flow of Water at High Speeds, pp. 219-223, NAUKA, Moscow.

Hunter, C. (1960) On the collapse of an empty cavity in water. J. Fluid Mech. 8, 241-263.

Welsh, R. L. (1967) Imploding shocks and detonations. J. Fluid Mech. 29, 61-79.

Crack Propagation in Burning Solid Propellants

J. G. Siefert* and K. K. Kuo†
The Pennsylvania State University
University Park, Pennsylvania

Abstract

Crack propagation in a burning composite propellant subjected to rapid pressurization in the order of 10 GPa/s was investigated experimentally. A visual record of the event was made using a high-speed (35,000 pps) movie camera. The transient pressurization process was recorded by high-frequency pressure transducers in the combustion chamber. The effect of pressurization rate on both crack propagation velocity and time variation of crack shape was studied. Experimental results indicate that the crack velocity increases as the pressurization rate is raised. The measured dependence of crack propagation velocity on pressurization rate was found to be stronger than that based only upon viscoelastic material property variation at different loading rates. A characteristic difference of crack geometry variation during propagation was observed at different pressurization rates. At higher pressurization rates, the geometric transformation of the crack tip was preceded by the emergence of a fan region immediately above the crack tip. It is believed that the fan region is caused by microstructure damage in the propellant at the crack tip, which allows the bright combustion gases to penetrate the voids ahead of the tip region and spread in a radial fashion. The actual geometric change, a transformation of the crack shape from a smooth triangular contour into a square contour with a jagged leading edge, is believed to be the growth of small fractures into major crack branches.

Presented at the 9th ICODERS, Poitiers, France, July 3-8, 1983. Copyright © American Institute of Aeronautics and Astronautics, Inc., 1984. All rights reserved.
*Graduate Student, Department of Mechanical Engineering.
†Professor, Department of Mechanical Engineering.

Introduction

At present, there is a lack of experimental and theoretical knowledge in the area of solid-propellant grain fracture during rocket motor firing. This area is of special interest to investigators of rocket motor grain integrity, where high-speed crack growth and propellant fragmentation may provide sufficient additional burning surface area to enhance the possibility of deflagration-to-detonation transition (DDT) in solid-propellant rocket motors.

In the past, numerous studies, based purely on solid mechanics considerations, have been conducted to investigate crack propagation. Even though there is no experimental evidence to support the validity of applying results of inert propellant crack propagation studies to live propellants under burning conditions, virtually all previous studies have employed inert propellants in non-burning environments. It is therefore apparent that an investigation of crack propagation under burning conditions would be useful for the interpretation of the mechanism of crack propagation in a combusting environment.

An investigation was undertaken to help bridge this technological gap. Specific objectives of this study were:

1) To develop an experimental technique to study crack propagation in a composite propellant under burning conditions.

2) To measure propagation velocities of the luminous crack front in solid-propellant samples.

3) To observe any abnormalities of the propagating crack.

4) To study the effects of chamber pressure, initial crack tip pressurization rate, and crack sample geometry on luminous crack front propagation velocity.

Experimental Setup

Crack propagation tests were conducted with a windowed chamber; the schematic diagram of the test rig is shown in Fig. 1. A propellant sample, with prefabricated crack, was internally pressurized by a small driving motor which produced high-temperature and high-pressure gases. The propellant sample configuration allowed for significant displacement of the crack walls, resulting in high stresses and strains at the crack tip, which in turn led to crack propagation. In the driving motor, an electric

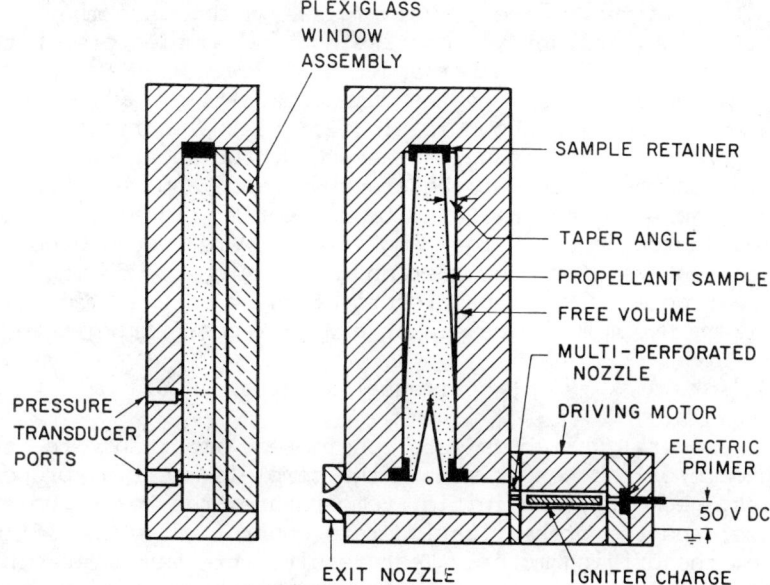

Fig. 1   Schematic diagram of crack propagation chamber.

primer (FA 874) was used to ignite the booster propellant and the igniter charge. Product gases flowed through a multiperforated nozzle into the main chamber, causing ignition, flame spreading, and mechanical deformation of exposed internal crack surfaces. The continued pressurization of the crack cavity is dominated by the output mass and energy fluxes of the driving motor. With this apparatus, it is possible to obtain pressures up to 50 MPa, and pressurization rates on the order of 100 GPa/s. For tests conducted in this study, pressurization rates ranged between 1 and 10 GPa/s.

The specific chamber pressurization rates used in these experiments are applicable to actual propellant rocket motor configurations. The lower limits of the pressurization rates tested are of the order of magnitude found during the ignition transient in certain rocket motors. The higher rates correspond to the rates during the early phase of the deflagration-to-detonation transition in damaged solid-propellant grains.

The propellant sample was machined to a desired geometry and a uniform thickness. A 3-mm deep slit of uniform width was cut with a sharp razor blade to facilitate initiation of crack growth. The initial slit angle of the crack sample is nearly zero. The propellant used was AP/HTPB 73/27 with 200-μm AP. Once the sample was installed in the chamber, a plexiglass window assembly was bolted on, slightly compressing the sample between the rear wall of the chamber and the window.

A block diagram of the data acquisition system used in the study is shown in Fig. 2. Piezoelectric transducers were used to measure pressure at two locations in the main chamber (one at the crack entrance and the other in the crack-tip region). The transducer's signals, amplified through a charge amplifier, were recorded on the transient waveform recorder and the high-speed magnetic tape recorder. A visual record of the event was made with a 16-mm Hycam movie camera at a 35,000-pps filming rate.

Experimental Results

Several variables were considered important to the outcome of the experiment: the sample's material properties, geometry, and initial temperature; the exit nozzle size; and the pressurization rate controlled by the efflux from the driving motor. Test results presented in this paper include only effects of pressurization rate and geometry; other variables were held constant.

A typical pressure vs time trace is presented in Fig. 3. From the plot, it can be seen that during the initial time interval (1 ms) of pressurization at the crack tip; the pressurization rate was approximately constant. Nearly constant pressurization rates observed for all test firings provided a convenient test parameter, varied from test to test by adjusting the size of the igniter charge loaded in the driving motor. In subsequent discussions,

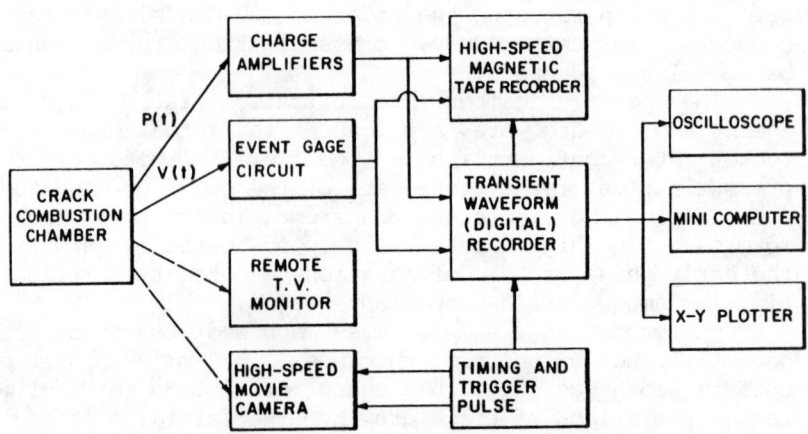

Fig. 2 Block diagram of data acquisition system.

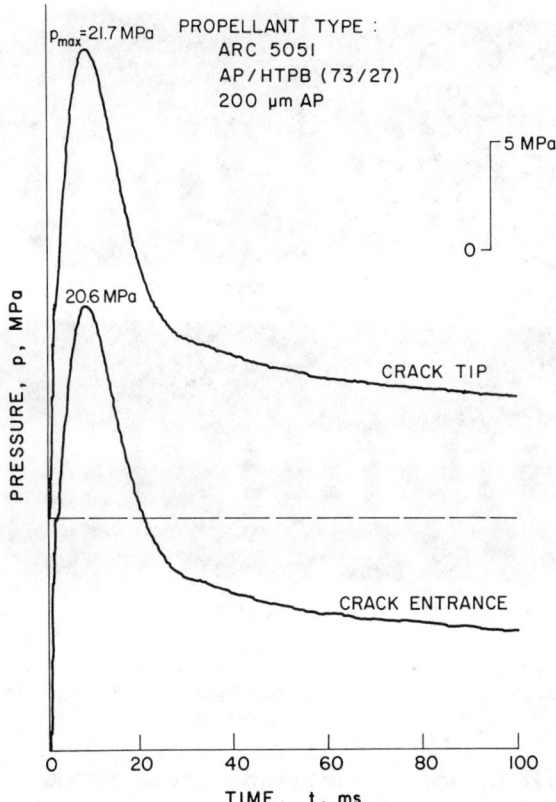

Fig. 3 Pressure-time trace for crack propagation test (Test No. DNCP1-7; dp/dt = 4x10³ MPa/s during measured crack propagation).

the pressurization rate is referred to the initial pressurization rate at the crack tip.

The bulk of data obtained in each test was recorded by means of high-speed photography. A typical film record is shown in Fig. 4. Hot gases from the driving motor illuminated the crack cavity and outlined the crack walls. In this figure, each vertical column of light is the outlined crack at a different time. The event proceeds from left to right, with a time interval of 29 μs between consecutive pictures.

After an initial period of irregular crack-tip displacement (first three pictures), the event shown in pictures 4-9 illustrates a typical mode I crack propagation with the crack-tip displacement in the vertical direction as expected. However, in picture 10, a dramatic change takes place in the propagation mechanism. A fan region of light emerges just above the luminous front of

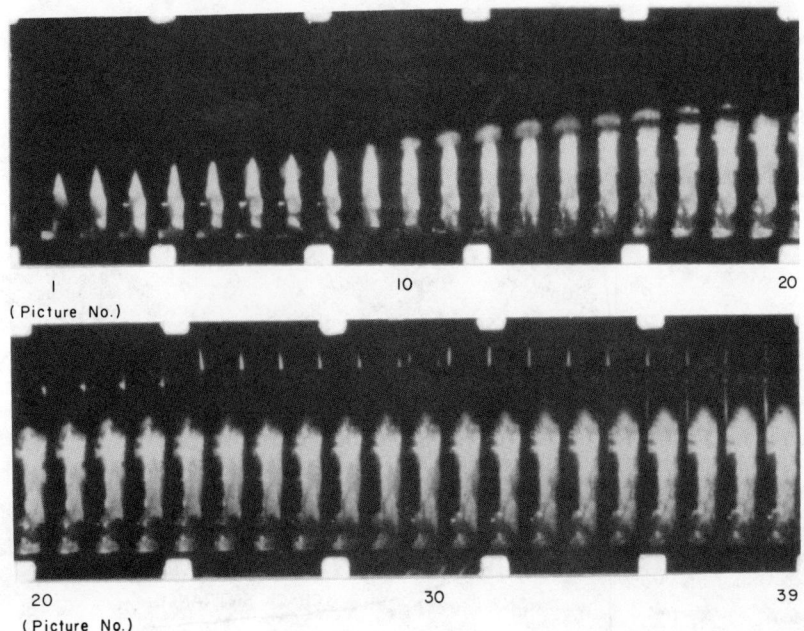

Fig. 4 Photograph of luminous crack-tip front propagation in a burning solid propellant with high pressurization rate (35,000 pps).

the crack tip, and is accompanied by a change in the geometry of the crack. After the fan region appeared, the crack walls became almost parallel, and the crack tip itself lost its definite triangular shape and became a jagged horizontal line. One explanation for the fan region and the corresponding change in crack geometry is the creation of micropores by the material fracture near the crack tip. This leads to branching of the crack, which allows crack walls to displace outward into the square-shaped contour described above. From picture 20 on, the crack front is highly irregular with multiple burning branches. The growth of the bulk crack cavity has essentially stopped, due to leaks to the outer cavities (see Fig. 1). These leaks create an equal pressure situation stopping further crack fracture.

The luminous front of the crack tip for each test was plotted as a function of time. An example is presented in Fig. 5. If the tests were conducted in a nonburning environment, one could easily associate crack-tip displacement with crack propagation velocity. However, in the case of a burning environment, mechanical displacement, material regression due to burning, and the possibility of flame

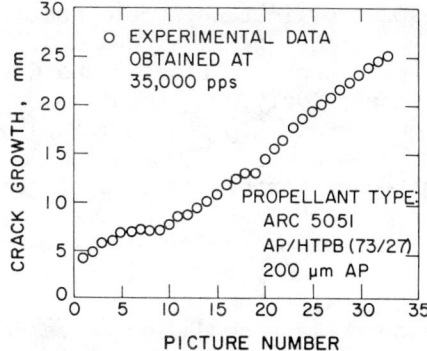

Fig. 5 Measured displacement of the luminous front of the crack tip in a burning solid propellant.

spreading between the sample and window must be considered. Burning rate calculations reveal that the effect of deflagration accounts for only 0.1% of the displacement experienced, and therefore may be considered negligible. The sample's vertical displacement at the crack tip due to pressurization was calculated with a finite-element program. From this analysis, it was determined that the mechanical displacement could also be considered negligible when compared to the total displacement. The effect of possible flame spreading between the sample and window above the crack tip can be determined by interpretation of the film. While it is not obvious from the black-and-white reproductions of the photographs included in this paper, the actual geometry of the luminous region of the crack cavity can be identified in most cases by inspection of the films with a motion analyzer.

Once the displacement plot shown in Fig. 5 was established as the approximate luminous crack-tip front propagation plot, average front velocities were calculated by a least-square fit. All tests revealed nearly constant propagation velocities after a short initial period of irregular crack growth. These velocities ranged from 20 to 80 m/s. A major objective of this study was to determine the effect of pressurization rate on crack velocities. Since an approximate constant velocity and pressurization rate were observed for each test, each experiment provided a data point for the plot of crack propagation velocity vs pressurization rate for a fixed initial crack geometry. This type of plot will be discussed in a later section. One definite trend observed from experimental results is that the higher the pressurization rate, the higher the propagation velocity of the luminous front.

Another aspect of increasing pressurization rates is illustrated by comparing Figs. 4 and 6. Pressurization rates for the tests shown in Figs. 4 and 6 were 15,000 and 3000 MPa/s, respectively. At the lower pressurization rate, the luminous crack-tip region maintained its original triangular shape with only small deformations, while the test conducted at the higher pressurization rate revealed a square crack shape with a ragged tip region, an indication of crack branching. The geometric transformation was accompanied by an increase in combustion intensity (see Fig. 4). The pictures on the right side of Fig. 6 show some flame spreading across the face of the sample. In this case, it is possible to distinguish between the flame spreading and the triangular crack contour. The pictures in Fig. 4, on the other hand, show limited flame spreading and the clearly outlined square shaped crack.

Discussion of Experimental Results

Determination of the relationship between vertical luminous crack-tip front displacement and chamber pressure was achieved by combining displacement vs. time and pressure vs time data obtained in this study. Examination of this relationship illustrates the influence of chamber pressurization rate and crack sample geometry on propagation velocity of the luminous front. Two explanations for the tendencies exhibited by this relationship were considered, and were found to provide new insight into the principal mechanism of crack propagation encountered in this study.

Displacement vs pressure curves for three tests, having identical geometries but different pressurization

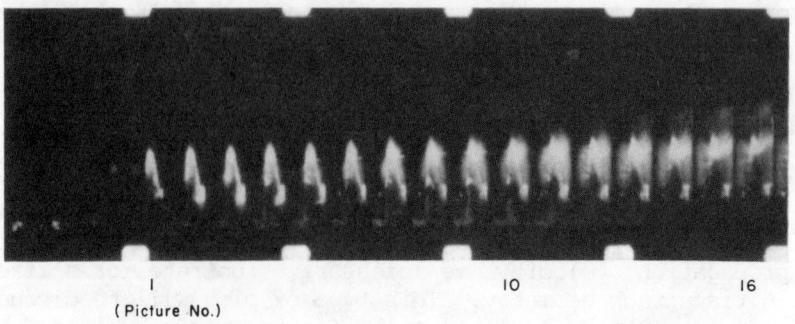

Fig. 6 Photograph of luminous crack-tip front propagation in a burning solid propellant with low pressurization rate (35,000 pps).

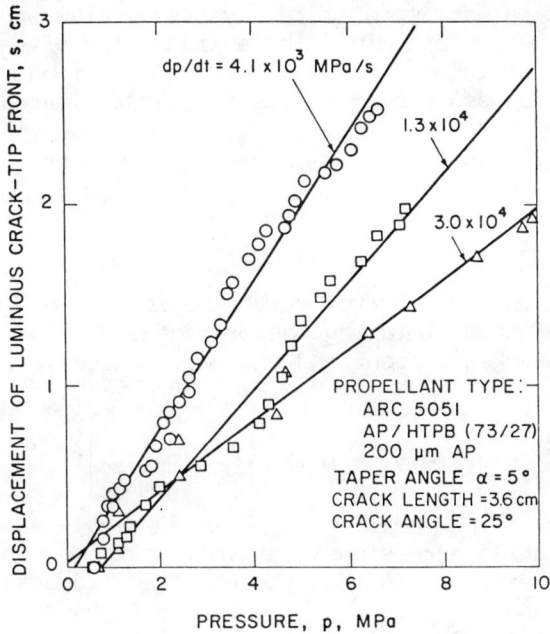

Fig. 7 Measured luminous crack-tip front displacement vs chamber pressure for crack samples with the same initial geometries.

rates, are shown in Fig. 7. The straight lines drawn through the data points represent the best linear fit for each test. It is clear from this figure that the tests resulted in three approximately linear relationships between displacement and pressure. The average pressurization rate measured in these tests is shown in Fig. 7. It can be seen from the figure that the slopes of the displacement vs pressure curves decrease with increasing pressurization rate. If only pressurization rate and sample geometry were varied, the relationship between the slope of the displacement-pressure curves and these parameters can be generally written as

$$\frac{ds}{dp} = F(\frac{dp}{dt}, \text{geometry}) \qquad (1)$$

For those tests shown in Fig. 7, geometry was held constant; therefore, the expression above reduces to

$$\frac{ds}{dp} = f(\frac{dp}{dt}) \qquad (2)$$

This function was approximated by a power law whose coefficients were determined by a linear fit of the log-log plot of ds/dt for the three tests with fixed geometry. The function was found to be closely approximated by

$$\frac{ds}{dp} = 0.086 \left(\frac{dp}{dt}\right)^{-0.363} \qquad (3)$$

This equation was used to derive the following experimental correlation relating the propagation velocity of the luminous crack-tip front and chamber pressurization rate:

$$\frac{ds}{dt} = V_c \text{ (m/s)} = 0.086 \left[\frac{dp}{dt} \text{ (Mpa/s)}\right]^{0.637} \qquad (4)$$

The correlation agrees qualitatively with the observation discussed earlier in that the observed propagation velocity increases as chamber pressurization rate becomes higher. It should be noted that this correlation is valid only for the sample geometry considered and for the particular type of propellant studied. The crack angle of 25 deg marked on Fig. 7 is the initial angle of the triangular cavity. The real crack angle is nearly 0 deg, created by the 3-mm deep slit cut by a razor blade.

To illustrate the effect of different geometries on observed luminous crack-tip front propagation velocities, three crack-tip displacement vs pressure curves are shown in Fig. 8. Approximately the same pressurization rate of 15,000 MPa/s was measured during three separate tests with different sample geometries. With the same initial contact surface between the sample and chamber wall, the sample geometry differed only in the taper angle of the external surfaces of the propellant sample, which in turn changes the free volume adjacent to the sample (see Fig. 1). This observation remains consistent for the two remaining curves. Since these tests were conducted at the same average pressurization rate, the steeper the slope of the curve in Fig. 8, the higher the average crack propagation velocity.

Data obtained from several tests with different geometries and pressurization rates are shown in Fig. 9. The solid line represents the correlation given by Eq. (4), and the dash lines represent the conjectured relationships at different taper angles. Since the data base for these curves is limited, the usefulness of the figure lies in the qualitative trends indicated by the plot. It can be seen from the figure that propagation velocity of the luminous crack-tip front is a function of both pressurization rate and sample geometry.

To understand crack propagation velocity dependence on pressurization rate and sample geometry, two propagation mechanisms were considered. Swanson (1979) first assumes that the crack is propagating at its terminal velocity V_t, which is approximately equal to one-third the speed of sound in the propellant. As pressurization rate is increased, the dynamic and relaxation moduli of the viscoelastic propellant increase. This results in a higher terminal velocity in the propellant, according to the following formula:

$$V_t = \frac{1}{3}\sqrt{E/\rho} \qquad (5)$$

where E is the dynamic modulus and ρ is the density of the sample. However, over the range of pressurization rates tested, changes in moduli are approximately 10%; there-

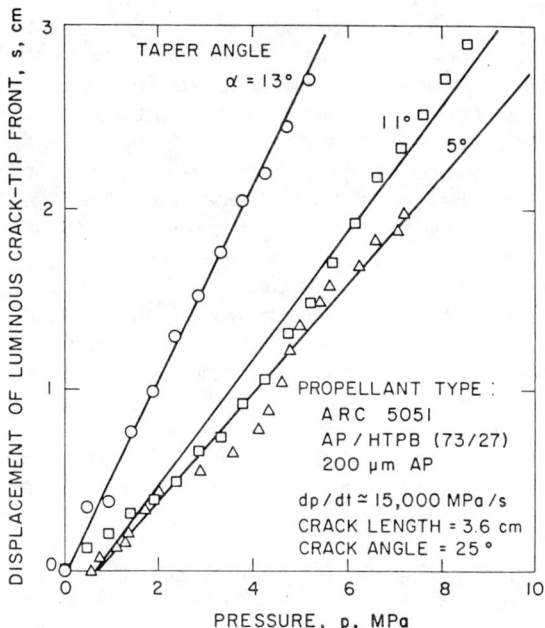

Fig. 8 Measured luminous crack-tip front displacement vs chamber pressure for crack samples with different initial geometries and pressurization rates.

fore, these small changes in moduli do not result in significant increase in crack propagation velocity as measured in this study. This observation has been made by Gent and Marteny (1982) in studying natural rubbers under rapid loading rates. In addition, this mechanism does not predict changes in crack propagation velocity due to different sample geometries.

The second crack propagation mechanism considered assumes that the velocity of the running crack is limited by the speed of the deforming crack contour. In order for propagation to progress, a sufficient stress state must be maintained at the crack tip. For geometries tested in this study, the large deformation of the crack walls accounted for the high stress at the crack tip. Therefore, the crack tip cannot progress at a faster rate than the crack contour.

To identify the significance of the second mechanism, one can test its usefulness in interpreting the test results shown in Figs. 7 and 8. Over the range of pressurization rates tested, it appears that the second mechanism satisfactorily explains the experimental trends. The quasistatic application of this mechanism would predict that the crack propagation velocity is proportional to pressurization rate to the first power. This is due to the fact that a static pressure load would correspond to a given displacement of the crack contour; therefore, under this limiting condition, the curves in Fig. 7 would collapse into one. However, experimental results in Fig. 7 show significant dependence of ds/dp on dp/dt. The changes in slope at different pressurization rates shown in Fig. 7 are believed to be caused by the delay of the sample leg displacement at higher loading rates. This delay is due to the effect of inertia during the acceleration of the sample mass. Therefore, the dynamic application of this mechanism would show weaker dependence of V_c on dp/dt than the static or quasistatic application. The exponent of dp/dt in the dynamic application would be lower than one; this is consistent with the experimental finding of 0.637, which is less than one. It should be noted that the time delay mentioned above does not refer to stress wave propagation times but rather the time needed to accelerate and displace the actual mass of the sample legs.

Similarly, the changes in slope at different geometries shown in Fig. 8 are believed to be caused by the decrease in resistance of the sample leg displacement for larger taper angles. This implies that the greater the free volume, the quicker the response of the propellant sample to the pressure load, which in turn would result in

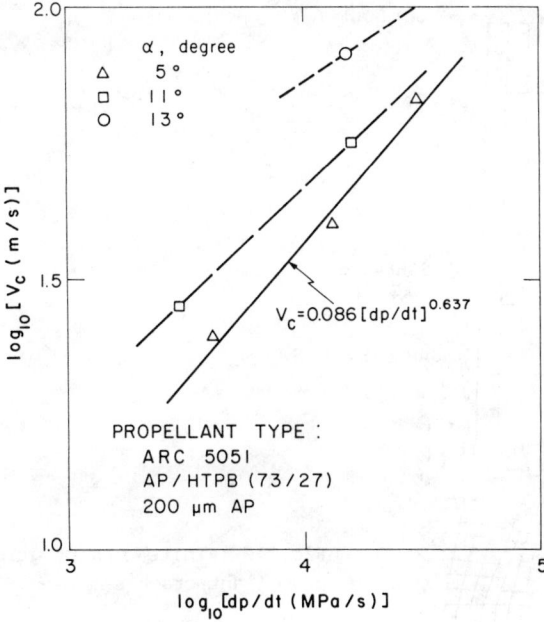

Fig. 9 Combined effect of pressurization rate and sample geometry on propagation velocity of luminous crack-tip front.

higher crack propagation velocities. The approximate magnitudes and the tendencies of curves shown in Figs. 7 and 8 were simulated by analytically solving a model consisting of a Voigt-Kelvin parallel spring-damper element in series with a point mass. The coefficients for the system were approximated from results of the static finite-element analysis and sample geometry. Results of the simple analysis produce qualitatively the same changes in ds/dp, due to different loading rates and geometries.

It is believed that at higher pressurization rates than those tested in this study, the theoretical maximum velocity would be reached. Under these conditions, the influence of pressurization rate and sample geometry on crack propagation velocity would be small, and the first crack propagation mechanism would be important.

It is interesting to note that the maximum observed luminous crack-tip front propagation velocity of 80 m/s is of the same order as that reported by Swanson (1979) and Gent when testing inert propellants and rubber, respectively. Swanson reported maximum terminal velocities of approximately 40 m/s, while the maximum velocities reported by Gent and Marteny (1982) were 80 m/s in specimens

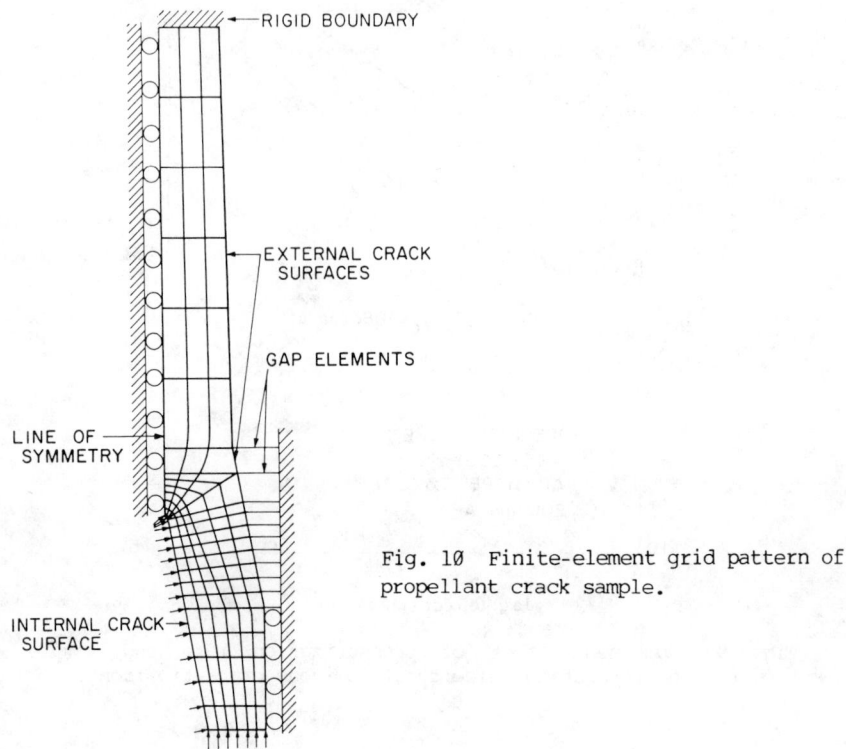

Fig. 10 Finite-element grid pattern of propellant crack sample.

with no preimposed strains in the crack direction. It should be noted that the actual propagation velocity of the crack tip ahead of the luminous front may be much higher than 80 m/s. The real crack propagation velocity, however, cannot be measured by present high-speed photographic methods.

Finite-Element Analysis

An analytic stress analysis of the propellant sample was performed, using a finite-element technique. Objectives of the analysis were threefold: 1) to gain insight into how the sample would deform in a cold environment under uniform internal pressure loading; 2) to predict the manner in which one could expect the propellant sample to fracture; and 3) to determine the contribution of mechanical deformation associated with the observed crack tip displacement.

The sample geometry modeled was similar to that shown in Fig. 1. The figure shows that the propellant sample is symmetric with respect to the crack axis. A finite crack-tip radius of 1.78 mm was selected in order to keep the number of mesh elements to a manageable size. The finite radius approximation reduces the accuracy of the solution near the tip; however, the solution still yields a good qualitative picture of crack wall deformation and stress contours in the vicinity of the crack tip.

The finite-element grid used in the analysis was generated by using an interactive mesh generation program (STAB) developed by Fitzgerald et al. (1971) and is shown in Fig. 10. The final grid density was determined by several test runs in which the density of the grid was increased in each successive run. The final result was a compromise between accuracy and computation time. The assumed boundary conditions for the model are also shown in Fig. 10. The gap elements in the figure allow for unrestricted displacement of the outer surface of the sample legs until contact is made with the chamber wall.

Material properties assumed for the model were an elastic modulus E equal to 55 MPa, and Poisson's ratio ν equal to 0.500. Although elastic behavior was assumed, the viscoelastic character of the sample was approximated by substituting the relaxation modulus for the elastic modulus. According to the correspondence principle, selection of the relaxation modulus was based upon the test pressurization rate and initial temperature of the propellant. This is a common procedure used in solid-propellant grain deformation analysis.

A static analysis of the model described was solved with a nonlinear finite-element code (ABAQUS-1981). The program allowed incompressible material behavior and large deformations which were needed in order to solve the problem. Results of the study can be divided into three areas, all of which provide insight into the mechanism of crack propagation: (1) displacement of the crack tip along the crack centerline; (2) contour deformation; and (3) stress and strain states near the crack tip.

Crack-Tip Displacement

Figure 11 shows the displacement of the finite elements as the result of a static pressure of 5.2 MPa. This figure demonstrates that pressure loading on the inner crack walls results in a mechanical displacement along the crack centerline. Even though the material is incompressible, vertical displacement of the crack tip is possible due to the fact that the free boundaries of the sample allowed for the horizontal expansion shown in the figure. The calculated displacement of the crack-tip node was determined for several pressures, and results indicated the nonlinearity of the problem. Calculated displacement vs pressure results can be easily converted to displacement vs time data, assuming that the pressure-time relationship is given and the quasisteady assumption is valid. Two mechanical displacement curves obtained in this manner are shown in Fig. 12; one corresponds to a low pressurization rate of 3000 MPa/s, and the other to a high dp/dt of 30,000 MPa/s. To determine the contribution of mechanical deformation to the overall observed crack-tip displacement, the experimentally measured displacements (under these two pressurization rates) were plotted on the same figure. In the case of low pressurization rate, it is obvious that the calculated mechanical displacement was considerably smaller than the overall displacement; however, in the case of higher pressurization rate, there appears to be a more significant contribution of mechanical deformation to the overall crack-tip displacement. That is, the higher the pressurization rate, the larger the percentage of crack-tip displacement due to mechanical deformation.

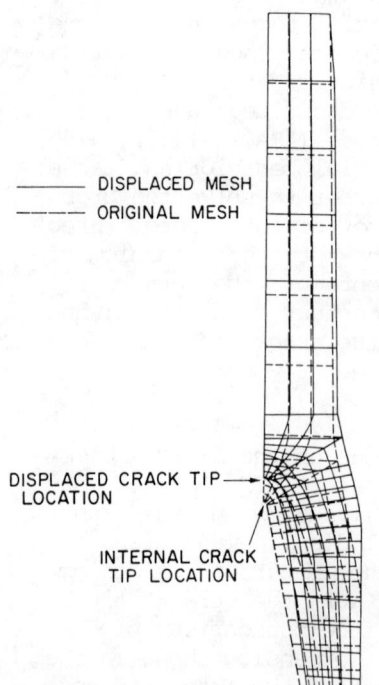

Fig. 11 Finite-element displacement plot due to static pressure of 5.2 MPa.

To evaluate the accuracy of this observation, it becomes necessary to re-examine the quasisteady assumption made in the finite-element analysis. A simple lumped-mass analysis demonstrates that time lags due to inertia effects are on the same order of magnitude as the duration of the experiment. Therefore, the higher the loading rate, the greater the errors due to the quasisteady assumption. Also, the static solution gives an upper bound of the displacement at a given time, since the inertia effect causes actual displacement to lag behind the calculated value. The contribution of mechanical deformation to overall crack-tip displacement is estimated as small (<15%) even at high pressurization rates.

Contour Deformation

Another interesting aspect of the finite-element displacement plot shown in Fig. 11 is the shape of the crack cavity after pressure is applied. This figure shows that when fracture at the crack tip is not allowed, the mechanical deformation tends to slightly round the original triangular shape of the crack inner walls. However, the degree of deformation is small when compared with the ob-

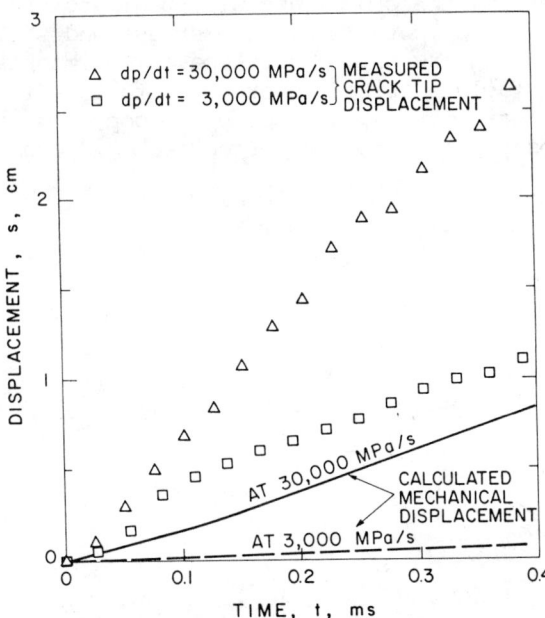

Fig. 12 Calculated and measured crack-tip displacement.

served abrupt crack contour changes shown in Fig. 4. This implies that in order for the crack contour to deform in the manner shown in Fig. 4, crack propagation must be caused by fracture.

Another observation which can be made from the plot in Fig. 11 is the extent to which the outer surface of the sample's leg deforms against the chamber wall. Even at a static pressure of 5.2 MPa, the crack sample leg is fairly rigid, and only the lower portion deforms to a position flat against the chamber wall. If one or more cracks were propagating, the leg would be free to deform out towards the chamber wall and create drastic changes in the crack cavity contour similar to changes seen in Fig. 4.

Stress and Strain States Near the Crack Tip

To gain some insight into the mode of failure of the propellant sample, the finite-element results were analyzed in conjunction with a typical propellant grain failure criterion. An experimentally derived failure envelope, based upon the maximum principal deviatoric stress and the corresponding strain, is a common criterion for determining a loading rate independent failure of solid-propellant grains subjected to internal pressurization. According to this criterion, the propellant will begin to crack at the surface near a point where the stress and strain states exceed the boundary of the envelope. Two important points of the failure criterion are: 1) failure is dependent upon deviatoric stress, and not upon hoop or maximum principal stress; and 2) the criterion is based upon the level

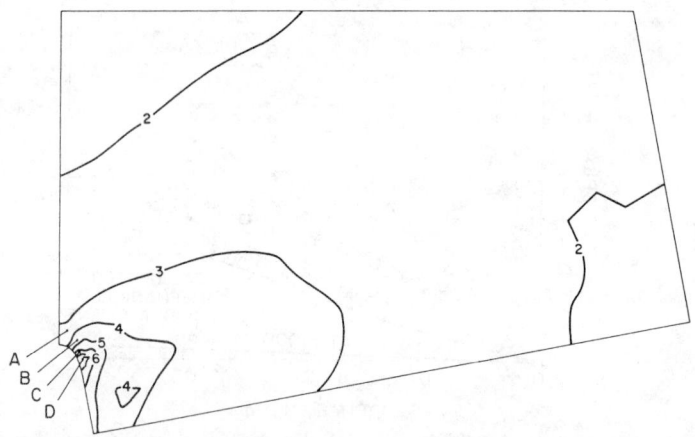

Fig. 13 Isopleths for deviatoric stresses.

of the combined state of stress and strain. Both of these points derive from the rubberlike character of composite solid propellants.

Deviatoric stress contours near the crack-tip radius are shown in Fig. 13. This plot clearly shows that the maximum deviatoric stress concentration is near the shoulder of the crack-tip radius. Four locations on the crack radius are indicated on the figure as points A, B, C, and D. A plot of the maximum deviatoric stress vs the corresponding strains for these four points is shown in Fig. 14. It can be seen that the maximum deviatoric stress occurs at D, on the shoulder of the crack tip, not at the tip zenith. In addition, the maximum strain occurs at C. This implies that, for the geometry and load conditions modeled by the finite-element analysis, the most likely area for failure would be in the vicinity of C and D, and not at the crack tip.

When the finite-element calculations are compared to the experimental data, it is necessary to keep in mind the simplifying assumptions made in the analysis. If the small crack radius caused by the razor blade is ignored, the finite-element results did not predict the large stress gradients which exist in the actual situation.

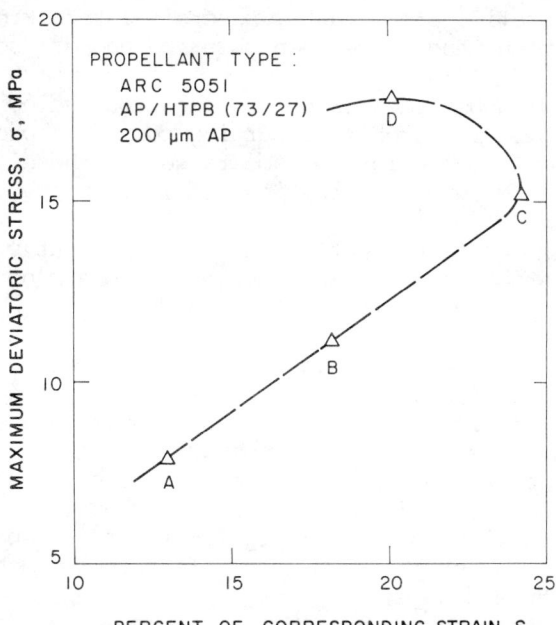

Fig. 14 Maximum deviatoric stress vs corresponding strain for crack-tip elements at a pressure of 5.17 MPa.

However, it is believed that the basic pattern of relative magnitudes of stress and strain states around the inner radius of the crack tip would not change significantly with the reduction of radius. This implies that the likelihood of failure occurring on the shoulder of the crack tip, as opposed to the exact apex of the crack tip, still exists. The symmetry of the model implies that failure at the shoulder would, in the actual sample, create two running cracks. If this were to occur, the two cracks would be extremely close together at the onset of failure because of the small radius of the crack tip.

Considering the possibility of a dual fracture in the propellant helps to explain the transition to a square-shaped geometry as observed during high pressurization rate tests. A multiple fracture of this type would allow dramatic displacement of the crack cavity walls outward towards the chamber wall, and create the rapid transition from the original triangular-shaped cavity to the square-shaped cavity shown in Fig. 4.

Summary and Conclusions

Even though the study presented in this paper is only a first step in the investigation of crack propagation in a burning solid propellant, several interesting facts about the phenomenon have been observed and are summarized below

1) Luminous crack front propagation velocities of 20-80 m/s were measured in a burning composite propellant. These velocities are in the same range as those reported in nonburning dynamic testing of inert propellants and rubbers.

2) It was found that luminous crack front propagated at a constant velocity when subjected to a constant rate of initial pressurization at the crack tip.

3) It was observed that there is a characteristic difference in crack propagation processes between low and high rates of pressurization. The crack-tip geometry maintained its original contour for low dp/dt. However, a fan region above the crack tip (accompanied by a squaring off of the tip region) was observed for high pressurization rates in the order of 2×10^4 MPa/s. The uneven multiple front of the luminous crack tip and the appearance of the fan region suggest the occurrence of crack branching and microstructure damage.

4) A tentative experimental correlation showed a stronger dependence of luminous crack-tip front propagation velocity on initial pressurization rate than

predictions based solely on the elastic-glassy transition behavior of viscoelastic material during dynamic loading.

5) Two crack propagation mechanisms were tested against experimental data. It was found that for the test configuration and range of pressurization rates tested, crack propagation velocity appears to be limited by the speed at which the crack cavity walls can deform. This limitation of the propagation velocity accounts for the strong dependence of V_c on dp/dt and sample geometry.

Acknowledgement

This research represents a part of results obtained under Contract No. N00014-79-C-0762, sponsored by the Power Program, Office of Naval Research, Arlington, Va., under the management of Dr. R. S. Miller. The participation of Dr. M. Kumar in the early portion of this study is acknowledged.

References

ABAQUS Finite Element Program (1981) Version 3-21-8 . Hibbitt, Karlson, and Sorenson, Inc.

Fitzgerald, J. E. and Hufferd, W. L. (1971) Handbook for the engineering structural analysis of solid propellants. Technical Report to ONR, CPIA Publication No. 214, University of Utah, Salt Lake City, Utah.

Gent, A. N. and Marteny, P. (1982) The effect of strain upon the velocity of sound and the velocity of free retraction for natural rubber. Technical Report No. 16, to ONR from the University of Akron, Akron, Ohio.

Swanson, S. R. (1979) An experimental study of dynamic crack propagation in a filled rubber. 16th Annual Meeting Soc. Eng. Sci., Northwestern University, Evanston, Illinois.

Author Index for Volume 94

Adams, P. R. 221
Arisoy, A. 221
Bakke, J. R. 504
Bauer, P. 118
Benedick, W. B. 546
Boiko, V. M. 293
Book, D. L. 449
Boris, J. P. 429
Borisov, A. A. 332
Bouriannes, R. 343
Brochet, C. 118, 343, 416
Brossard, J. 556
Cheret, R. 343
Clarke, J. F. 175
Cowperthwaite, M. 387
DeBoni, T. M. 405
Desbordes, D. 556
Djomehri, M. J. 569
Dupré, G. 201
Elsworth, J. E. 130
Eyre, J. A. 523
Fomin, N. A. 201
Fry, M. A. 449
Fuhre, K. 504
Fujiwara, T. 186, 309, 320
Funk, J. W. 55
Garnier, J. L. 556
Geiger, W. 474, 491
Gel'fand, B. E. 332
Guirao, C. 23
Hamada, L. 343
Hardy, J. R. 405
Harigaya, Y. 3
Hasegawa, T. 309
Hendrickx, S. 556
Hjertager, B. H. 504
Holt, M. 569
Kailasanath, K. 38
Karo, A. M. 405
Kassoy, D. R. 175
Kato, Y. 416
Kauffman, C. W. 221, 241
Khomic, S. V. 332
Knystautas, R. 23, 546
Kuo, K. K. 575
Lannoy, A. 556
Lee, D. 241

Lee, J. H. 23, 80, 546
Leyer, J. C. 556
Maker, B. N. 221
Moen, I. O. 55
Murray, S. B. 80
Nicholls, J. A. 221, 241
Ohyagi, S. 3
Oran, E. S. 38, 429
Paillard, C. 201
Papyrin, A. N. 293
Parker, S. J. 504
Peraldi, O. 302
Perrot, J. 556
Picone, J. M. 429
Pitz, W. J. 151
Presles, H. N. 118, 343
Rude, G. M. 55
Saint-Cloud, J. P. 302, 556
Schildknecht, M. 474, 491
Selman, J. R. 277
Setchell, R. E. 350
Shuff, P. J. 130, 523
Sichel, M. 241
Siefert, J. G. 575
Stock, M. 474, 491
Sugimura, T. 320
Sulmistras, A. 23
Taki, S. 186
Tarver, C. M. 369
Taylor, P. A. 350
Thibault, P. A. 55
Timofeev, E. I. 332
Tokita, K. 320
Tsyganov, S. A. 332
Tulis, A. J. 277
Ungut, A 130, 523
Urtiew, P. A. 151, 369
Vandermeiren, M. 104
Van Tiggelen, P. J. 104
Veyssière, B. 264
Walker, F. E. 405
Ward, S. A. 55
Westbrook, C. K. 151
Wolański, P. 221, 241, 293
Woliński, M. 293
Yoshihashi, T. 3
Young, T. R. 429

PROGRESS IN ASTRONAUTICS AND AERONAUTICS SERIES VOLUMES

VOLUME TITLE/EDITORS

*1. Solid Propellant Rocket Research (1960)
Martin Summerfield
Princeton University

*2. Liquid Rockets and Propellants (1960)
Loren E. Bollinger
The Ohio State University
Martin Goldsmith
The Rand Corporation
Alexis W. Lemmon Jr.
Battelle Memorial Institute

*3. Energy Conversion for Space Power (1961)
Nathan W. Snyder
Institute for Defense Analyses

*4. Space Power Systems (1961)
Nathan W. Snyder
Institute for Defense Analyses

*5. Electrostatic Propulsion (1961)
David B. Langmuir
Space Technology Laboratories, Inc.
Ernst Stuhlinger
NASA George C. Marshall Space Flight Center
J.M. Sellen Jr.
Space Technology Laboratories, Inc.

*6. Detonation and Two-Phase Flow (1962)
S.S. Penner
California Institute of Technology
F.A. Williams
Harvard University

*7. Hypersonic Flow Research (1962)
Frederick R. Riddell
AVCO Corporation

*8. Guidance and Control (1962)
Robert E. Roberson
Consultant
James S. Farrior
Lockheed Missiles and Space Company

*9. Electric Propulsion Development (1963)
Ernst Stuhlinger
NASA George C. Marshall Space Flight Center

*10. Technology of Lunar Exploration (1963)
Clifford I. Cummings and Harold R. Lawrence
Jet Propulsion Laboratory

*11. Power Systems for Space Flight (1963)
Morris A. Zipkin and Russell N. Edwards
General Electric Company

*12. Ionization in High-Temperature Gases (1963)
Kurt E. Shuler, Editor
National Bureau of Standards
John B. Fenn, Associate Editor
Princeton University

*13. Guidance and Control—II (1964)
Robert C. Langford
General Precision Inc.
Charles J. Mundo
Institute of Naval Studies

*14. Celestial Mechanics and Astrodynamics (1964)
Victor G. Szebehely
Yale University Observatory

*15. Heterogeneous Combustion (1964)
Hans G. Wolfhard
Institute for Defense Analyses
Irvin Glassman
Princeton University
Leon Green Jr.
Air Force Systems Command

*16. Space Power Systems Engineering (1966)
George C. Szego
Institute for Defense Analyses
J. Edward Taylor
TRW Inc.

*17. Methods in Astrodynamics and Celestial Mechanics (1966)
Raynor L. Duncombe
U.S. Naval Observatory
Victor G. Szebehely
Yale University Observatory

*18. Thermophysics and Temperature Control of Spacecraft and Entry Vehicles (1966)
Gerhard B. Heller
NASA George C. Marshall Space Flight Center

*19. Communication Satellite Systems Technology (1966)
Richard B. Marsten
Radio Corporation of America

*Out of print.

*20. **Thermophysics of Spacecraft and Planetary Bodies: Radiation Properties of Solids and the Electromagnetic Radiation Environment in Space** (1967)
Gerhard B. Heller
NASA George C. Marshall Space Flight Center

*21. **Thermal Design Principles of Spacecraft and Entry Bodies** (1969)
Jerry T. Bevans
TRW Systems

*22. **Stratospheric Circulation** (1969)
Willis L. Webb
Atmospheric Sciences Laboratory, White Sands, and University of Texas at El Paso

*23. **Thermophysics: Applications to Thermal Design of Spacecraft** (1970)
Jerry T. Bevans
TRW Systems

24. **Heat Transfer and Spacecraft Thermal Control** (1971)
John W. Lucas
Jet Propulsion Laboratory

25. **Communication Satellites for the 70's: Technology** (1971)
Nathaniel E. Feldman
The Rand Corporation
Charles M. Kelly
The Aerospace Corporation

26. **Communication Satellites for the 70's: Systems** (1971)
Nathaniel E. Feldman
The Rand Corporation
Charles M. Kelly
The Aerospace Corporation

27. **Thermospheric Circulation** (1972)
Willis L. Webb
Atmospheric Sciences Laboratory, White Sands, and University of Texas at El Paso

28. **Thermal Characteristics of the Moon** (1972)
John W. Lucas
Jet Propulsion Laboratory

29. **Fundamentals of Spacecraft Thermal Design** (1972)
John W. Lucas
Jet Propulsion Laboratory

30. **Solar Activity Observations and Predictions** (1972)
Patrick S. McIntosh and Murray Dryer
Environmental Research Laboratories, National Oceanic and Atmospheric Administration

31. **Thermal Control and Radiation** (1973)
Chang-Lin Tien
University of California at Berkeley

32. **Communications Satellite Systems** (1974)
P.L. Bargellini
COMSAT Laboratories

33. **Communications Satellite Technology** (1974)
P.L. Bargellini
COMSAT Laboratories

34. **Instrumentation for Airbreathing Propulsion** (1974)
Allen E. Fuhs
Naval Postgraduate School
Marshall Kingery
Arnold Engineering Development Center

35. **Thermophysics and Spacecraft Thermal Control** (1974)
Robert G. Hering
University of Iowa

36. **Thermal Pollution Analysis** (1975)
Joseph A. Schetz
Virginia Polytechnic Institute

37. **Aeroacoustics: Jet and Combustion Noise; Duct Acoustics** (1975)
Henry T. Nagamatsu, Editor
General Electric Research and Development Center
Jack V. O'Keefe, Associate Editor
The Boeing Company
Ira R. Schwartz, Associate Editor
NASA Ames Research Center

38. **Aeroacoustics: Fan, STOL, and Boundary Layer Noise; Sonic Boom; Aeroacoustic Instrumentation** (1975)
Henry T. Nagamatsu, Editor
General Electric Research and Development Center
Jack V. O'Keefe, Associate Editor
The Boeing Company
Ira R. Schwartz, Associate Editor
NASA Ames Research Center

39. **Heat Transfer with Thermal Control Applications** (1975)
M. Michael Yovanovich
University of Waterloo

SERIES LISTING

40. **Aerodynamics of Base Combustion** (1976)
S.N.B. Murthy, Editor
Purdue University
J.R. Osborn, Associate Editor
Purdue University
A.W. Barrows and J.R. Ward, Associate Editors
Ballistics Research Laboratories

41. **Communications Satellite Developments: Systems** (1976)
Gilbert E. LaVean
Defense Communications Agency
William G. Schmidt
CML Satellite Corporation

42. **Communications Satellite Developments: Technology** (1976)
William G. Schmidt
CML Satellite Corporation
Gilbert E. LaVean
Defense Communications Agency

43. **Aeroacoustics: Jet Noise, Combustion and Core Engine Noise** (1976)
Ira R. Schwartz, Editor
NASA Ames Research Center
Henry T. Nagamatsu, Associate Editor
General Electric Research and Development Center
Warren C. Strahle, Associate Editor
Georgia Institute of Technology

44. **Aeroacoustics: Fan Noise and Control; Duct Acoustics; Rotor Noise** (1976)
Ira R. Schwartz, Editor
NASA Ames Research Center
Henry T. Nagamatsu, Associate Editor
General Electric Research and Development Center
Warren C. Strahle, Associate Editor
Georgia Institute of Technology

45. **Aeroacoustics: STOL Noise; Airframe and Airfoil Noise** (1976)
Ira R. Schwartz, Editor
NASA Ames Research Center
Henry T. Nagamatsu, Associate Editor
General Electric Research and Development Center
Warren C. Strahle, Associate Editor
Georgia Institute of Technology

46. **Aeroacoustics: Acoustic Wave Propagation; Aircraft Noise Prediction; Aeroacoustic Instrumentation** (1976)
Ira R. Schwartz, Editor
NASA Ames Research Center
Henry T. Nagamatsu, Associate Editor
General Electric Research and Development Center
Warren C. Strahle, Associate Editor
Georgia Institute of Technology

47. **Spacecraft Charging by Magnetospheric Plasmas** (1976)
Alan Rosen
TRW Inc.

48. **Scientific Investigations on the Skylab Satellite** (1976)
Marion I. Kent and Ernst Stuhlinger
NASA George C. Marshall Space Flight Center
Shi-Tsan Wu
The University of Alabama

49. **Radiative Transfer and Thermal Control** (1976)
Allie M. Smith
ARO Inc.

50. **Exploration of the Outer Solar System** (1976)
Eugene W. Greenstadt
TRW Inc.
Murray Dryer
National Oceanic and Atmospheric Administration
Devrie S. Intriligator
University of Southern California

51. **Rarefied Gas Dynamics, Parts I and II (two volumes)** (1977)
J. Leith Potter
ARO Inc.

52. **Materials Sciences in Space with Application to Space Processing** (1977)
Leo Steg
General Electric Company

53. **Experimental Diagnostics in Gas Phase Combustion Systems** (1977)
Ben T. Zinn, Editor
Georgia Institute of Technology
Craig T. Bowman, Associate Editor
Stanford University
Daniel L. Hartley, Associate Editor
Sandia Laboratories
Edward W. Price, Associate Editor
Georgia Institute of Technology
James G. Skifstad, Associate Editor
Purdue University

54. **Satellite Communications: Future Systems** (1977)
David Jarett
TRW Inc.

55. **Satellite Communications: Advanced Technologies** (1977)
David Jarett
TRW Inc.

56. **Thermophysics of Spacecraft and Outer Planet Entry Probes** (1977)
Allie M. Smith
ARO Inc.

57. **Space-Based Manufacturing from Nonterrestrial Materials** (1977)
Gerard K. O'Neill, Editor
Princeton University
Brian O'Leary, Assistant Editor
Princeton University

58. **Turbulent Combustion** (1978)
Lawrence A. Kennedy
State University of New York at Buffalo

59. **Aerodynamic Heating and Thermal Protection Systems** (1978)
Leroy S. Fletcher
University of Virginia

60. **Heat Transfer and Thermal Control Systems** (1978)
Leroy S. Fletcher
University of Virginia

61. **Radiation Energy Conversion in Space** (1978)
Kenneth W. Billman
NASA Ames Research Center

62. **Alternative Hydrocarbon Fuels: Combustion and Chemical Kinetics** (1978)
Craig T. Bowman
Stanford University
Jorgen Birkeland
Department of Energy

63. **Experimental Diagnostics in Combustion of Solids** (1978)
Thomas L. Boggs
Naval Weapons Center
Ben T. Zinn
Georgia Institute of Technology

64. **Outer Planet Entry Heating and Thermal Protection** (1979)
Raymond Viskanta
Purdue University

65. **Thermophysics and Thermal Control** (1979)
Raymond Viskanta
Purdue University

66. **Interior Ballistics of Guns** (1979)
Herman Krier
University of Illinois at Urbana-Champaign
Martin Summerfield
New York University

67. **Remote Sensing of Earth from Space: Role of "Smart Sensors"** (1979)
Roger A. Breckenridge
NASA Langley Research Center

68. **Injection and Mixing in Turbulent Flow** (1980)
Joseph A. Schetz
Virginia Polytechnic Institute and State University

69. **Entry Heating and Thermal Protection** (1980)
Walter B. Olstad
NASA Headquarters

70. **Heat Transfer, Thermal Control, and Heat Pipes** (1980)
Walter B. Olstad
NASA Headquarters

71. **Space Systems and Their Interactions with Earth's Space Environment** (1980)
Henry B. Garrett and Charles P. Pike
Hanscom Air Force Base

72. **Viscous Flow Drag Reduction** (1980)
Gary R. Hough
Vought Advanced Technology Center

73. **Combustion Experiments in a Zero-Gravity Laboratory** (1981)
Thomas H. Cochran
NASA Lewis Research Center

74. **Rarefied Gas Dynamics, Parts I and II (two volumes)** (1981)
Sam S. Fisher
University of Virginia at Charlottesville

75. **Gasdynamics of Detonations and Explosions** (1981)
J.R. Bowen
University of Wisconsin at Madison
N. Manson
Université de Poitiers
A.K. Oppenheim
University of California at Berkeley
R.I. Soloukhin
Institute of Heat and Mass Transfer, BSSR Academy of Sciences

76. **Combustion in Reactive Systems** (1981)
J.R. Bowen
University of Wisconsin at Madison
N. Manson
Université de Poitiers
A.K. Oppenheim
University of California at Berkeley
R.I. Soloukhin
Institute of Heat and Mass Transfer, BSSR Academy of Sciences

77. **Aerothermodynamics and Planetary Entry** (1981)
A.L. Crosbie
University of Missouri-Rolla

78. **Heat Transfer and Thermal Control** (1981)
A.L. Crosbie
University of Missouri-Rolla

SERIES LISTING

79. **Electric Propulsion and Its Applications to Space Missions** (1981)
Robert C. Finke
NASA Lewis Research Center

80. **Aero-Optical Phenomena** (1982)
Keith G. Gilbert and Leonard J. Otten
Air Force Weapons Laboratory

81. **Transonic Aerodynamics** (1982)
David Nixon
Nielsen Engineering & Research, Inc.

82. **Thermophysics of Atmospheric Entry** (1982)
T.E. Horton
The University of Mississippi

83. **Spacecraft Radiative Transfer and Temperature Control** (1982)
T.E. Horton
The University of Mississippi

84. **Liquid-Metal Flows and Magnetohydrodynamics** (1983)
H. Branover
Ben-Gurion University of the Negev
P.S. Lykoudis
Purdue University
A. Yakhot
Ben-Gurion University of the Negev

85. **Entry Vehicle Heating and Thermal Protection Systems: Space Shuttle, Solar Starprobe, Jupiter Galileo Probe** (1983)
Paul E. Bauer
McDonnell Douglas Astronautics Company
Howard E. Collicott
The Boeing Company

86. **Spacecraft Thermal Control, Design, and Operation** (1983)
Howard E. Collicott
The Boeing Company
Paul E. Bauer
McDonnell Douglas Astronautics Company

87. **Shock Waves, Explosions, and Detonations** (1983)
J.R. Bowen
University of Washington
N. Manson
Université de Poitiers
A.K. Oppenheim
University of California at Berkeley
R.I. Soloukhin
Institute of Heat and Mass Transfer, BSSR Academy of Sciences

88. **Flames, Lasers, and Reactive Systems** (1983)
J.R. Bowen
University of Washington
N. Manson
Université de Poitiers
A.K. Oppenheim
University of California at Berkeley
R.I. Soloukhin
Institute of Heat and Mass Transfer, BSSR Academy of Sciences

89. **Orbit-Raising and Maneuvering Propulsion: Research Status and Needs** (1984)
Leonard H. Caveny
Air Force Office of Scientific Research

90. **Fundamentals of Solid-Propellant Combustion** (1984)
Kenneth K. Kuo
The Pennsylvania State University
Martin Summerfield
Princeton Combustion Research Laboratories, Inc.

91. **Spacecraft Contamination: Sources and Prevention** (1984)
J.A. Roux
The University of Mississippi
T.D. McCay
NASA Marshall Space Flight Center

92. **Combustion Diagnostics by Nonintrusive Methods** (1984)
T.D. McCay
NASA Marshall Space Flight Center
J.A. Roux
The University of Mississippi

93. **The INTELSAT Global Satellite System** (1984)
Joel Alper
COMSAT Corporation
Joseph Pelton
INTELSAT

94. **Dynamics of Shock Waves, Explosions, and Detonations** (1984)
J.R. Bowen
University of Washington
N. Manson
Universite de Poitiers
A.K. Oppenheim
University of California
R.I. Soloukhin
Institute of Heat and Mass Transfer, BSSR Academy of Sciences

95. **Dynamics of Flames and Reactive Systems** (1984)
J.R. Bowen
University of Washington
N. Manson
Universite de Poitiers
A.K. Oppenheim
University of California
R.I. Soloukhin
Institute of Heat and Mass Transfer, BSSR Academy of Sciences

(Other Volumes are planned.)